AF812613

Smart Agriculture

Volume 13

Series Editors

Zhao Zhang, Key Laboratory of Smart Agriculture System Integration Research, Ministry of Education, China Agricultural University, Beijing, China

Yiannis Ampatzidis, UF/IFAS Southwest Florida Research and Education Center, University of Florida, Immokalee, USA

Paulo Flores, Department of Agricultural and Biosystems Engineering, North Dakota State University, Fargo, USA

Yuanjie Wang, Agricultural Information Institute, Chinese Academy of Agricultural Sciences, Beijing, China

The book series Smart Agriculture presents progress of smart agricultural technologies, which includes, but not limited to, specialty crop harvest robotics, UAV technologies for row crops, innovative IoT applications in plant factories, and big data for optimizing production process. It includes both theoretical study and practical applications, with emphasis on systematic studies. AI technologies in agricultural productions will be emphasized, consisting of innovative algorithms and new application domains. Additionally, new crops are emerging, such as hemp in U.S., and covered as well. This book series would cover regions worldwide, such as U.S., Canada, China, Japan, Korea, and Brazil.

The book series Smart Agriculture aims to provide an academic platform for interdisciplinary researchers to provide their state-of-the-art technologies related to smart agriculture. Researchers of different academic backgrounds are encouraged to contribute to the book, such as agriculture engineers, breeders, horticulturist, agronomist, and plant pathologists. The series would target a very broad audience – all having a professional related to agriculture production. It also could be used as textbooks for graduate students.

Imran · Jiyu Li

UAV Aerodynamics and Crop Interaction

Revolutionizing Modern Agriculture with Drone

Springer

Imran
College of Engineering
South China Agricultural University
Guangzhou, Guangdong, China

Jiyu Li
College of Engineering
South China Agricultural University
Guangzhou, Guangdong, China

ISSN 2731-3476 ISSN 2731-3484 (electronic)
Smart Agriculture
ISBN 978-981-96-8401-4 ISBN 978-981-96-8402-1 (eBook)
https://doi.org/10.1007/978-981-96-8402-1

© The Editor(s) (if applicable) and The Author(s), under exclusive license to Springer Nature Singapore Pte Ltd. 2025

This work is subject to copyright. All rights are solely and exclusively licensed by the Publisher, whether the whole or part of the material is concerned, specifically the rights of translation, reprinting, reuse of illustrations, recitation, broadcasting, reproduction on microfilms or in any other physical way, and transmission or information storage and retrieval, electronic adaptation, computer software, or by similar or dissimilar methodology now known or hereafter developed.
The use of general descriptive names, registered names, trademarks, service marks, etc. in this publication does not imply, even in the absence of a specific statement, that such names are exempt from the relevant protective laws and regulations and therefore free for general use.
The publisher, the authors and the editors are safe to assume that the advice and information in this book are believed to be true and accurate at the date of publication. Neither the publisher nor the authors or the editors give a warranty, expressed or implied, with respect to the material contained herein or for any errors or omissions that may have been made. The publisher remains neutral with regard to jurisdictional claims in published maps and institutional affiliations.

This Springer imprint is published by the registered company Springer Nature Singapore Pte Ltd.
The registered company address is: 152 Beach Road, #21-01/04 Gateway East, Singapore 189721, Singapore

If disposing of this product, please recycle the paper.

Competing Interests The authors have no competing interests to declare that are relevant to the content of this manuscript.

Contents

About the Authors

Dr. Imran is an Agriculture Officer in the Extension Wing of the Ministry of Agriculture, Government of Khyber Pakhtunkhwa, Pakistan. He holds a Ph.D. in Agronomy and Climate Change from The University of Agriculture, Peshawar. Renowned for his commitment to agro-meteorology and climate resilience in agriculture, he was awarded the Swiss Development Cooperation Fellowship during his doctoral studies and later received the prestigious Turk Presidential Fellowship for his postdoctoral research. He has served as a visiting professor at the Rhizosphere Laboratory, University of Çukurova, Turkey, and is currently pursuing postdoctoral research at South China Agricultural University, Guangzhou, China.

He has authored over 106 research and review articles and contributed 40 book chapters on topics such as climate change, crop production, and sustainable soil management. He is also the editor and author of three academic books. Beyond research, his professional experience includes roles as a soil ecologist, lecturer in agronomy, and consultant for international organizations including USAID, IAO Italiana, and PDMA. His research interests encompass crop nutrition, micro climate engineering through UAV aerodynamics and its relation with agronomy, degraded soil management, crop competition, and the development of sustainable agricultural practices in challenging environments. He also possesses specialized expertise in climate forecasting and agro-meteorology.

Prof. Dr. Jiyu Li is a distinguished Doctor of Engineering, Professor, and Doctoral Supervisor at South China Agricultural University. He serves on the Editorial Committee of Agriculture, Youth Editorial Committee of the Journal of Agricultural Machinery, and Youth Editorial Committee of the Journal of South China Agricultural University. Additionally, he holds key positions, including Deputy Director of the Guangdong Provincial Agricultural Aviation Applied Engineering Technology Research Center, Founder of the UAV Association of South China Agricultural University, Director of the Department of Agricultural Engineering, Robotic Engineering Director, and Head of AOPA College Training Institution. Prof. Li's research focuses on the effect and efficiency of precise operation of rotor aircraft, developing new field and airborne sensors, airflow target flight control systems, configuring UAV energy, load, and architecture, and exploring the operation performance of large-load UAVs. He has presided over 15 projects, including National Natural Science Foundation surface projects and key special projects of the national key research and development plan. His work has resulted in 85 published papers, with 49 included in SCI/EI, and 59 effective invention patents, of which 49 have been authorized. He has also transformed 7 achievements and received notable awards, including the Guangdong Science and Technology Progress Award and the Dabei Agricultural Science and Technology Progress Award. In addition to his research accomplishments, Prof. Li teaches national first-class undergraduate courses, including "Basics of Small Aircraft" and "Nature and Flight." He has guided excellent graduate students and postgraduate students in various competitions, leading to national awards.

Chapter 1
Overview of UAV Technology: History and Evolution

Abstract This chapter provides a comprehensive overview of Unmanned Aerial Vehicle (UAV) technology, tracing its development from early conceptual prototypes to today's sophisticated systems widely used across various sectors. The historical trajectory begins with initial military applications in the early twentieth century, progressing through pivotal moments in World War I and II, the Cold War, and the technological boom of the twenty-first century. The evolution of UAVs is explored through advances in design, propulsion, control systems, payload capabilities, and autonomy, driven by innovations in artificial intelligence, GPS, sensor fusion, and data processing. Beyond defense, UAVs have found transformative applications in civilian domains. A key area is agriculture, where drones are revolutionizing farm management by enhancing productivity, reducing input costs, and promoting sustainability. Despite challenges like cost and regulatory barriers, continuous innovation in AI, automation, and data analytics is cementing UAVs as essential tools for precision farming and smart agriculture. The chapter further highlights the critical role of academic and research institutions in advancing UAV technology, particularly for agricultural use. Their contributions drive the development of cutting-edge drone solutions aimed at optimizing resource use and minimizing environmental impact. In conclusion, the evolution of UAV technology—from early experimental designs to AI-driven systems—has transformed aerial operations across industries. As adoption grows, UAVs will continue to shape military strategies, commercial practices, environmental stewardship, and sustainable agriculture, making them indispensable in modern society.

Keywords UAV technology · Environmental conservation · Future of smart agriculture · Aerial technology · Drones in agriculture

© The Author(s), under exclusive license to Springer Nature Singapore Pte Ltd. 2025
J. Li, *UAV Aerodynamics and Crop Interaction*, Smart Agriculture 13,
https://doi.org/10.1007/978-981-96-8402-1_1

1.1 Introduction

1.1.1 Overview of UAV Technology: History and Evolution

Unmanned Aerial Vehicles (UAVs), more commonly known as drones, have become a revolutionary technology shaping various aspects of modern life—from defense and surveillance to precision agriculture and environmental conservation. The rise of UAVs is a testament to the rapid progression of technological innovation, interdisciplinary engineering, and the growing demand for automated, efficient, and cost-effective solutions across industries. What once began as rudimentary remote-controlled flying machines for military purposes has evolved into complex, autonomous aerial systems capable of executing highly specialized tasks with remarkable precision. The journey of UAV technology is not only one of technical evolution but also of social and economic transformation. UAVs have transitioned from niche military tools to essential assets in civilian applications, spurred by advancements in artificial intelligence (AI), navigation systems, sensor integration, and data processing. This introduction aims to provide a comprehensive overview of the historical development of UAVs, outline the key technological milestones, and explore their expanding role in modern society. It serves as a foundation for understanding the current state of UAV innovation and offers a contextual backdrop for future explorations in the field.

Unmanned Aerial Vehicles (UAVs) are revolutionizing agriculture by providing real-time data, precision farming solutions, and automation. Equipped with high-resolution cameras, multispectral sensors, GPS, and AI-powered analytics, UAVs enable farmers to monitor crops, analyze soil conditions, optimize irrigation, and manage pests more efficiently than traditional methods. Key applications of UAVs in agriculture include real-time crop monitoring and health analysis, soil analysis and field mapping, precision spraying of pesticides and fertilizers, irrigation and water management, automated planting and seeding, pest and disease detection, livestock monitoring, and increased efficiency and productivity [1]. Benefits of UAVs in agriculture include increased efficiency and productivity, cost reduction and resource optimization, environmental sustainability, risk mitigation and disaster management, and early warnings for crop diseases, pest outbreaks, and drought conditions. However, challenges in UAV adoption include high initial costs, regulatory restrictions, technical skills and training, data management and integration issues, and limited technical expertise in rural regions. The future of UAVs in agriculture will be driven by advancements in AI, automation, and IoT (Internet of Things), with key trends including autonomous UAV swarms, AI-driven crop intelligence, integration with smart farming systems, and affordable UAV solutions [2]. Autonomous UAV swarms involve multiple drones collaborating for large-scale planting, spraying, and monitoring. AI-driven crop intelligence provides predictive analytics for yield forecasting and disease prevention, while integration with smart farming systems connects with IoT sensors, automated tractors, and farm management platforms for real-time decision-making. As technology evolves, cost-effective drone solutions will become more accessible to small-scale farmers worldwide. In UAV technology is rapidly transforming agriculture by enhancing productivity, reducing costs, and promoting sustainability [3]. While challenges such as high costs and regulations remain, continuous innovations in AI, automation, and data analytics make UAVs an essential tool for modern precision farming. As adoption increases globally, UAVs will play a crucial role in ensuring food security, environmental conservation, and the future of smart agriculture.

1.1.2 The Origins of UAV Technology

The history of UAVs dates back more than a century, with early concepts inspired by the desire to remove the human pilot from dangerous aerial missions. The first recorded use of an unmanned flying device can be traced to 1849, when the Austrian army launched bomb-laden balloons against Venice—an extremely primitive form of UAVs. However, the foundational era of UAV development began in the early twentieth century, particularly during World War I, when both British and American military engineers began experimenting with pilotless aircraft. One of the first serious attempts was the Kettering Bug developed by Charles Kettering and the U.S. Army in 1918. It was an early cruise missile prototype—a small biplane equipped with a gyroscopic guidance system and programmed to follow a predetermined flight path. Though it never saw combat, the Kettering Bug laid the groundwork for future UAV innovations by demonstrating the feasibility of automated flight. The interwar years saw minimal development, but the outbreak of World War II catalyzed new efforts. During this period, UAVs like the Radioplane OQ-2 were introduced for training anti-aircraft gunners. This UAV was mass-produced and flown via radio control, representing a key transition from concept to practical military use. Moreover, this era marked the entry of Hollywood actor Ronald Reagan into the UAV story—he was assigned to photograph Radioplane manufacturing facilities while serving in the U.S. Army Air Forces.

The Kettering Bug. picture from Encyclopedia of Astrobiology, Astronomy, and Space Flight

1.2 Historical Evolution of UAV Technology

Unmanned Aerial Vehicles (UAVs) have a long history, dating back to ancient China. The first significant UAV development occurred during World War I with the Kettering Bug, a rudimentary cruise missile. During World War II, UAVs were limited, but during the Cold War, they advanced for surveillance and reconnaissance. The 1990s and 2000s saw a revolution in UAV development, with the MQ-1 Predator and MQ-9 Reaper becoming central to U.S. military operations in Iraq and Afghanistan [6]. Civilian and commercial UAV applications expanded to agriculture, disaster response, delivery services, and aerial photography. Recent advancements include AI-powered drones, swarming technology, and autonomous UAVs. The historical evolution of UAVs reflects a progression from simple pilotless aircraft to highly sophisticated, multi-purpose systems [7]. Today, UAVs are equipped with advanced features like high-resolution cameras, multispectral sensors, and AI-powered analytics, enabling widespread adoption in precision farming and environmental monitoring.

1.3 UAVs in the Cold War Era

The Cold War period, particularly from the 1950s through the 1980s, was instrumental in shaping UAV technology. The geopolitical tensions between superpowers fueled the demand for intelligence-gathering technologies, leading to the development of high-altitude, long-endurance UAVs for reconnaissance missions. Notably, the U.S. developed systems such as Ryan Model 147 "Lightning Bug" drones, which were used extensively for surveillance during the Vietnam War and over China and North Korea. This era saw advancements in flight control, propulsion, and remote communication systems. UAVs grew larger and more capable of carrying high-resolution cameras and radar equipment. However, these early surveillance drones still relied heavily on ground-based pilots and were limited in autonomy and intelligence.

Simultaneously, advances in miniaturization and electronics improved payload capabilities. UAVs began to include early forms of data recorders and sensors, enabling more detailed information gathering over hostile or inaccessible territories. Despite these improvements, UAVs were still confined primarily to military applications, hindered by cost, size, and limited regulatory frameworks for civilian use.

1.3.1 Early Innovations

The history of Unmanned Aerial Vehicles (UAVs) dates back to the early twentieth century, with key milestones in the evolution of drone technology. During World War I and II, the first UAVs were developed as target drones for military training, such as the Kellering Bug (1918). During World War II, UAVs like the Radioplane OQ-2 were used for target practice and reconnaissance missions [8]. The Cold War saw significant technological advancements in UAV technology, particularly for surveillance, with the Ryan Firebee (1950s) being a notable example. Navigation systems, such as gyroscopes and autopilots, improved the stability and accuracy of UAVs, making them more reliable for military applications. In the late twentieth century, GPS integration and miniaturization revolutionized UAVs, enabling precise navigation and autonomous flight. Advances in electronics and materials science led to smaller, lighter, and more affordable drones, paving the way for civilian and commercial applications. Early agricultural applications saw drones being used for aerial photography and mapping, providing farmers with new ways to monitor crops and land. Research and development focused on testing their potential for crop monitoring, spraying, and data collection, setting the stage for their widespread adoption in the twenty-first century [9]. Key innovations in early UAV technology included remote control systems, lightweight materials, and basic sensors. The military legacy of early UAVs demonstrated the potential for remote-controlled and autonomous aircraft, and the technological advancements made during this period laid the groundwork for sophisticated drones used in agriculture and other industries today.

1.3.2 The UAV Technological Boom of the 21st Century

The turn of the twenty-first century ushered in a period of unprecedented technological development that significantly altered the landscape of UAV systems. Several converging technologies—including Global Positioning Systems (GPS), miniaturized sensors, lithium-polymer batteries, real-time data transmission, and artificial intelligence—catalyzed the creation of more compact, affordable, and versatile UAV platforms. Perhaps the most transformative development came with the proliferation of multi-rotor UAVs, such as quadcopters and hexacopters, which offered excellent maneuverability, vertical takeoff and landing (VTOL) capabilities, and ease of use. The emergence of companies like DJI in China democratized UAV technology by introducing consumer-grade drones that were not only affordable but also capable of high-resolution photography, autonomous navigation, and app-based controls.

At the same time, military UAVs also experienced significant evolution. Systems like the MQ-1 Predator and its successor the MQ-9 Reaper became synonymous with modern aerial warfare, offering persistent surveillance and precision-strike capabilities without endangering pilots. These UAVs were remotely operated over long distances and equipped with advanced imaging, thermal sensors, and weapons, proving critical in the Global War on Terror. This era also marked the beginning of regulatory efforts aimed at governing the growing presence of UAVs in civilian airspace. Governments and aviation authorities such as the Federal Aviation Administration (FAA) in the U.S., EASA in Europe, and CAAC in China began developing rules to ensure the safe integration of UAVs into national airspaces. The convergence of technological readiness and regulatory clarity led to an explosion in commercial UAV applications.

1.3.3 Drones in Transition: From Combat Zones to Crop Fields

The evolution of Unmanned Aerial Vehicles (UAVs) from military applications to agriculture has significantly improved efficiency, sustainability, and productivity in modern agriculture. Several UAV technologies first developed for military purposes have been successfully repurposed for agriculture, including aerial surveillance and crop monitoring, GPS and autonomous navigation, targeted airstrikes vs. precision spraying, remote sensing and data collection, swarm technology and automated farming, and disaster management and risk reduction. The benefits of military drone technology in agriculture include increased efficiency and productivity, reduced costs and resource optimization, environmental sustainability, and disaster management and risk reduction [10]. UAVs allow rapid and real-time monitoring of crops, reducing the need for manual field inspections and covering thousands of acres in a fraction of the time required by traditional methods. Automated drones can cover thousands of acres in a fraction of the time required by traditional methods.

Future trends in advanced UAV technologies in agriculture include AI and machine learning, autonomous farming drones, IoT and smart farming, and hyper-precision agriculture. AI-powered precision spraying, soil mapping, and real-time growth monitoring will be used in next-generation UAVs. The transition of UAV technology from military applications to agriculture has revolutionized modern agriculture by leveraging high-tech drone innovations [11]. By leveraging high-tech drone innovations, farmers can now increase yields, reduce costs, and promote sustainable farming practices. As drone technology continues to advance, fully autonomous farms may soon become a reality, ensuring global food security and environmental sustainability.

Academic institutions have played a central and multifaceted role in the advancement of Unmanned Aerial Vehicle (UAV) technology, especially in its agricultural applications. Serving as innovation hubs, universities foster multidisciplinary collaboration that integrates aeronautical engineering, robotics, computer science, agriculture, and environmental sciences. This collaborative environment provides the foundation for cutting-edge research and the development of UAV technologies tailored to address modern agricultural challenges. Research conducted within universities spans a wide range of UAV-related topics. These include the refinement of drone designs, improvements in flight stability and obstacle avoidance, optimization of energy efficiency, and the advancement of autonomous navigation and decision-making systems [20]. Through experimental designs, field trials, and advanced software development, academic researchers contribute significantly to the evolution of UAV capabilities. UAV systems are increasingly developed and tested in academic settings to solve domain-specific problems—such as plant phenotyping, precision crop monitoring, and atmospheric data collection for climate analysis [20].

Field trials conducted by university research teams are essential for validating UAV performance in real agricultural conditions. These trials encompass tasks like crop health monitoring, aerial spraying, soil analysis, and yield estimation, helping to ensure that new technologies meet both agronomic needs and operational feasibility. Data generated from these trials inform iterative improvements in UAV design and functionality, fostering a cycle of innovation grounded in practical application [20]. Beyond research, universities play a crucial role in education and workforce development. Academic programs dedicated to UAV technology, precision agriculture, and environmental monitoring equip students with both theoretical knowledge and hands-on experience. These include undergraduate and postgraduate degree programs, specialized coursework, field-based learning, and lab-based experiments. Students gain practical skills through internships, capstone projects, and drone deployment in real-world settings, preparing them for careers in the rapidly expanding drone industry [21]. Universities also engage with industry and government agencies to co-develop UAV solutions that are commercially viable and socially relevant. Joint research initiatives with drone manufacturers and agricultural technology companies accelerate the translation of laboratory findings into market-ready products. These collaborations often result in the development of innovative UAV platforms and sensor technologies tailored to meet the needs of smallholder farmers, agribusinesses, and environmental monitoring agencies [21].

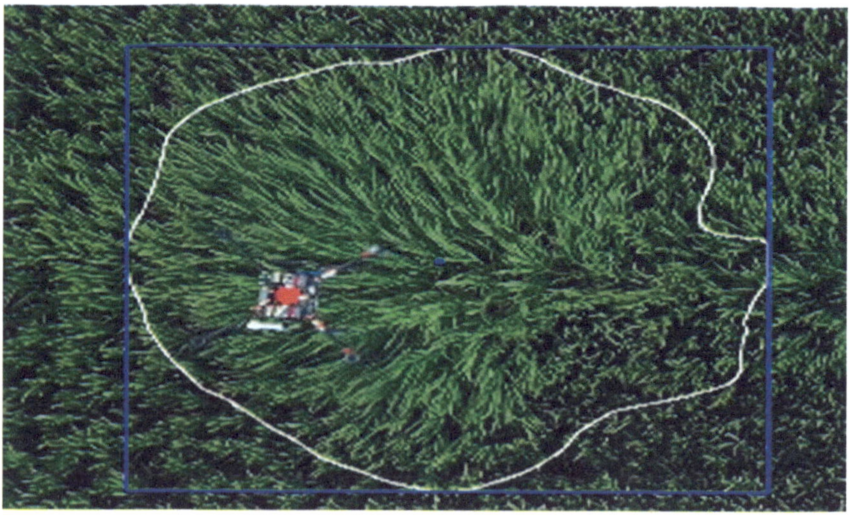

In addition to driving innovation and training professionals, academic institutions contribute to policy development and regulatory frameworks that govern UAV usage. Working alongside government bodies, universities provide research-based insights that inform regulations on drone flight operations, safety standards, data privacy, and environmental impact. This regulatory engagement helps ensure the responsible and scalable integration of UAVs into agriculture and other sectors [21]. A key focus for many university-led UAV programs is addressing challenges related to cost, accessibility, and sustainability. By designing affordable drone systems and open-source software solutions, academic institutions aim to make UAV technology accessible to small-scale farmers, particularly in developing countries. These efforts are vital for promoting equitable access to precision agriculture tools and bridging the digital divide in rural regions. Looking ahead, the role of universities in shaping the future of UAV technology will only intensify. Emerging areas of focus include the integration of artificial intelligence (AI) and machine learning into UAV platforms to enable real-time decision-making and predictive analytics. Research in this domain is expected to further enhance UAV autonomy, adaptability, and data analysis capabilities. In addition, universities are increasingly embracing global partnerships and collaborative networks to co-develop UAV applications that address transboundary agricultural issues, such as food security, climate resilience, and environmental conservation [22]. Open-source development is another future trajectory where universities are poised to make significant contributions. By promoting transparency and community-based innovation, open-source UAV projects reduce barriers to adoption and encourage the customization of drone solutions for diverse agroecological contexts. This democratization of drone technology aligns with broader goals of sustainability, resilience, and inclusive development [22].

1.3.4 Warfare to Workforce: UAVs Finding New Purpose in Agriculture

Unmanned Aerial Vehicles (UAVs), or drones, have revolutionized modern farming by enhancing precision, efficiency, and productivity. They provide real-time data, monitor crop health, and optimize resource management. UAVs equipped with multispectral and thermal cameras enable farmers to detect issues like nutrient deficiencies, diseases, and pest infestations, providing high-resolution images for early intervention. Precisely applying fertilizers, pesticides, and water is another benefit of UAVs. GPS-guided spraying systems ensure even distribution, minimizing overuse and runoff. Soil and field analysis is crucial for effective farming, with UAVs generating detailed 3D maps that assess soil conditions, moisture levels, and nutrient content [4]. This data aids farmers in making informed decisions about planting patterns, irrigation schedules, and fertilization plans. Irrigation management is another benefit of UAVs. Thermal imaging sensors can identify areas experiencing water stress, enabling farmers to adjust irrigation accordingly, improving crop health and sustainability. Livestock monitoring is another benefit of UAVs. Infrared cameras can detect sick animals early, allowing timely intervention. Yield estimation and forecasting are also benefits of UAVs. By analyzing plant growth and density, UAVs contribute to accurate yield prediction, helping farmers plan for storage, marketing, and financial management. This reduces post-harvest losses and improves overall farm profitability. Disaster and risk management are also significant applications of UAVs [5]. Quick aerial surveys help farmers and authorities take prompt action to mitigate losses and apply for insurance claims efficiently. As technology advances, UAV applications in agriculture will continue to expand, making them an essential component of future farming systems.

1.3.5 Leading Contributors to UAV Innovation in Agriculture

The adoption of Unmanned Aerial Vehicles (UAVs) in agriculture has been driven by pioneering companies, researchers, and innovators who have developed drone technology specifically for precision farming, crop monitoring, and resource optimization. Key companies leading Agricultural UAV development include DJI Agriculture, Yamaha Motor Co., Ltd., Parrot Drones, SenseFly, Trimble Agriculture, Dr. José Luis Vázquez-Arellano, Dr. Craig Lobsey, and Dr. Salah Sukkarieh. DJI Agriculture developed the first commercial agricultural drones for precision spraying, focusing on AI-driven autonomous flight for crop monitoring and spraying. Yamaha Motor Co., Ltd. pioneered UAV crop spraying technology in the 1990s, using the RMAX drone for aerial pesticide application in rice fields. Parrot Drones introduced lightweight, fixed-wing drones for large-scale crop surveying and developed multispectral imaging technology for plant health monitoring [12]. SenseFly created fully autonomous fixed-wing drones optimized for large farm surveys and developed high-resolution aerial mapping solutions for soil analysis. Trimble Agriculture pioneered drone-based precision agriculture by integrating UAVs with GPS, GIS, and AI-driven analytics.

Influential researchers and innovators in Agricultural UAVs include Dr. José Luis Vázquez-Arellano, Dr. Craig Lobsey, and Dr. Salah Sukkarieh. Dr. Vázquez-Arellano developed UAV-based remote sensing systems for crop stress detection and AI-powered UAV solutions for real-time agricultural decision-making. Dr. Lobsey worked on robotics and drone automation for precision farming and developed autonomous UAV swarms for large-scale crop monitoring. Dr. Sukkarieh developed AI-powered UAVs for fruit and crop detection and real-time sensor integration for farm automation. Future pioneers and innovations in Agricultural UAVs include AI-powered UAVs capable of identifying plant diseases and optimizing farm inputs, swarm drone technology, nano-drones for pollination, and space-based UAV monitoring [13]. With continuous advancements in AI, automation, and data analytics, UAVs will play an even greater role in shaping the future of precision agriculture and global food security.

1.3.6 From Film to AI: The Evolution of Sensors and Imaging

The evolution of sensors and imaging technologies has significantly impacted the capabilities of Unmanned Aerial Vehicles (UAVs), particularly in agriculture. Early imaging technologies included basic cameras capable of capturing low-resolution aerial photographs, while digital imaging introduced high-resolution cameras, allowing for clearer and more detailed aerial photographs [14]. This allowed for real-time data transmission, enabling farmers and operators to view images and videos instantly. Multispectral and hyperspectral imaging revolutionized crop monitoring by capturing data across specific wavelengths, such as visible and near-infrared. Hyperspectral imaging captured data across hundreds of narrow spectral bands, allowing for more detailed analysis of crop conditions and soil properties. Thermal imaging enabled farmers to identify water stress, irrigation issues, and pest infestations, making it useful for monitoring large fields and optimizing water use. Light Detection and Ranging (LiDAR) technology provided 3D mapping of terrain and vegetation, which was invaluable for topographic surveys, soil analysis, and forest management [15]. LiDAR data helped farmers understand field topography, plan irrigation systems, and manage land more effectively.

Advancements in sensor integration have led to multi-sensor platforms, miniaturization, AI and machine learning, data analytics, automation, and applications in agriculture. Multispectral and hyperspectral imaging provide detailed insights into plant health, helping farmers detect issues early and take corrective actions. Soil analysis using sensors like LiDAR and ground-penetrating radar (GPR) helps analyze soil structure, moisture levels, and nutrient content. Irrigation management using thermal imaging and soil moisture sensors optimizes practices, conserving water and improving crop yields. Future trends in sensor technology include enhanced resolution, cost reduction, and integration with IoT devices. As sensor technology becomes more affordable, it will become accessible to a wider range of farmers, including small-scale operations [16]. The combination of drone sensors with IoT devices will enable real-time monitoring and decision-making, further enhancing precision agriculture. The evolution of sensors and imaging technologies has significantly enhanced the capabilities of drones in agriculture, enabling farmers to monitor crops, manage resources, and optimize yields with unprecedented precision.

1.3.7 The Global Shift: Adoption Trends of Precision Agriculture Technologies

The global adoption of Unmanned Aerial Vehicles (UAVs) in agriculture is rapidly increasing due to advancements in precision farming, automation, and sustainability initiatives. Governments, agricultural industries, and research institutions are promoting UAVs for crop monitoring, spraying, and resource optimization, leading to a shift in global farming practices. Key drivers of UAV adoption include precision agriculture, labor shortages, climate change, and government regulations and subsidies. Regional adoption trends include North America, Europe, Asia–Pacific, Latin America, Africa, and the Middle East. North America leads in precision agriculture with regulatory support from the FAA and Transport Canada, large-scale farms, and tech investments from companies like John Deere, Trimble, and AgEagle [17]. Europe is leading adoption with strict environmental laws, government-funded projects, and AI and robotics integration.

The Asia–Pacific region sees rapid growth in agricultural drones, with China dominating the market with DJI's Agras series, Japan promoting spraying drones due to limited agricultural labor, India promoting drone subsidies for small-scale farmers through initiatives like PM Kisan Drone Yojana, and Australia using large-scale UAVs for livestock monitoring, irrigation management, and weed control. Latin America boosts agricultural productivity with precision agriculture growth, Brazil, Argentina, and Mexico supporting drone-based spraying and soil analysis, and private sector investments from companies like XMobots in Brazil [18]. Africa and the Middle East are emerging markets for UAVs, with South Africa, Kenya, UAE, and Israel leading UAV adoption in the region. Key trends include water management, drought monitoring, small-scale farmer initiatives, and agricultural drone startups. Challenges in UAV adoption include regulatory barriers, high initial costs, training and awareness, and data management issues. Future trends include AI and machine learning, swarm drones, 5G and IoT integration, and autonomous UAVs [19]. While challenges remain, the future of smart, AI-powered drone farming looks promising, transforming global food production and resource management.

1.4 Global Trends in Technology Adoption: From Innovation to Integration

The adoption of drones in agriculture has significantly transformed traditional farming practices by enhancing efficiency, precision, and productivity. Drones, also known as Unmanned Aerial Vehicles (UAVs), have revolutionized modern agriculture by providing real-time data, reducing manual labor, and improving decision-making. Early use and experimentation focused on aerial photography and crop monitoring, but battery life, sensor quality, and costs restricted widespread use. Technological advancements, including improved sensors, artificial intelligence (AI), and automation, have accelerated adoption of drone technology [23]. Key developments include multispectral and thermal imaging, precision spraying, GPS and AI integration, and crop monitoring and health assessment. Key applications in agriculture include crop monitoring and health assessment, precision agriculture and variable rate application (VRA), aerial seeding and planting, livestock monitoring, soil and field analysis, and IoT integration. Challenges remain, such as high initial costs, regulatory issues, and data management.

The future of drones in agriculture is promising, with AI-driven analytics, autonomous swarm technology, and IoT integration connecting drones with smart farming systems for real-time decision-making. Machine learning will improve predictive analytics for farming, autonomous swarm technology, and IoT integration. The adoption of drones in agriculture continues to grow, offering farmers innovative solutions to improve efficiency, sustainability, and profitability. Factors driving adoption include technological advancements, precision farming needs, cost efficiency, government support, environmental concerns, and increased efficiency. Drones equipped with multispectral and thermal cameras provide detailed insights into crop health, helping farmers detect diseases, nutrient deficiencies, and pest infestations early [24]. Precision spraying allows drones to precisely apply fertilizers, pesticides, and herbicides, reducing chemical use and environmental harm. Thermal imaging from drones identifies areas with water stress, allowing for efficient irrigation practices. Planting and seeding involve specialized drones dispersing seeds or seedlings, particularly in hard-to-reach or large areas, improving planting efficiency. Livestock monitoring helps track livestock movements, monitor health, and even assist in herding, saving time and labor. Yield prediction is achieved by analyzing crop growth patterns, helping farmers predict yields and plan harvests more accurately. Benefits of drone adoption include increased efficiency, cost savings, improved yields, sustainability, and data-driven decisions [25]. However, challenges include high initial costs, regulatory restrictions, technical expertise, limited battery life, and data management. Future prospects for drone adoption in agriculture include advancements in AI-powered analytics, longer battery life, and autonomous swarms of drones. As governments and organizations promote sustainable farming practices, drones will play an increasingly important role in addressing global challenges like food security and climate change. With ongoing improvements in affordability and accessibility, drones are set to become an integral part of modern agriculture, transforming the way farmers manage their land and crops. The adoption of drones in agriculture has revolutionized farming practices, offering numerous benefits in terms of efficiency, sustainability, and productivity [26]. While challenges remain, the future of drone technology in agriculture is promising, with the potential to drive significant advancements in the industry.

1.5 Technological Contributions and Agronomic Advantages of UAV

The integration of Unmanned Aerial Vehicles (UAVs) in agriculture has revolutionized traditional farming practices by providing accurate, real-time data, reducing costs, and enhancing productivity. UAVs play a critical role in precision agriculture, helping farmers make data-driven decisions that lead to higher yields and more sustainable farming methods. The adoption of drones in agriculture has grown significantly over the past decade, driven by their ability to enhance efficiency, reduce costs, and promote sustainable farming practices [27]. Factors driving adoption include technological advancements, precision farming needs, cost efficiency, government support, environmental concerns, and applications of drones in agriculture. Technological advancements have made drones more accessible and effective for agricultural use, with improvements in drone hardware and software. Precision farming needs have fueled the adoption of drones, as they require accurate data to optimize inputs like water, fertilizers, and pesticides. Cost efficiency is achieved through automated tasks like crop monitoring, spraying, and mapping, while government support has introduced subsidies, training programs, and relaxed regulations to encourage the use of drones in agriculture. Drones also help minimize the environmental impact of farming by enabling targeted application of resources, reducing chemical runoff, and conserving water [28]. Applications of drones in agriculture include crop monitoring, precision spraying, soil and field analysis, irrigation management, planting and seeding, livestock monitoring, and yield prediction. Benefits of drone adoption include increased efficiency, cost savings, improved yields, sustainability, and data-driven decisions. However, challenges in adoption include high initial costs, regulatory restrictions, technical expertise, limited battery life, and data management. Advanced drones with specialized sensors can be expensive, posing a barrier for small-scale farmers. Regulatory restrictions on drone usage, including flight permissions and licensing, can hinder adoption [29]. Technical expertise may require farmers to operate drones and interpret the data they collect, and drones have short flight times, requiring frequent recharging or battery swaps.

Future prospects for drone adoption include AI and machine learning integration, swarm technology, and autonomous drones. These self-operating drones are capable of completing tasks without human intervention, making farming more efficient, cost-effective, and environmentally sustainable. The adoption of drones in agriculture has revolutionized farming practices, offering numerous benefits in terms of efficiency, sustainability, and productivity. While challenges remain, the future of drone technology in agriculture is promising, with the potential to drive significant advancements in the industry. As technology continues to advance, innovations such as AI-powered analytics, longer battery life, and autonomous swarms of drones will further enhance their capabilities [30]. Additionally, as governments and organizations promote sustainable farming practices, drones will play an increasingly important role in addressing global challenges like food security and climate change. With ongoing improvements in affordability and accessibility, drones are set to become an integral part of modern agriculture, transforming the way farmers manage their land and crops.

1.6 Enhanced Precision in Crop Management

Precision agriculture relies on the use of advanced technologies, such as Unmanned Aerial Vehicles (UAVs), to improve farm operations and optimize crop growth. UAVs provide real-time data, allowing farmers to make data-driven decisions that optimize crop growth, reduce resource wastage, and increase overall yields. They enable farmers to apply inputs like water, fertilizers, and pesticides only where and when they are needed, leading to higher efficiency, reduced costs, and environmental sustainability. UAVs enhance precision in crop management through high-resolution imaging, real-time analytics, and automation, leading to site-specific interventions that improve plant health and productivity [31]. Crop health monitoring uses multi-spectral and thermal imaging to detect crop stress, nutrient deficiencies, and diseases before they are visible to the human eye. Variable Rate Application (VRA) of inputs allows for precise application of fertilizers, pesticides, and herbicides by analyzing variations in soil and plant health. This reduces overuse of chemicals, prevents nutrient runoff, and lowers input costs. Irrigation optimization and water conservation are also achieved through thermal imaging from UAVs, which detects variations in soil moisture, helping farmers optimize irrigation schedules and identify dry spots and over-watered areas. Pest and disease control is another advantage of UAVs, as they can identify pest-infested areas and implement localized pest control measures instead of widespread pesticide application [32].

The advantages of enhanced precision in crop management include increased yield and productivity, cost reduction, environmental sustainability, real-time decision making and data-driven farming, and the integration of AI and machine learning. Autonomous swarm UAVs work together to scan, spray, and monitor large farms simultaneously, while IoT and smart farming systems integrate drones with soil sensors, weather stations, and automated irrigation systems. Enhancing precision in crop management using UAVs optimizes farm productivity, reduces resource wastage, and promotes sustainability [33]. With continuous advancements in drone technology, smart farming practices will become even more efficient, ensuring global food security and agricultural resilience.

1.7 Enhancing Agricultural Sustainability Through Cost-Effective UAV Technologies

The integration of drones in agriculture has significantly improved cost efficiency and sustainability. By leveraging advanced technologies, drones help farmers reduce operational costs, optimize resource use, and adopt eco-friendly practices. Cost efficiency is achieved through automation of tasks such as crop monitoring, spraying, and mapping, which traditionally require significant manual effort. Precision application of inputs, faster data collection, and reduced operational costs are some of the benefits of drones. Sustainability is another benefit of drones [34]. They reduce chemical use, water conservation, soil health preservation, lower carbon footprint, biodiversity protection, and sustainable land management. Drones provide high-resolution maps and data that help farmers plan land use more effectively, ensuring sustainable management of agricultural resources.

Examples of cost efficiency and sustainability in action include precision agriculture, water management, pest control, and soil conservation. Precision agriculture can reduce fertilizer use by 20–30%, saving costs and minimizing environmental harm. Water management can be optimized by optimizing irrigation in water-scarce regions, while pest control can be achieved by detecting pest infestations early and applying targeted pesticides. Soil conservation can be achieved by identifying erosion-prone areas and implementing conservation measures like contour plowing and terracing. However, there are challenges to achieving cost efficiency and sustainability. High initial investment, technical expertise, and regulatory barriers can be barriers for small-scale farmers [35]. Solutions include government subsidies, leasing options, and shared drone services. Technical expertise can be addressed through training programs and user-friendly drone software. Regulations on drone usage can be simplified and provided clear guidelines. Innovations like AI-powered analytics, longer battery life, and autonomous swarms will further enhance their role in promoting cost efficiency and sustainability. As global awareness of environmental issues grows, drones will play an increasingly important role in helping farmers adopt sustainable practices while maintaining profitability [36]. In summary, drones are transforming agriculture by making farming more cost-efficient and sustainable, ensuring a balance between productivity and ecological preservation.

1.7.1 Cost-Efficient Approaches for Achieving Long-Term Sustainability

Unmanned Aerial Vehicles (UAVs) are transforming agriculture by providing cost-effective solutions that enhance productivity, reduce expenses, and optimize resource management. By integrating precision agriculture technologies, UAVs help farmers make data-driven decisions, ultimately lowering input costs and improving profitability. One of the most significant cost-saving aspects of UAVs is their ability to reduce input costs, particularly fertilizers, pesticides, and water. Precision spraying technology ensures that chemicals are applied only where needed, reducing waste by up to 50% [37]. This targeted application not only lowers costs but also minimizes environmental impact. Similarly, UAVs equipped with multispectral imaging help assess soil moisture levels, allowing for efficient irrigation management and preventing unnecessary water use. By avoiding overuse of resources, farmers can significantly cut expenses while maintaining high crop yields.

UAVs also contribute to lower labor and machinery costs. Traditional farming methods require a large workforce for manual tasks like crop scouting, spraying, and monitoring. UAVs automate these tasks, reducing the need for human labor and mitigating labor shortages. Additionally, UAVs replace heavy machinery such as tractors and sprayers, which consume large amounts of fuel and require frequent maintenance. Since UAVs run on rechargeable batteries, they provide a more energy-efficient alternative, cutting down fuel costs and reducing greenhouse gas emissions.

Another major financial benefit of UAVs is increased crop yield and profitability. Drones equipped with advanced sensors can detect crop diseases, pest infestations, and nutrient deficiencies early, allowing farmers to take preventive measures before major damage occurs [38]. Early detection minimizes crop losses and ensures higher yields, directly contributing to greater profits. Furthermore, UAVs provide accurate field data, helping farmers make better decisions regarding planting schedules, harvesting times, and resource allocation Beyond direct cost savings, UAVs also aid in risk mitigation and insurance benefits. Drones can assess damage caused by natural disasters, such as floods or droughts, helping farmers document losses for insurance claims. Some insurance companies even offer lower premiums for farms that utilize UAV technology, as drones provide accurate risk assessments and monitoring capabilities. Low-cost UAV models, government subsidies, and drone rental services are increasing accessibility. Additionally, AI-driven UAVs and IoT-integrated smart farming solutions are further enhancing efficiency by automating data analysis and farm management processes [39]. As UAV technology continues to evolve, farmers will benefit from increased affordability, higher efficiency, and improved sustainability.

1.7.1.1 Smart Agriculture for Reduced Input Costs: Enhancing Efficiency in Fertilizer, Pesticide, and Water Use

UAVs play a crucial role in minimizing agricultural input costs by optimizing the application of fertilizers, pesticides, and water. Traditional farming methods often lead to overuse of these resources, increasing expenses and causing environmental harm. UAVs, equipped with precision spraying technology and advanced sensors, ensure that inputs are applied only where necessary, improving efficiency and reducing waste One of the major advantages of UAVs is their ability to reduce chemical waste by up to 30–50% through precision spraying [40]. Traditional spraying methods often result in excessive application, leading to unnecessary expenses and potential soil degradation. UAVs, however, use targeted spraying techniques to apply the right amount of fertilizers and pesticides, minimizing costs while maintaining crop health. This precision approach not only saves money but also reduces the environmental impact of chemical runoff. Similarly, UAVs enhance water conservation through targeted irrigation mapping. By using thermal and multispectral imaging, UAVs can detect moisture levels in the soil, helping farmers determine which areas need watering and which do not. This prevents overwatering, reducing water consumption costs and preserving vital water resources. Efficient irrigation not only lowers expenses but also promotes sustainable water management, a critical factor in regions facing water scarcity. By integrating UAV technology, farmers can achieve higher productivity at lower costs, ensuring sustainable and profitable agricultural practices [41]. As precision agriculture continues to advance, the cost-saving benefits of UAVs in fertilizer, pesticide, and water management will become even more significant.

1.7.1.2 Lower Labor and Machinery Expenses

UAVs significantly reduce labor and machinery costs in agriculture by automating essential tasks such as field scouting, pest control, and spraying. Traditionally, these operations require a large workforce and expensive ground equipment, increasing farming expenses. However, UAVs streamline these processes, reducing reliance on manual labor and costly machinery. By replacing tractors, boom sprayers, and ground-based scouting vehicles, UAVs help cut fuel expenses and lower maintenance costs. Heavy farming machinery requires frequent servicing, fuel, and repairs, which add to operational expenses. In contrast, UAVs operate on rechargeable batteries, making them a cost-effective and eco-friendly alternative. The reduced need for multiple passes across the field also minimizes soil compaction, preserving soil health and improving long-term productivity [42]. Additionally, UAVs address the issue of rising agricultural labor shortages by automating repetitive tasks. With fewer workers needed for scouting, monitoring, and spraying, farmers can allocate labor resources more efficiently, reducing wage-related costs. Automated drone operations also enable continuous farm monitoring, improving efficiency while lowering dependency on seasonal labor.

1.7.1.3 Improved Yield and Profitability

UAVs play a crucial role in boosting crop yield and farm profitability by providing early detection of diseases, pests, and nutrient deficiencies. Traditional monitoring methods can be time-consuming and inefficient, leading to delayed interventions and crop losses. UAVs, equipped with multispectral and thermal imaging sensors, enable farmers to identify plant stress at an early stage, allowing for quick corrective actions before major damage occurs. Accurate and real-time field data collection further enhances decision-making in farm management. UAVs provide precise insights into soil health, irrigation needs, and crop growth patterns, helping farmers optimize input application and planting schedules. By reducing uncertainty and improving efficiency, farmers can minimize losses and maximize productivity. Higher crop yields directly translate into greater profitability [43]. By preventing losses and optimizing resource use, UAVs help farmers achieve higher returns on investment (ROI). As UAV technology becomes more affordable and accessible, its ability to increase productivity and profitability makes it an essential tool for modern, sustainable agriculture.

1.7.1.4 Reduced Fuel and Energy Costs

Traditional farming heavily depends on fuel-powered machinery such as tractors, sprayers, and scouting vehicles, leading to high fuel consumption and operational costs. The rising costs of diesel and gasoline add to the financial burden on farmers, making fuel efficiency a key concern in modern agriculture. UAVs offer a cost-effective and sustainable alternative by operating on electricity or rechargeable batteries, significantly reducing fuel expenses. Unlike traditional ground equipment, drones require minimal energy to cover large fields, making them a more efficient and eco-friendly solution. The transition from fuel-powered machinery to battery-operated UAVs also contributes to lower greenhouse gas emissions, promoting sustainable farming practices. Additionally, automated flight paths and GPS navigation ensure minimal energy wastage during field surveys and crop monitoring [44]. By optimizing drone movements, farmers can maximize coverage while minimizing power consumption, further enhancing cost savings. By replacing fuel-intensive machinery with UAV technology, farmers can lower energy expenses, improve efficiency, and reduce their environmental footprint, making agriculture more profitable and sustainable.

1.7.1.5 Insurance and Risk Mitigation

UAVs play a vital role in reducing financial risks for farmers by providing accurate data on crop health, soil conditions, and potential weather threats. Traditional methods of risk assessment rely on manual inspections, which can be time-consuming and less precise. UAVs, equipped with advanced imaging and real-time monitoring capabilities, offer detailed insights that help farmers secure better insurance coverage by providing concrete evidence of field conditions In the event of natural disasters such as floods, droughts, or pest outbreaks, UAVs enable rapid post-disaster assessments. By capturing high-resolution aerial images, farmers can quickly document damage, making the insurance claim process more efficient and transparent. This ensures faster compensation, helping farmers recover financially and restore productivity [45]. By integrating UAV technology for risk assessment and insurance documentation, farmers can minimize financial losses, improve resilience against unforeseen events, and secure more favorable insurance policies, ensuring greater stability in agricultural operations.

1.7.2 Sustainability Benefits of UAVs in Agriculture

UAVs promote sustainable farming by reducing environmental impact, conserving natural resources, and minimizing chemical use.

1.7.2.1 Reduced Chemical Pollution and Soil Contamination

UAVs contribute significantly to reducing chemical pollution and soil contamination by ensuring precise application of herbicides, pesticides, and fertilizers. Traditional farming methods often involve broadcast spraying, which can lead to over-application, soil degradation, and contamination of surrounding ecosystems. In contrast, UAVs use precision spraying technology to apply chemicals only where needed, minimizing environmental impact. One of the biggest concerns in conventional agriculture is chemical runoff, where excess fertilizers and pesticides wash into rivers, lakes, and groundwater sources, leading to water pollution and ecosystem damage [46]. UAVs reduce this risk by targeting specific areas, preventing unnecessary chemical spread and promoting sustainable farming practices. By optimizing chemical application, UAVs help maintain soil health, preserve biodiversity, and ensure long-term agricultural sustainability, making them a crucial tool for eco-friendly farming.

1.7.2.2 Water Conservation and Efficient Irrigation Management

UAV technology is transforming water management in agriculture by enabling precise and efficient irrigation. Drones equipped with thermal and multispectral imaging cameras can detect variations in soil moisture levels, allowing farmers to apply water only where needed. This targeted approach helps prevent overwatering and under-watering, ensuring crops receive the optimal amount of moisture for healthy growth. Smart drone-assisted irrigation management is especially valuable in drought-prone regions, where water conservation is critical [47]. By identifying water-stressed areas, UAVs help farmers prioritize irrigation, reducing water wastage and promoting sustainable resource use. This technology also supports precision agriculture practices, improving crop yield while lowering water consumption costs.

1.7.2.3 Soil Health Preservation

UAVs play a crucial role in maintaining soil health and fertility by eliminating the negative impact of heavy agricultural machinery. Traditional farming equipment, such as tractors and boom sprayers, compacts the soil, reducing aeration and affecting root development. In contrast, UAVs operate above the ground, preventing soil compaction and preserving its natural structure for better water infiltration and nutrient absorption. Additionally, UAVs enable Variable Rate Application (VRA) of fertilizers, ensuring that nutrients are evenly distributed across the field [48]. This precision nutrient management prevents over-fertilization and soil degradation, maintaining soil fertility for long-term sustainability. By applying fertilizers only where needed, UAVs reduce chemical buildup, minimizing the risk of soil acidification and contamination.

1.7.2.4 Reduction in Greenhouse Gas Emissions

UAVs contribute to lowering greenhouse gas emissions by replacing traditional fuel-powered agricultural machinery such as tractors, sprayers, and scouting vehicles. These conventional machines consume large amounts of diesel and gasoline, releasing carbon dioxide (CO_2) and other greenhouse gases into the atmosphere. In contrast, UAVs operate on electricity or rechargeable batteries, significantly reducing fossil fuel dependency and emissions. Additionally, UAVs enhance resource efficiency by optimizing the use of fertilizers, pesticides, and water, reducing unnecessary applications that contribute to environmental pollution [49]. By ensuring precision agriculture practices, UAVs help lower the overall carbon footprint of farming operations, making agriculture more climate-friendly and sustainable.

1.7.2.5 Enhanced Biodiversity and Eco-Friendly Farming

UAV technology supports biodiversity and sustainable farming practices by enabling precise monitoring of crop rotations, soil diversity, and pollinator-friendly environments. By analyzing vegetation patterns and soil conditions, UAVs help farmers implement diverse planting strategies that maintain soil health and encourage beneficial organisms. Additionally, UAVs reduce chemical exposure by ensuring targeted application of pesticides and fertilizers, minimizing the impact on surrounding ecosystems. This supports wildlife conservation, protecting essential species like bees, butterflies, and other pollinators that play a vital role in agriculture [50]. With fewer chemical residues in the environment, natural pest control mechanisms, such as predator insects and birds, can thrive, reducing reliance on synthetic pesticides.

1.7.2.6 Climate Adaptation and Disaster Response

UAVs play a crucial role in helping farmers adapt to climate change and respond effectively to natural disasters. By providing real-time climate data, UAVs enable farmers to monitor temperature shifts, rainfall patterns, and drought conditions, allowing them to make informed decisions about crop selection, irrigation, and planting schedules. This proactive approach helps mitigate the risks associated with unpredictable weather changes. In the aftermath of natural disasters such as floods, droughts, or storms, UAVs assist in rapid damage assessment. Equipped with high-resolution imaging and mapping capabilities, drones can quickly survey affected farmland, identifying damaged crops, soil erosion, and areas in need of urgent intervention [28]. This data helps farmers and authorities develop efficient recovery and restoration plans, minimizing economic losses and ensuring faster rehabilitation of agricultural land.

1.7.3 Future Outlook: Cost-Effective and Sustainable UAV Innovations

The future of UAVs in agriculture will focus on making drone technology more affordable, energy-efficient, and eco-friendly.

1.7.3.1 Lower-Cost UAVs for Small Farmers

Advancements in drone manufacturing and technology are making UAVs more affordable and accessible for small and mid-sized farms. As production costs decrease and more efficient, compact models are developed, farmers with limited budgets can integrate UAV technology into their operations without heavy financial burdens. Additionally, government subsidies, financial incentives, and UAV rental services are playing a key role in expanding drone adoption. Many governments and agricultural organizations are offering grants or tax breaks to encourage small-scale farmers to use precision agriculture technologies [51]. Drone rental services also provide an alternative for farmers who cannot afford to purchase UAVs outright, allowing them to benefit from aerial monitoring, precision spraying, and data analysis at a lower cost.

1.7.3.2 AI and IoT Integration for Smarter Resource Management

The integration of Artificial Intelligence (AI) and the Internet of Things (IoT) with UAV technology is transforming agriculture by enabling smarter, data-driven resource management. AI-powered UAVs can automatically analyze crop health, predict yields, and optimize irrigation plans by processing vast amounts of aerial data. Advanced machine learning algorithms help identify pest infestations, nutrient deficiencies, and disease outbreaks with high accuracy, allowing farmers to take timely corrective actions and increase productivity. The use of 5G and IoT-connected UAVs ensures real-time farm monitoring, allowing seamless data transfer between drones, sensors, and farm management systems. These smart UAVs can work autonomously, adjusting operations based on weather conditions, soil moisture levels, and crop needs, ensuring efficient use of water, fertilizers, and pesticides [52]. This reduces resource waste, lowers costs, and minimizes environmental impact, leading to sustainable and precision-driven agriculture.

1.7.3.3 Solar-Powered and Autonomous UAVs

The future of agricultural UAVs is shifting toward solar-powered and fully autonomous drones, enhancing energy efficiency and operational capabilities. Solar-powered UAVs harness renewable energy, reducing dependence on frequent battery charging and extending flight durations. This innovation is particularly beneficial for large-scale farms, where continuous monitoring and coverage are required over vast areas. By minimizing energy consumption, solar-powered drones contribute to sustainable farming practices and lower operational costs. Additionally, advancements in autonomous UAV technology are enabling the deployment of drone swarms—groups of UAVs working together with minimal human intervention [53]. These drones can coordinate tasks such as crop monitoring, spraying, planting, and soil analysis, significantly reducing labor requirements and improving efficiency. Equipped with AI and machine learning, these UAVs can adapt to real-time field conditions, optimizing resource use and increasing farm productivity.

1.7.3.4 Government Policies Promoting Sustainable UAV Usage

Governments worldwide are recognizing the environmental and economic benefits of UAV technology in agriculture and are implementing policies to promote its sustainable use. By introducing eco-friendly UAV regulations, authorities aim to ensure that drone operations align with green agriculture initiatives, reducing chemical runoff, optimizing water usage, and minimizing carbon footprints. To encourage farmers to adopt UAV technology, many governments offer tax incentives, subsidies, and grants [54]. These financial supports help offset the high initial costs of drone adoption, making advanced UAV solutions more accessible to small and medium-scale farmers. Additionally, governments are investing in training programs and research to enhance farmers' technical knowledge and encourage the responsible use of UAVs.

1.8 Conclusion and Summary

UAV technology is revolutionizing agriculture by enhancing productivity, reducing costs, and promoting sustainability. Despite challenges like high costs and regulations, continuous innovations in AI, automation, and data analytics are making UAVs an essential tool for modern precision farming. As adoption increases globally, UAVs will play a crucial role in ensuring food security, environmental conservation, and the future of smart agriculture. The integration of UAV technology in modern farming has transformed traditional practices, offering numerous benefits in terms of efficiency, sustainability, and productivity. As technology continues to advance, their role in agriculture will only grow, helping farmers meet the demands of a rapidly changing world while ensuring food security and environmental sustainability. Enhanced precision in crop management using UAVs optimizes farm productivity, reduces resource wastage, and promotes sustainability. With continuous advancements in drone technology, smart farming practices will become even more efficient, ensuring global food security and agricultural resilience.

The evolution of UAV technology has transformed aerial technology, evolving from early experimental designs to highly advanced, AI-driven systems used across multiple industries. As technology progresses, UAVs will continue to shape military strategies, commercial applications, and environmental conservation efforts, making them an indispensable tool in modern society. The transition of UAV technology from military applications to farming has revolutionized modern agriculture by enabling precision farming, efficient resource management, and sustainable practices. As drone technology continues to advance, fully autonomous farms may soon become a reality, ensuring global food security and environmental sustainability. The evolution of sensors and imaging technologies has significantly enhanced the capabilities of drones in agriculture, from basic cameras to advanced multispectral, thermal, and LiDAR sensors. As technology continues to advance, the role of drones in agriculture will only grow, driving further improvements in efficiency, sustainability, and productivity. The global adoption of UAVs in agriculture is expanding rapidly, driven by technological advancements, sustainability goals, and regulatory support.

Universities are essential drivers of research and innovation in UAV technology and its applications in agriculture. Through their research, education, and collaboration efforts, they contribute to the development of advanced drone solutions that enhance farming efficiency, sustainability, and productivity. Their role will continue to be critical as the agricultural sector embraces new technologies to meet the challenges of the future. In conclusion, UAVs are revolutionizing agriculture by providing cost-effective and sustainable solutions. By reducing input costs, optimizing resource use, and minimizing environmental impact, UAVs help farmers achieve higher profitability while promoting eco-friendly farming practices. As automation, AI, and renewable energy continue to evolve, UAVs will play a critical role in shaping the future of precision farming and global food security.

References

1. El Alaoui M, Amraoui KE, Masmoudi L et al (2024) Unleashing the potential of IoT, artificial Intelligence, and UAVs in contemporary agriculture: a comprehensive review. J Terrramech 115:100986
2. Gouiza N, Jebari H, Reklaoui K (2024) Integration of Iot-enabled technologies and artificial intelligence in diverse domains: recent advancements and future trends. J Theor Appl Inf Technol 102:1975–2029
3. Sarfraz S, Ali F, Hameed A, et al (2023) Sustainable agriculture through technological innovations. In: Sustainable agriculture in the era of the OMICs revolution. Springer, pp 223–239
4. Vellingiri A, Kokila R, Nisha P, et al (2025) Harnessing GPS, sensors, and drones to minimize environmental impact: precision agriculture. In: Designing sustainable internet of things solutions for smart industries. IGI Global, pp 77–108
5. Chandel NS, Chakraborty SK, Jat D, Chouhan P (2024) Smart farming management system: pre and post-production interventions. In: Artificial intelligence techniques in smart agriculture. Springer, pp 67–82
6. Vincent S (2019) Innovation, technology and security: the emergence of unmanned aerial vehicles before and after 9/11
7. Kunertova D (2024) Learning from the Ukrainian battlefield: tomorrow's drone warfare, today's innovation challenge. ETH Zurich
8. Orgeira-Crespo P, García-Luis U (2024) Brief introduction to unmanned aerial systems. In: Applying drones to current societal and industrial challenges. Springer, pp 1–22
9. Messina G, Modica G (2022) The role of remote sensing in olive growing farm management: a research outlook from 2000 to the present in the framework of precision agriculture applications. Remote Sensing 14:5951
10. Emimi M, Khaleel M, Alkrash A (2023) The current opportunities and challenges in drone technology. Int J Electr Eng and Sustain 74–89
11. Singh R, Kumar S (2025) A comprehensive insights into drones: history, classification, architecture, navigation, applications, challenges, and future trends. arXiv preprint arXiv:2501 10066
12. Gulak AM (2024) Implementation of drone for spraying herbicide and pesticide
13. Agrawal J, Arafat MY (2024) Transforming farming: a review of AI-powered UAV technologies in precision agriculture. Drones (2504–446X) 8
14. Olson D, Anderson J (2021) Review on unmanned aerial vehicles, remote sensors, imagery processing, and their applications in agriculture. Agron J 113:971–992

15. Guo Q, Su Y, Hu T et al (2020) Lidar boosts 3D ecological observations and modelings: a review and perspective. IEEE Geoscience and Remote Sensing Magazine 9:232–257
16. Dhillon R, Moncur Q (2023) Small-scale farming: a review of challenges and potential opportunities offered by technological advancements. Sustainability 15:15478
17. Krishna KR (2018) Agricultural drones: a peaceful pursuit. Apple Academic Press
18. Reis JG, Rios S, Fernandes JAC, da Motta Veiga P (2024) Reorganization of global value chains: are there opportunities for Brazil?
19. Javed S, Hassan A, Ahmad R et al (2024) State-of-the-art and future research challenges in UAV swarms. IEEE Internet Things J 11:19023–19045
20. Padhiary M, Kumar R, Sethi LN (2024) Navigating the future of agriculture: a comprehensive review of automatic all-terrain vehicles in precision farming. J Instit Eng (India): Series A 105:767–782
21. Njoku O, Gideon NM (2024) Adoption of precision agricultural techniques in vocational agriculture training programmes in universities in Abia state: a pathway to sustainable workforce development. J Assoc Vocational Tech Educators Nigeria (Javten) 29:241–252
22. Narne H (2024) Sustainable agricultural practices through technological innovations: the role of artificial intelligence and machine learning
23. Schaefer L (2023) An emerging era of artificial intelligence research in agriculture. J Robot Spectrum 1:036–046
24. Vashishth TK, Sharma V, Sharma KK, et al (2024) Integration of unmanned aerial vehicles (UAVs) and IoT for crop monitoring and spraying. In: Internet of things applications and technology. Auerbach Publications, pp 95–117
25. Titirmare S, Margal PB, Gupta S, Kumar D (2024) AI-powered predictive analytics for crop yield optimization. In: Agriculture 4.0. CRC Press, pp 89–110
26. Khan N, Ray RL, Sargani GR et al (2021) Current progress and future prospects of agriculture technology: gateway to sustainable agriculture. Sustainability 13:4883
27. Mihret YC, Takele MM, Mintesinot SM (2025) Advancements in Agriculture 4.0 and the needs for introduction and adoption in Ethiopia: a review. advances in agriculture 2025:8828400
28. Auma JA (2023) The application of drone technology in micro-watershed evaluation. J Geograph Assoc Tanzania 43:47–66
29. Singh P, Singh P (2023) Drones in Indian agriculture: trends, challenges, and policy implications
30. Rashid AB, Kausik AK (2024) AI revolutionizing industries worldwide: a comprehensive overview of its diverse applications. Hybrid Adv 100277
31. Upadhyay A, Zhang Y, Koparan C et al (2024) Advances in ground robotic technologies for site-specific weed management in precision agriculture: a review. Comput Electron Agric 225:109363
32. Lochan K, Khan A, Elsayed I, et al (2024) Advancements in precision spraying of agricultural robots: a comprehensive review. IEEE Access
33. Sabir RM, Mehmood K, Sarwar A, et al (2024) Remote sensing and precision agriculture: a sustainable future. In: Transforming agricultural management for a sustainable future: climate change and machine learning perspectives. Springer, pp 75–103
34. Maghazei O, Netland T (2020) Drones in manufacturing: exploring opportunities for research and practice. J Manuf Technol Manag 31:1237–1259
35. Mizik T (2021) Climate-smart agriculture on small-scale farms: a systematic literature review. Agronomy 11:1096
36. Rai AK, Bana SR, Sachan DS, Singh B (2023) Advancing sustainable agriculture: a comprehensive review for optimizing food production and environmental conservation. Int J Plant Soil Sci 35:417–425
37. Taseer A, Han X (2024) Advancements in variable rate spraying for precise spray requirements in precision agriculture using Unmanned aerial spraying Systems: a review. Comput Electron Agric 219:108841
38. Abbas A, Zhang Z, Zheng H et al (2023) Drones in plant disease assessment, efficient monitoring, and detection: a way forward to smart agriculture. Agronomy 13:1524

39. Sutikno T, Ardiansyah A, Pujiyanta A, Handayani L (2024) Artificial intelligence technologies in urban smart waste management. High-Tech Innovative Series 1:55–66
40. Sahni RK, Kumar SP, Thorat D, et al (2024) Drone spraying system for efficient agrochemical application in precision agriculture. In: Applications of computer vision and drone technology in agriculture 4.0. Springer, pp 225–244
41. Yadav N, Sidana N (2023) Precision agriculture technologies: analysing the use of advanced technologies, such as drones, sensors, and GPS, in precision agriculture for optimizing resource management, crop monitoring, and yield prediction. J Adv Zool 44
42. Ali G, Mijwil MM, Adamopoulos I, Ayad J (2025) Leveraging the Internet of things, remote sensing, and artificial intelligence for sustainable forest management. Babylonian J Internet Things 2025:1–65
43. Mgendi G (2024) Unlocking the potential of precision agriculture for sustainable farming. Discover Agriculture 2:87
44. Xie B, Jin Y, Faheem M et al (2023) Research progress of autonomous navigation technology for multi-agricultural scenes. Comput Electron Agric 211:107963
45. Munkombwe J (2024) Understanding the diffusion of weather index insurance among smallholder farmers in Zambia: case of four (4) selected districts
46. Zahoor I, Mushtaq A (2023) Water pollution from agricultural activities: a critical global review. Int J Chem Biochem Sci 23:164–176
47. Ahmed Z, Gui D, Murtaza G et al (2023) An overview of smart irrigation management for improving water productivity under climate change in drylands. Agronomy 13:2113
48. He L (2023) Variable rate technologies for precision agriculture. In: Encyclopedia of digital agricultural technologies. Springer, pp 1533–1542
49. Yablokova A, Kovalev D, Kovalev I, et al (2024) Environmental safety problems of swarm use of UAVs in precision agriculture. EDP Sciences, p 04018
50. Mujeeb SA, Divya P, Chaitanya G, Manisha B (2023) The vital role of pollinators in agriculture and ecosystems. Integrated Publications TM New Delhi 25
51. Thompson A (2024) A review and prospects on the use of unmanned aerial systems (UAS) in precision agriculture for the mid-Atlantic United States
52. Almalki FA, Soufiene BO, Alsamhi SH, Sakli H (2021) A low-cost platform for environmental smart farming monitoring system based on IoT and UAVs. Sustainability 13:5908
53. Campion M, Ranganathan P, Faruque S (2018) UAV swarm communication and control architectures: a review. J Unmanned Vehicle Syst 7:93–106
54. Agbelusi J, Arowosegbe OB, Alomaja OA et al (2024) Strategies for minimizing carbon footprint in the agricultural supply chain: leveraging sustainable practices and emerging technologies. World J Adv Res Rev 23:2625–2646

Chapter 2
Understanding UAV Aerodynamics in Agriculture

Abstract Understanding UAV aerodynamics is crucial for optimizing performance, efficiency, and stability in various applications. Key principles include lift, drag, stability, and control mechanisms in fixed-wing, rotary-wing, and hybrid UAVs. Factors like airfoil design, propeller dynamics, and environmental influences impact UAV efficiency and maneuverability. Computational fluid dynamics (CFD) plays a role in UAV design, along with challenges like wind resistance, structural limitations, and power constraints. Newton's laws of motion explain UAVs' stability and control, with the First Law describing how UAVs resist changes unless acted upon by external forces, the Second Law determining forces influencing acceleration and movement, and the Third Law describing lift and thrust generation. Mastering these principles allows for improved UAV applications in military operations, environmental monitoring, disaster management, agriculture, and commercial logistics. Bernoulli's principle is essential for understanding lift generation in fixed-wing UAVs. Continuous improvement in aerodynamics can lead to higher efficiency, longer endurance, and enhanced capabilities for various applications.

Keywords UAV aerodynamics · Lift-to-drag ratio · Airfoil design · Propeller dynamics · CFD in UAVs · Flight stability · UAV efficiency

2.1 Introduction

Aerodynamics is the study of how air interacts with moving objects, and it plays pivotal role in the design, performance, and operation of Unmanned Aerial Vehicles (UAVs). For drones, aerodynamics determines how efficiently they can generate lift, overcome drag, and maintain stability during flight. Whether it's a fixed-wing drone soaring over vast agricultural fields or a multirotor drone hovering precisely to spray crops, the principles of aerodynamics are fundamental to their functionality. UAVs rely on a delicate balance of four primary forces—lift, thrust, drag, and weight—to achieve stable and efficient flight [1]. Lift allows the drone to rise and stay airborne, thrust propels it forward, drag resists its motion, and weight pulls it downward.

© The Author(s), under exclusive license to Springer Nature Singapore Pte Ltd. 2025

J. Li, *UAV Aerodynamics and Crop Interaction*, Smart Agriculture 13,
https://doi.org/10.1007/978-981-96-8402-1_2

Understanding and optimizing these forces is critical for designing UAVs that can perform complex tasks, such as crop monitoring, precision spraying, and livestock management, with accuracy and efficiency. In addition to the forces of flight, UAV aerodynamics involves the design of wings, rotors, and propulsion systems, as well as the use of advanced materials and technologies to enhance performance [2]. For instance, lightweight materials like carbon fiber reduce weight without compromising strength, while advanced sensors and software improve stability and control. Aerodynamic optimization also extends to reducing energy consumption, improving payload capacity, and ensuring adaptability to varying environmental conditions, such as wind and altitude. As UAV technology continues to evolve, aerodynamics remains at the heart of innovation. From AI-driven flight control systems to sustainable materials and hybrid designs, advancements in aerodynamics are enabling drones to become more efficient, versatile, and environmentally friendly. This chapter explores the fundamental principles of UAV aerodynamics, their applications in agriculture, and the future trends shaping this dynamic field [3]. By understanding these concepts, we can unlock the full potential of UAVs to revolutionize farming and address global challenges like food security and resource management.

2.2 Principles of Aerodynamics

Aerodynamics is the cornerstone of UAV (Unmanned Aerial Vehicle) design and operation. It governs how air interacts with the UAV, influencing its ability to generate lift, overcome drag, and maintain stable flight. This section provides a detailed exploration of the principles of aerodynamics, including definitions, Newton's laws, Bernoulli's principle, and the four forces of flight, along with their applications and significance in UAV technology [4].

1. **Definition of Aerodynamics**

   Aerodynamics is the study of the behavior of air as it interacts with solid objects, such as UAVs. It involves analyzing the forces generated by air movement and how they affect the motion, stability, and efficiency of flying objects.
   Airflow: The movement of air around the UAV.
   Pressure Distribution: Variations in air pressure across the UAV's surfaces.
   Forces: Lift, drag, thrust, and weight, which determine flight dynamics.

2. **Importance in UAVs**

   Aerodynamics is the study of how air interacts with objects in motion, particularly UAVs, influencing their efficiency, stability, and control. It plays a crucial role in UAV performance by determining how effectively they generate lift while overcoming drag, optimizing energy consumption, and enhancing flight stability. Newton's laws of motion explain the fundamental principles of flight—objects remain in motion unless acted upon by external forces, acceleration depends on force and mass, and every action has an equal and opposite reaction, as seen in UAV propulsion [5].

Bernoulli's principle further explains lift generation, where faster airflow over a curved wing surface creates lower pressure above the wing, allowing higher pressure beneath to push it upward. UAV flight is governed by four key forces: lift, which counteracts gravity; weight, which pulls the UAV downward; thrust, which propels it forward; and drag, which resists motion. Efficient UAV design relies on balancing these forces to achieve stable and energy-efficient flight, making aerodynamics a critical factor in UAV development and operation [6].

Newton's First Law (Law of Inertia)

Newton's First Law states that an object at rest will stay at rest, and an object in motion will remain in motion with the same speed and direction unless acted upon by an external force. This principle applies to UAVs in several ways. A drone in steady flight will continue moving in the same direction unless an external force—such as wind, pilot input, or obstacles—changes its motion. Similarly, a hovering multirotor drone remains stationary until a force, such as a throttle adjustment or air disturbance, alters its state [7]. UAVs are equipped with stability systems, such as gyroscopes and accelerometers, to counteract unintended movements caused by external forces, ensuring smooth and controlled flight.

Newton's Second Law (Law of Acceleration)

Newton's Second Law states that the acceleration of an object is directly proportional to the net force acting on it and inversely proportional to its mass ($F = maF = ma$). This law explains how UAVs generate movement. When a UAV's motors increase thrust, it accelerates upward or forward depending on the direction of force. The force required to accelerate a UAV depends on its mass—heavier drones require more thrust to achieve the same acceleration as lighter drones. This principle also governs lift generation: for a fixed-wing UAV, increasing airspeed over the wings generates more lift, enabling it to climb. For a multirotor drone, increasing motor speed increases the force generated by the rotors, lifting the drone into the air [8]. The balance between applied force and mass determines how efficiently a UAV can change its altitude, speed, or direction.

Newton's Third Law (Law of Action and Reaction)

Newton's Third Law states that for every action, there is an equal and opposite reaction. This principle is fundamental in UAV propulsion and lift. In a multirotor drone, the propellers push air downward (action), generating an equal and opposite force that lifts the UAV upward (reaction). Similarly, in a fixed-wing UAV, as air is pushed backward by the propulsion system, the aircraft moves forward due to the reaction force. This law also explains why UAVs need counter-rotating propellers—when one set of propellers spins in one direction, an opposite force tries to spin the UAV in the other direction [9]. To maintain stability, drones use multiple propellers that rotate in opposite directions to counteract unwanted rotational forces.

Newton's laws of motion explain how UAVs achieve stable and controlled flight. The First Law describes how UAVs resist changes in motion unless acted upon by an external force, the Second Law determines how forces influence UAV acceleration and movement, and the Third Law explains how UAVs generate lift and thrust through action-reaction forces [10]. Understanding these principles helps in designing UAVs with better stability, efficiency, and control, making them effective for various applications such as surveillance, agriculture, and logistics.

3. **Bernoulli's Principle and Its Role in UAV Flight**

Bernoulli's principle is a fundamental concept in aerodynamics that explains how differences in air pressure contribute to lift generation, which is essential for UAV flight. This principle is based on the relationship between airspeed and pressure, stating that as the speed of a fluid (such as air) increases, its pressure decreases [11].

Application in UAVs

In UAVs, particularly fixed-wing drones, Bernoulli's principle plays a key role in producing lift. The wings of a UAV are designed with an airfoil shape, meaning the upper surface is curved while the lower surface remains relatively flat. As air moves over the wing, it travels faster over the curved upper surface than it does under the wing. Due to Bernoulli's principle, the increase in airflow speed over the top reduces the air pressure above the wing, while the slower-moving air below maintains higher pressure [12]. This pressure difference generates an upward force known as lift, which counteracts gravity and keeps the UAV airborne.

Example in UAV Flight

A fixed-wing UAV, such as those used for aerial surveillance or mapping, relies on this principle for sustained flight. As the UAV moves forward, the wings force air to move over and under them at different speeds, creating the necessary pressure difference for lift [13]. Even in multirotor drones, while lift is primarily generated by rotor thrust, Bernoulli's principle influences airflow dynamics around the propellers, contributing to overall efficiency and stability.

Limitations and Considerations

While Bernoulli's principle is crucial for lift generation, it works alongside Newton's Third Law of Motion, which states that air being deflected downward by the wing also contributes to lift through an equal and opposite reaction. UAV designers must balance these aerodynamic principles to optimize wing shape, rotor efficiency, and overall flight stability [14].

Bernoulli's principle is essential in understanding how UAVs, particularly fixed-wing types, generate lift. By utilizing differences in airspeed and pressure, UAV wings create an upward force that allows them to fly efficiently. This principle, combined with Newtonian physics, helps engineers design UAVs that are both aerodynamically efficient and capable of stable flight in various conditions [15].

4. **The Four Forces of Flight in UAVs**

The flight of an Unmanned Aerial Vehicle (UAV) is influenced by four fundamental aerodynamic forces: lift, weight, thrust, and drag. These forces act simultaneously, determining whether a UAV can take off, maintain altitude, maneuver, or land [16]. Understanding the interaction between these forces is essential for optimizing UAV performance and efficiency.

Lift is the upward force that counteracts gravity and allows a UAV to remain in the air. It is generated through the movement of air over wings in fixed-wing UAVs or by the spinning rotors in multirotor drones. Lift is produced by the difference in air pressure between the upper and lower surfaces of an airfoil, as described by Bernoulli's Principle. Additionally, Newton's Third Law of Motion contributes to lift, as the downward deflection of air by the wing or rotors results in an equal and opposite upward force [17]. Several factors influence lift, including air density, velocity, wing surface area, and the angle of attack. UAVs must maintain sufficient lift to counteract their weight and sustain controlled flight.

Weight is the force of gravity pulling the UAV downward. It depends on the mass of the UAV and its payload, including onboard components such as cameras, sensors, and batteries. Weight must be carefully managed to ensure a UAV remains light enough for efficient flight while still carrying the necessary equipment. If the weight exceeds the lift force, the UAV will not be able to take off or will struggle to stay in the air. Engineers use lightweight materials such as carbon fiber and aluminum to reduce overall weight while maintaining structural integrity [7]. Proper weight distribution is also crucial for stability during flight.

Thrust is the forward force that propels the UAV through the air. It is generated by motors and propellers in multirotor drones and by engines or propeller systems in fixed-wing UAVs. Thrust is required to overcome drag and maintain forward motion. The amount of thrust produced depends on factors such as motor power, propeller efficiency, and energy availability from the UAV's battery or fuel source. In fixed-wing UAVs, thrust not only moves the aircraft forward but also contributes to lift by increasing airflow over the wings [18]. In multirotor drones, thrust is directed downward to provide lift, and adjustments in thrust control altitude and maneuverability.

Drag is the resistive force that opposes thrust as the UAV moves through the air. It is caused by air resistance and is influenced by the UAV's shape, surface roughness, and speed. Drag can be classified into two types: parasitic drag and induced drag. Parasitic drag results from the UAV's body shape and external components, such as antennas and landing gear, which create resistance as air flows around them. Induced drag is a byproduct of lift and occurs when airflow around the wings creates vortices, reducing aerodynamic efficiency. UAV designers minimize drag by using streamlined shapes, smooth surfaces, and efficient airflow management techniques to enhance performance and conserve energy [1].

Balancing these four forces is essential for stable and efficient UAV flight. Lift must be sufficient to counteract weight, while thrust must be strong enough to overcome drag. By optimizing these forces through careful design, UAVs can achieve better flight stability, longer endurance, and improved maneuverability, making them more effective for various applications such as surveillance, delivery, and aerial mapping (El Adawy et al., 2023).

5. **Practical Applications in UAVs**

Understanding the four forces of flight—lift, weight, thrust, and drag—is crucial for designing and operating UAVs efficiently. These principles directly influence UAV performance, maneuverability, energy consumption, and operational capabilities across various applications.

Fixed-Wing Drones rely on wings to generate lift through forward motion, following Bernoulli's Principle. Unlike multirotor drones, they do not need to continuously expend energy to stay aloft, making them highly efficient for long-range missions such as aerial surveying, mapping, and agricultural monitoring. Their thrust is typically generated by propellers or jet engines, which must overcome drag to maintain steady flight [19]. However, they require runways or launch mechanisms for takeoff and landing, limiting their operational flexibility in confined spaces.

Multirotor Drones generate both lift and thrust through their rotating propellers, allowing them to hover, ascend, descend, and move in any direction. They are highly maneuverable and capable of vertical takeoff and landing (VTOL), making them ideal for applications like aerial photography, package delivery, and search-and-rescue operations. However, their energy-intensive nature means they have shorter flight durations compared to fixed-wing UAVs. By adjusting the rotational speed of individual rotors, they achieve precise control over altitude, direction, and stability, even in complex environments [20].

Hybrid Drones combine elements of both fixed-wing and multirotor designs, offering a balance between endurance and maneuverability. These UAVs can take off and land vertically like multirotor drones but transition to forward flight for greater efficiency, similar to fixed-wing aircraft. This makes them suitable for applications requiring both extended range and operational flexibility, such as military reconnaissance, environmental monitoring, and large-area surveillance [21]. Their ability to switch flight modes enhances versatility, though the added mechanical complexity may increase maintenance requirements.

By applying aerodynamic principles effectively, UAV designers and operators can optimize performance, reduce energy consumption, and enhance mission success across various industries, from agriculture and logistics to defense and disaster response.

6. **Importance of Aerodynamics in UAV Design**

Aerodynamics plays a crucial role in UAV design, influencing efficiency, stability, performance, and sustainability. A well-optimized aerodynamic design ensures that UAVs can operate effectively while minimizing energy consumption and maximizing flight capabilities [22].

Efficiency is a key factor in UAV operation, as improved aerodynamics reduce drag and optimize lift generation, leading to lower energy consumption. This is especially important for battery-powered drones, where extended flight time is crucial for mission success. By designing streamlined airframes and optimizing wing or rotor configurations, UAV manufacturers can enhance efficiency and increase operational endurance [23].

Stability is essential for UAVs to maintain controlled flight, particularly in unpredictable environmental conditions such as wind turbulence. Understanding aerodynamic forces and airflow interactions helps engineers design UAVs with enhanced stability mechanisms, such as gyroscopes, stabilizers, and adaptive control systems [24]. Multirotor drones, for example, use rapid adjustments in rotor speeds to maintain balance, while fixed-wing drones rely on carefully shaped wings and control surfaces for stability.

Performance improvements in aerodynamics lead to increased speed, higher payload capacity, and better maneuverability. Reducing drag allows UAVs to move faster with less energy, while optimized lift generation enables them to carry heavier payloads without compromising efficiency. Additionally, improved aerodynamics contribute to better handling and responsiveness, making UAVs more effective in applications such as aerial photography, surveying, and emergency response [25].

Sustainability is a growing concern in UAV development, and aerodynamic efficiency plays a significant role in reducing environmental impact. By designing UAVs that require less power to achieve flight and maintain stability, energy consumption is minimized, leading to lower emissions in fuel-based UAVs and extended battery life in electric models [26]. Advancements in materials and design techniques further contribute to creating UAVs that are both high-performing and environmentally friendly.

Incorporating aerodynamic principles into UAV design is essential for achieving superior flight performance, operational reliability, and energy efficiency. By continuously improving aerodynamic efficiency, UAVs can be adapted for a wide range of applications, from commercial deliveries and scientific research to military surveillance and disaster management. The principles of aerodynamics—rooted in Newton's laws, Bernoulli's principle, and the four forces of flight—are essential for understanding how UAVs operate [27]. These principles guide the design and optimization of UAVs, ensuring they can generate lift, overcome drag, and maintain stable flight. By mastering these concepts, engineers and operators can develop UAVs that are efficient, versatile, and capable of performing complex tasks in agriculture and beyond. Aerodynamics is not just a theoretical concept but a practical tool that drives innovation and sustainability in UAV technology [28].

2.3 UAV Flight Mechanics

UAV flight mechanics encompass the principles and forces that dictate how unmanned aerial vehicles (UAVs) achieve lift, maintain stability, maneuver, and respond to environmental factors. These mechanics vary based on the UAV type (fixed-wing or rotary-wing) and influence performance, endurance, and operational capabilities. A deep understanding of UAV flight mechanics is essential for optimizing aerodynamics, reducing energy consumption, and enhancing flight control [29].

2.3.1 *Fixed-Wing vs. Rotary-Wing UAVs*

Unmanned Aerial Vehicles (UAVs) are broadly classified into fixed-wing and rotary-wing types, each designed with distinct aerodynamic properties suited for specific applications. Fixed-wing UAVs resemble traditional airplanes, using stationary wings to generate lift through forward motion, making them highly efficient for long-range missions, high-speed operations, and carrying larger payloads. In contrast, rotary-wing UAVs, such as quadcopters and helicopters, rely on rotating blades to generate lift and thrust, allowing for vertical takeoff and landing (VTOL), hovering, and superior maneuverability in confined spaces. While fixed-wing drones offer greater endurance and speed, rotary-wing drones provide enhanced agility and flexibility, making them ideal for tasks requiring precision and close-range operations [30]. The choice between these UAV types depends on the operational requirements, including flight duration, maneuverability, and environmental conditions.

2.3.2 *Fixed-Wing UAVs*

Fixed-wing UAVs are designed to operate similarly to conventional airplanes, utilizing stationary wings to generate lift. Unlike rotary-wing UAVs, which rely on spinning rotors, fixed-wing drones require continuous forward motion to stay airborne. This motion is usually powered by propellers, jet engines, or electric motors. As the UAV moves forward, air flows over and under the wings, creating a pressure difference based on Bernoulli's principle—the faster-moving air above the wings results in lower pressure, while the relatively slower air below generates higher pressure, producing lift [31].

Advantages of fixed-wing UAVs

One of the key benefits of fixed-wing UAVs is their longer endurance. Since they do not need to continuously generate lift through rotor movement, they consume less energy and can stay in the air for extended periods. This makes them ideal for applications requiring long-range flight, such as aerial mapping, surveillance, and environmental monitoring. Additionally, fixed-wing UAVs have a greater range, allowing them to cover large distances efficiently [32]. Their aerodynamic efficiency also enables them to fly at higher speeds compared to rotary-wing drones, making them suitable for missions that demand quick response times.

Disadvantages of fixed-wing UAVs

Despite these advantages, fixed-wing UAVs have several limitations. A major drawback is their need for a runway or a launching mechanism (such as a catapult or hand launch) for takeoff and landing, which can be a challenge in areas with limited space. Unlike rotary-wing UAVs, they lack the ability to hover, making them unsuitable for operations that require stationary flight, such as close-up inspections or precise aerial photography [33]. Furthermore, maneuverability is limited since fixed-wing UAVs must maintain a certain speed to generate lift, preventing them from making sharp turns or hovering in place.

Applications of fixed-wing UAVs

Due to their endurance, range, and speed, fixed-wing UAVs are commonly used in applications that demand long-duration flights. These include military reconnaissance, disaster response, agriculture, cargo transport, and infrastructure monitoring. Their ability to carry heavier payloads, such as advanced cameras and sensors, makes them valuable for surveying and mapping large areas [34]. However, their operational constraints must be considered when choosing the right UAV for specific missions.

2.3.3 Rotary-Wing UAVs

Rotary-wing UAVs, including quadcopters, hexacopters, and helicopters, generate lift through rotating blades (propellers) rather than relying on fixed wings and forward motion. Unlike fixed-wing UAVs, they do not need continuous movement to remain airborne, allowing them to hover, take off, and land vertically. This Vertical Takeoff and Landing (VTOL) capability makes them highly versatile for operations in confined spaces or urban environments. By adjusting the speed and angle of their rotors, these UAVs achieve precise control over their movement, making them ideal for applications that require agility and stability [35].

Advantages of rotary-wing UAVs

One of the biggest advantages of rotary-wing UAVs is their VTOL capability, allowing them to operate in areas without runways or open spaces. This feature is especially useful in urban settings, emergency response situations, and military operations. Additionally, they offer excellent maneuverability, as they can hover, change direction quickly, and perform complex movements, making them superior to fixed-wing UAVs in close-range operations [36]. Their precise control makes them ideal for tasks such as aerial photography, inspections, search and rescue, and surveillance.

Disadvantages of rotary-wing UAVs

Despite their versatility, rotary-wing UAVs come with some limitations. Their flight time is significantly shorter than fixed-wing UAVs because they require continuous energy to keep their rotors spinning, leading to higher battery consumption. This makes them less suitable for long-duration missions. Additionally, rotary-wing UAVs typically have lower speeds, as their design prioritizes stability over high-speed flight [37]. Another drawback is their higher maintenance requirements due to the presence of multiple moving parts, such as rotors, motors, and control mechanisms, which are more prone to wear and tear.

Applications of rotary-wing UAVs

Given their ability to hover and maneuver in tight spaces, rotary-wing UAVs are commonly used in law enforcement, firefighting, industrial inspections, delivery services, and environmental monitoring. They are particularly valuable for real-time surveillance and disaster response, where precise positioning and agility are essential [38]. However, their shorter flight time and lower speeds must be considered when planning missions that require extensive coverage or long-range travel.

2.3.4 *Hybrid UAVs*

Hybrid UAVs integrate the capabilities of both fixed-wing and rotary-wing UAVs, combining the endurance and efficiency of fixed-wing flight with the vertical takeoff and landing (VTOL) capabilities of rotary-wing drones. These UAVs can lift off and land without a runway, like a multirotor, and then transition into fixed-wing forward flight for improved range and speed. This unique combination makes them highly adaptable for a variety of applications, especially in areas where long-range travel is needed but traditional runways are unavailable [39].

Advantages of Hybrid UAVs

The most significant advantage of hybrid UAVs is that they offer the best of both worlds—the ability to take off and land in confined spaces while also achieving efficient, long-duration flight. This makes them suitable for missions requiring both precision hovering and long-distance coverage, such as disaster response, military reconnaissance, search and rescue, and large-scale environmental monitoring [40]. Their versatility allows them to operate effectively in challenging environments where either a fixed-wing or rotary-wing UAV alone would be insufficient.

Disadvantages of Hybrid UAVs

Despite their advantages, hybrid UAVs come with certain challenges. Their complex design requires sophisticated control systems and advanced flight software to ensure smooth transitions between hovering and forward flight. This added complexity makes them more expensive to develop, produce, and maintain compared to traditional UAVs. Additionally, the integration of both flight mechanisms often results in a heavier structure, which can impact overall efficiency and payload capacity [28].

2.3.5 *Applications of Hybrid UAVs*

Hybrid UAVs are widely used in commercial, military, and scientific operations where both VTOL capability and long-range efficiency are needed. They are ideal for mapping large areas, delivering cargo to remote locations, conducting aerial surveys, and supporting emergency response efforts. As UAV technology continues to evolve, hybrid designs are expected to become more efficient and widely adopted across various industries [41].

2.4 Explanation of Stability and Control in UAVs

Stability and control are crucial for ensuring safe and efficient UAV operation. Stability refers to the UAV's ability to maintain a steady flight path without excessive deviations, while control refers to the mechanisms that allow pilots or onboard systems to maneuver the UAV effectively. Both fixed-wing and rotary-wing UAVs rely on aerodynamic principles and advanced flight control systems to achieve smooth and predictable flight [42].

2.4.1 Fixed-Wing UAV Stability and Control

Fixed-wing UAVs maintain stability and maneuverability using control surfaces that influence movement along three primary axes:

- **Pitch (Up/Down Movement)**: Controlled by the elevator on the tail. Raising or lowering the elevator changes the aircraft's angle of attack, causing it to climb or descend.
- **Yaw (Left/Right Rotation)**: Controlled by the rudder on the vertical stabilizer. The rudder helps the UAV turn left or right without banking the wings.
- **Roll (Tilting Left/Right)**: Controlled by the ailerons on the wings. Tilting the ailerons in opposite directions causes the UAV to roll, allowing it to bank and make turns.

To enhance stability, autopilot systems and gyroscopes help compensate for environmental disturbances, such as wind turbulence, by making automatic adjustments to the control surfaces. This ensures smoother flights, particularly in autonomous or long-distance missions [43].

2.4.2 Rotary-Wing UAV Stability and Control

Unlike fixed-wing UAVs, rotary-wing UAVs (e.g., quadcopters) achieve stability and control by adjusting individual rotor speeds rather than using fixed aerodynamic surfaces. Their movement is controlled as follows:

- **Vertical Motion (Lift/Descent)**: Increasing or decreasing power to all rotors simultaneously lifts or lowers the UAV.

- **Yaw (Rotating Left/Right)**: Adjusting the speed of opposite rotors creates rotational torque, allowing the UAV to spin left or right.
- **Pitch & Roll (Tilting Forward/Backward/Sideways)**: Increasing rotor speed on one side while decreasing it on the other tilts the UAV, enabling directional movement.

To ensure stability, rotary-wing UAVs rely on gyroscopes, accelerometers, and GPS sensors, which constantly monitor and adjust flight dynamics. These sensors provide real-time data to onboard flight controllers, allowing precise positioning, even in windy or unstable environments.

2.4.3 *Importance of Stability and Control in UAVs*

Both fixed-wing and rotary-wing UAVs require precise stability and control systems to perform their intended tasks efficiently. Fixed-wing UAVs need stability for long-range efficiency, while rotary-wing UAVs require high maneuverability for close-range precision operations. Advanced stabilization technologies, such as autopilot systems, inertial measurement units (IMUs), and GPS-guided navigation, continue to enhance UAV performance, making them more reliable in diverse applications, from aerial surveillance to cargo transport and disaster response [44].

2.4.4 *Explanation of Lift and Drag Coefficients in UAVs*

Understanding lift and drag coefficients is essential for optimizing UAV aerodynamics. The lift coefficient (C_L) determines how effectively a UAV's airfoil (wing or rotor) generates lift, while the drag coefficient (C_D) quantifies the resistance force opposing motion. These coefficients help engineers design UAVs that maximize efficiency and stability in flight [45].

Lift Coefficient (CLC_LCL)

The lift coefficient (C_L) measures how efficiently an airfoil produces lift based on its shape, flight conditions, and airflow dynamics. A higher C_L indicates a more effective lift generation. Several factors influence the lift coefficient [46]:

- **Wing Shape**: More curved airfoils (such as cambered wings) create greater pressure differences, increasing lift.
- **Angle of Attack**: Increasing the angle of attack (the angle between the wing and the airflow) enhances lift, but exceeding a critical angle leads to stall, where lift rapidly decreases.
- **Air Density and Velocity**: Higher air density and greater flight speed contribute to increased lift generation.

The lift force (L) is given by the equation:

$$L = \frac{1}{2}\rho v^2 S C_L$$

where:

L is the lift force (N)
ρ is the air density (kg/m^3)
v is the velocity of the air (m/s)
S is the wing (or rotor) surface area (m^2)
C_L is the lift coefficient (dimensionless)

Optimizing the lift coefficient ensures that UAVs can generate sufficient lift for takeoff, maintain altitude, and perform stable flight operations.

Drag Coefficient (C_D)

Drag is the aerodynamic resistance opposing a UAV's forward motion, and the drag coefficient (C_D) determines how much drag a UAV experiences under specific flight conditions. Reducing drag improves UAV efficiency by lowering energy consumption and increasing flight duration [47]. The drag coefficient depends on several factors:

- **Surface Roughness**: Smooth surfaces reduce friction drag, while rough surfaces increase resistance.
- **Shape**: Aerodynamic, streamlined designs minimize drag by reducing airflow disruptions.
- **Speed**: Higher speeds increase parasitic drag, as air resistance grows with velocity.

The total drag force (D) is calculated using the formula:

$$D = \frac{1}{2}\rho v^2 S C_D$$

where:

D = Drag force (N).
ρ = Air density (kg/m^3).
v = Velocity of the object relative to air (m/s).
S = Reference area (m^2).
C_D = Drag coefficient (dimensionless).

Importance of Lift and Drag Coefficients in UAVs

For UAVs, balancing lift and drag is crucial for maximizing performance:

- A higher lift-to-drag ratio ensures better efficiency, enabling longer flight times and improved payload capacity.
- Reducing drag minimizes power consumption, extending battery life in electric UAVs.

- Adjusting wing design, airfoil shape, and flight angles can enhance both stability and maneuverability.

By carefully optimizing C_L and C_D, UAV designers can develop aircraft that achieve greater endurance, fuel efficiency, and aerodynamic performance, making them more effective for surveillance, mapping, and cargo transport applications.

2.4.5 Explanation of the Effect of Air Density on UAV Performance

Air density (ρ) plays a crucial role in UAV aerodynamics, influencing lift generation, drag resistance, and propulsion efficiency. Since air density changes with altitude, temperature, and humidity, UAVs must adjust their flight characteristics accordingly to maintain stability and performance [48].

Effects of Altitude on UAV Performance

At higher altitudes, air density decreases, which directly impacts UAV operation in the following ways:

- **Reduced Lift**: Since lift depends on air density (ρ) in the equation lower density results in decreased lift. To compensate, UAVs must increase airspeed or have larger wings to generate sufficient lift.
- **Thinner Air**: Propellers and rotors become less effective because they push against less air, requiring higher rotational speeds to maintain thrust.

At lower altitudes, air density is higher, leading to:

- **Increased Lift**: More air molecules interact with the wings or rotors, generating greater lift, which is beneficial for takeoff and maneuvering.
- **Higher Drag**: While increased air density improves lift, it also creates greater air resistance, requiring more thrust to maintain speed.

Effects of Temperature on UAV Performance

- **Warm air is less dense**, reducing lift and making propellers less efficient. UAVs operating in hot environments may require increased propeller speeds or larger wing surfaces to compensate for the lower air density.
- **Cold air is denser**, improving lift efficiency but also increasing aerodynamic drag. While UAVs experience better lift, they must overcome higher resistance, which can slightly reduce overall efficiency.

Effects of Humidity on UAV Performance

- Moist air is less dense than dry air because water vapor is lighter than oxygen and nitrogen molecules. This reduces lift, which means UAVs may need more thrust to maintain altitude in humid conditions.

- In high humidity, UAV motors and propellers may need additional power output to counteract the slight loss in lift.

Compensating for Air Density Changes

To maintain stable flight under varying air density conditions, UAVs use several adaptation mechanisms:

- **Adjusting Angle of Attack**: Increasing the angle of attack helps generate more lift in low-density air. However, excessive angles can lead to aerodynamic stall.
- **Increasing Propeller Speeds**: UAVs adjust rotor speeds to compensate for changes in thrust efficiency due to air density variations.
- **Automatic Flight Control Systems**: Advanced UAVs use altitude sensors, barometers, and GPS data to automatically adjust flight parameters in real-time, ensuring stability across different atmospheric conditions.

Importance of Air Density Considerations in UAV Design

Understanding how air density affects UAV performance is crucial for designing aircraft that can adapt to different altitudes, temperatures, and humidity levels. Engineers must carefully optimize wing shape, propeller efficiency, and onboard sensors to ensure that UAVs operate reliably in diverse environments, whether at high altitudes, in hot desert conditions, or in humid coastal regions. UAV flight mechanics involve a balance of aerodynamic forces, stability mechanisms, and environmental adaptations [49]. Understanding how fixed-wing and rotary-wing UAVs achieve flight, how stability is maintained, and how lift and drag coefficients impact performance is essential for optimizing UAV efficiency. Additionally, environmental factors such as air density must be considered for effective UAV operation in various conditions. By leveraging these principles, UAV designers and operators can enhance endurance, maneuverability, and overall flight performance across different applications [50].

2.5 UAV Design and Aerodynamic Considerations

Designing an efficient UAV requires careful consideration of aerodynamic principles to optimize lift, thrust, stability, and energy consumption. Key factors influencing UAV aerodynamics include wing shape, propeller and rotor dynamics, airfoil selection, and Computational Fluid Dynamics (CFD) simulations.

Wing Shapes and Their Effects

The shape of a UAV's wing plays a crucial role in determining its aerodynamic efficiency, affecting lift generation, drag resistance, and overall flight performance. Different wing configurations are chosen based on mission requirements, speed, endurance, and maneuverability [51]. Each wing type has its advantages and limitations, making it important to select the right design for specific applications.

Straight wings are commonly used in slow-speed UAVs due to their ability to provide high lift and stability. These wings create a large amount of lift at lower speeds, making them ideal for long-endurance missions such as surveillance, reconnaissance, and agricultural monitoring. The simple design of straight wings also reduces manufacturing and maintenance costs [52]. However, they generate higher drag at high speeds, limiting their efficiency for fast-moving UAVs.

Swept wings, on the other hand, are designed for high-speed UAVs, as they reduce aerodynamic drag and improve performance at greater velocities. These wings are often found in military and performance-oriented drones that require fast and efficient flight. While swept wings allow UAVs to travel longer distances with better fuel efficiency, they require more power to generate lift at low speeds and may be less stable during takeoff and landing [53].

Delta wings are triangular in shape and provide high maneuverability, making them suitable for UAVs designed for agility and supersonic capabilities. Their design minimizes drag at high speeds, allowing for rapid movements and sharp turns. Delta wings are often used in stealth UAVs and high-performance drones that need to operate in dynamic environments. However, they generate less lift at lower speeds, requiring powerful propulsion systems to maintain stable flight [54].

A more advanced design is the blended-wing body (BWB), which merges the fuselage and wings into a single, aerodynamic structure. This configuration significantly improves fuel efficiency, enhances payload capacity, and reduces drag. BWB designs are often used in large UAVs meant for cargo transport, military operations, or long-endurance surveillance. Despite their advantages, these UAVs require sophisticated flight control systems to maintain stability due to their unique shape. The selection of wing shape for a UAV depends on the intended mission. While straight wings excel in endurance and low-speed stability, swept and delta wings are ideal for high-speed applications [55]. The blended-wing body design offers exceptional efficiency but requires advanced control mechanisms. Understanding these aerodynamic principles helps engineers design UAVs that optimize performance for various operational needs.

Propeller and Rotor Dynamics

Propellers and rotors play a vital role in UAV thrust generation and maneuverability. Their design determines the efficiency of rotary-wing and fixed-wing propulsion systems. The design of a UAV's propellers and rotors plays a crucial role in determining its flight efficiency, thrust generation, and energy consumption. Several factors influence how effectively a UAV can maneuver and sustain flight, including propeller pitch, the number of blades, rotor size, and the use of coaxial rotors. These elements must be carefully optimized to balance performance and energy efficiency. Propeller pitch refers to the angle at which the blades are set relative to the airflow. A higher pitch increases the amount of air pushed backward with each rotation, generating more forward thrust [56]. However, this also requires more power from the UAV's motor, which can lead to higher energy consumption and reduced battery life. Lower-pitch propellers, on the other hand, generate less thrust but allow for more efficient energy use, making them suitable for UAVs focused on endurance rather than speed.

The number of blades on a propeller also affects UAV performance. More blades generate higher thrust, which is useful for lifting heavier payloads or operating in environments with high air resistance. However, additional blades also increase aerodynamic drag and reduce efficiency. UAVs designed for long-endurance missions often use two-blade propellers to minimize drag and optimize fuel or battery efficiency, whereas high-power UAVs might use three or four blades to achieve greater lifting capacity [57]. Rotor size is another important consideration, particularly for rotary-wing UAVs. Larger rotors generate more lift with less power, improving efficiency and reducing the strain on the UAV's propulsion system. However, large rotors require precise balancing to maintain stability and can limit the UAV's agility. Smaller rotors provide better maneuverability but require higher rotational speeds to generate sufficient lift, increasing energy consumption. Coaxial rotors, where two rotors are mounted on the same axis but rotate in opposite directions, are commonly used in multirotor UAVs. This design enhances lift capacity while keeping the UAV compact. Coaxial rotors help counteract torque effects, improving stability and control. They are particularly useful in UAVs that need to hover for extended periods or operate in confined spaces. However, they add complexity to the propulsion system and may require more advanced flight control mechanisms [58]. The choice of propeller and rotor design significantly impacts UAV performance. Optimizing propeller pitch, blade count, rotor size, and rotor configuration ensures a balance between thrust generation, efficiency, and stability. Engineers carefully analyze these factors to design UAVs that meet specific operational requirements, whether for endurance, speed, payload capacity, or maneuverability.

Airfoil Selection for UAVs

The airfoil shape of UAV wings or rotors determines how efficiently they generate lift. Key factors in airfoil selection include:

The selection of an airfoil shape is a crucial factor in UAV design, as it directly impacts lift generation, drag resistance, and overall flight performance. Various airfoil characteristics, such as camber, thickness, leading-edge shape, and symmetry, determine how efficiently a UAV can operate under different flight conditions. Camber refers to the curvature of an airfoil. A more curved airfoil generates greater lift because it increases the pressure difference between the upper and lower surfaces of the wing. However, higher camber also increases drag, which can reduce overall efficiency. UAVs designed for slow-speed, high-lift applications, such as surveillance and agricultural drones, often use airfoils with significant camber to maximize lift while operating at lower speeds [59]. The thickness of an airfoil plays a role in both aerodynamics and structural integrity. Thicker airfoils provide greater structural strength, which is beneficial for UAVs carrying heavy payloads. However, increased thickness also leads to greater drag, which can limit speed and efficiency. Designers must balance airfoil thickness to ensure sufficient strength without compromising performance.

The shape of the leading edge influences a UAV's ability to handle stall conditions and aerodynamic efficiency. A rounded leading edge improves stall resistance by allowing smoother airflow over the wing, making it ideal for UAVs that require stable flight characteristics. On the other hand, a sharper leading edge reduces drag at high speeds, making it suitable for UAVs designed for fast, efficient travel, such as military reconnaissance drones [60]. Airfoil symmetry also affects UAV performance. Symmetrical airfoils, which have identical upper and lower surfaces, generate equal lift in both directions. These are commonly used in aerobatic drones and UAVs that require inverted flight capabilities. Asymmetrical airfoils, which have a more curved upper surface and a flatter lower surface, generate more lift and are commonly used in conventional UAVs that prioritize stable and efficient flight. Airfoil selection is a critical design consideration in UAVs. Factors such as camber, thickness, leading-edge shape, and symmetry must be optimized based on the intended mission and flight characteristics of the UAV [61]. Proper airfoil design ensures a balance between lift, drag, stability, and efficiency, allowing UAVs to perform effectively in diverse operational environments.

2.5.1 Computational Fluid Dynamics (CFD) in UAV Design

CFD simulations play a crucial role in UAV design by allowing engineers to analyze airflow, lift, and drag forces before physical testing. Benefits of CFD in UAV design include:

Computational Fluid Dynamics (CFD) plays a vital role in UAV design by enabling engineers to analyze and optimize aerodynamic performance before physical testing. By simulating airflow over wings, propellers, and fuselage components, CFD helps reduce drag and increase efficiency, leading to improved flight characteristics and energy savings. One of the key benefits of CFD is its ability to predict UAV performance under different environmental conditions. By modeling how air interacts with the UAV at various altitudes, temperatures, and wind speeds, designers can ensure optimal performance in real-world scenarios [62]. This is particularly important for UAVs operating in challenging environments, such as high-altitude surveillance drones or agricultural UAVs flying in turbulent weather. CFD is also instrumental in enhancing UAV stability. By identifying areas of turbulence, pressure changes, or airflow separation, engineers can refine wing and rotor designs to maintain smooth, controlled flight. This leads to better maneuverability, reduced risk of stalling, and more efficient propulsion, which is essential for both fixed-wing and rotary-wing UAVs [63].

Another major advantage of CFD is cost reduction in UAV development. Traditional wind tunnel testing and physical prototypes can be expensive and time-consuming. CFD allows designers to perform virtual tests, minimizing the need for multiple iterations of physical models. This not only accelerates the development process but also ensures that UAVs are aerodynamically optimized before they even take flight. CFD is a powerful tool for UAV design, allowing for precise aerodynamic optimization, accurate performance predictions, enhanced stability, and significant cost savings. By leveraging advanced simulations, engineers can create UAVs that are more efficient, stable, and adaptable to various flight conditions [64]. By integrating advanced aerodynamic design principles, UAVs can achieve higher efficiency, longer endurance, and improved flight performance, making them suitable for diverse applications in surveillance, cargo transport, and scientific research.

2.5.2 UAV Performance and Efficiency

UAV Performance and Efficiency

The performance and efficiency of UAVs depend on several aerodynamic and mechanical factors, including the power-to-weight ratio, glide ratio, endurance, and energy consumption. Optimizing these factors ensures better flight capabilities, extended operational time, and enhanced overall performance [65].

Power-To-Weight Ratio

The power-to-weight ratio is a crucial metric in UAV design, representing the relationship between the thrust produced by the propulsion system and the UAV's total weight. A higher power-to-weight ratio allows for better acceleration, agility, and lift capabilities. Fixed-wing UAVs typically require less power to sustain flight due to their efficient lift generation, whereas rotary-wing UAVs, which rely on continuous thrust for lift, demand more power. Optimizing this ratio ensures UAVs can carry payloads efficiently while maintaining stable flight [66].

Glide Ratio and Endurance

The glide ratio determines how far a UAV can travel horizontally compared to its vertical descent, which is particularly important for fixed-wing UAVs. A higher glide ratio means the UAV can cover more distance with minimal power usage, improving efficiency and endurance. Endurance, or the ability to remain airborne for extended periods, depends on factors such as battery capacity, fuel efficiency, and aerodynamic design. UAVs designed for long-range missions, such as surveillance or environmental monitoring, require optimized aerodynamics and lightweight materials to maximize endurance [67].

Aerodynamic Efficiency and Energy Consumption

Aerodynamic efficiency directly influences energy consumption, as reducing drag and optimizing airflow around the UAV helps minimize power requirements. Streamlined designs, efficient airfoils, and optimized propeller or rotor configurations contribute to lower energy consumption, allowing UAVs to operate longer on limited power sources. Fixed-wing UAVs generally have higher aerodynamic efficiency compared to rotary-wing UAVs, making them ideal for extended missions. Additionally, advanced energy management systems, such as regenerative power sources or solar panels, further enhance UAV efficiency and operational time. UAV performance and efficiency depend on maintaining an optimal power-to-weight ratio, improving glide ratio and endurance, and enhancing aerodynamic efficiency to reduce energy consumption [68]. These factors collectively contribute to better flight stability, extended operational range, and overall effectiveness in UAV applications.

2.5.3 UAV Stability and Control Systems

UAV Stability and Control Systems

Stability and control systems are critical in UAV design to ensure smooth flight, maneuverability, and responsiveness to external factors such as wind and turbulence. These systems rely on aerodynamic control surfaces, automated flight control mechanisms, and onboard sensors to maintain balance and precision during flight [69].

Role of Ailerons, Elevators, and Rudders

Fixed-wing UAVs use three primary control surfaces to maintain stability and execute maneuvers:

- **Ailerons** (located on the wings) control roll, allowing the UAV to tilt left or right for turning.
- **Elevators** (positioned on the horizontal tail) control pitch, enabling the UAV to climb or descend.
- **Rudders** (mounted on the vertical stabilizer) control yaw, helping the UAV change direction while maintaining level flight.

By adjusting these control surfaces, UAVs can maintain stability in different flight conditions and execute precise movements for navigation.

Flight Control Systems (Autopilot and Stability Augmentation)

Modern UAVs incorporate autopilot and stability augmentation systems (SAS) to enhance control and reduce pilot workload.

- **Autopilot systems** allow UAVs to follow pre-programmed flight paths, automatically adjusting thrust, altitude, and navigation without manual input.
- **Stability Augmentation Systems (SAS)** use feedback from onboard sensors to correct disturbances, ensuring smooth flight by counteracting external forces like turbulence.

These systems enable UAVs to perform complex missions with high accuracy while improving flight safety and reliability.

Sensors and Their Impact on Stability

UAV stability is highly dependent on sensors that provide real-time data for flight adjustments. Key sensors include:

- **Gyroscopes**—Detect angular velocity and help maintain orientation.
- **Accelerometers**—Measure linear acceleration and assist in stabilizing movement.
- **GPS Modules**—Provide precise location data for navigation and autopilot functions.
- **Barometric Altimeters**—Measure altitude and help regulate UAV flight levels.
- **IMUs (Inertial Measurement Units)**—Combine gyroscopes and accelerometers to provide comprehensive motion data for stability corrections.

By integrating these sensors, UAVs can autonomously adjust their flight characteristics, ensuring stable operation in varying environmental conditions. UAV stability and control systems rely on aerodynamic control surfaces, automated flight mechanisms, and advanced sensors to maintain smooth and precise flight. These components work together to enhance maneuverability, improve safety, and enable UAVs to operate effectively in diverse applications [70].

2.6 Advantages of Understanding UAV Aerodynamics

Understanding UAV aerodynamics is crucial for optimizing performance, stability, energy efficiency, and overall reliability. Aerodynamic principles dictate how UAVs generate lift, counteract drag, maintain stability, and maneuver effectively. By leveraging these principles, engineers and operators can enhance UAV capabilities for various applications, such as surveillance, delivery, agriculture, and disaster response [71]. Below are the key benefits of understanding UAV aerodynamics in greater detail.

1. **Enhanced Performance**

Aerodynamics directly influence a UAV's flight performance, including speed, endurance, and range. By designing UAVs with efficient aerodynamic profiles, engineers can reduce air resistance (drag) and improve lift generation. Fixed-wing UAVs, for instance, rely on streamlined wings to maximize their flight range and minimize energy consumption. Rotary-wing UAVs, like quadcopters, benefit from optimized rotor blade shapes to improve lift efficiency and stability [72]. An aerodynamically efficient UAV can carry heavier payloads while maintaining longer flight durations, making it more effective for critical missions.

2. **Better Stability**

Stability is essential for safe UAV operations, especially in adverse weather conditions. UAVs maintain stability through a combination of aerodynamic design and flight control systems. Fixed-wing UAVs achieve stability using control surfaces like ailerons, elevators, and rudders, which adjust airflow and balance forces. Rotary-wing UAVs maintain stability by adjusting rotor speeds, aided by onboard gyroscopes and accelerometers. Without proper aerodynamic stability, UAVs are prone to turbulence, loss of control, and inefficient maneuvering, leading to reduced reliability in real-world applications [73].

3. **Energy Efficiency**

One of the primary concerns in UAV operation is energy consumption. UAVs powered by batteries, fuel, or hybrid systems must optimize their aerodynamic efficiency to conserve energy and extend flight time. A high lift-to-drag ratio allows UAVs to stay airborne with minimal power usage. For example, reducing drag through streamlined body shapes and efficient propeller designs minimizes energy wastage. Improved aerodynamic efficiency directly translates into longer flight durations, reducing the need for frequent battery replacements or refueling, which is critical for long-range UAV missions [74].

4. **Improved Maneuverability**

Aerodynamics play a fundamental role in a UAV's ability to maneuver effectively. The shape of the wings, propellers, and fuselage impacts how a UAV responds to control inputs. Fixed-wing UAVs use control surfaces to execute precise turns and altitude adjustments, whereas rotary-wing UAVs depend on differential rotor thrust to change direction. Understanding airflow behavior around UAV surfaces helps optimize control response, allowing UAVs to navigate tight spaces, perform agile movements, and adapt to changing environmental conditions with greater precision [75].

5. **Safety and Reliability**

A UAV must withstand various environmental factors, such as wind gusts, temperature fluctuations, and humidity changes. Aerodynamic stability ensures that UAVs remain operational under these conditions without excessive power consumption or structural stress. For instance, turbulence can disrupt flight paths, but well-designed aerodynamic surfaces, coupled with stabilization systems, help UAVs maintain smooth flight. Additionally, aerodynamics contribute to structural integrity, preventing excessive vibration or stress on UAV components. This leads to fewer mechanical failures, reducing the risk of crashes and extending the UAV's lifespan [76]. Understanding UAV aerodynamics is fundamental for improving flight efficiency, stability, energy conservation, maneuverability, and overall safety. Engineers use aerodynamic principles to design UAVs that can operate in diverse conditions while maximizing flight performance. Operators benefit from aerodynamic knowledge by making better decisions on flight paths, payload distribution, and power management. By continuously improving UAV aerodynamics, the industry can achieve higher efficiency, longer endurance, and enhanced capabilities for a wide range of applications [77].

2.7 Importance of UAV Aerodynamics

UAV aerodynamics is a critical aspect of drone design and operation, impacting flight efficiency, stability, energy consumption, and overall mission success. Understanding aerodynamics allows engineers to design UAVs that can fly longer, carry heavier payloads, operate in harsh conditions, and complete tasks with precision. In various industries, from defense and agriculture to commercial logistics and environmental research, aerodynamics plays a fundamental role in ensuring UAVs perform at their best [78].

1. **The Role of Aerodynamics in Military UAVs**

The military relies extensively on UAVs for reconnaissance, surveillance, target tracking, and combat operations. Aerodynamics plays a crucial role in enhancing their endurance, stealth, speed, and precision, making them more effective for various defense applications. One of the most critical aspects of military UAVs is extended flight time. By optimizing lift-to-drag ratios and reducing aerodynamic resistance, UAVs can remain airborne for long periods with minimal energy consumption [79]. This extended endurance allows them to conduct prolonged surveillance missions over enemy territory without requiring frequent refueling or battery replacements.

Stealth capabilities are another key advantage of aerodynamically optimized military UAVs. Many UAVs are designed with smooth, low-profile bodies and blended-wing configurations that reduce their radar cross-section. These aerodynamic enhancements make them less detectable by enemy radar and air defense systems, allowing for covert operations in hostile environments. For combat and reconnaissance UAVs, high-speed performance is essential. Military UAVs often need to travel at high velocities while maintaining stability and maneuverability. Advanced aerodynamic designs minimize drag and optimize propulsion efficiency, enabling UAVs to sustain high-speed flight without excessive energy loss [80]. Finally, precision targeting is a crucial function of military UAVs used in combat or missile guidance. UAVs designed for strike missions require agile maneuverability to track and engage targets accurately. Proper aerodynamic balancing allows for rapid directional changes while maintaining flight stability, ensuring accurate target engagement in dynamic combat situations. By leveraging advanced aerodynamics, military UAVs achieve superior endurance, stealth, speed, and agility, making them invaluable assets for modern defense strategies [58]. These design optimizations enhance their effectiveness in surveillance, reconnaissance, and combat missions, providing a significant tactical advantage on the battlefield.

2. **The Role of UAV Aerodynamics in Agriculture**

The agricultural sector has increasingly adopted UAV technology to enhance efficiency in crop monitoring, irrigation management, and pesticide spraying. The aerodynamic design of UAVs plays a crucial role in ensuring energy efficiency, stability, and precision in these applications [81].

One of the primary benefits of aerodynamically optimized UAVs in agriculture is energy efficiency. Fixed-wing UAVs, commonly used for surveying large farmland areas, benefit from natural lift generated by their wings. This reduces the reliance on motor power, allowing for extended flight times and greater coverage per mission. Their streamlined design minimizes drag, making them ideal for tasks that require long-duration aerial observations, such as monitoring crop health and mapping field conditions. For hovering-based applications, such as pesticide and fertilizer spraying, multirotor UAVs rely on precise aerodynamic control to maintain stability. Effective aerodynamic design ensures that these UAVs can hover steadily over crops, providing uniform spraying while minimizing drift caused by wind [82]. This reduces chemical waste and enhances the effectiveness of targeted spraying, leading to better crop yields and reduced environmental impact.

Additionally, aerodynamics plays a key role in precision farming by allowing UAVs to operate at low altitudes with minimal turbulence. UAVs equipped with optimized airfoil shapes and controlled airflow mechanisms can capture high-resolution imagery of farmland, enabling farmers to analyze soil health, detect crop diseases, and assess irrigation needs. This level of precision helps in making data-driven decisions that improve agricultural productivity and sustainability. By integrating advanced aerodynamic designs, agricultural UAVs enhance efficiency, reduce operational costs, and contribute to sustainable farming practices [83]. Their ability to perform long-range surveys, hover with stability, and collect precise data makes them invaluable tools for modern agriculture.

3. **Disaster Management and Search & Rescue Operations**

In emergency response and disaster management, UAVs play a critical role in assessing damage, locating survivors, and providing real-time data to rescue teams. The effectiveness of these operations heavily depends on the UAV's aerodynamic design, which ensures stability, efficiency, and adaptability in extreme conditions. One of the primary challenges of disaster response is operating in harsh conditions. UAVs deployed in disaster zones must endure strong winds, airborne debris, and rapidly changing atmospheric conditions. Aerodynamic stability is essential to maintaining reliable flight performance in these unpredictable environments [84]. Fixed-wing UAVs, designed for endurance, and rotary-wing drones, built for maneuverability, must both have optimized aerodynamic features to handle these challenges effectively. Extended surveillance capabilities are another key advantage of aerodynamically efficient UAVs. Fixed-wing UAVs, with high lift-to-drag ratios, can remain airborne for extended periods, allowing them to cover vast areas affected by natural disasters such as floods, earthquakes, or wildfires. This prolonged flight time enables emergency responders to gather continuous, high-resolution aerial imagery and assess damage with greater accuracy [85].

In contrast, close-proximity operations require UAVs to navigate confined and hazardous spaces where human access is limited. Rotary-wing UAVs, designed with precise aerodynamic control, can hover and maneuver through collapsed buildings, rugged mountainous terrains, and dense urban environments. Their ability to operate in tight spaces makes them invaluable for search-and-rescue missions, where they can locate trapped individuals and deliver emergency supplies. By leveraging advanced aerodynamic designs, UAVs enhance the speed, efficiency, and safety of disaster response efforts [86]. Their ability to operate in challenging environments, conduct long-duration surveillance, and navigate complex terrains makes them indispensable tools in modern emergency management.

4. **UAVs in Commercial Logistics and Delivery Services**

With the rapid growth of e-commerce, UAVs are transforming package delivery, medical supply transport, and logistics. Aerodynamic advancements play a crucial role in enhancing UAV efficiency, allowing them to carry heavier payloads, travel longer distances, and operate effectively in urban environments. One of the most significant advantages of aerodynamically optimized UAVs is increased payload capacity. By reducing unnecessary drag, UAVs can carry heavier loads without requiring excessive power. This is particularly important for medical supply deliveries, where drones must transport critical items such as vaccines, blood samples, or emergency medications to remote or disaster-affected areas [87]. Aerodynamics also contributes to longer flight range, especially in fixed-wing UAVs used for long-distance deliveries. These drones generate lift efficiently, enabling them to travel farther while consuming minimal energy. This makes UAV-based delivery a viable option for reaching rural or geographically isolated regions where traditional transportation is costly or impractical.

For urban package delivery, maneuverability is a key factor. Rotary-wing UAVs, such as quadcopters, are designed for last-mile delivery, where precise navigation through dense urban landscapes is essential. Advanced aerodynamic balancing allows these drones to hover, avoid obstacles, and land safely in tight spaces, ensuring reliable and efficient delivery services. Finally, energy efficiency and sustainability are major benefits of aerodynamic UAV design. By minimizing air resistance and optimizing lift-to-drag ratios, UAVs require less power to operate. This translates to lower energy consumption, reducing operational costs and making drone-based delivery a more environmentally friendly alternative to fuel-powered transportation methods [88]. As UAV technology continues to evolve, aerodynamic innovations will play a crucial role in shaping the future of e-commerce and logistics. By improving efficiency, stability, and range, drones are set to become an integral part of modern delivery networks.

5. **The Role of UAV Aerodynamics in Environmental Monitoring**

Unmanned Aerial Vehicles (UAVs) play a crucial role in climate monitoring, wildlife tracking, and environmental protection. Their aerodynamic design directly impacts their ability to perform these tasks efficiently, ensuring stability, endurance, and minimal disruption to ecosystems. One key advantage of UAVs in environmental monitoring is their ability to conduct long-range surveillance. Fixed-wing UAVs, optimized for low drag and high lift-to-drag ratios, can fly at high altitudes and cover vast geographic areas. This capability allows researchers to collect climate data, track deforestation, monitor ocean temperatures, and survey remote or inaccessible locations with minimal resource consumption [89]. Aerodynamics also plays a vital role in silent operations, which is especially important for wildlife monitoring. Many UAVs used for tracking endangered species or studying animal behavior are designed to minimize noise and turbulence. Aerodynamically efficient airframes and propeller designs help reduce sound emissions, ensuring that animals are not disturbed during observations. This allows scientists to gather accurate behavioral data without influencing wildlife activities.

Additionally, UAVs often operate in challenging environmental conditions, such as strong winds, extreme temperatures, or varying air densities. Advanced aerodynamic designs help maintain stability and control in these conditions, ensuring that UAVs can continue their missions without disruption. Features such as adaptive flight control systems, streamlined fuselages, and efficient wing designs allow UAVs to resist turbulence, maintain steady flight paths, and capture precise environmental data. Overall, aerodynamics plays a fundamental role in optimizing UAV performance for environmental applications [2]. By enhancing endurance, reducing noise, and improving stability, aerodynamic advancements enable UAVs to be powerful tools in scientific research and conservation efforts worldwide.

Key Factors Affecting UAV Aerodynamics and Performance

Understanding the fundamental aerodynamic principles of UAVs is essential for optimizing their efficiency, stability, and overall performance. Several key factors influence how UAVs generate lift, reduce drag, and maintain energy-efficient flight.

1. **Lift-To-Drag Ratio (L/D Ratio)**

The lift-to-drag ratio is a crucial indicator of UAV aerodynamic efficiency. A higher L/D ratio means that the UAV generates more lift relative to the drag it experiences, allowing it to stay airborne for longer durations while consuming less power. Fixed-wing UAVs, designed with streamlined bodies and efficient wings, generally have a higher L/D ratio compared to rotary-wing UAVs, making them more suitable for long-endurance missions [90].

2. **Airfoil Shape and Wing Design**

The design of a UAV's wings and airfoil plays a crucial role in its ability to generate lift and maintain efficient flight. Various factors, including airfoil curvature, thickness, and angle of attack, influence how air moves over the surface, ultimately determining the UAV's aerodynamic performance. These design considerations directly affect the aircraft's stability, speed, and energy efficiency. A highly curved (cambered) airfoil is designed to maximize lift production at lower speeds. The increased curvature allows air to travel faster over the top surface than the bottom, creating a significant pressure difference that generates lift [91]. This makes cambered airfoils ideal for UAVs that need to maintain steady flight at lower speeds, such as agricultural drones and surveillance UAVs. However, the added curvature also increases drag, which can reduce overall flight efficiency and speed. On the other hand, thin, streamlined airfoils are optimized to minimize drag, making them well-suited for high-speed UAVs. These airfoils allow for smoother airflow, reducing resistance and enabling the UAV to achieve greater speeds with less power consumption. High-speed UAVs, such as those used for reconnaissance or military applications, often incorporate this design to improve fuel efficiency and extend operational range [92].

In addition to airfoil shape, wing sweep and aspect ratio significantly impact a UAV's aerodynamic performance. Swept wings are commonly used in high-speed UAVs to reduce drag and improve stability during fast-forward flight. Meanwhile, a higher aspect ratio—meaning longer, narrower wings—reduces induced drag and enhances glide efficiency. This design is commonly seen in fixed-wing UAVs designed for long-endurance missions, such as weather monitoring and aerial mapping. By carefully selecting the right airfoil and wing configuration, UAV designers can optimize performance based on the intended mission [93]. Whether prioritizing high lift for slow-speed stability, reducing drag for high-speed efficiency, or improving glide capabilities, wing and airfoil design remain fundamental to UAV aerodynamics.

3. **Propeller Efficiency**

Propellers and rotors are essential components in UAV propulsion systems, directly influencing thrust generation, energy efficiency, and flight performance. Their design and operational characteristics determine how effectively a UAV can generate lift, maintain stability, and maneuver through various flight conditions. Several key factors impact their efficiency, including size, blade shape, and pitch angle. The **size** of a propeller significantly affects its thrust output and power consumption. Larger propellers can generate more lift while operating at lower revolutions per minute (RPM), which improves energy efficiency and reduces mechanical stress on the UAV's motor. This makes them ideal for applications requiring extended flight durations, such as aerial surveillance and mapping [94]. However, larger propellers also increase weight and require a more powerful motor, which can affect overall UAV design.

The **blade shape** is another critical factor in propeller performance. Aerodynamically optimized blades are designed to minimize energy loss and maximize lift generation. Features such as curved leading edges and tapered tips help reduce drag and enhance overall propulsion efficiency. Well-designed blades allow UAVs to operate more smoothly, consume less power, and achieve better control during flight. The **pitch angle** of a propeller determines how much air is displaced with each rotation. A higher pitch increases forward thrust, allowing the UAV to achieve greater speeds [95]. However, it also demands more power from the motor, which can lead to higher energy consumption and shorter flight times. On the other hand, a lower pitch improves control at lower speeds, making it ideal for tasks requiring precision, such as aerial photography, inspection, and close-range surveillance. By carefully selecting and optimizing propellers and rotors, UAV designers can enhance thrust efficiency, improve flight stability, and maximize operational performance. Understanding these factors helps in designing UAVs that meet specific mission requirements, whether for endurance-focused applications or agile maneuverability [96]. Optimized propeller design ensures effective thrust generation while minimizing energy consumption, contributing to better endurance and flight stability.

4. **Weight Distribution**

Proper weight distribution is a critical factor in ensuring UAV stability, maneuverability, and overall flight efficiency. An imbalanced UAV can experience a range of performance issues, including instability, increased power consumption, and reduced control responsiveness. Engineers and UAV operators must carefully manage weight placement to optimize flight performance and energy efficiency. When the weight of a UAV is unevenly distributed, it can lead to unstable flight dynamics. The UAV may struggle to maintain level flight, causing it to tilt or drift unintentionally. This instability is particularly problematic for both fixed-wing and multirotor UAVs, as they rely on precise aerodynamic forces and thrust control to maintain smooth operation. Improper balance can make takeoff, landing, and mid-flight adjustments more difficult, increasing the risk of operational failure [67].

In addition to stability concerns, poor weight distribution results in higher power consumption. A UAV that is constantly making corrections to counterbalance an uneven load will use more energy, reducing flight time and efficiency. Multirotor UAVs, in particular, are highly affected by imbalanced weight, as they depend on equal thrust distribution across all rotors. If one side of the UAV is heavier than the other, some rotors will have to work harder to compensate, leading to faster battery depletion and reduced endurance. Control responsiveness is also significantly impacted by weight distribution. UAVs with uneven weight placement may experience sluggish or delayed responses to pilot inputs, making them less maneuverable. This issue is especially critical in applications requiring precise navigation, such as aerial photography, surveillance, or delivery drones [97]. In extreme cases, a severely imbalanced UAV may become uncontrollable, leading to crashes or mission failures. To prevent these issues, UAV designers and operators use advanced techniques such as center-of-gravity calculations, strategic placement of components, and automated flight stabilization systems. Ensuring proper weight distribution allows UAVs to achieve smoother flight, better energy efficiency, and improved maneuverability across various operational environments [98]. By carefully positioning payloads, batteries, and sensors, designers ensure that the UAV maintains an optimal center of gravity, improving both flight efficiency and stability.

5. **Impact of Environmental Conditions on UAV Aerodynamics**

Environmental factors such as air density, humidity, and wind resistance significantly influence UAV performance and stability. These conditions affect lift generation, drag, propulsion efficiency, and overall maneuverability. Understanding these effects is essential for designing UAVs that can adapt to varying atmospheric conditions and maintain optimal flight performance. Air density plays a crucial role in UAV aerodynamics and varies with altitude, temperature, and atmospheric pressure. At lower altitudes, where air density is higher, UAVs experience greater lift, which improves flight stability. However, the increased air resistance also results in higher drag, requiring more thrust to maintain speed. In contrast, at higher altitudes, the air becomes thinner, reducing lift and making it necessary for UAVs to either fly at higher speeds or have larger wings to compensate [99]. Additionally, lower air density at high altitudes can reduce engine efficiency, particularly for combustion-powered UAVs. To counteract these effects, modern UAVs utilize adjustable flight control systems that modify wing angles or rotor speeds to maintain stable flight.

Humidity also influences UAV flight by affecting air density. Moist air contains water vapor, which is less dense than dry air, leading to a slight reduction in lift. This means that UAVs operating in high-humidity environments may require additional thrust to maintain altitude. On the other hand, dry air increases density, providing more lift but also generating more drag. Although humidity's impact on UAV performance is relatively minor compared to air density or wind conditions, it is still an important factor to consider, particularly for UAV operations in tropical or coastal regions. Wind resistance is another significant factor affecting UAV stability and efficiency (Silva 2014). Strong winds can push UAVs off course, necessitating real-time adjustments from onboard control systems. Fixed-wing UAVs generally handle windy conditions better than rotary-wing UAVs due to their streamlined design. Additionally, turbulence caused by irregular airflow can lead to sudden altitude and directional changes, making precise control challenging. Advanced stabilization technologies, such as gyroscopes, accelerometers, and GPS-based autopilot systems, help UAVs counteract these disturbances. Crosswinds, in particular, can be challenging for takeoff and landing, requiring UAVs to adjust their flight path dynamically to maintain stability [92]. By considering these environmental factors in UAV design and operation, engineers and pilots can optimize performance, improve stability, and ensure energy efficiency in different atmospheric conditions.

Advanced UAVs use real-time sensors and adaptive flight control systems to adjust for environmental changes, ensuring consistent performance across varying conditions. Optimizing UAV aerodynamics involves balancing multiple factors to achieve efficiency, stability, and performance. By enhancing the lift-to-drag ratio, selecting the right airfoil design, optimizing propeller efficiency, ensuring proper weight distribution, and accounting for environmental factors, UAVs can operate with greater endurance, reliability, and energy efficiency [92]. These considerations are essential for improving UAV applications across military, commercial, agricultural, and scientific fields. Aerodynamics is the foundation of UAV performance across multiple sectors, impacting everything from flight endurance and maneuverability to payload efficiency and stability. Whether in military surveillance, precision agriculture, disaster response, commercial logistics, or environmental research, understanding and optimizing UAV aerodynamics leads to better efficiency, cost savings, and mission success [100]. As UAV technology continues to evolve, advancements in aerodynamics will further enhance their capabilities, making them more reliable, energy-efficient, and versatile in real-world applications.

2.8 Challenges and Considerations in UAV Aerodynamics

Despite advancements in UAV technology, several challenges must be addressed to ensure optimal performance, safety, and efficiency. Aerodynamic design plays a crucial role in overcoming these obstacles, but various factors still pose limitations. One of the primary challenges is the effect of wind and turbulence. UAVs must be designed to handle unpredictable wind conditions, which can cause instability, affect maneuverability, and impact mission success. Advanced stabilization systems, aerodynamic shaping, and real-time flight adjustments are necessary to counteract the effects of turbulence, especially for UAVs operating in extreme weather conditions [101].

Structural limitations also influence UAV aerodynamics. The choice of materials impacts both weight and durability. Lightweight materials improve flight efficiency and reduce energy consumption, but they must also provide sufficient strength to withstand aerodynamic forces and environmental stress. Engineers must balance weight reduction with structural integrity to ensure safe and reliable UAV operations. Another significant factor is battery and power constraints. Aerodynamics directly affect energy consumption and battery life. UAVs with poor aerodynamic designs experience higher drag, requiring more power to sustain flight. Optimizing lift-to-drag ratios and incorporating energy-efficient propulsion systems help maximize endurance and operational range while minimizing power usage [102].

Lastly, UAV designs must comply with regulatory constraints. Aviation regulations and airspace restrictions dictate various aerodynamic considerations, such as flight altitude, speed limits, and noise reduction. Compliance with these rules ensures UAVs can operate safely in shared airspace without interfering with manned aircraft or violating legal restrictions. Addressing these challenges requires a combination of advanced aerodynamic engineering, material innovations, power management strategies, and regulatory adherence. Overcoming these limitations will enable UAVs to operate more efficiently and effectively across diverse applications. Understanding UAV aerodynamics is essential for designing efficient, stable, and high-performance drones [103]. By analyzing lift, drag, thrust, and control mechanisms, engineers can optimize UAVs for various applications. Continuous advancements in aerodynamics, materials, and computational modeling will further enhance UAV capabilities in the future.

2.9 Conclusion and Summary

The study of UAV aerodynamics is crucial for enhancing flight efficiency, stability, and performance across various applications. Fixed-wing UAVs use streamlined airfoil designs for long-range efficiency, while rotary-wing UAVs rely on rotor dynamics for maneuverability and vertical takeoff and landing. Hybrid UAVs combine both capabilities to maximize versatility. Key aerodynamic factors, including lift, drag, propeller efficiency, and air density effects, directly influence UAV flight characteristics. Advances in computational fluid dynamics (CFD) aid in optimizing UAV designs by reducing drag and improving stability. However, challenges such as wind resistance, structural integrity, power constraints, and regulatory compliance must be addressed for optimal UAV functionality. Understanding these principles allows for improved UAV applications in military operations, environmental monitoring, disaster management, agriculture, and commercial logistics. Newton's laws of motion explain how UAVs achieve stable and controlled flight. The First Law describes how UAVs resist changes in motion unless acted upon by an external force, the Second Law determines how forces influence UAV acceleration and movement, and the Third Law explains how UAVs generate lift and thrust through action-reaction forces. Understanding these principles helps in designing UAVs with better stability, efficiency, and control, making them effective for various applications such as surveillance, agriculture, and logistics.

The principles of aerodynamics—rooted in Newton's laws, Bernoulli's principle, and the four forces of flight—are essential for understanding how UAVs operate. These principles guide the design and optimization of UAVs, ensuring they can generate lift, overcome drag, and maintain stable flight. By mastering these concepts, engineers and operators can develop UAVs that are efficient, versatile, and capable of performing complex tasks in agriculture and beyond. Bernoulli's principle is essential in understanding how UAVs, particularly fixed-wing types, generate lift. By utilizing differences in airspeed and pressure, UAV wings create an upward force that allows them to fly efficiently. This principle, combined with Newtonian physics, helps engineers design UAVs that are both aerodynamically efficient and capable of stable flight in various conditions. UAV flight mechanics involve a balance of aerodynamic forces, stability mechanisms, and environmental adaptations. Understanding how fixed-wing and rotary-wing UAVs achieve flight, how stability is maintained, and how lift and drag coefficients impact performance is essential for optimizing UAV efficiency. UAV performance and efficiency depend on maintaining an optimal power-to-weight ratio, improving glide ratio and endurance, and enhancing aerodynamic efficiency to reduce energy consumption. By continuously improving UAV aerodynamics, the industry can achieve higher efficiency, longer endurance, and enhanced capabilities for a wide range of applications. As UAV technology continues to evolve, advancements in aerodynamics will further expand their capabilities, enabling more efficient and innovative uses across different sectors.

References

1. Joy IM (2024) computational fluid dynamics and aerodynamics-system identification of a real hexacopter for controlling the flight within urban environments
2. Telli K, Kraa O, Himeur Y et al (2023) A comprehensive review of recent research trends on unmanned aerial vehicles (UAVS). Systems 11:400
3. Li W, Wu B (2024) Computational fluid dynamics investigation of aerodynamics for agricultural drones. Comput Electron Agric 227:109528

4. Muraleedharan N, Cohen DS, Estrela VV et al (2020) Modelling and simulation of UAV systems. Imag Sens Unmanned Aircraft Syst 1:101–121
5. Orgeira-Crespo P, García-Luis U (2024) Brief introduction to unmanned aerial systems. In: Applying drones to current societal and industrial challenges. Springer, pp 1–22
6. Sharma S (2024) Drone development from concept to flight: design, assemble, and discover the applications of unmanned aerial vehicles. Packt Publishing Ltd
7. Arshad HMS (2024) Development of heavy lift drone platform for various payloads
8. Bai S, He Q, Chirarattananon P (2022) A bioinspired revolving-wing drone with passive attitude stability and efficient hovering flight. Science Robotics 7:eabg5913
9. Giurato M (2020) Design, integration and control of multirotor UAV platforms
10. Goodrich P, Betancourt O, Arias AC, Zohdi T (2023) Placement and drone flight path mapping of agricultural soil sensors using machine learning. Comput Electron Agric 205:107591
11. Landell-Mills MN (2023) The Bernoulli paper experiment explained using Newtonian mechanics
12. Wild G (2023) Misunderstanding flight part 1: a century of flight and lift education literature. Educ Sci 13:762
13. Ayele W (2024) Design and dynamics of light unmanned aircraft wing structures subject to extreme flow turbulence
14. Sanguino T de JM, Domínguez JML (2024) Design and stabilization of a Coandă effect-based UAV: comparative study between fuzzy logic and PID control approaches. Robot Autonomous Systems 175:104662
15. Vepa R (2023) Flight dynamics, simulation, and control: for rigid and flexible aircraft. CRC Press
16. Elmeseiry N, Alshaer N, Ismail T (2021) A detailed survey and future directions of unmanned aerial vehicles (uavs) with potential applications. Aerospace 8:363
17. Lee C, Kim S, Chu B (2021) A survey: flight mechanism and mechanical structure of the UAV. Int J Precis Eng Manuf 22:719–743
18. Shams OA, Alturaihi MH, Mustafa MAS, Majdi HS (2023) Enhancement of drones' control and guidance systems channels: a review. J Européen des Systèmes Automatisés 56:201–212
19. Garg P (2022) Characterisation of fixed-wing versus multirotors UAVs/Drones. J Geomatics 16:152–159
20. Hofer N, Abdulrahim M (2025) An exploration of passive lift device integration and applications in multirotor UAVs, p 0116
21. Febrianto R, Yeoh JC, Putra IGAK, et al (2024) Design and development of a hybrid tricopter fixed-wing UAV for precision agriculture. Indonesian J Comput Sci 13
22. Mieloszyk J, Tarnowski A, Goetzendorf-Grabowski T (2024) Designing aerodynamic devices for UAV–lessons learned. Aircr Eng Aerosp Technol 96:73–83
23. Al-Haddad LA, Jaber AA, Giernacki W et al (2024) Quadcopter unmanned aerial vehicle structural design using an integrated approach of topology optimization and additive manufacturing. Designs 8:58
24. Zuo Z, Liu C, Han Q-L, Song J (2022) Unmanned aerial vehicles: control methods and future challenges. IEEE/CAA J Automatica Sinica 9:601–614
25. Harvey C, Gamble LL, Bolander CR et al (2022) A review of avian-inspired morphing for UAV flight control. Prog Aerosp Sci 132:100825
26. Wan K, Gao X, Hu Z, Wu G (2020) Robust motion control for UAV in dynamic uncertain environments using deep reinforcement learning. Remote sensing 12:640
27. Nahiyoon SA, Ren Z, Wei P et al (2024) Recent development trends in plant protection UAVs: a journey from conventional practices to cutting-edge technologies—a comprehensive review. Drones 8:457
28. Mohsan SAH, Othman NQH, Li Y et al (2023) Unmanned aerial vehicles (UAVs): practical aspects, applications, open challenges, security issues, and future trends. Intel Serv Robot 16:109–137
29. Grubesic TH, Nelson JR (2020) Uavs and urban spatial analysis. Springer

30. Muhammed M, Virk MS (2023) Ice accretion on rotary-wing unmanned aerial vehicles—a review study. Aerospace 10:261
31. Szekely BJ (2024) Fixed-wing UAV design
32. Zhang Y, Zhao Q, Mao P et al (2024) Design and control of an ultra-low-cost logistic delivery fixed-wing UAV. Appl Sci 14:4358
33. Ramesh P, Jeyan JML (2022) Comparative analysis of fixed-wing, rotary-wing and hybrid mini unmanned aircraft systems (UAS) from the applications perspective. INCAS Bulletin 14:137–151
34. Farajijalal M, Eslamiat H, Avineni V et al (2025) Safety systems for emergency landing of civilian unmanned aerial vehicles—a comprehensive review. Drones 9:141
35. Özbek E, Ekici S, Karakoc TH (2023) An evaluation on landing gear configurations of fixed-wing, rotary-wing, and hybrid UAVs. In: Unmanned aerial vehicle design and technology. Springer, pp 153–165
36. Haisley S (2021) Designing and developing stable body flight in drones
37. Ucgun H, Yuzgec U, Bayilmis C (2021) A review on applications of rotary-wing unmanned aerial vehicle charging stations. Int J Adv Rob Syst 18:17298814211015864
38. Sabour M, Jafary P, Nematiyan S (2023) Applications and classifications of unmanned aerial vehicles: a literature review with focus on multi-rotors. Aeronautical J 127:466–490
39. Ekbote A (2020) Conceptual design of an autonomous quad plane hybrid unmanned aerial vehicle
40. Bergonti F (2023) Motion and hardware intelligence for the adaptive morphology of flying multibody robots
41. Santoso F, Garratt MA, Anavatti SG (2017) State-of-the-art intelligent flight control systems in unmanned aerial vehicles. IEEE Trans Autom Sci Eng 15:613–627
42. Mohsan SAH, Khan MA, Noor F et al (2022) Towards the unmanned aerial vehicles (UAVs): a comprehensive review. Drones 6:147
43. Petritoli E, Leccese F, Ciani L (2018) Reliability and maintenance analysis of unmanned aerial vehicles. Sensors 18:3171
44. Abdelmaksoud SI, Mailah M, Abdallah AM (2020) Control strategies and novel techniques for autonomous rotorcraft unmanned aerial vehicles: a review. IEEE Access 8:195142–195169
45. Wood AD (2004) A model for the design and field evaluation of the structural stability of ballasted rootwads in streambanks. The Pennsylvania State University
46. Khatouri H, Benamara T, Breitkopf P, Demange J (2022) Metamodeling techniques for CPU-intensive simulation-based design optimization: a survey. Adv Model Simulat Eng Sci 9:1
47. Ferdousi SA (2009) Decolorization of wastewater by adsorption on activated carbon in a liquid solid fluidized bed
48. Marins JL (2020) A closed-form energy model for quadcopters based on the dynamic of the movement applied to calculate wind parameters aiming to extend the flight time
49. Bacanli SS, Elgeldawi E, Turgut B, Turgut D (2022) UAV charging station placement in opportunistic networks. Drones 6:293
50. Song L, Di B, Zhang H, Han Z (2023) Aerial access networks: integration of UAVs, HAPs, and satellites. Cambridge University Press
51. Panagiotou P, Yakinthos K (2020) Aerodynamic efficiency and performance enhancement of fixed-wing UAVs. Aerosp Sci Technol 99:105575
52. Levin JM, Paranjape AA, Nahon M (2018) Sideslip and slipstream in extreme maneuvering with fixed-wing unmanned aerial vehicles. J Guid Control Dyn 41:1610–1616
53. Harvey C, Inman DJ (2021) Aerodynamic efficiency of gliding birds vs comparable UAVs: a review. Bioinspir Biomim 16:031001
54. Aleisa H, Kontis K, Pirlepeli B, Nikbay M (2023) Conceptual design of a nonconstant swept flying wing unmanned combat aerial vehicle. J Aircr 60:1872–1888
55. Throneberry G, Hassanalian M, Abdelkefi A (2019) Insights into sensitivity of wing shape and kinematic parameters relative to aerodynamic performance of flapping wing nano air vehicles. Drones 3:49

56. Andria G, Di Nisio A, Lanzolla A, et al (2018) Design and performance evaluation of drone propellers. IEEE, pp 407–412
57. Doerffer P, Doerffer K, Ochrymiuk T, Telega J (2019) Variable size twin-rotor wind turbine. Energies 12:2543
58. Alotaibi A, Chatwin C, Birch P (2023) Ubiquitous unmanned aerial vehicles (UAVs): a comprehensive review. Shanlax Int J Arts, Sci Human 11:62–90
59. Kapoulas IK, Statharas J, Hatziefremidis A, Baldoukas A (2022) Fast airfoil selection methodology for small unmanned aerial vehicles. Appl Sci 12:9328
60. Ceruti A, Fiorini T, Boggi S, Mischi L (2018) Engineering optimization based on dynamic technique for order preference by similarity to ideal solution fitness: application to unmanned aerial vehicle wing airfoil geometry definition. J Multi-Criteria Decision Analysis 25:88–100
61. Hieu NK, Loc HT (2016) Airfoil selection for fixed wing of small unmanned aerial vehicles. Springer, pp 881–890
62. Basri EI, Basri AA, Ahmad KA (2023) Computational fluid dynamics analysis in biomimetics applications: a review from aerospace engineering perspective. Biomimetics 8:319
63. Khan FA, Islam A, Abbas H, et al (2023) Computational analysis of a UAV drone with varying angle of attacks. IEEE, pp 1–11
64. Haoran L, Zhang Y, Haixin C (2022) Optimization design of airfoils under atmospheric icing conditions for UAV. Chin J Aeronaut 35:118–133
65. Moustafa MH, Khashwain HS, Dol SS, Al Ramahi A (2024) Enhancing aerodynamic performance in UAV design: CFD analysis
66. Bajwa OI, Baluch HA, Saeed HA (2024) Machine learning approach for predicting key design parameters in UAV conceptual design. Ain Shams Eng J 15:102932
67. Gholami A (2023) Analyzing fixed-wing drone design and evaluating financial viability in unmanned aerial vehicle. J Multidisciplinary Eng Sci Stud (JMESS) 9:5186–5193
68. Firoozi AA, Hejazi F, Firoozi AA (2024) Advancing wind energy efficiency: a systematic review of aerodynamic optimization in wind turbine blade design. Energies 17:2919
69. Bravo-Mosquera PD, Cerón-Muñoz HD, Catalano FM (2022) Design, aerodynamic analysis and optimization of a next-generation commercial airliner. J Braz Soc Mech Sci Eng 44:609
70. Chai R, Guo Y, Zuo Z et al (2024) Cooperative motion planning and control for aerial-ground autonomous systems: methods and applications. Prog Aerosp Sci 146:101005
71. Misiorowski MP (2019) Understanding and improvement of eVTOL UAV performance through high-fidelity analysis of interactional aerodynamics. Rensselaer Polytechnic Institute
72. Lim SKY (2024) Comprehensive multi-disciplinary design of an electric vertical take-off and landing aircraft
73. Shukla P, Shukla S, Singh AK (2024) Trajectory-Prediction techniques for unmanned aerial vehicles (UAVs): a comprehensive survey. IEEE Communications Surveys & Tutorials
74. MahmoudZadeh S, Yazdani A, Kalantari Y et al (2024) Holistic review of UAV-centric situational awareness: applications, limitations, and algorithmic challenges. Robotics 13:117
75. Løw-Hansen B, Hann R, Gryte K, et al (2025) Modeling and identification of a small fixed-wing UAV using estimated aerodynamic angles. CEAS Aeronautical J 1–23
76. Wang Y (2024) Optimizing PID and sliding mode control for quadcopter UAV stabilization. Atlantis Press, pp 481–493
77. Hage C, Sophy T, Aglzim E-H (2023) Investigating UAV propellers performances near moving obstacles: CFD study, thrust control, and battery energy management. IEEE Open J Vehicular Technol 4:590–609
78. Bhat SS, Anavatti SG, Garratt M, Ravi S (2024) Review of autonomous outdoor blimps and their applications. Drone Syst Appl 12:1–21
79. Criollo L, Mena-Arciniega C, Xing S (2024) Classification, military applications, and opportunities of unmanned aerial vehicles. Aviation 28:115–127
80. Lyu C, Zhan R (2022) Global analysis of active defense technologies for unmanned aerial vehicle. IEEE Aerosp Electron Syst Mag 37:6–31
81. Toscano F, Fiorentino C, Capece N, et al (2024) Unmanned aerial vehicle for precision agriculture: a review. IEEE access

82. Gulak AM (2024) Implementation of drone for spraying herbicide and pesticide
83. Chen H, Lan Y, Fritz BK et al (2021) Review of agricultural spraying technologies for plant protection using unmanned aerial vehicle (UAV). Int J Agri Biol Eng 14:38–49
84. Alsamhi SH, Shvetsov AV, Kumar S et al (2022) UAV computing-assisted search and rescue mission framework for disaster and harsh environment mitigation. Drones 6:154
85. Damaševičius R, Bacanin N, Misra S (2023) From sensors to safety: internet of emergency services (IoES) for emergency response and disaster management. J Sens Actuator Netw 12:41
86. Khan A, Gupta S, Gupta SK (2022) Emerging UAV technology for disaster detection, mitigation, response, and preparedness. J Field Robot 39:905–955
87. Nguyen H-T, Ghazali I (2024) Trends of unmanned aerial vehicles in logistics delivery. Bincang Sains dan Teknologi 3:52–59
88. Abiodun TF (2021) Public support and prospects of drones or unmanned aerial vehicles (UAVs) technologies for effective transport and logistics delivery in Nigeria. J DOI 7
89. Betti Sorbelli F (2024) UAV-based delivery systems: a systematic review, current trends, and research challenges. J Autonom Transp Syst 1:1–40
90. Karunathilake E, Le AT, Heo S et al (2023) The path to smart farming: Innovations and opportunities in precision agriculture. Agriculture 13:1593
91. Liang H, Lee S-C, Bae W et al (2023) Towards UAVs in construction: advancements, challenges, and future directions for monitoring and inspection. Drones 7:202
92. Munkombwe J (2024) Understanding the diffusion of weather index insurance among smallholder farmers in Zambia: case of four (4) selected districts
93. den Besten N, Steele-Dunne S, de Jeu R, van der Zaag P (2021) Towards monitoring waterlogging with remote sensing for sustainable irrigated agriculture. Remote Sensing 13:2929
94. Joshi R, Pandey K (2024) IoT-enabled UAV: a comprehensive review of technological change in Indian farming. Unmanned Aircraft Systems 93–135
95. Pajares G (2015) Overview and current status of remote sensing applications based on unmanned aerial vehicles (UAVs). Photogramm Eng Remote Sens 81:281–330
96. Balestrieri E, Daponte P, De Vito L et al (2021) Sensors and measurements for UAV safety: an overview. Sensors 21:8253
97. Watson R (2023) Unmanned aerial vehicles for contact-based inspection
98. Chandana G, Gowda GS, Sayeeganesh A, Vannarth RR (2024) High payload fraction UAV design and performance evaluation. SAE Technical Paper
99. Scanavino M (2021) Design and testing methodologies for UAVs under extreme environmental conditions
100. Auma JA (2023) The application of drone technology in micro-watershed evaluation. J Geogr Assoc Tanzania 43:47–66
101. Sepulveda E, Smith H (2017) Technology challenges of stealth unmanned combat aerial vehicles. The Aeronautical Journal 121:1261–1295
102. Abbas N, Abbas Z, Liu X et al (2023) A survey: future smart cities based on advance control of unmanned aerial vehicles (UAVs). Appl Sci 13:9881
103. Naveed M (2024) The adoption of 4.0 agriculture for wine production in order to improve efficiency, sustainability and competitiveness

Chapter 3
Fundamentals and Applications of UAV Aerodynamics in Precision Agriculture

Abstract Unmanned Aerial Vehicles (UAVs) have become indispensable tools in modern agriculture, enabling high-resolution data collection, precise input application, and real-time field monitoring. The efficiency of these aerial systems is closely tied to a comprehensive understanding of their aerodynamic behavior. This chapter explores the foundational principles of UAV aerodynamics, with particular emphasis on rotorcraft and fixed-wing platforms commonly deployed in precision agriculture. Core aerodynamic elements such as lift, thrust, drag, and rotor-induced airflow patterns—including downwash—are analyzed in the context of flight stability, maneuverability, and payload delivery. Special focus is given to how aerodynamic interactions affect crop canopies, revealing both advantageous effects, like enhanced pesticide dispersion, and potential drawbacks such as crop stress or structural damage. Environmental factors including wind speed, temperature, and terrain are also examined for their influence on UAV performance in agricultural settings. While UAVs hold the potential to transform agriculture by increasing efficiency, reducing labor, and enhancing sustainability, widespread adoption remains constrained by high initial costs, regulatory barriers, and the need for technical expertise. By linking aerodynamic science with field-level challenges and innovations, this chapter presents a holistic view of UAV integration in agriculture. It highlights how improvements in drone design, autonomy, and data processing will continue to drive their utility in transforming traditional farming into smart, sustainable systems.

Keywords AVs airflow and downwash · Precision farming · Drone based smart agriculture

3.1 Introduction

Unmanned Aerial Vehicles (UAVs), commonly known as drones, have emerged as transformative tools in modern agriculture, introducing a new era of precision, efficiency, and sustainability in farming practices. As the global demand for increased agricultural productivity intensifies, UAV technology offers an advanced solution

© The Author(s), under exclusive license to Springer Nature Singapore Pte Ltd. 2025
J. Li, *UAV Aerodynamics and Crop Interaction*, Smart Agriculture 13,
https://doi.org/10.1007/978-981-96-8402-1_3

by enabling site-specific management, real-time monitoring, and optimized input application. Among the key drivers of this technological advancement is a growing understanding of UAV aerodynamics—how lift, drag, thrust, and airflow patterns interact with agricultural environments and influence operational outcomes [1]. In agricultural applications, UAVs perform diverse tasks such as high-resolution field mapping, multispectral imaging, soil analysis, and precision spraying. Their ability to cover large agricultural landscapes swiftly and accurately has made them indispensable in the implementation of precision farming. However, the performance and reliability of UAVs in such tasks depend not only on onboard sensors or data-processing algorithms but also on aerodynamic characteristics that govern stability, maneuverability, and payload capacity. For instance, rotor-induced airflow—particularly the downwash generated by rotary-wing drones—can influence plant canopies in beneficial ways, such as improving pesticide penetration, but may also cause mechanical stress or damage to delicate crops [2].

Understanding UAV aerodynamics is thus critical for both optimizing drone design and maximizing their effectiveness in field operations. Equipped with multispectral and thermal sensors, UAVs are widely used for assessing plant health, identifying pest infestations, and detecting nutrient deficiencies. The way these drones interact aerodynamically with the crop environment, especially during low-altitude or close-proximity operations, significantly affects the quality and accuracy of data collection. These aerodynamic interactions are influenced by flight altitude, speed, rotor configuration, and environmental factors such as wind turbulence and terrain variability. Beyond crop surveillance, UAVs contribute significantly to soil and water management strategies. Drones equipped with LiDAR and infrared sensors analyze topographical variations and monitor soil moisture distribution. These functions, while often treated as purely sensor-driven tasks, are also influenced by the flight stability and aerodynamic efficiency of the UAV platform. For example, fixed-wing UAVs, which typically have superior glide ratios and longer endurance, are often preferred for large-scale mapping and water management due to their aerodynamic design, whereas multirotor drones are more suitable for localized applications and vertical take-off due to their hovering capabilities [3].

In animal husbandry, UAVs are used to monitor livestock movements and assess pasture conditions. Although aerodynamics may seem peripheral in these applications, flight endurance, stability under gusty conditions, and maneuverability in complex terrains play crucial roles in determining the operational success of UAV-based livestock management systems. Despite these advancements, several challenges remain that hinder widespread UAV adoption in agriculture. Regulatory restrictions, high capital costs, limited battery life, and the need for skilled operation and data interpretation are among the most pressing obstacles. Notably, aerodynamic limitations such as excessive energy consumption during hover or instability in turbulent wind conditions directly affect mission planning and data quality. Training in UAV aerodynamics, therefore, becomes essential for operators aiming to use drones efficiently and safely. Improvements in aerodynamic design—such as enhanced rotor configurations, reduced drag surfaces, and energy-efficient propulsion—are increasingly being developed to overcome these operational barriers [4].

UAVs in agriculture lies in their evolution into fully autonomous systems with integrated aerodynamic intelligence. As artificial intelligence (AI) and machine learning algorithms are combined with aerodynamic modeling, new-generation UAVs are expected to dynamically adjust their flight behavior in response to crop height, wind patterns, or canopy resistance. These innovations will not only extend flight times and increase payload efficiency but also minimize the environmental impact of UAV operations by ensuring targeted, low-disturbance field interventions.

UAVs are reshaping agricultural practices through their multifunctional capabilities and operational flexibility. However, to fully realize their potential, a deeper understanding of UAV aerodynamics is essential. This chapter explores the fundamental aerodynamic principles of rotary and fixed-wing UAVs and their influence on flight dynamics, crop interaction, and environmental performance. By bridging the gap between aeronautical science and agricultural application, we aim to provide a scientific foundation for improving drone-based agricultural systems and fostering sustainable, data-driven farming.

3.2 Fundamentals of UAV Aerodynamics

Drone Flight Dynamics: The Science of Aerodynamics and Control

Drones, or unmanned aerial vehicles (UAVs), have surged in popularity across a spectrum of applications, from capturing breathtaking aerial footage to performing intricate industrial inspections. To harness the full potential of these versatile machines, it's crucial to understand the intricate dance of aerodynamics and Drone Flight Dynamics that govern their behavior in the skies. This comprehensive guide delves into the principles that ensure efficient and stable drone flight. Understanding the aerodynamics and Drone Flight Dynamics of drones is fundamental for anyone looking to fully exploit their capabilities, whether for leisure or professional purposes. As drone technology evolves, so too will the sophistication of their design and control systems, promising even greater achievements in UAV performance. By mastering these principles, drone operators can ensure safer, more efficient, and more productive flights across a myriad of applications (Table 3.1).

Table 3.1 Key aerodynamic factors

Factor	Role in flight	Impact on drone performance
Lift	Opposes gravity	Essential for takeoff and altitude maintenance
Thrust	Propels forward	Determines speed and acceleration
Drag	Resists motion	Reduces efficiency; needs minimization
Weight	Downward force	Must be balanced by lift for stable flight

Drone Propulsion and Movement

Drones manipulate these forces through sophisticated propulsion systems and agile design:

- **Fixed-Wing Drones**: These drones generate lift through static wings and thrust via propellers or jet engines. They are ideal for long distances and high-altitude flights.
- **Multirotor Drones**: Commonly seen in quadcopters, these drones use multiple rotors to lift and propel themselves, offering excellent vertical takeoff and landing capabilities.

Stability and Control

Achieving and maintaining stability is crucial for effective drone operation. This involves a delicate balance of design, weight distribution, and real-time control adjustments:

- **Design**: Aerodynamic shapes and balanced weight distribution aid in natural stability.
- **Control Systems**: Modern drones incorporate advanced gyroscopes and accelerometers to monitor and adjust their position, orientation, and motion.

  Control in drones is achieved through varying the speed of the rotors:

- **Roll**: Controlled by adjusting the speed of rotors on opposite sides.
- **Pitch**: Managed by altering the speed between front and back rotors.
- **Yaw**: Adjusted by changing the relative speed of rotors spinning in opposite directions.

Effects of Wind on Drone Flight

Wind plays a significant role in drone operations, affecting everything from trajectory to battery efficiency:

- **Headwinds** increase drag and reduce speed, demanding more power and reducing efficiency.
- **Tailwinds** can boost speed but may compromise control and stability.
- **Crosswinds** introduce lateral forces that can cause drift and require adjustments to maintain the intended flight path.

Experienced pilots must account for these conditions, adjusting their flight strategies to maintain control and efficiency.

Rotoraircraft Aerodynamics in Agricultural Applications

Unmanned aerial vehicles, especially agricultural unmanned helicopters (AUH), are nowadays extensively used in precision agriculture. AUHs have recently become responsible for spraying fertilizers and pesticides for crop yields. The strong downward rotating flow produced by the main rotor helps very uniform crop spraying which determines that how important is the aerodynamics of rotor blade in AUH. In this work, the aerodynamic performance of AUH rotor blades is evaluated and an efficient blade is obtained by numerically investigating the influence of design variables on the aerodynamics of rotor blades. The design variables consist of airfoil shape, pitch settings, and twist angle. The limited power available in hover and aerodynamic requirements (lift and drag) are the aerodynamic constraints. This analysis only considers the hovering flight condition at a constant rotational speed. The aerodynamically efficient rotor blade which is based on gradually varying and linearly twisted airfoil shapes, show a significant improvement in hover performance with relatively uniform blade loading. After testing, the optimum blade can be used as the main rotor in the AUH to perform precision farming.

A desire to increase productivity and profitability in the agriculture has stimulated interest in innovative technologies. Advances in information technology and their application in crop production, which is labeled as precision agriculture (PA), is creating potential for substantial changes in management and decision making in the farming sector. Farmers use PA practices by applying precise and correct amounts of seed, water, pesticides, and herbicides at the right time, which benefits the entire crop cycle. Consequently, precision agriculture management has turned out to be a keystone of sustainable agriculture.

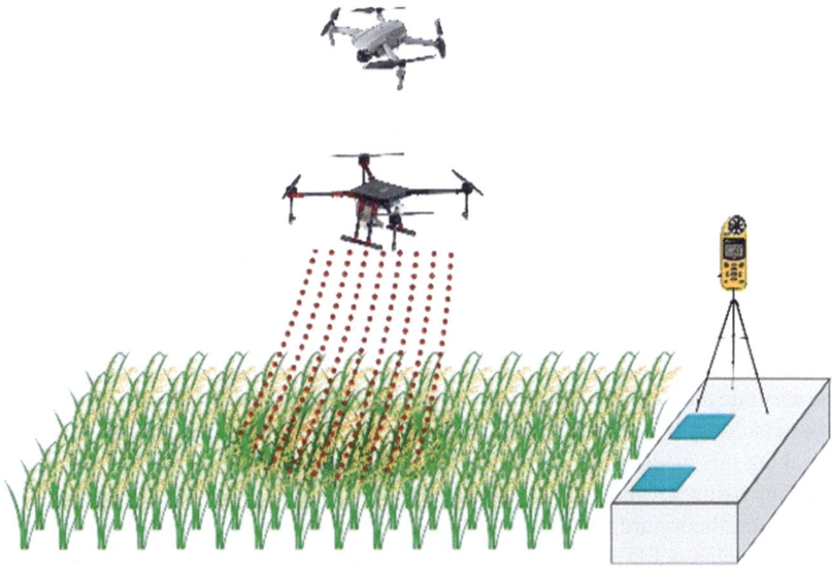

Indeed, PA is not based on a single technology, but an integration of various modern techniques. These include high precision positioning systems (e.g. GPS), geographical information systems (GIS), grain yield monitoring devices, sensors and remote sensing, variable-rate technologies, and crop management. Aerial systems such as satellites, air- craft, balloons, unmanned aerial systems (UAS), and rotorcraft are the common platforms used in data acquisition and field variability maps to support decision making and management practices. The application of these airborne platforms significantly reduces the time necessary for surveying the land area, and consequently, allows farmers to adjust their agricultural practices. However, it is believes that the cost, availability, and processing of crop imagery from satellites and aircraft have made their applications limiting and impractical. As a result, UAS could be an economical, feasible, and potential substitute for high-resolution photography and spraying pesticides.

In recent years, the application of UAS has flourished, ranging from fixed-wing aircraft to multicopters. Specifically, the AUH has been applied to agricultural use because heli copters have numerous benefits over fixed wing air craft, such as vertical take-off and landing (VTOL), hovering, low runway requirements, and the ability to fly at low speeds and altitude. AUHs are smaller than full-scale manned helicopters and readily used as a tool for crop status sensing. In contrast to sophisticated applications of unmanned helicopters, AUHs have also been used for chemical spraying. Applying pesticides manually is a very labor intensive, inefficient, and hazardous job for the operators. However, aerial application of chemicals with an AUH ensures accurate and timely spraying, decreases the drift potential and evaporation, and reduces the environmental impact. In addition, the rotor downwash generated while hovering results deep penetration of the chemical and ensures uniform application.

Helicopters are available in a range of shapes and sizes, but most share the same major parts. All parts of a helicopter are important, but main rotors are the most important because they develop aerodynamic lift. Accordingly, the aerodynamicists take special care in the aerodynamic design of the main rotor to ensure the optimum performance both in hover and cruise flight conditions. For aerial applications, the agricultural helicopters operate at a relatively low speed (4–7 m/s) and altitude; therefore, the aerodynamics of AUHs could be considered as nearly hovering.

3.3 Aerodynamics Computationalmethodology

The accurate and reliable computation of the helicopter rotor performance continues to be a challenge in rotorcraft design. A variety of numerical techniques can be utilized in making this assessment, ranging from the simplified momentum theory to direct modeling using advanced computational fluid dynamics. The methods based on the momentum theory are low-fidelity methods while for better accuracy, high-fidelity methods such as CFD are desired. CFD has matured as a general predictive tool with sensible reliability in the industry. Since the first application of CFD to the rotary wing, recent solutions of Reynolds-Averaged Navier–Stokes (RANS) equations offer substantial developments in the computation of complex aero-dynamics around the rotorcraf) and the similar applications. The details of the applied method are discussed in the subsequent sections.

Table 3.2 Characteristics of the test bed (AgroHeli-4G)

Parameters	Symbols	Values
Number of blades	N_b	2
Rotor blade radius	R	1.510 m
Blade chord length	c	0.135 m
Nominal rotor speed	Ω	890 rpm
Blade tip Mach number	M_{tip}	0.414
Blade aspect ratio	AR	10.0
Rotor solidity	$\sigma = \frac{N_b c}{\pi R}$	0.0569
Rated power available	p_{av}	25 kW
Total dead weight	W_d	650N

3.3.1 Agricultural Unmanned Helicopter

The AUH employed in the performance evaluation is a two-bladed AgroHeli-4G (AH-4G), a prototype designed and constructed by Kyungpook National University. The total lift capacity of AH-4G is up to 1176 N with an engine power of 25 kW. To improve the performance of the AH-4G; airfoil shape, collective pitch, and blade twist are the parameters considered for the simulation. The characteristics of the AUH test bed are listed in Table 3.2.

Flow Physics Solver

The computations are performed using the Ansys Fluent Release 16.2 solver. A multiple reference frame (MRF) method is use to simulate the isolated rotating blades. The isolated rotor in hovering flight exhibits a steady-state flow in the moving reference frame. The MRF method is a steady state approximation, and the flow in each moving cell zone is solved using the moving reference frame equations while the stationary equations are used to solve the stationary zone ($\Omega = 0$). A local reference frame transformation is performed at the interfaces between adjacent cell zones to enable flow variables. The AUH rotor configurations are evaluated by computing the aerodynamic coefficients, including lift, drag, and moment. These aerodynamic quantities are then used to determine the thrust, power, and torque required for a particular flight regime. In hovering, the thrust is entirely related to the lift force, as it overcomes the helicopter weight and the drag is mainly induced drag incurred while the blades are producing lift. An important consequence of providing thrust is torque. The torque is directly related to the amount of engine power being used to turn the main rotor system. The rotor design pursues a maximal lift for a corresponding power source. Oftentimes the required power level is limited by a choice of aeronautical power supply due to its availability; therefore, the power limit should be considered in the design of the rotor, as well as the performance. Moreover, the hover performance for the configurations is calculated by plugging the aerodynamic parameters. Koo has already discussed the two-dimensional aerodynamic characteristics of the airfoil sections, but three-dimensional aerodynamic characteristics are yet to be analyzed. These characteristics are deliberated subsequently (Table 3.3).

Technological Breakthroughs in UAV Hardware

Unmanned Aerial Vehicles (UAVs) have seen significant advancements in hardware technology over the past decade, enhancing flight endurance, payload capacity, navigation accuracy, and automation. Key innovations include hybrid-electric and solar-powered UAVs, high-performance batteries, wireless charging pads, AI-driven autopilot systems, RTK GPS, and Lidar-based navigation. Advanced flight control and navigation systems include AI-driven autopilot systems, RTK GPS, and Lidar-based navigation. Aerodynamics and airframes have been improved with lightweight composite materials, morphing wings technology, and VTOL designs. High-resolution multispectral and hyperspectral cameras are used in agriculture to detect crop health, soil conditions, and water stress [5]. Thermal imaging sensors

Table 3.3 Summary of optimum performance parameters for rotor blade configurations

Parameters	C1		C2		C3	
	Untwisted	Twisted	Untwisted	Twisted	Untwisted	Twisted
Peak FM	0.4153	0.4526	0.4531	0.4820	0.4486	0.4753
CT at peak FM	0.0070	0.0075	0.0080	0.0085	0.0078	0.0083
$C_{Q,\text{opt}} = C_{P,opt}$ $(\times 10^{-2})$	0.0995	0.1023	0.113	0.1146	0.1090	0.1116
θ_{opt} at Peak FM	12°	13°	12°	13°	12°	13°
$\theta_{\max}$ *at* $C_{T,\max}$	20°	18°	18°	18°	20°	20°

are useful for search-and-rescue missions, precision agriculture, and infrastructure inspections. Gas and chemical sensors detect pollutants, making them valuable for environmental monitoring and disaster response.

Swarm technology and AI integration have led to drone swarming algorithms, autonomous coordination of multiple UAVs, and edge computing in UAVs. Onboard AI processes data in real time, reducing latency and improving decision-making. High-speed 5G connectivity enables real-time data transmission for UAV fleets, improving remote operation and live-streaming capabilities. Satellite-enabled UAVs allow for beyond-line-of-sight operations in remote areas. Technological advancements in UAV hardware continue to push the boundaries of performance, making drones more efficient, autonomous, and versatile [6]. As these breakthroughs progress, UAVs will play an even more significant role in various industries, including agriculture, defense, logistics, and environmental monitoring.

Case Studies of Improved Yield:

The integration of Unmanned Aerial Vehicles (UAVs) in agriculture has led to significant improvements in crop yields by enabling precision farming practices. Drones equipped with advanced imaging, artificial intelligence (AI), and automated spraying systems help farmers optimize resources, detect problems early, and implement targeted interventions. Several real-world case studies demonstrate how UAVs have successfully enhanced agricultural productivity and sustainability. One notable case is the use of drones in rice farming in Japan. With an aging farming population and labor shortages, Japanese farmers have increasingly adopted UAVs for crop monitoring and spraying. In a study conducted by the National Agriculture and Food Research Organization (NARO), drones equipped with multispectral cameras were used to assess rice plant health [7]. The data helped farmers detect nitrogen deficiencies, allowing them to apply fertilizers precisely where needed. This resulted in a 15% increase in yield while reducing fertilizer usage by 30%. Additionally, UAV-assisted pesticide spraying reduced labor costs and ensured even coverage, preventing disease outbreaks that could harm the crop.

Another successful application of UAVs is seen in vineyard management in France. Wine producers in Bordeaux have adopted drones to optimize irrigation and monitor vine health. By using thermal imaging, UAVs identified areas experiencing water stress, allowing farmers to adjust irrigation schedules accordingly. This precision approach led to an improvement in grape quality and yield, with some vineyards reporting a 20% increase in production [8]. Additionally, drones helped detect early signs of fungal infections, enabling targeted fungicide application and reducing chemical use by 40%. This not only improved sustainability but also enhanced the overall quality of the wine. A third case study involves maize and wheat farmers in Kenya, where drought conditions often threaten crop production. The Kenyan Agricultural Research Institute (KARI) introduced UAVs to monitor crop health and soil moisture levels. Drones helped identify drought-stressed areas and guided farmers in adjusting irrigation and fertilizer application accordingly. As a result, maize farmers reported a 25% increase in yield while reducing water consumption by 35%. This initiative demonstrated the potential of UAVs to improve food security in regions facing climate challenges [9, 10].

Enhancing Yield Efficiency:

Increasing agricultural yield efficiency is a top priority for farmers and researchers worldwide, as growing populations and climate change continue to threaten food security [11]. Traditional farming methods often lead to resource inefficiencies, uneven crop growth, and increased environmental degradation [12]. However, the adoption of Unmanned Aerial Vehicles (UAVs) in agriculture has provided innovative solutions that improve crop monitoring, optimize input distribution, and maximize productivity [13]. By leveraging UAV technology, farmers can make data-driven decisions that enhance yield efficiency while reducing costs and environmental impact. One of the key ways UAVs enhance yield efficiency is through precision crop monitoring and health assessment. Unlike manual inspections, which are time-consuming and prone to errors, UAVs equipped with high-resolution cameras and multispectral sensors provide real-time data on plant health, soil conditions, and growth patterns. These drones can detect early signs of disease, nutrient deficiencies, and water stress, enabling farmers to address issues before they affect yield. For example, in a study conducted on wheat farms, UAV-based crop monitoring helped farmers increase yields by 20% by identifying and correcting nitrogen deficiencies before they became widespread [14].

Another critical factor in yield efficiency is optimized resource distribution, particularly for water, fertilizers, and pesticides. UAVs help achieve variable rate application (VRA) by mapping field conditions and identifying areas that require specific treatments. This targeted approach ensures that inputs are applied only where needed, reducing waste and preventing soil degradation. Farmers using UAV-guided fertilization and pesticide application have reported reductions in input costs by up to 30% while maintaining or even improving crop yields. Additionally, UAV-assisted irrigation management has been shown to improve water-use efficiency by identifying dry zones and adjusting irrigation schedules accordingly. UAVs also play a crucial role in yield estimation and harvest planning. Traditional methods of predicting yields often rely on historical data and manual sampling, which can be inaccurate [15]. UAVs, however, provide real-time aerial imagery and AI-driven analysis to estimate crop production with high accuracy. This information helps farmers plan harvest schedules, manage labor efficiently, and optimize storage and transportation logistics. In large-scale farming operations, the use of UAVs for yield prediction has led to a 15% increase in post-harvest efficiency by reducing losses and ensuring timely market supply [16].

UAV Aerodynamics and Micro Climate Engineering and Adaptation

Climate change poses one of the most significant threats to global agriculture, affecting crop yields, soil health, water availability, and overall food security. Rising temperatures, unpredictable rainfall patterns, increased frequency of droughts, and extreme weather events make traditional farming methods less effective. However, Unmanned Aerial Vehicles (UAVs), commonly known as drones, are emerging as a vital tool for farmers to adapt to climate change. UAVs offer real-time data, precision monitoring, and automated solutions that help mitigate the effects of climate variability and enhance the resilience of agricultural systems. One of the primary ways UAVs contribute to climate change adaptation is through real-time environmental monitoring and early warning systems [17]. Equipped with multispectral, thermal, and hyperspectral sensors, drones can assess soil moisture, vegetation health, and temperature variations across large agricultural fields. This data allows farmers to anticipate drought conditions, detect water stress in crops, and implement timely interventions such as adjusting irrigation schedules or selecting drought-resistant crop varieties [18]. For example, in regions facing recurrent droughts, UAV-assisted water management has helped farmers reduce water consumption by 30% while maintaining crop productivity.

Another critical application of UAVs in climate adaptation is disaster response and recovery. Extreme weather events, such as floods, hurricanes, and wildfires, can cause severe damage to farmlands. UAVs provide rapid post-disaster assessment, enabling farmers and policymakers to evaluate crop losses, assess soil erosion, and develop strategies for land rehabilitation. In flood-affected areas, drones have been used to map submerged farmland, helping farmers determine which crops can be salvaged and which areas require replanting. Additionally, UAVs play a crucial role in reseeding efforts, such as dispersing native plant seeds in degraded lands to restore vegetation and prevent further soil degradation. UAVs also support sustainable farming practices that help mitigate climate change effects. By facilitating precision agriculture, drones minimize excessive use of chemical inputs, reducing greenhouse gas emissions from fertilizers and pesticides. UAVs optimize the use of bio-fertilizers and organic farming methods by delivering inputs only where needed, promoting soil health and reducing dependency on synthetic chemicals [19]. Additionally, UAVs contribute to carbon sequestration projects by assisting in afforestation and reforestation efforts, helping farmers transition towards more sustainable land management practices.

Supporting Small Farmers

Small-scale farmers form the backbone of global food production, yet they often face significant challenges such as limited access to technology, unpredictable weather conditions, high input costs, and difficulties in monitoring crop health. Traditional farming methods can be labor-intensive and inefficient, making it difficult for small farmers to maximize yields and profitability. However, Unmanned Aerial Vehicles (UAVs), commonly known as drones, are emerging as a transformative tool that empowers small farmers by providing cost-effective solutions for precision farming, resource management, and climate resilience. One of the biggest challenges small farmers face is crop monitoring and disease detection. In traditional farming, identifying plant stress, pest infestations, and nutrient deficiencies is often based on visual inspections, which can be inaccurate and time-consuming. UAVs equipped with multispectral and thermal imaging sensors can quickly assess large areas of farmland, identifying issues before they become critical [20]. This allows small farmers to take timely action, reducing crop losses and improving overall productivity. Studies have shown that drone-assisted crop monitoring can increase yields by up to 20% by enabling early detection of plant health issues.

Another crucial benefit of UAVs is precision agriculture and cost savings. Small farmers often struggle with the overuse or underuse of water, fertilizers, and pesticides, leading to increased expenses and environmental degradation. UAVs enable variable rate application (VRA), where inputs are applied only where needed [21]. This precision reduces waste, lowers input costs, and improves soil health. For instance, smallholder farmers using UAV-guided irrigation have reported up to 30% reductions in water usage, helping them manage resources more efficiently, especially in regions with water scarcity. UAVs also play a significant role in market access and financial inclusion for small farmers. By providing accurate yield predictions and real-time farm data, drones help farmers secure better prices for their produce and access financial support from banks and insurance companies. Some financial institutions now use UAV-generated farm data to assess loan eligibility, making it easier for small farmers to obtain credit for investment in better seeds, equipment, and sustainable practices [22]. Additionally, UAV-based data collection supports participation in government subsidy programs and climate-smart agriculture initiatives, ensuring that small farmers benefit from modern advancements in agriculture.

Reducing Human Labor

Labor shortages are a growing challenge in agriculture, especially as rural populations decline and fewer people enter the farming profession. Traditional farming methods rely heavily on manual labor for tasks such as planting, monitoring crop health, applying fertilizers and pesticides, and harvesting. However, these processes can be time-consuming, physically demanding, and costly. Unmanned Aerial Vehicles (UAVs), commonly known as drones, are revolutionizing farming by automating many labor-intensive tasks, increasing efficiency, and reducing reliance on human workers [23]. One of the most significant ways UAVs reduce human labor is through automated crop monitoring and data collection. In traditional farming, monitoring plant health requires extensive field inspections, which are labor-intensive and prone to human error. UAVs equipped with high-resolution cameras and multispectral sensors can scan entire fields in a fraction of the time, identifying issues such as water stress, pest infestations, and nutrient deficiencies. This automation allows farmers to focus on strategic decision-making rather than spending hours walking through their fields. Studies have shown that drone-assisted crop monitoring reduces labor hours by up to 40%, leading to increased productivity and cost savings. Another area where UAVs reduce manual labor is in precision spraying of fertilizers and pesticides. Traditional methods, such as manual spraying or tractor-mounted sprayers, require significant labor and often result in excessive chemical use, leading to environmental pollution. UAVs equipped with precision spraying systems can target specific areas of a field, reducing waste and ensuring that chemicals are applied only where needed [24]. This not only minimizes labor costs but also improves worker safety by reducing exposure to harmful chemicals. In some cases, farms using UAVs for spraying have reported a 50% reduction in labor requirements and a 30% decrease in chemical use. UAVs also play a role in automated planting and seeding, particularly for cover crops and reforestation projects. Some drones are designed to drop seeds directly into the soil, eliminating the need for manual planting [25]. This technology is especially beneficial in large-scale farming operations and areas where traditional machinery is impractical, such as steep terrains or flood-affected lands. By automating the seeding process, UAVs help reduce the demand for human labor while improving planting efficiency and crop establishment rates.

Long-Term Economic Benefits of UAV Technology in Agriculture

The adoption of Unmanned Aerial Vehicles (UAVs) in agriculture is not only revolutionizing farming practices but also providing substantial long-term economic benefits. As global food demand increases and farming costs continue to rise, UAVs offer solutions that improve efficiency, reduce waste, and enhance productivity. By integrating UAV technology into agricultural operations, farmers can optimize resource management, increase profitability, and ensure long-term sustainability. The economic advantages of UAVs extend beyond individual farms, contributing to overall agricultural sector growth and food security. One of the key long-term economic benefits of UAVs is cost savings through precision agriculture. Traditional farming methods often rely on uniform application of water, fertilizers, and pesticides, leading to inefficient resource use and increased expenses. UAVs enable Variable Rate Application (VRA) by identifying specific areas of a field that require treatment, reducing input costs while maintaining or even improving crop yields [26]. Studies have shown that farms using UAV-assisted precision agriculture can reduce fertilizer costs by up to 30% and pesticide costs by 40%, resulting in significant savings over time. These reductions in input costs allow farmers to allocate resources more efficiently and reinvest in farm improvements.

Another major economic benefit is increased yield and profitability. UAVs provide real-time data on crop health, soil conditions, and climate variability, allowing farmers to make data-driven decisions that enhance productivity. Early detection of diseases, pest infestations, and nutrient deficiencies enables timely interventions, preventing major crop losses. In large-scale studies, farms that utilized UAV monitoring systems reported yield increases of 10–20%, leading to higher profitability and improved financial stability [27]. By ensuring consistent and optimized production, UAVs contribute to long-term revenue growth for farmers. UAVs also reduce labor costs and operational expenses, addressing the challenges of labor shortages and high wages in the agricultural sector. Traditional methods of farm management require extensive human labor for field inspections, planting, spraying, and harvesting. UAVs automate many of these processes, reducing the need for manual labor and minimizing overall operational costs. For instance, farms using UAVs for aerial spraying have reported up to a 50% reduction in labor costs, making large-scale farming more financially viable [28]. Additionally, the ability to monitor vast areas of land remotely improves time efficiency and reduces fuel and machinery expenses.

UAV Adoption Around the World

The use of Unmanned Aerial Vehicles (UAVs) in agriculture is rapidly increasing worldwide due to the need for increased efficiency, sustainability, and productivity.

North America is a global leader in UAV adoption, with companies like John Deere and DJI promoting solutions for precision farming. Canada is also experiencing growth in UAV adoption, particularly in the Prairie provinces, where large-scale farming dominates. Europe is seeing an increase in UAV adoption, supported by the EU's Common Agricultural Policy (CAP), with countries like France, Germany, and Spain leading in precision agriculture. The European Union Aviation Safety Agency (EASA) provides a framework for safe UAV operations. The UK has seen significant growth in UAV adoption, particularly for crop monitoring and spraying. Post-Brexit agricultural policies emphasize technology-driven farming, further encouraging UAV use [29]. Asia is also increasingly adopting UAV technology in agriculture, driven by government support and major UAV manufacturers like DJI. Latin America is a leading adopter of UAVs, driven by large-scale soybean, sugarcane, and corn farms. Future trends include increased automation, integration with IoT, expansion in developing countries, and a focus on sustainability.

North America: The Early Adopters of UAV Technology in Agriculture

North America has been at the forefront of agricultural innovation, with farmers and agribusinesses quickly embracing Unmanned Aerial Vehicles (UAVs) to enhance productivity and sustainability. The region's advanced technological infrastructure, strong research institutions, and supportive regulatory environment have contributed to the widespread adoption of drones in farming. The United States and Canada, in particular, have led the way in integrating UAVs into precision agriculture, setting a global precedent for modern farming practices [30]. The early adoption of UAVs in North American agriculture has resulted in increased efficiency, reduced costs, and improved yield management. One of the primary reasons for the rapid adoption of UAVs in North America is the availability of advanced agricultural technology and infrastructure. The region has long been a pioneer in precision agriculture, with farmers utilizing GPS-guided tractors, automated irrigation systems, and AI-powered analytics. The integration of UAVs into this ecosystem has provided real-time aerial data, enabling farmers to optimize crop management strategies. Major agribusiness corporations, such as John Deere and Bayer, have invested heavily in drone technology, further accelerating its adoption [31]. In the U.S. Corn Belt, for example, UAVs are widely used for crop health monitoring, allowing farmers to detect nutrient deficiencies and pest infestations early, thereby reducing losses.

Another significant factor driving UAV adoption is government support and regulatory advancements. The Federal Aviation Administration (FAA) in the U.S. and Transport Canada have developed clear guidelines for commercial drone use in agriculture, making it easier for farmers to legally operate UAVs. Additionally, research institutions such as the University of Illinois and the University of Guelph have conducted extensive studies on UAV applications in agriculture, providing farmers with valuable insights on how to integrate drones into their farming operations. Government incentives and funding programs have further encouraged small and large-scale farmers to invest in UAV technology, ensuring broader accessibility across the agricultural sector. The economic benefits of UAV technology have also played a crucial role in its widespread adoption in North America. Farmers have been able to significantly reduce costs associated with labor, fertilizers, pesticides, and water consumption. UAV-guided precision agriculture ensures that inputs are applied only where needed, reducing waste and maximizing efficiency. For instance, soybean farmers in the Midwest have reported a 25% reduction in pesticide use by employing UAVs for targeted spraying [32]. These cost savings, combined with increased yields, have improved the profitability of North American farms, making UAVs a valuable long-term investment.

The Rise of UAV Technology in Agriculture : Emerging Markets in Asia and Africa:

Agriculture is the backbone of many economies in Asia and Africa, providing livelihoods for millions of farmers. However, challenges such as low productivity, inefficient resource use, and climate variability have historically hindered agricultural growth in these regions. In recent years, the adoption of Unmanned Aerial Vehicles (UAVs), or drones, has begun transforming farming practices, offering a cost-effective solution to improve efficiency and sustainability. While UAV adoption in Asia and Africa is still in its early stages compared to North America and Europe, rapid technological advancements, supportive policies, and growing private sector investments are accelerating drone usage in agriculture across these regions. One of the key drivers of UAV adoption in Asia and Africa is the need to improve smallholder farming efficiency. Unlike large commercial farms in developed countries, agriculture in these regions is dominated by small-scale farmers who often lack access to modern technology. UAVs provide an affordable way to bridge this gap by offering precision agriculture solutions tailored to small farms [33]. In countries like India, China, and Kenya, drones are being used to monitor crop health, optimize pesticide application, and improve irrigation management. For example, in India, the government has launched initiatives to provide subsidies for drone technology in agriculture, making it more accessible to farmers.

Another significant factor contributing to UAV growth is climate resilience and disaster management. Asia and Africa are highly vulnerable to extreme weather events such as droughts, floods, and locust infestations, which can devastate agricultural production. UAVs are playing a crucial role in monitoring climate-related risks and enabling farmers to respond proactively. In Africa, organizations like the United Nations Food and Agriculture Organization (FAO) have deployed drones to track locust swarms and guide pest control efforts, preventing massive crop losses [34]. Similarly, in Southeast Asia, UAVs are being used to map flood-prone areas and help farmers develop adaptive strategies to minimize damage. The rise of agritech startups and government initiatives has further accelerated UAV adoption in these regions. In Asia, countries like China and Japan have developed domestic drone manufacturing industries, producing cost-effective UAV solutions specifically designed for agriculture. In Africa, startups such as Aerobotics (South Africa) and Uhurulabs (Tanzania) are providing drone-based agricultural services, offering farmers access to UAV technology without the high costs of ownership. Governments in both continents are also recognizing the benefits of UAVs and are working to develop regulatory frameworks that facilitate their safe and effective use in farming [35].

Europe's Sustainability-Focused Use of UAV Technology in Agriculture:

Europe has been at the forefront of sustainable agricultural practices, and the adoption of Unmanned Aerial Vehicles (UAVs) is a key part of this movement. With strict environmental regulations, a focus on reducing chemical usage, and an emphasis on precision farming, European countries are leveraging drone technology to make farming more sustainable. UAVs help European farmers optimize resources, reduce waste, and minimize the environmental impact of agriculture. By integrating UAVs with data analytics and artificial intelligence, Europe is setting a global example of how advanced technology can support sustainable and eco-friendly farming [36]. One of the most significant ways UAVs contribute to sustainability in Europe is reducing chemical use through precision agriculture. European Union regulations impose strict limits on the use of pesticides and fertilizers due to concerns about environmental pollution and biodiversity loss. UAVs equipped with multispectral sensors and artificial intelligence allow farmers to apply chemicals only where needed, significantly reducing overall usage. For example, studies in Germany and France have shown that drone-guided pesticide application can cut chemical use by up to 40%, benefiting both farmers and the environment. By ensuring that chemicals are used efficiently, UAVs help protect soil health and prevent water contamination. Another major advantage of UAV technology in Europe is enhancing biodiversity and soil conservation. With increasing awareness of the importance of protecting natural ecosystems, European farmers are using UAVs to monitor field boundaries, hedgerows, and wildlife corridors [37]. This helps them maintain biodiversity while still optimizing land use for agricultural production. In the Netherlands, for example, drones are used to identify areas where cover crops should be planted to prevent soil erosion and improve carbon sequestration. Such practices align with the European Green Deal's objectives of promoting regenerative agriculture and carbon–neutral farming. UAVs also play a crucial role in water management and irrigation efficiency, which is essential for sustainable farming. Climate change has led to irregular rainfall patterns and increased drought risks in Southern Europe, making water conservation a priority. UAVs equipped with thermal imaging cameras help farmers detect moisture variations in fields, allowing them to adjust irrigation systems accordingly. In Spain and Italy, vineyards and olive farms have successfully used drones to reduce water waste, with some farms reporting a 30% reduction in irrigation needs [38]. By improving water efficiency, UAVs help European farmers adapt to climate change while preserving natural resources.

Key Challenges in UAV Adoption for Aerodynamics:

Unmanned Aerial Vehicles (UAVs) have revolutionized agriculture by enabling precision farming, improving efficiency, and reducing costs. However, the widespread adoption of UAVs faces several challenges, including regulatory restrictions, high costs, technical limitations, and farmer acceptance. Navigating the complex regulatory environment is one of the most significant challenges in UAV adoption. In the United States, the FAA requires farmers to obtain special permits to operate

UAVs commercially, while the European Union has introduced regulations to standardize drone operations across member states [39]. Compliance can be complicated for small farmers, and inconsistent or unclear drone policies further hinder adoption in developing regions like Africa and Asia. High initial costs and limited access to financing are also significant barriers to UAV adoption. Advanced agricultural drones equipped with multispectral cameras, GPS, and AI-driven analytics can cost thousands of dollars, making them inaccessible to many farmers. Additionally, ongoing costs such as software subscriptions, maintenance, and operator training further increase financial burdens. Although some governments and agritech startups offer subsidies and financing options, many farmers still struggle to afford UAV technology, slowing its widespread adoption. Technical limitations and infrastructure gaps are another challenge. UAVs depend on strong digital infrastructure, including GPS signals, reliable internet connectivity, and data processing capabilities [40]. In many rural areas, poor network coverage and weak technological infrastructure make drone operations difficult. Battery life and payload capacity limitations restrict the area a drone can cover in a single flight, adding to operational costs. Lastly, lack of awareness and farmer training is another challenge. Farmers must be trained in drone operation, data analysis, and interpretation of aerial imagery. However, in many regions, especially traditional farming communities, awareness of UAV technology remains low, and there is limited access to training programs that teach farmers how to integrate UAVs into their daily operations [41].

High Initial Costs: A Major Barrier to UAV Adoption in Agriculture:

The adoption of Unmanned Aerial Vehicles (UAVs) in agriculture has the potential to revolutionize farming practices, offering benefits such as improved efficiency, reduced resource use, and higher yields. However, one of the most significant challenges hindering widespread UAV adoption is the high initial cost associated with purchasing, maintaining, and operating drone technology. For many farmers, especially smallholders and those in developing countries, the financial burden of investing in UAVs remains a major obstacle [9]. Addressing this challenge requires strategic solutions, such as subsidies, financing options, and technological advancements that make drones more affordable and accessible.

The Cost of UAV Hardware and Equipment

One of the primary contributors to the high initial cost of UAV adoption is the price of the drone itself. Agricultural drones range from simple, low-cost models priced around \$1000 to advanced drones equipped with multispectral cameras, LiDAR sensors, and AI-driven analytics, which can cost anywhere between \$10,000 and \$50,000. These high-end drones provide essential data for precision agriculture, but their cost is prohibitive for many farmers. Additionally, UAVs require additional accessories such as high-capacity batteries, GPS modules, and storage devices, further increasing upfront expenses [42].

Software and Data Processing Costs

Beyond hardware, specialized software is required to analyze UAV-captured data and translate it into actionable farming decisions. Advanced drone software for mapping, plant health monitoring, and yield predictions often requires subscription-based models, with prices ranging from $500 to several thousand dollars annually. For farmers unfamiliar with digital tools, investing in these software solutions—and the necessary training to use them effectively—adds to the financial burden. Without affordable software solutions, many farmers may struggle to maximize the benefits of UAV technology [43].

Training and Operational Expenses

Even if farmers can afford UAVs, they must also invest in training and certification to operate them legally and effectively. In many countries, pilots of commercial drones need a license, which involves training courses that cost hundreds to thousands of dollars. Furthermore, maintaining drones—such as repairing damaged propellers, updating firmware, and ensuring compliance with local regulations—adds ongoing costs that many farmers are not prepared for. Hiring a professional drone service provider is an alternative, but this also comes at a significant cost, making it an impractical solution for long-term use [44].

Limited Financing and Lack of Subsidies

Unlike other agricultural investments such as tractors or irrigation systems, financing options for drones remain limited. Many financial institutions do not yet recognize UAVs as essential farming equipment, making it difficult for farmers to obtain loans for drone purchases. Government subsidies for agricultural technology adoption also tend to focus on traditional machinery rather than UAVs. In contrast, countries like China and India have begun providing incentives for farmers to adopt drone technology, setting an example for other nations to follow [12]. Expanding financial assistance programs could significantly lower the cost barrier and encourage more farmers to invest in UAVs.

The Challenge for Small Farmers and Developing Nations

While large commercial farms in the United States, Canada, and Europe can often justify UAV investments due to their high production volumes, smallholder farmers in developing countries face a much steeper challenge. In regions like Africa and South Asia, where agriculture is dominated by small farms with limited capital, spending thousands of dollars on UAV technology is not feasible. This creates a technology gap, where wealthier farms benefit from modern innovations while smaller farms struggle to compete [45]. Initiatives such as drone leasing programs and cooperatives could help bridge this gap, allowing multiple farmers to share the cost of UAV services.

Unmanned Aerial Vehicles (UAVs) have revolutionized agriculture by enabling precision farming, improving efficiency, and reducing costs. However, widespread adoption faces challenges such as regulatory restrictions, high costs, technical limitations, and farmer acceptance. Compliance can be difficult for small farmers, and inconsistent drone policies hinder adoption in developing regions. High initial costs and limited financing access are significant barriers to UAV adoption. Technical limitations and infrastructure gaps are another challenge, as UAVs require strong digital infrastructure, but poor network coverage and weak technological infrastructure in rural areas make operations difficult. Additionally, lack of awareness and farmer training is another challenge, as farmers need to be trained in drone operation, data analysis, and aerial imagery interpretation. Addressing this requires strategic solutions like subsidies, financing options, and technological advancements to make drones more affordable and accessible.

Training Needs for Farmers in UAV Adoption

The integration of Unmanned Aerial Vehicles (UAVs) in agriculture is revolutionizing farming practices by improving efficiency, optimizing resource use, and increasing yields. However, to maximize the benefits of UAV technology, farmers must receive proper training in drone operation, data analysis, and regulatory compliance [46]. Without adequate education, the potential of drones in precision agriculture remains underutilized. Training programs tailored to different levels of farming expertise are essential to ensure successful adoption.

Understanding UAV Operations and Flight Control

One of the fundamental training requirements for farmers is learning how **to** operate and control UAVs**.** Flying a drone requires knowledge of basic aerodynamics, maneuvering techniques, and troubleshooting common technical issues. Many agricultural drones come with automated flight features, but manual intervention is sometimes necessary to adjust flight paths, avoid obstacles, or respond to unexpected weather conditions [47]. Farmers must be trained on pre-flight checks, safety procedures, and emergency handling to prevent crashes and equipment damage.

Data Collection and Analysis Skills

A key advantage of UAVs in agriculture is their ability to collect valuable aerial data using multispectral cameras, thermal sensors, and LiDAR technology. However, many farmers lack the technical skills needed to interpret this data. Training programs must focus on how to process drone images, analyze crop health indicators, and use Geographic Information Systems (GIS) for mapping. Without proper knowledge, farmers may struggle to make informed decisions about irrigation, fertilization, and pest control based on UAV-generated insights [48].

Regulatory Compliance and Licensing Requirements

Many countries require UAV operators to obtain a license or permit before using drones for commercial agricultural purposes. Farmers must be trained on local aviation laws, no-fly zones, privacy regulations, and drone registration procedures**.** In regions where UAV regulations are still evolving, ongoing education is necessary to keep farmers updated on changing policies. Compliance training ensures that farmers can legally and safely integrate UAVs into their operations without risking fines or penalties [49].

Maintenance and Troubleshooting of UAVs

Drones **are** high-tech machines that require regular maintenance to function optimally. Farmers need to be trained in battery management, sensor calibration, firmware updates, and minor repairs. Understanding how to troubleshoot common problems—such as connectivity issues, GPS errors, or camera malfunctions—can prevent costly downtime. Without this technical knowledge, farmers may rely on external service providers, increasing operational costs and limiting the efficiency of UAV adoption [50].

Accessible Training Programs for Small Farmers

While large agribusinesses may have access to specialized drone training programs, smallholder farmers often lack formal education in UAV technology. Governments, universities, and agricultural organizations should develop affordable and accessible training programs tailored to small-scale farmers. Mobile-based learning platforms, community workshops, and hands-on field demonstrations can help bridge the knowledge gap. In developing countries, subsidized training initiatives can ensure that resource-limited farmers are not excluded from UAV advancements [51].

The integration of Unmanned Aerial Vehicles (UAVs) in agriculture is revolutionizing farming practices by improving efficiency, optimizing resource use, and increasing yields. However, to fully benefit from UAV technology, farmers need proper training in drone operation, data analysis, and regulatory compliance. Training programs should be tailored to different levels of farming expertise, focusing on understanding UAV operations, flight control, data collection, and analysis skills. Farmers should be trained on pre-flight checks, safety procedures, and emergency handling to prevent crashes and equipment damage. Regulatory compliance and licensing requirements are also crucial, as many countries require UAV operators to obtain a license or permit before using drones for commercial agricultural purposes. Farmers should be trained on local aviation laws, no-fly zones, privacy regulations, and drone registration procedures. Maintenance and troubleshooting of UAVs are also crucial, including battery management, sensor calibration, firmware updates, and minor repairs. Affordable and accessible training programs should be developed by governments, universities, and agricultural organizations for small-scale farmers.

Regulatory Hurdles and Airspace Laws in UAV Adoption

The widespread adoption of Unmanned Aerial Vehicles (UAVs) in agriculture has introduced significant benefits, including improved efficiency, precision farming, and cost reduction. However, regulatory hurdles and airspace laws remain a major challenge for farmers seeking to integrate drones into their operations. Different countries have varying policies regarding drone usage, licensing, and airspace restrictions, which can complicate UAV adoption. Striking a balance between safety, privacy, and agricultural innovation is essential for ensuring that UAVs can be effectively used in farming without violating legal requirements [52].

Strict Licensing and Certification Requirements:

One of the biggest regulatory challenges for farmers using UAVs is obtaining the necessary licenses and certifications. Many countries require drone operators to pass training courses and register their UAVs before using them for commercial purposes. In the United States, for instance, the Federal Aviation Administration (FAA) mandates that drone operators obtain a Part 107 certification for commercial drone use [53]. Similarly, the European Union has introduced standardized drone regulations that require remote pilots to complete training and pass an exam. These licensing requirements add costs and time constraints, making it difficult for small farmers to adopt UAV technology.

Airspace Restrictions and No-Fly Zones:

Many agricultural areas are located near restricted airspace, airports, or military zones, limiting where drones can legally operate. Governments have imposed no-fly zones to prevent drones from interfering with manned aircraft, but these restrictions can make it difficult for farmers to fully utilize UAVs. In some cases, farmers must apply for special permissions to operate drones in regulated airspace, a process that can be time-consuming and bureaucratic. Countries with complex airspace laws may require UAV operators to obtain real-time flight approvals, further complicating drone usage [54].

Privacy and Data Protection Concerns

Drones equipped with high-resolution cameras and sensors can capture detailed images of private properties, raising concerns about surveillance and data security. Many countries have strict privacy laws that regulate how aerial data can be collected and used. In the European Union, for example, the General Data Protection Regulation (GDPR) imposes strict rules on data collection and storage. Farmers using UAVs must ensure that they do not violate privacy laws by inadvertently capturing images of neighboring properties or individuals without consent. Failing to comply with privacy regulations can result in legal penalties and restrictions on UAV usage [55].

Import and Export Restrictions on UAV Technology:

In some regions, the availability of high-quality agricultural drones is limited due to import and export restrictions on UAV technology. Governments impose these restrictions due to national security concerns, limiting the supply of advanced drones for farming. In India, for example, the government has placed restrictions on importing foreign drones to encourage local manufacturing, but this has led to supply shortages and increased costs for farmers. Trade policies and geopolitical tensions can also impact the availability of UAV technology, making it difficult for farmers to access the latest innovations.

Need for Policy Reforms and International Standards:

To facilitate the widespread adoption of UAVs in agriculture, governments need to develop clearer, more flexible policies that support innovation while ensuring safety and compliance. The lack of standardized regulations across countries creates confusion for drone operators, especially those involved in cross-border agricultural trade. Developing internationally recognized drone regulations could help harmonize policies, making it easier for farmers to adopt UAVs without facing legal uncertainties [56]. Additionally, governments should consider providing subsidies and simplified licensing processes for agricultural drone users to encourage more widespread adoption.

The use of Unmanned Aerial Vehicles (UAVs) in agriculture faces regulatory challenges such as strict licensing requirements, airspace restrictions, privacy and data protection concerns, import and export restrictions, and trade policies. The FAA's Part 107 certification and the European Union's standardized regulations add costs and time constraints to farmers' adoption. Governments impose airspace restrictions and no-fly zones to prevent drone interference, while farmers must ensure they don't violate privacy laws. Import and export restrictions limit the availability of high-quality agricultural drones, such as India's restrictions on importing foreign drones. Trade policies and geopolitical tensions also impact the availability of UAV technology. To facilitate widespread adoption, governments need to develop clearer, more flexible policies that support innovation while ensuring safety and compliance. Internationally recognized drone regulations could help harmonize policies, making it easier for farmers to adopt UAVs without facing legal uncertainties. Additionally, governments should consider providing subsidies and simplified licensing processes for agricultural drone users. Unmanned Aerial Vehicles (UAVs) have become a transformative tool in modern agriculture, offering significant benefits such as increased yields, optimized resource use, and reduced labor dependency. However, challenges such as high initial costs, regulatory barriers, and the need for specialized training hinder widespread adoption. This chapter examines these factors while highlighting case studies of successful UAV applications in farming. By addressing these challenges and leveraging advancements in UAV technology, agriculture can move toward a more efficient and sustainable future. As military advancements continue to advance, civilian applications will likely follow, further transforming industries and improving everyday life. This interplay between military advancements and civilian adaptation underscores the importance of continued investment in research and technology. As technology advances, UAVs are expected to become more autonomous, with improved battery life, payload capacity, and data processing capabilities. Integration with other technologies such as artificial intelligence, machine learning, and the Internet of Things (IoT) will further enhance their capabilities, making UAVs an indispensable tool in modern agriculture. Technological advancements in UAV hardware continue to push the boundaries of performance, making drones more efficient, autonomous, and versatile.

UAV technology is revolutionizing traditional farming by evolving into a data-driven, efficient, and sustainable practice. Traditional agriculture faces major challenges such as inefficiency in resource use, climate change impacts, and pest outbreaks, which contribute to declining productivity, increased costs, and environmental degradation. Addressing these challenges requires modern solutions such as precision agriculture, climate-resilient farming techniques, and sustainable pest management strategies to ensure long-term food security and economic stability. UAV technology is also revolutionizing small-scale farming by providing cost-effective solutions for crop monitoring, precision agriculture, and financial inclusion. By enabling early disease detection, optimizing resource use, and improving access to financial and market opportunities, drones help small farmers increase productivity and profitability. The integration of UAV technology is playing a crucial role in ensuring financial sustainability, food security, and overall economic growth in the farming industry. As UAV adoption continues to grow, North America remains a global leader in this field due to advanced technology, supportive regulations, and economic incentives. Europe's sustainability-focused use of UAV technology is transforming agriculture by reducing chemical use, enhancing biodiversity, and improving water management. However, the widespread adoption of UAV technology faces challenges such as regulatory restrictions, high costs, technical limitations, and a lack of farmer awareness. Addressing these issues through policy improvements, financial incentives, infrastructure development, and education programs is essential for successful integration of UAVs into global farming practices.

References

1. Yadav N, Sidana N (2023) Precision agriculture technologies: analysing the use of advanced technologies, such as drones, sensors, and GPS, in precision agriculture for optimizing resource management, crop monitoring, and yield prediction. J Adv Zool 44
2. Vashishth TK, Sharma V, Sharma KK, et al (2024) Integration of unmanned aerial vehicles (UAVs) and IoT for crop monitoring and spraying. In: Internet of Things applications and technology. Auerbach Publications, pp 95–117
3. Dayoub M, Shnaigat S, Tarawneh RA et al (2024) Enhancing animal production through smart agriculture: possibilities, hurdles, resolutions, and advantages. Ruminants 4:22–46
4. Emimi M, Khaleel M, Alkrash A (2023) The current opportunities and challenges in drone technology. Int J Electr Eng and Sustain 74–89
5. Toscano F, Fiorentino C, Capece N, et al (2024) Unmanned aerial vehicle for precision agriculture: a review. IEEE access
6. Telli K, Kraa O, Himeur Y et al (2023) A comprehensive review of recent research trends on unmanned aerial vehicles (uavs). Systems 11:400
7. Njane SN, Tsuda S, van Marrewijk BM et al (2023) Effect of varying UAV height on the precise estimation of potato crop growth. Front Plant Sci 14:1233349
8. Naveed M (2024) The adoption of 4.0 agriculture for wine production in order to improve efficiency, sustainability and competitiveness
9. McCarthy C, Nyoni Y, Kachamba DJ et al (2023) Can drones help smallholder farmers improve agriculture efficiencies and reduce food insecurity in Sub-Saharan Africa? Local perceptions from Malawi. Agriculture 13:1075

10. Maddikunta PKR, Hakak S, Alazab M et al (2021) Unmanned aerial vehicles in smart agriculture: applications, requirements, and challenges. IEEE Sens J 21:17608–17619
11. Zhang J, Yu F, Zhang Q et al (2024) Advancements of UAV and deep learning technologies for weed management in Farmland. Agronomy 14:494
12. Joshi R, Pandey K (2024) IoT-Enabled UAV: a comprehensive review of technological change in Indian farming. Unmanned Aircraft Systems 93–135
13. Getahun S, Kefale H, Gelaye Y (2024) Application of precision agriculture technologies for sustainable crop production and environmental sustainability: a systematic review. Scientific World J 2024:2126734
14. Silva L, Conceição LA, Lidon FC, Maçãs B (2023) Remote monitoring of crop nitrogen nutrition to adjust crop models: a review. Agriculture 13:835
15. Bisht B (2023) Yield prediction using spatial and temporal deep learning algorithms and data fusion
16. Kar D, Dhal SB (2025) Advancing food security through drone-based hyperspectral imaging: applications in precision agriculture and post-harvest management. Environ Monit Assess 197:1–26
17. Ewane EB, Mohan M, Bajaj S et al (2023) Climate-change-driven droughts and tree mortality: assessing the potential of UAV-derived early warning metrics. Remote Sensing 15:2627
18. Xing Y, Wang X (2024) Precision agriculture and water conservation strategies for sustainable crop production in arid regions. Plants 13:3184
19. Sharma P, Sharma P, Thakur N (2024) Sustainable farming practices and soil health: a pathway to achieving SDGs and future prospects. Discover Sustainability 5:250
20. Ndlovu HS, Odindi J, Sibanda M, Mutanga O (2024) A systematic review on the application of UAV-based thermal remote sensing for assessing and monitoring crop water status in crop farming systems. Int J Remote Sens 45:4923–4960
21. Abdalla A, Mirzakhani Nafchi A (2024) Development and evaluation of an affordable variable rate applicator controller for precision agriculture. AgriEngineering 6:4639–4657
22. Duporge I (2021) Advancing remote sensing methods to monitor wildlife
23. Rai AK, Kumar N, Katiyar D et al (2023) Unlocking productivity potential: the promising role of agricultural robots in enhancing farming efficiency. Int J Plant Soil Sci 35:624–633
24. Taseer A, Han X (2024) Advancements in variable rate spraying for precise spray requirements in precision agriculture using unmanned aerial spraying systems: a review. Comput Electron Agric 219:108841
25. Bazargani K, Deemyad T (2024) Automation's impact on agriculture: opportunities, challenges, and economic effects. Robotics 13:33
26. Pagliai A (2023) Evaluation of proximal sensing techniques and testing of variable-rate kit for sprayers to improve environmental and economic sustainability of crop protection stages
27. Amarasingam N, Salgadoe ASA, Powell K et al (2022) A review of UAV platforms, sensors, and applications for monitoring of sugarcane crops. Remote Sensing Appl: Soc Environ 26:100712
28. Dhillon R, Moncur Q (2023) Small-scale farming: a review of challenges and potential opportunities offered by technological advancements. Sustainability 15:15478
29. Knierim A, Herrera B, Paulus M (2024) Analysis of digital ecosystems
30. Orzel EC (2024) Sowing sustainable agriculture: investigating "sustainability" in the academic discourse on precision agriculture
31. McFadden J, Njuki E, Griffin T (2023) Precision agriculture in the digital era: recent adoption on US farms
32. Thompson A (2024) A review and prospects on the use of unmanned aerial systems (UAS) in PRECISION AGRICULTURE FOR The Mid-Atlantic United States
33. Júnior MRB, de Almeida Moreira BR, dos Santos CV et al (2024) Precision agriculture in the United States: a comprehensive meta-review inspiring further research, innovation, and adoption. Comput Electron Agric 221:108993
34. Ochieng VA (2022) Assessment of the efficiency of drones in surveillance and control of desert locust, Schistocerca Gregaria

35. Tyshchuk VV (2024) A review of legal regulation regarding the use of unmanned aerial vehicles for border security and the impact of global technologies. International Comparative Jurisprudence 10:61–81
36. Akter J, Kamruzzaman M, Hasan R, et al (2024) Artificial intelligence in American agriculture: a comprehensive review of spatial analysis and precision farming for sustainability. IEEE, pp 1–7
37. Pirbasti MA, McArdle G, Akbari V (2024) Hedgerows monitoring in remote sensing: a comprehensive review. IEEE Access
38. Marques P, Pádua L, Sousa JJ, Fernandes-Silva A (2024) Advancements in remote sensing imagery applications for precision management in olive growing: a systematic review. Remote Sensing 16:1324
39. Lee D, Hess DJ, Heldeweg MA (2022) Safety and privacy regulations for unmanned aerial vehicles: a multiple comparative analysis. Technol Soc 71:102079
40. Khan A, Gupta S, Gupta SK (2022) Emerging UAV technology for disaster detection, mitigation, response, and preparedness. J Field Robot 39:905–955
41. Mishra S (2023) Internet of things enabled deep learning methods using unmanned aerial vehicles enabled integrated farm management. Heliyon 9
42. Harder A, Vilter R, Doctor AC (2024) Advancements in the connection between unmanned systems (UxS) and geospatial technologies
43. Srivastava A, Prakash J (2023) Techniques, answers, and real-world UAV implementations for precision farming. Wireless Pers Commun 131:2715–2746
44. Navarro Lafuente A (2024) Business modelling of 5G-based Drone-as-a-service solution
45. Salihu H, Jibril H, Roper S, et al (2025) Mapping the innovation and commercialisation infrastructure for agricultural technology in the UK
46. Westbrooke V, Lucock X, Greenhalgh I (2023) Drone use in on-farm environmental compliance: an investigation of regulators' perspectives. Sustainability 15:2153
47. Young S, Ancel E, Moore A, et al (2020) Architecture and information requirements to assess and predict flight safety risks during highly autonomous urban flight operations
48. Dorbu FE (2024) Data-Driven approach for processing remote sensing time-series imagery for precision agriculture
49. Matalonga S, White S, Hartmann J, Riordan J (2022) A review of the legal, regulatory and practical aspects needed to unlock autonomous beyond visual line of sight unmanned aircraft systems operations. J Intell Rob Syst 106:10
50. Parmaksiz O, Cinar G (2023) Technology acceptance among farmers: examples of agricultural unmanned aerial vehicles. Agronomy 13:2077
51. Heiss N, Meier J, Gessner U, Kuenzer C (2025) A review: potential of earth observation (EO) for mapping small-scale agriculture and cropping systems in West Africa. Land 14:1–47
52. Al-Dosari K, Fetais N (2023) A new shift in implementing unmanned aerial vehicles (UAVs) in the safety and security of smart cities: a systematic literature review. Safety 9:64
53. Marshall DM (2021) UAS regulations, standards, and guidance. In: Introduction to unmanned aircraft systems. CRC Press, pp 101–136
54. McTegg SJ, Tarsha Kurdi F, Simmons S, Gharineiat Z (2022) Comparative approach of unmanned aerial vehicle restrictions in controlled airspaces. Remote Sensing 14:822
55. Stöcker C, Bennett R, Nex F et al (2017) Review of the current state of UAV regulations. Remote sensing 9:459
56. Askerbekov D, Garza-Reyes JA, Ghatak RR et al (2024) Embracing drones and the Internet of drones systems in manufacturing—an exploration of obstacles. Technol Soc 78:102648

Chapter 4
Drone Airflow Dynamics and Crop Interaction

Abstract Drones, particularly rotor-based Unmanned Aerial Vehicles (UAVs), have become integral to precision agriculture, enabling efficient crop monitoring, spraying, and data collection. However, the aerodynamic interactions between drone-generated airflow and plant canopies remain a critical yet underexplored aspect influencing agricultural outcomes. This chapter investigates the dynamics of drone airflow—especially rotor-induced downwash—and its interaction with various crop structures. It focuses on how airflow intensity, angle, and turbulence affect canopy penetration, pesticide and nutrient deposition, and microclimate modification. The role of flight parameters such as altitude, speed, and tilt angle in shaping airflow behavior and their impact on crop morphology, stress responses, and yield formation is also examined. Additionally, environmental variables like wind speed, temperature, and crop height are analyzed to understand their modulation of drone-crop aerodynamic interactions. To support this analysis, the chapter integrates foundational principles of UAV aerodynamics—lift, drag, thrust, and weight—and examines how different UAV configurations (rotary-wing, fixed-wing, and hybrid designs) influence airflow characteristics. By linking aerodynamic behavior to real-world agricultural performance, this study offers practical guidance for optimizing drone-based operations, enhancing input delivery, minimizing crop stress, and promoting sustainable farming practices.

Keywords UAV aerodynamics · Fixed-Wing UAV · Rotary-Wing UAV · Airfoil design · Flight efficiency · Energy consumption

4.1 Introducation

Unmanned Aerial Vehicles (UAVs), commonly known as drones, rely on fundamental aerodynamic principles to achieve stable flight, maneuver efficiently, and optimize performance for various applications. The aerodynamics of UAVs is governed by four primary forces: lift, weight, thrust, and drag. A thorough understanding of these forces is essential for designing UAVs that can perform tasks ranging from surveillance

© The Author(s), under exclusive license to Springer Nature Singapore Pte Ltd. 2025

J. Li, *UAV Aerodynamics and Crop Interaction*, Smart Agriculture 13,
https://doi.org/10.1007/978-981-96-8402-1_4

and agriculture to cargo transport and military operations. Lift is the upward force that counteracts the weight of the UAV and is primarily generated by air moving over wings in fixed-wing UAVs or by rotating blades in rotary-wing UAVs such as quadcopters [1]. The amount of lift depends on factors such as airfoil shape, angle of attack, airspeed, and air density. In rotary-wing UAVs, lift is generated dynamically as rotors spin, whereas in fixed-wing UAVs, forward motion through the air is necessary to maintain lift. Weight is the downward force due to gravity and includes the UAV's structural weight, onboard equipment, and payload. A UAV must generate sufficient lift to overcome its weight to achieve and sustain flight. Heavier UAVs require more powerful propulsion systems and efficient aerodynamic designs to ensure optimal performance.

Thrust is the forward force that propels the UAV through the air. In fixed-wing UAVs, thrust is produced by propellers or jet engines, while in rotary-wing UAVs, thrust is generated by rotating blades that provide both lift and directional control. The efficiency of the propulsion system impacts the UAV's speed, endurance, and ability to carry payloads. Drag is the aerodynamic resistance experienced by the UAV as it moves through the air. It is classified into two main types: parasitic drag and induced drag [2]. Parasitic drag results from air resistance against the UAV's body, including its fuselage, wings, and other components. Induced drag is generated as a byproduct of lift, particularly at high angles of attack. Minimizing drag is essential for improving UAV efficiency, reducing energy consumption, and increasing flight endurance. Stability is crucial for ensuring that a UAV maintains a steady flight path without excessive oscillations. Fixed-wing UAVs achieve stability through control surfaces such as ailerons, elevators, and rudders, which adjust pitch, roll, and yaw. Rotary-wing UAVs, such as quad copters, rely on differential rotor speeds to maintain stability and change direction. Advanced UAVs incorporate gyroscopes, accelerometer, and flight control software to enhance stability and enable autonomous flight capabilities. Aerodynamic efficiency plays a critical role in UAV performance [3]. Factors such as wing shape, aspect ratio, airfoil design, and weight distribution influence how efficiently a UAV can generate lift and minimize drag. Streamlined body shapes and lightweight materials help reduce drag and improve flight endurance. Additionally, optimizing propeller or rotor design enhances thrust efficiency and reduces energy consumption By applying aerodynamic principles, UAV designers can create more efficient, stable, and versatile aircraft suitable for a wide range of applications, from environmental monitoring to logistics and defense operations [4]. Understanding UAV aerodynamics is key to developing innovative solutions that maximize flight performance and operational efficiency.

The Four Forces of Flight:

Flight is governed by four fundamental aerodynamic forces: lift, weight, thrust, and drag. These forces interact to determine an aircraft's ability to take off, maneuver, and sustain flight. Whether discussing traditional manned aircraft, Unmanned Aerial Vehicles (UAVs), or even birds, the same principles apply [5]. Each force plays a distinct role in flight dynamics, and understanding their interplay is crucial for aeronautical engineers, pilots, and drone designers.

Lift: The Upward Force

Lift is the force that enables an aircraft to rise and stay in the air. It counteracts weight and is primarily generated by the wings or rotors. Lift is produced when air moves over an airfoil—a specially shaped wing or blade—creating a pressure difference between the upper and lower surfaces. This phenomenon is explained by Bernoulli's Principle, which states that as the velocity of a fluid (air in this case) increases, its pressure decreases. The curved shape of a wing allows air to travel faster over the top surface than underneath, creating a lower pressure area above the wing and a higher pressure area below. This pressure differential generates the upward force known as lift. The amount of lift produced depends on several factors, including wing shape, angle of attack, airspeed, and air density [6]. The angle of attack refers to the angle between the wing and the oncoming airflow. Increasing this angle can enhance lift to a certain extent, but beyond a critical point, excessive angle causes turbulent airflow separation, leading to a stall where lift dramatically decreases. Aircraft designers carefully balance wing design, aspect ratio (the ratio of wingspan to chord length), and airfoil characteristics to maximize lift while minimizing drag. Fixed-wing aircraft require continuous forward motion to generate lift, as their wings rely on airflow to function effectively. In contrast, rotary-wing aircraft such as helicopters and drones generate lift dynamically through rotating blades [7]. These blades act as miniature wings, adjusting their pitch cyclically to control altitude and direction. Efficient lift generation is crucial for maximizing fuel efficiency, increasing payload capacity, and ensuring stable flight in various atmospheric conditions.

Weight: The Downward Force

Weight, or gravity, acts as the opposing force to lift, pulling the aircraft toward the Earth. It includes not only the aircraft's own structural weight but also the weight of fuel, passengers, cargo, and onboard equipment. To achieve flight, an aircraft must generate lift equal to or greater than its weight. Heavier aircraft require larger wings or more powerful engines to generate sufficient lift. Weight distribution is another critical factor in aerodynamics. Engineers carefully balance the center of gravity to maintain stability and control. If an aircraft's center of gravity is too far forward or backward, it can become difficult to maneuver, increasing the risk of stalls or unintentional nose-dives. In UAVs, weight distribution affects battery efficiency, payload handling, and flight endurance [8]. For instance, drones designed for aerial photography need to balance the camera weight with the drone's center of gravity to avoid tilting or instability. Fuel consumption is also linked to weight. As an aircraft burns fuel, its weight decreases, affecting lift requirements and flight characteristics. Pilots and autopilot systems adjust control surfaces and power settings to maintain equilibrium throughout a flight [9]. For commercial aviation, reducing weight through lightweight composite materials and fuel-efficient designs has become a key focus to enhance performance and reduce emissions.

Thrust: The Forward Force

Thrust is the force that propels an aircraft forward. It is generated by engines, propellers, or jet turbines, depending on the type of aircraft. In fixed-wing planes, thrust overcomes drag to maintain motion, while in rotary-wing aircraft like helicopters and drones, thrust contributes to both lift and movement. The amount of thrust required depends on the aircraft's weight, aerodynamic efficiency, and mission requirements. Propeller-driven aircraft generate thrust by pulling air backward, creating forward motion through Newton's Third Law of Motion, which states that every action has an equal and opposite reaction [10]. In jet engines, high-speed exhaust gases push the aircraft forward, allowing for greater speeds and efficiency at higher altitudes. UAVs use different propulsion systems based on their design and intended use. Small consumer drones typically use electric motors with propellers, offering quiet and efficient operation, while larger UAVs designed for military or long-range applications may use internal combustion or hybrid engines for extended endurance [11]. The efficiency of a propulsion system directly impacts range, payload capacity, and energy consumption. Optimizing thrust-to-weight ratios ensures that an aircraft can accelerate, climb, and maneuver effectively.

Drag: The Resisting Force

Drag is the aerodynamic force that opposes thrust and slows down an aircraft. It occurs due to air resistance and is classified into two main types: parasitic drag and induced drag. Parasitic drag includes three subtypes: form drag, skin friction, and interference drag. Form drag results from the aircraft's shape and size; larger, bulkier designs create more resistance. Skin friction drag arises from air molecules interacting with the aircraft's surface, which is why smoother, streamlined designs reduce drag. Interference drag occurs when airflow from different components, such as wings and fuselage, collide and create turbulence. Induced drag is a byproduct of lift and occurs due to vortices forming at the wingtips. These vortices create swirling air currents that reduce lift efficiency. To minimize induced drag, aircraft designers use winglets—upward extensions at the wingtips that help redirect airflow and reduce vortex formation. Reducing drag is essential for improving flight efficiency, conserving fuel, and increasing speed [12]. Aircraft achieve this through streamlined designs, retractable landing gear, and advanced materials that reduce surface roughness. In UAVs, efficient aerodynamic shaping allows drones to extend battery life, improve stability, and achieve smoother flight.

The Balance of Forces in Flight

For an aircraft to maintain steady, level flight, lift must equal weight, and thrust must equal drag. If lift exceeds weight, the aircraft climbs; if weight overcomes lift, it descends. Similarly, if thrust is greater than drag, the aircraft accelerates; if drag is greater, it slows down. Different flight phases require adjustments in these forces. During takeoff, maximum thrust is applied to overcome drag and generate sufficient lift to become airborne [13]. In cruising flight, pilots or autopilot systems adjust power settings and control surfaces to maintain equilibrium. During landing, pilots reduce thrust, allowing drag and weight to lower the aircraft safely.

Applications of the Four Forces in UAVs

UAVs, ranging from small quadcopters to large autonomous aircraft, must carefully balance the four forces of flight. In drone technology, optimizing lift is crucial for increasing payload capacity while maintaining flight efficiency. Engineers experiment with wing designs, rotor configurations, and lightweight materials to enhance performance. Quadcopters adjust thrust and lift dynamically by varying rotor speeds, allowing for precise hover control, agile maneuvering, and stable flight even in turbulent conditions. Fixed-wing UAVs require continuous forward motion to sustain lift, making them ideal for long-range missions such as agricultural mapping, environmental monitoring, and military reconnaissance [14]. Advancements in aerodynamics, such as the use of computational fluid dynamics (CFD) simulations and wind tunnel testing, help refine UAV designs for improved performance. Future developments focus on reducing drag, enhancing thrust efficiency, and incorporating hybrid propulsion systems to extend flight times and expand operational capabilities.

The four forces of flight—lift, weight, thrust, and drag—form the foundation of all aerial motion. Understanding their interactions allows engineers to design efficient aircraft and UAVs that meet specific mission requirements. Whether developing commercial airliners, military jets, or autonomous drones, balancing these forces ensures safe, stable, and efficient flight [15]. As aerospace technology advances, continuous improvements in aerodynamics and propulsion systems will push the boundaries of flight performance, leading to more capable and versatile aerial vehicles in the future.

4.2 Types of UAV Configurations

Unmanned Aerial Vehicles (UAVs) come in various configurations to suit different mission requirements, operational constraints, and environmental conditions. These configurations influence aerodynamic performance, stability, endurance, and maneuverability. The primary UAV configurations include fixed-wing, rotary-wing, hybrid, and unconventional designs. Each configuration follows fundamental aerodynamic principles, governed by forces like lift, thrust, drag, and weight [16]. Understanding these configurations is crucial for optimizing UAV performance across various applications, including military reconnaissance, agriculture, environmental monitoring, and commercial delivery services. Each configuration has its unique advantages and trade-offs.

4.2.1 Fixed-Wing UAVs

Fixed-wing UAVs resemble traditional aircraft with a rigid wing structure that provides lift through aerodynamic forces. These UAVs require forward motion to generate lift and sustain flight. Fixed-wing drones are known for their efficiency, long-range capabilities, and extended endurance, making them ideal for missions requiring extensive coverage, such as surveillance, mapping, and aerial surveying. They typically feature a fuselage, wings, control surfaces (such as rudders, elevators, and ailerons), and a propulsion system, which may consist of propellers, internal combustion engines, or jet turbines. Fixed-wing UAVs can be launched via runways, catapults, or hand launches, while landing methods may include runways, parachutes, or belly landings [17]. Their primary limitation is that they require a takeoff and landing space, unlike vertical takeoff and landing (VTOL) systems, and they cannot hover in place.

4.2.1.1 Aerodynamic Principles and Equations

Lift (L) is generated when air moves over the wing, creating a pressure difference. The lift equation is:

$$L = \frac{1}{2}\rho \upsilon^2 S C_L$$

where:

L = Lift (N)
ρ = Air density (kg/m^3)
υ = Airspeed (m/s)
S = Wing area (m^2)
C_L = Coefficient of lift (dimensionless)

For stable flight, lift must balance weight ($W = mgW = mg$, where mm is the mass and gg is gravitational acceleration).

4.2.1.2 Example: The RQ-4 Global Hawk

The RQ-4 Global Hawk is a high-altitude, long-endurance (HALE) fixed-wing UAV used for reconnaissance. With a wingspan of 39.9 m, it can stay airborne for over 30 h at altitudes exceeding 18,000 m [18]. The large wing area and optimized airfoil design enhance lift and fuel efficiency.

4.2.1.3 Experimental Validation

Wind tunnel tests are commonly conducted to analyze fixed-wing UAV aerodynamics. In one study, researchers tested different airfoil shapes to measure lift and drag coefficients. By varying airspeed and angle of attack, they optimized wing geometry for maximum efficiency. Computational Fluid Dynamics (CFD) simulations further validated the results, helping improve UAV performance [19].

4.2.2 Rotary-Wing UAVs

Rotary-wing UAVs, also known as rotorcraft, generate lift and thrust through rotating blades. This category includes single-rotor helicopters, multi-rotor drones, and coaxial rotor configurations. The most common type is the multi-rotor drone, which includes quadcopters (four rotors), hexacopters (six rotors), and octocopters (eight rotors). These UAVs offer vertical takeoff and landing (VTOL) capabilities, making them highly versatile for operations in confined spaces, urban environments, and locations without runways [20]. Rotary-wing UAVs excel in hovering, precise maneuvering, and low-speed stability, making them ideal for applications such as aerial photography, inspection, search and rescue, and delivery services. However, their endurance and range are limited due to high energy consumption and battery constraints, making them less efficient than fixed-wing UAVs for long-duration missions. Single-rotor helicopters, while more complex, provide better endurance and lifting capacity compared to multi-rotor systems, often used in heavy-lift transport and military reconnaissance [21].

4.2.2.1 Aerodynamic Principles and Equations

Energy consumption is a major limitation of multi-rotors, as power required for hover increases with mass.

For a hovering rotorcraft, the thrust (T) must equal the weight (W):

$$T = W = mg$$

where:

T = Thrust force (N).
W = Weight (N).
m = Mass of the rotorcraft (kg).
g = Gravitational acceleration ($9.81\ m/s^2$).

The thrust generated by a rotor is given by:

$$T = C_T \rho A R^2 \omega^2$$

Where:

C_T = Thrust coefficient (dimensionless).
ρ = Air density (kg/m^3).
A = Rotor disk area (m^2).
R = Rotor radius (m).
ω = Angular velocity (rad/s).

Note: In some cases, $A = \pi R^2$, so you can substitute if needed.

Hover Power Equation

In a hovering multi-rotor aircraft, the required power increases as the thrust increases. Since the thrust is proportional to the mass of the aircraft ($T = mg$), power consumption will increase with the aircraft's mass as well. The hover power equation is typically derived from the relationship between thrust and power, with an approximation for the power required to maintain a hover:

$$P = \frac{T^{3/2}}{\sqrt{2\rho A}}$$

where:

P = Power required for hover (W).
T = Thrust (N).
ρ = Air density (kg/m^3).
A = Rotor disk area (m^2).

Since $\boldsymbol{T}$ is proportional to mg, as the mass of the rotorcraft increases, the thrust requirement $\boldsymbol{T}$ increases. This causes the power requirement $\boldsymbol{P}$ to increase significantly due to the $\boldsymbol{T}^{3/2}$ term.

Example: DJI Phantom 4

The DJI Phantom 4 is a quadcopter UAV widely used for photography, surveying, and mapping. With four rotors, it achieves stable flight through differential thrust control. Each rotor adjusts speed independently to maintain balance, control pitch, roll, and yaw.

Experimental Validation

Researchers have tested quadcopters in controlled environments to analyze stability. In one experiment, a quadcopter was flown with varying payloads, and power consumption was measured. Results showed that adding weight increased rotor speed requirements, confirming theoretical predictions [22]. Motion capture systems were also used to validate real-time stabilization algorithms.

4.2.3 Hybrid UAVs

Hybrid UAVs combine features of both fixed-wing and rotary-wing configurations to leverage the advantages of both designs. These UAVs typically feature VTOL capabilities for easy takeoff and landing while also having the ability to transition into forward flight for improved efficiency and endurance. Hybrid UAVs use tilt-rotor, tilt-wing, or tail-sitter mechanisms to switch between hover and forward flight modes. One common example is the tilt-rotor UAV, which has rotors that tilt forward to act as propellers during horizontal flight and rotate upward for vertical lift during takeoff and landing. Hybrid UAVs are particularly useful for missions requiring extended range, speed, and hover capabilities, such as logistics, surveillance, and medical supply transport [23]. Their complexity and cost, however, are higher than conventional UAV designs due to the need for sophisticated transition mechanisms and control systems.

4.2.3.1 Aerodynamic Principles and Equations

During vertical takeoff, a hybrid UAV operates like a rotary-wing drone, where thrust equals weight. When transitioning to horizontal flight, lift is generated aerodynamically, reducing the need for rotor thrust. The transition phase is modeled using equations that balance lift and thrust.

A simplified transition equation is:

$$\mathrm{L} + \mathrm{T}_{\mathrm{forward}} = \mathrm{W}$$

where:

L = **Aerodynamic lift** (N)
$\mathrm{T}_{\mathrm{forward}}$ = **Forward thrust** from rotors or propellers (N)
W = **Weight** of the UAV (N)

Explanation:

In vertical takeoff, the UAV is primarily relying on the rotor's thrust (T) to balance its weight (W), with no significant aerodynamic lift. During the transition to horizontal flight, the UAV starts generating lift aerodynamically (L) with its wings, and rotor thrust is reduced because the aerodynamic lift (L) helps balance some or all of the weight. The remaining forward thrust ($T_{forward}$) from the rotors or propellers helps maintain forward motion.

The simplified equation $L + T_{forward} = W$ describes the balance between the aerodynamic lift, forward thrust, and the weight of the UAV during this transition phase.

Notes:

The transition phase is critical because the rotorcraft moves from a vertical takeoff mode (where the rotors provide all the thrust) to a more efficient horizontal flight mode, where lift is mostly provided by wings, and rotors are mainly for propulsion.

4.2.3.2 Example: V-22 Osprey (Tilt-Rotor UAV Concept)

The V-22 Osprey is a manned tilt-rotor aircraft, but similar UAV concepts exist. Tilt-rotor UAVs transition from rotary to fixed-wing flight by rotating their engines. This design enables extended range and endurance compared to pure rotary-wing UAVs.

4.2.3.3 Experimental Validation

A study tested hybrid UAVs in a wind tunnel, measuring aerodynamic performance in vertical, transition, and cruise flight modes. Data showed efficiency improvements in forward flight compared to conventional multi-rotors. Engineers used this data to refine tilt-rotor control algorithms [24].

4.2.4 Unconventional UAV Configurations

In addition to standard UAV types, several unconventional configurations exist, developed to address specific challenges or improve efficiency in niche applications. Some notable designs include flapping-wing UAVs (ornithopters), ducted fan UAVs, and solar-powered UAVs [25]. Ornithopters mimic the flight mechanics of birds and insects by using wing-flapping motions to generate lift and thrust, making them highly maneuverable and capable of stealth operations. These UAVs are often used in research and surveillance applications where silent flight is beneficial. Ducted fan UAVs use an enclosed rotor system, improving safety and reducing noise, often applied in urban drone delivery and compact surveillance systems. Solar-powered

UAVs integrate photovoltaic cells into their wings, enabling ultra-long-endurance flight by harnessing solar energy [26]. These UAVs are used for high-altitude, long-endurance (HALE) applications such as atmospheric research, disaster monitoring, and communication relay systems, operating for weeks or even months without refueling.

4.2.4.1 Flapping-Wing UAVs (Ornithopters)

Ornithopters mimic birds and insects by flapping their wings to generate both lift and thrust. The lift equation for an ornithopter, accounting for the oscillatory motion of the wings, is:

$$L = \frac{1}{2}\rho V^2 S C_{L_{\text{oscillatory}}}$$

where:

- L = Lift generated by the wings (N).
- ρ = Air density (kg/m^3).
- V = Velocity of the airflow relative to the ornithopter (m/s).
- S = Wing area (m^2).
- $CL_{\text{oscillatory}}$ = Oscillatory lift coefficient (dimensionless), which accounts for the unsteady aerodynamic effects due to wing flapping.

Explanation:

The equation is similar to the traditional lift equation used for fixed-wing aircraft but with an added component: the oscillatory lift coefficient $C_{L\text{oscillatory}}$, which models the unsteady aerodynamics caused by the flapping motion. The lift generated by the ornithopters wings depends on the velocity of the air, the wing area, and the oscillatory effects that arise from the wings' motion. This is crucial for modeling the dynamic behavior of flapping-wing UAVs (ornithopters).

Notes:

The unsteady aerodynamic effects can be quite complex and are often modeled using more advanced techniques (like Theodorsen's theory, or dynamic stall models), but this simplified equation captures the basic idea.

Example: Festo SmartBird

The Festo SmartBird replicates seagull flight, using articulated wings to generate efficient lift and propulsion.

Experimental Validation

High-speed cameras and airflow sensors were used to analyze wing motion. Results showed that synchronized wing flapping improved lift-to-drag ratios compared to fixed-wing equivalents.

4.2.4.2 Solar-Powered UAVs

Solar-powered UAVs integrate photovoltaic cells into their wings, enabling high-altitude, long-endurance (HALE) missions.

General Equation for Power from Photovoltaic Cells:

The power generated by a solar-powered UAV can be expressed as:

$$P_{solar} = A \cdot G \cdot \eta$$

where:

P_{solar} = Power generated by the solar cells (W)
A = Area of the photovoltaic cells on the UAV's wings (m^2)
G = Solar irradiance (W/m^2), typically varying with altitude and time of day
H = Efficiency of the photovoltaic cells (dimensionless, typically between 0.15 and 0.25)

Energy Balance for Solar-Powered UAV:

To maintain long-endurance flight, the **energy balance** must be considered:

$$P_{solar} + P_{battery} = P_{total} = P_{required}$$

where:

$P_{battery}$ = Power supplied by the battery (W).
P_{total} = Total power required by the UAV (W) (includes power for propulsion, avionics, etc.)
$P_{required}$ = Total power required for maintaining flight, including both solar and battery contributions.

During **daylight hours**, the solar cells contribute to the total power required, reducing the load on the battery. At night or in conditions with low solar irradiance, the battery would provide the necessary power.

Example of UAV Power Needs:

Propulsion power is often modeled as:

$$P_{prop} = T \cdot v$$

where:

P_{prop} = Propulsion power (W)
T = Thrust (N)
v = Velocity (m/s)

So, to maintain flight, the total power required to operate the UAV should be split between the solar power and battery power.

Example: Zephyr S

The Zephyr S, a solar UAV, operates for months in the stratosphere, serving as a pseudo-satellite.

Experimental Validation

Laboratory tests measured solar panel efficiency at different altitudes, confirming power sustainability for continuous flight.

UAV configurations are designed to optimize aerodynamics, endurance, and maneuverability based on mission needs. Fixed-wing UAVs excel in range and efficiency, rotary-wing UAVs provide superior control and VTOL capabilities, hybrid UAVs offer versatility, and unconventional UAVs push technological boundaries. Theoretical equations, real-world examples, and experimental research validate UAV performance across diverse applications [27]. As UAV technology advances, engineers continue refining configurations for improved efficiency, autonomy, and adaptability in real-world operations.

4.3 Stability and Control in Flight

Stability and control in flight are essential for ensuring safe and efficient aircraft operation. Stability refers to an aircraft's ability to return to its original flight path after being disturbed by external forces, while control involves the pilot's or automated system's ability to maneuver the aircraft as needed. Achieving stability and control requires a combination of physical design, mechanical systems, and advanced software-based technologies [28]. The three key factors that contribute to stability and control in flight include gyroscopic stabilization mechanisms, wind and weather adaptability, and software-driven stability enhancements.

4.3.1 Gyroscopic Stabilization Mechanisms

Gyroscopic stabilization is based on the principle of angular momentum, where a spinning object resists changes to its orientation. In aviation, this principle is applied through gyroscopes, which help maintain the aircraft's stability and provide critical navigational information [29].

Gyroscopes in Flight Instruments

Gyroscopes are fundamental components in various flight instruments, helping pilots maintain aircraft stability and orientation, especially in conditions where visibility is limited. These instruments rely on the principle of gyroscopic inertia, meaning that a spinning object resists changes to its orientation [30]. The three main flight instruments that incorporate gyroscopes are the attitude indicator, heading indicator, and turn coordinator.

Attitude Indicator (Artificial Horizon):

The attitude indicator provides the pilot with a real-time visual representation of the aircraft's orientation relative to the horizon. It displays pitch (nose-up or nose-down attitude) and bank (left or right tilt of the wings). The internal gyroscope within this instrument remains stable, while the aircraft moves around it, allowing the display to accurately reflect the aircraft's attitude [31]. This is crucial when flying in low visibility conditions, such as fog or nighttime operations, where external references like the horizon are not visible.

Heading Indicator (Directional Gyro):

The heading indicator helps pilots maintain a precise heading by displaying the aircraft's direction relative to magnetic north. Unlike a traditional magnetic compass, which can be affected by acceleration and turning errors, the gyroscopic heading indicator provides a stable and accurate directional reference [32]. It needs periodic realignment with the magnetic compass due to minor drift over time, but it remains an essential tool for navigation.

Turn Coordinator:

The turn coordinator is designed to show the rate of turn and whether the aircraft is executing a coordinated turn. It helps pilots avoid slips and skids, which occur when the aircraft is not correctly aligned with the direction of movement. The internal gyroscope measures the rate of rotation about the vertical axis, providing feedback on how much rudder and aileron input is needed for smooth turns. This instrument is particularly useful during instrument flight conditions, where external visual cues are limited. Together, these gyroscopic flight instruments ensure that pilots can maintain precise control over the aircraft, even when external visibility is poor [33]. They are especially critical in instrument flight rules (IFR) conditions, where pilots rely on cockpit instruments rather than visual references to navigate and maintain stability.

Gyroscopic Precession and Stability

Gyroscopic precession is a key principle that affects the stability and control of an aircraft. It occurs when an external force is applied to a spinning gyroscope, causing the reaction to manifest 90 degrees in the direction of rotation rather than at the point of force application [34]. This characteristic is crucial in aviation, as it influences the behavior of rotating components such as propellers, helicopter rotors, and jet engine turbines.

Effect on Propellers and Helicopter Rotors:

In aircraft with propellers, gyroscopic precession plays a role in handling characteristics. When a pilot applies pitch control (raising or lowering the nose), the resulting precessional effect causes a slight yawing motion. This is particularly noticeable in tail-dragger aircraft during takeoff, where the rapid pitch change causes the aircraft to yaw left or right, requiring counteracting rudder input. In helicopters, gyroscopic precession affects the rotor disc's response to control inputs. When the cyclic control is used to tilt the rotor forward, the actual response occurs 90 degrees later in the rotation cycle [35]. This is why helicopter controls must compensate for precession when maneuvering.

Jet Engine and Turbine Stability:

Jet engines also experience gyroscopic effects due to the high-speed rotation of turbine blades. These effects must be carefully managed during high-speed maneuvers, as rapid attitude changes can introduce unexpected forces on the engine mounts and airframe. Aircraft designers account for these forces when designing engine placement and mounting systems to ensure smooth and stable flight [36].

Contribution to Overall Stability:

Gyroscopic effects help maintain stability by resisting sudden changes in aircraft orientation. This contributes to smoother flight characteristics and helps prevent excessive rolling or pitching due to external disturbances such as turbulence. By understanding and managing gyroscopic precession, pilots and engineers can optimize flight control strategies for enhanced stability and maneuverability [37].

Application in Unmanned Aerial Vehicles (UAVs) and Drones

Modern drones and UAVs rely heavily on electronic gyroscopes, particularly Micro-Electro-Mechanical Systems (MEMS) gyroscopes, which are compact, lightweight, and highly sensitive to changes in orientation. These gyroscopes play a vital role in autonomous flight stabilization, allowing drones to hover, maintain position, and respond to external disturbances without constant human input [38].

Orientation and Balance Control:

Gyroscopes in drones continuously monitor the aircraft's roll, pitch, and yaw and send data to the flight control system. If a drone encounters wind gusts or sudden movements, the onboard gyroscopes detect the changes and signal the flight controller to adjust motor speeds accordingly [39]. This ensures that the drone remains level and stable during flight.

Hovering and Stability in Windy Conditions:

In quadcopters and multirotor drones, stability is a key challenge due to their lightweight frames. MEMS gyroscopes work alongside accelerometers to detect unwanted tilting or drifting. The flight controller then compensates by adjusting the thrust output of individual rotors, allowing the drone to hover steadily even in windy conditions [1].

Autonomous Navigation and Obstacle Avoidance: Many modern UAVs integrate gyroscopes with GPS, cameras, and AI-driven software to enable advanced features such as:

Autonomous waypoint navigation: The drone follows a pre-programmed flight path while maintaining stability.

- **Obstacle avoidance systems:** Gyroscopic data helps adjust flight paths when obstacles are detected.
- **Precision landings:** Drones use gyroscopes to stabilize themselves during descent and landing, ensuring soft and controlled landings.

Military and Commercial Applications:

Military UAVs and high-end commercial drones use advanced triple-axis gyroscopes and inertial measurement units (IMUs) to achieve high-precision stability. This is essential for applications such as surveillance, mapping, search and rescue, and aerial photography, where stable flight and camera positioning are critical. By incorporating gyroscopic stabilization, UAVs and drones can operate with remarkable accuracy, efficiency, and resilience against environmental disturbances [40]. These advancements continue to push the boundaries of autonomous aerial technology, making drones more capable and reliable for various industries.

Gyroscopes play an indispensable role in modern aviation, from essential flight instruments in manned aircraft to advanced stabilization systems in UAVs and drones. The principles of gyroscopic precession and angular stability contribute to smooth and controlled flight, helping pilots and automated systems navigate safely. Whether in commercial airliners, military jets, helicopters, or drones, gyroscopes remain a foundational technology that enhances flight stability, control, and performance [41].

4.3.2 Wind and Weather Adaptability

Aircraft must be designed to withstand a wide range of weather conditions, including strong winds, turbulence, wind shear, and rapid atmospheric changes. Since weather conditions can be unpredictable, aircraft stability and control systems play a crucial role in ensuring smooth flight performance. Several aerodynamic features and onboard control systems allow modern aircraft to adapt to changing weather conditions, improving safety and passenger comfort [42]. These include specialized wing designs, adjustable control surfaces, turbulence adaptation systems, and advanced autopilot features.

Dihedral and Anhedral Wings for Stability

The shape and orientation of an aircraft's wings significantly influence its stability and maneuverability in different weather conditions. Two common wing configurations, dihedral and anhedral, serve different purposes based on the type of aircraft and its operational needs.

Dihedral Wings (Upward-Angled Wings):

Dihedral wings are designed with an upward tilt, meaning the wingtips are positioned higher than the wing roots where they connect to the fuselage. This design enhances roll stability, making the aircraft naturally return to level flight after experiencing turbulence or wind-induced tilting. When an aircraft with dihedral wings encounters crosswinds or slight roll disturbances, aerodynamic forces automatically generate a restoring force that stabilizes the aircraft without excessive pilot input [43]. This design is commonly found in commercial airliners and general aviation aircraft, where stability and smooth flight are prioritized.

Anhedral Wings (Downward-Angled Wings):

Anhedral wings, in contrast, are designed with a downward tilt, meaning the wingtips are positioned lower than the wing roots. This configuration enhances maneuverability by making the aircraft more responsive to pilot inputs, which is beneficial for high-speed fighter jets and military aircraft that require quick roll control. However, anhedral wings make the aircraft less inherently stable, meaning the pilot or onboard stability control systems must constantly make small adjustments to maintain level flight. The choice between dihedral and anhedral wings depends on the intended role of the aircraft [44]. Commercial airliners prioritize stability for passenger comfort, while military jets favor maneuverability for combat and tactical operations.

Control Surfaces for Weather Adaptability

Aircraft are equipped with several adjustable control surfaces that help maintain stability and counteract the effects of wind disturbances. These surfaces allow pilots to manipulate the aircraft's movement along different axes, ensuring a controlled and steady flight even in changing weather conditions.

Ailerons (Roll Control):

Ailerons are hinged flight control surfaces located at the trailing edge of the wings. They control the roll motion of the aircraft, which helps counteract lateral wind disturbances. When wind causes an aircraft to tilt or roll to one side, the pilot can adjust the ailerons to correct the roll and stabilize the wings (Collinson, 2023a). This is especially useful in strong crosswinds, where roll stability is critical for maintaining the flight path.

Elevators (Pitch Control):

The elevators, located on the horizontal stabilizer at the tail of the aircraft, control pitch movement (nose-up or nose-down). During turbulent weather or sudden wind gusts, rapid changes in airflow can cause unwanted nose movements. The pilot adjusts the elevators to counteract these forces and maintain level flight. Elevators are especially important during takeoff and landing, where precise control of pitch is necessary for a safe descent and approach [46].

Rudder (Yaw Control):

The rudder, located on the vertical stabilizer, helps control yaw (left or right movement of the aircraft's nose). It is particularly useful in crosswind landings, where strong side winds can push the aircraft off its intended path. By adjusting the rudder, pilots can counteract these lateral forces and align the aircraft properly with the runway during landing [47]. Together, these control surfaces allow aircraft to remain stable in unpredictable weather conditions, giving pilots the ability to adjust to turbulence, gusts, and other environmental challenges.

Turbulence Adaptation Systems

Turbulence is a common atmospheric phenomenon that occurs due to variations in wind speed and direction at different altitudes. While turbulence itself is usually not dangerous, it can cause discomfort to passengers and momentarily reduce pilot control. Modern aircraft are equipped with advanced turbulence detection and adaptation systems that help minimize its effects [48].

Turbulence Detection Sensors:

Advanced aircraft are fitted with turbulence detection sensors that monitor wind speed variations and air pressure changes in real-time. These sensors can predict areas of severe turbulence, allowing pilots to adjust their altitude or flight path to avoid rough air. Some commercial jets use LIDAR (Light Detection and Ranging) and Doppler radar to detect turbulence several kilometers ahead [49].

Gust Alleviation Systems:

Modern aircraft are designed with gust alleviation technology, which automatically adjusts the aircraft's control surfaces to counteract turbulence. These systems detect sudden shifts in airflow and instantly make small corrections to the wings and stabilizers, reducing the aircraft's movement and making the ride smoother for passengers [50].

Active Turbulence Damping:

Some advanced aircraft feature active turbulence damping systems, which use computer-controlled actuators to fine-tune the wing control surfaces in real-time. These systems can counteract sudden vertical or lateral movements caused by turbulence, reducing shaking and making the flight experience more stable [51]. Turbulence adaptation technology plays a significant role in enhancing flight comfort, reducing pilot workload, and maintaining overall aircraft stability in unpredictable weather conditions.

Wind Shear Warning and Autopilot Adjustments

Wind shear is one of the most dangerous weather phenomena for aircraft, particularly during takeoff and landing. It occurs when there is a sudden and drastic change in wind speed or direction over a short distance. Wind shear can cause a rapid loss of altitude and make it difficult for pilots to maintain control [52]. To counteract this threat, modern aircraft are equipped with wind shear detection and warning systems.

Predictive Wind Shear Warning Systems:

Many commercial aircraft are equipped with Doppler radar-based predictive wind shear warning systems. These systems scan the atmosphere ahead and detect sudden wind shifts that could pose a threat. If wind shear is detected, the aircraft issues an audible and visual warning, allowing pilots to take corrective action, such as increasing engine power or adjusting the flight path [53].

Autopilot Adjustments for Wind Shear:

Modern autopilot systems are designed to react quickly to wind shear conditions. If the aircraft encounters wind shear, the autopilot can automatically:

- **Increase engine thrust** to compensate for sudden loss of lift.
- **Adjust control surfaces** to stabilize the aircraft and counteract sudden wind shifts.
- **Modify approach angles** to ensure a safer landing in strong wind conditions.

By combining predictive warning systems with automated control responses, modern aircraft can effectively handle wind shear conditions and reduce the risks associated with sudden atmospheric changes. Wind and weather adaptability is a critical aspect of modern aviation, ensuring that aircraft can handle unpredictable atmospheric conditions safely and efficiently. Through aerodynamic wing designs, adjustable control surfaces, turbulence adaptation systems, and wind shear warning technology, aircraft can maintain stability even in challenging weather. These advancements not only improve flight safety but also enhance passenger comfort by minimizing the effects of turbulence and sudden wind shifts [54]. As aviation technology continues to evolve, new innovations will further improve an aircraft's ability to adapt to complex weather patterns, making air travel safer and more reliable than ever before.

4.3.3 Software-Driven Stability Enhancements

With advancements in aviation technology, software-based systems play an essential role in improving flight stability, enhancing pilot control, and optimizing overall aircraft performance. These software-driven stability enhancements automate responses to disturbances, ensure smoother flights, and improve safety by integrating real-time data processing with advanced control mechanisms. Innovations such as Fly-By-Wire (FBW) technology, AI-driven autopilot systems, stability augmentation systems (SAS), and real-time data processing have significantly transformed how aircraft handle stability and control in various flight conditions [55].

Fly-By-Wire (FBW) Technology

Traditional aircraft control systems relied on mechanical linkages, such as cables, pulleys, and hydraulic systems, to transmit pilot inputs to the control surfaces. However, modern aircraft have replaced these mechanical systems with electronic Fly-By-Wire (FBW) technology, where pilot inputs are transmitted as electronic signals to the aircraft's flight control computers [45]. The computer then interprets these inputs and makes necessary adjustments to the control surfaces, improving both precision and safety.

Reduced Aircraft Weight and Improved Fuel Efficiency:

Since FBW eliminates bulky mechanical linkages, it significantly reduces an aircraft's overall weight, leading to improved fuel efficiency and operational cost savings.

Enhanced Response Time and Precision:

Electronic signals travel much faster than mechanical systems, improving the aircraft's response time and maneuverability. This makes FBW particularly useful for fighter jets and high-performance aircraft.

Automatic Stabilization During Disturbances:

FBW systems continuously monitor flight parameters and automatically make small corrections to stabilize the aircraft during turbulence, crosswinds, or sudden maneuvers. If a pilot accidentally inputs excessive control movements, software limits prevent overcorrections, ensuring smoother and safer flight operations. FBW technology is now standard in modern commercial aircraft such as the Airbus A320, Boeing 787 Dreamliner, and military jets like the F-35 Lightning II, where precise control is crucial [56].

Autopilot and Artificial Intelligence (AI) in Flight Control

Modern aircraft heavily rely on autopilot and AI-driven flight control systems to assist pilots in maintaining altitude, heading, and speed with minimal manual input. These systems continuously analyze real-time flight data and make adjustments to optimize flight stability.

Flight Management Systems (FMS):

- FMS helps optimize flight paths by adjusting navigation routes for better stability and efficiency.
- It reduces pilot workload by automating routine flight adjustments.

AI-Driven Autopilot Systems:

- AI-powered autopilot continuously detects anomalies in flight behavior and automatically corrects them to maintain stable flight.
- During emergency situations, AI can recommend corrective actions or even take control to prevent dangerous conditions such as stall or excessive bank angles.

AI in Unmanned Aerial Vehicles (UAVs) and Drones:

- AI-based stability algorithms help UAVs adjust to external forces such as wind gusts without human intervention.
- Some drones use machine learning models to predict and counteract disturbances, enabling them to hover more precisely and operate in complex environments.

AI-enhanced autopilot and flight control systems are becoming increasingly sophisticated, allowing pilots to focus more on high-level decision-making while the software manages routine stability adjustments [57].

Stability Augmentation Systems (SAS)

Stability Augmentation Systems (SAS) are designed to automatically adjust an aircraft's control surfaces based on real-time flight conditions, ensuring enhanced stability and smoother flight operations. SAS is particularly useful in aircraft that require high-speed maneuverability or experience significant environmental disturbances [58].

Application in Military Fighter Jets:

- Fighter jets often operate in extreme conditions, requiring **rapid and precise control adjustments**.
- SAS helps compensate for sudden aerodynamic changes, allowing pilots to perform complex maneuvers without losing stability.

Use in Commercial Aircraft for Passenger Comfort:

- SAS continuously **makes micro-corrections** to the aircraft's flight path, reducing turbulence-induced shaking and providing a more comfortable ride for passengers.
- It helps maintain a stable flight path, even when encountering strong crosswinds or gusty conditions.

SAS works in combination with FBW and autopilot systems to enhance aircraft control and responsiveness, making it an essential component of modern aviation.

Real-Time Data Processing for Stability

One of the most critical advancements in software-driven stability enhancements is real-time data processing. Modern aircraft use sensor fusion technology, where multiple onboard sensors work together to provide highly accurate flight data [59]. These include:

- **Gyroscopes and Accelerometers**: Measure angular velocity and linear acceleration to detect **sudden shifts in movement**.
- **GPS and Inertial Navigation Systems (INS)**: Provide precise positioning data to help maintain a stable trajectory.
- **Barometric Sensors**: Monitor altitude changes to ensure smooth altitude adjustments.

By processing data from these sensors in real time, aircraft flight control systems can detect stability issues before they become critical and make automatic corrections. This real-time processing capability is particularly valuable in:

- High-altitude flights, where atmospheric changes occur rapidly.
- Autonomous drones and UAVs, which require constant adjustments to maintain stability.
- Extreme weather conditions, where precise control is essential to counteract turbulence.

Software-driven stability enhancements have revolutionized modern aviation by providing pilots with advanced tools to maintain flight stability, safety, and efficiency. Technologies such as Fly-By-Wire (FBW), AI-driven autopilot, Stability Augmentation Systems (SAS), and real-time data processing have significantly improved the way aircraft handle flight control. These innovations allow aircraft to respond automatically to disturbances, optimize fuel efficiency, and enhance passenger comfort [42]. As aviation technology continues to evolve, software-based stability solutions will play an even greater role in ensuring safer and more reliable air travel.

4.4 UAV Propulsion Systems

Unmanned Aerial Vehicles (UAVs) depend on efficient propulsion systems to achieve stability, endurance, and performance. The type of propulsion system used in a UAV determines its ability to carry payloads, handle different environmental conditions, and perform specialized tasks. In recent years, electric propulsion has become dominant in agricultural applications, while advanced propeller designs have improved aerodynamic efficiency, and VTOL (Vertical Takeoff and Landing) technology has enhanced operational flexibility [60].

4.4.1 Electric Propulsion in Agriculture

Electric propulsion is the most widely used system in UAVs, especially in agricultural applications. Unlike fuel-powered drones, electric UAVs use battery-driven electric motors, which are lightweight, efficient, and environmentally friendly [61]. This allows drones to be used effectively in farming, reducing labor costs and increasing productivity.

Why is Electric Propulsion Important for Agriculture?

- Eco-Friendly Operation: Since electric UAVs do not produce emissions, they contribute to sustainable and green farming.

- Quiet and Low-Vibration Flight: Traditional fuel-powered drones produce significant noise and vibrations, but electric motors operate quietly, making them ideal for livestock monitoring and spraying pesticides without disturbing animals.
- High Energy Efficiency: Electric motors convert almost 90% of battery power into thrust, making them highly efficient for long-duration flights over farmland.

Applications of Electric UAVs in Agriculture:

- Aerial Crop Spraying: Electric drones carry fertilizers, pesticides, and herbicides, spraying them uniformly over fields. This method improves crop protection while reducing chemical wastage.
- Precision Agriculture & Crop Monitoring: Equipped with infrared and multispectral cameras, UAVs can analyze soil health, detect pest infestations, and optimize irrigation.
- Seed Sowing & Pollination: Some electric drones are designed to disperse seeds, helping in reforestation and large-scale plantation farming.

Because of these advantages, electric UAVs are revolutionizing modern agriculture by enhancing efficiency, reducing costs, and promoting sustainable farming practices.

4.4.2 Advanced Propeller Designs for UAV Efficiency

The efficiency of a UAV heavily depends on the design of its propellers, which determine factors such as thrust generation, noise levels, and power consumption. In recent years, researchers have developed advanced propeller technologies that improve UAV performance.

Key Innovations in UAV Propeller Designs:

Optimized blade shapes play a crucial role in improving UAV aerodynamics. Thin and curved airfoil designs help reduce drag and enhance lift, allowing UAVs to fly longer on the same battery charge. Additionally, variable pitch propellers can adjust blade angles based on flight conditions, optimizing energy efficiency. The use of lightweight composite materials further enhances UAV performance. Carbon fiber propellers, being both lightweight and strong, help reduce overall UAV weight while increasing thrust efficiency. Composite materials also aid in minimizing vibrations, which is particularly important for drones used in aerial photography or mapping. Moreover, advancements in low-noise propeller technology contribute to quieter UAV operations [62]. Serrated or trailing-edge modified propellers help reduce noise by minimizing turbulence at the blade tips. In specialized applications, some UAVs utilize stealth propellers, which are specifically designed for quiet operations, making them ideal for military surveillance and wildlife monitoring.

How Do Advanced Propellers Benefit UAVs?

- **Increase Flight Time:** Aerodynamically efficient propellers **reduce power consumption**, allowing UAVs to fly for extended durations.
- **Improve Stability:** Reduced vibrations lead to **smoother video recordings** for aerial photography and **better precision for mapping applications**.
- **Enhance Performance in Windy Conditions**: Advanced designs help UAVs remain stable **even in gusty winds**, making them more reliable for outdoor operations.

These innovations in propeller technology contribute to making UAVs **quieter, more energy-efficient, and more capable of handling various flight environments**.

4.4.3 VTOL Innovations for Takeoff Efficiency

One of the biggest limitations of traditional fixed-wing UAVs is that they require a runway for takeoff and landing. However, VTOL (Vertical Takeoff and Landing) technology allows UAVs to take off and land like a helicopter but fly like an airplane [63]. This innovation combines the efficiency of fixed-wing aircraft with the flexibility of multirotors, making UAVs more adaptable to different missions.

Types of VTOL UAVs:

Tilt-rotor UAVs feature rotors that tilt forward after takeoff, enabling a smooth transition from vertical lift to horizontal flight. This design enhances flight efficiency and maneuverability, similar to the Bell Boeing V-22 Osprey, which follows a comparable tilt-rotor mechanism. Another innovation in VTOL technology is hybrid fixed-wing VTOL drones, which integrate fixed wings with vertical lift motors. This combination allows them to take off vertically while still achieving efficient cruising, making them ideal for long-range surveying and mapping applications [64]. Additionally, coaxial VTOL systems employ stacked dual rotors to improve lift efficiency while reducing the overall drone size, enhancing stability and performance in compact UAV designs.

Advantages of VTOL UAVs:

- No Runway Required: VTOL drones can operate in urban areas, forests, or rugged terrain without needing a dedicated runway.
- Extended Flight Range: Hybrid VTOL drones transition into fixed-wing flight, reducing power consumption and increasing range.
- Better Stability in Windy Conditions: VTOL UAVs adjust their thrust dynamically, allowing them to hover steadily even in strong crosswinds.

Applications of VTOL UAVs:

- Search and Rescue Missions: VTOL UAVs can take off immediately in emergency situations, covering large areas quickly.
- Military Surveillance and Cargo Transport: Defense forces use VTOL drones for spying, reconnaissance, and supply deliveries.
- Environmental Monitoring and Disaster Response: Scientists use VTOL drones to monitor wildfires, floods, and climate change effects in remote regions.

The combination of vertical takeoff and long-range efficiency makes VTOL UAVs an ideal choice for both commercial and military applications. The evolution of UAV propulsion systems is transforming how drones operate in agriculture, surveillance, cargo transport, and scientific research [65].

- Electric propulsion is leading to eco-friendly, efficient, and quieter UAVs, making them ideal for precision farming and crop spraying.
- Advanced propeller designs are enhancing flight efficiency by reducing noise, vibration, and power consumption.
- VTOL technology is allowing UAVs to take off and land in confined spaces while covering long distances efficiently.

These advancements in propulsion technology are expanding the capabilities of UAVs, making them more versatile, efficient, and adaptable for future applications.

4.5 Flight Efficiency Optimization

Flight efficiency optimization is crucial for improving the performance and endurance of UAVs and aircraft. Key strategies include reducing energy consumption, improving payload-to-weight ratios, and refining aerodynamic design. These factors lead to longer flight durations, increased operational effectiveness, and better overall efficiency. By reducing energy consumption, improving payload-to-weight ratios, and optimizing aerodynamic design, UAVs can achieve longer flight times and higher payload capacities.

4.5.1 Reducing Energy Consumption

Reducing Energy Consumption is essential for maximizing UAV flight time and minimizing operational costs. One approach is to use energy-efficient propulsion systems, such as brushless electric motors, which offer higher efficiency and lower heat generation. Additionally, optimizing flight paths through AI-driven algorithms helps minimize unnecessary movements, reducing power usage [66]. Lightweight battery technologies, such as lithium-sulfur and solid-state batteries, also contribute to lower energy consumption by increasing energy density while maintaining a minimal weight.

Key Factors Influencing Energy Consumption:

Energy consumption in UAVs and aircraft is influenced by several factors, including propulsion system efficiency, optimized flight paths, and advanced battery technologies. Implementing improvements in these areas helps extend flight time, reduce operational costs, and enhance overall performance.

1. **Efficient Propulsion Systems**

   The propulsion system plays a crucial role in determining how much energy a UAV consumes. More efficient propulsion leads to less energy waste and longer flight durations.

- Brushless electric motors are commonly used in UAVs because they offer higher efficiency (80–90%) compared to traditional brushed motors. They generate less heat, require minimal maintenance, and improve overall power efficiency.
- For fuel-powered aircraft, turbofan engines with high bypass ratios improve efficiency by generating more thrust with less fuel consumption. These engines are commonly used in commercial jets to enhance fuel economy and extend range.

2. **Optimized Flight Paths**

   A well-planned flight path ensures that a UAV consumes the least amount of energy necessary to reach its destination.

- AI-driven flight control systems analyze terrain, weather conditions, and wind patterns to determine the most efficient flight route, minimizing unnecessary energy use.
- Example: Drones used in delivery services, such as Amazon Prime Air or Zipline medical drones, follow optimized routes to reduce power consumption while ensuring timely deliveries.

3. **Lightweight Battery Technologies**

   Battery efficiency is a critical factor in UAV endurance. High-energy–density batteries allow UAVs to fly longer without increasing overall weight.

- Lithium-sulfur (Li–S) batteries offer a higher energy density compared to traditional lithium-ion (Li-ion) batteries, meaning they store more energy per unit of weight. This technology is useful for long-endurance UAVs.
- Example: The Airbus Zephyr UAV, which relies on solar energy and Li–S batteries, can stay airborne for weeks, demonstrating the benefits of lightweight and high-capacity energy storage solutions.

By integrating efficient propulsion, optimized flight paths, and advanced battery technology, UAVs can significantly reduce energy consumption and improve flight performance, making them more effective for applications in agriculture, logistics, surveillance, and military operations [67].

Formula for Energy Consumption:

The total energy consumed by a UAV can be estimated using:

$$E = P \times t$$

where:

E = Total energy consumption (Joules or Watt-hours)
P = Power required for flight (Watts)
t = Flight duration (hours)

This formula assumes constant power consumption over time. It's commonly used to estimate energy use for electric UAVs, especially when computing battery requirements or mission endurance. By reducing PP (power requirement) through better aerodynamics and efficient propulsion, overall energy consumption decreases.

Example:

If a UAV consumes 200 W of power for 1.5 h:

$$E = 200 \quad W \times 1.5\,h = 300Wh$$

Or convert to joules (1 Wh = 3600 J):

$$E = 300 \quad Wh \times 3600 = 1,080,000\,J$$

4.5.2 Improving Payload-To-Weight Ratios

Improving Payload-to-Weight Ratios plays a significant role in enhancing UAV efficiency. Using lightweight composite materials, such as carbon fiber and advanced polymers, reduces the overall aircraft weight without compromising structural integrity. Optimized payload distribution ensures that weight is balanced effectively, improving stability and reducing excess energy expenditure. Furthermore, modular payload systems allow UAVs to carry only the necessary equipment, further optimizing the weight-to-efficiency ratio [68].

Strategies to Improve Payload-To-Weight Ratios:

Improving the payload-to-weight ratio is essential for enhancing UAV efficiency, as it allows drones to carry more useful equipment without excessive energy consumption. This is achieved through lightweight structural materials, modular payload systems, and optimized weight distribution.

1. **Lightweight Structural Materials**

Using lightweight yet strong materials reduces overall UAV weight while maintaining durability and performance.

- Carbon fiber composites are widely used in UAV construction because they are five times stronger than steel while being significantly lighter. This helps reduce the drone's weight, allowing it to carry heavier payloads without increasing energy consumption.
- Example: The DJI Matrice 300 RTK UAV utilizes a carbon fiber frame, enabling it to support heavy sensors while maintaining a lightweight structure, enhancing endurance and flight stability.

2. **Modular Payload Systems**

A customizable payload system ensures that UAVs carry only the necessary equipment, preventing excess weight and improving efficiency.

- Modular designs allow users to swap different sensors and tools based on specific mission needs. This reduces the overall burden on the UAV and optimizes energy consumption.
- Example: Some surveillance drones feature interchangeable payloads, such as cameras, LiDAR sensors, or infrared sensors, allowing operators to adapt to different operational requirements without adding unnecessary weight.

3. **Optimized Weight Distribution**

Proper weight distribution ensures flight stability and reduces drag, leading to lower energy consumption and better control.

- Strategic placement of payloads prevents instability, which can increase power usage due to constant flight corrections.
- Example: Fixed-wing UAVs distribute payloads closer to the center of gravity, which enhances flight stability and minimizes unnecessary energy loss due to imbalanced weight distribution.

By implementing these weight optimization strategies, UAVs can achieve higher efficiency, longer flight times, and better overall performance in applications such as agriculture, aerial mapping, and surveillance.

4.5.3 Aerodynamic Design Considerations

Aerodynamic Design Considerations are fundamental to reducing drag and enhancing flight stability. Streamlined airframes, blended wing designs, and optimized winglets help minimize air resistance, allowing UAVs to maintain higher speeds with lower energy consumption. Adaptive control surfaces, such as morphing wings and dynamic flaps, adjust to changing flight conditions to improve overall aerodynamic efficiency [69]. By integrating these design principles, UAVs and aircraft can achieve significant improvements in flight performance while maintaining energy efficiency.

Key Aerodynamic Design Strategies

Enhancing aerodynamic efficiency in UAVs and aircraft reduces drag, improves fuel economy, and extends flight time. Some key strategies include streamlined airframes, blended wing and winglet designs, and adaptive control surfaces.

1. **Streamlined Airframes**

   A smooth, aerodynamic body minimizes parasitic drag, allowing UAVs to move efficiently through the air.

- Sleek, curved fuselages reduce air resistance, improving overall flight performance.
- Example: The MQ-9 Reaper drone has a streamlined fuselage, enabling longer endurance and high-speed surveillance, making it highly effective for reconnaissance missions.

2. **Blended Wing and Winglet Designs**

   Wings play a crucial role in reducing drag and improving lift-to-drag ratios, leading to greater energy efficiency.

- Blended wing designs integrate the fuselage with the wings, reducing turbulence and improving fuel efficiency.
- Winglets minimize vortex drag at wingtips, reducing energy loss and improving stability.
- Example: The Boeing 787 Dreamliner uses raked wingtips, which enhance fuel efficiency by 5–7% by reducing induced drag.

3. **Adaptive Control Surfaces**

   Modern UAVs use dynamic control surfaces that adjust in response to wind conditions, optimizing flight stability and performance.

- Flaps and morphing wings help adapt to turbulence, improving energy efficiency.
- Example: NASA's MADCAT UAV features flexible, shape-changing wings, which optimize lift and drag for different flight conditions, leading to better performance in varying weather conditions.

By integrating streamlined airframes, advanced wing designs, and adaptive control surfaces, UAVs and aircraft achieve greater efficiency, longer endurance, and enhanced operational capability across industries like surveillance, cargo transport, and scientific research [70]. Minimizing C_D (drag coefficient) through aerodynamic design reduces the energy required to sustain flight, improving efficiency. Optimizing flight efficiency is essential for UAVs to maximize performance, reduce costs, and improve endurance.

- Reducing energy consumption through efficient motors, optimized flight paths, and advanced battery technologies extends UAV flight times.
- Improving payload-to-weight ratios by using lightweight materials and modular payloads ensures higher efficiency without excessive energy use.
- Aerodynamic design enhancements, including streamlined airframes, winglets, and adaptive control surfaces, help reduce drag and increase flight stability.

By integrating these strategies, UAVs can achieve longer flight durations, higher payload efficiency, and better overall performance in various applications, from agriculture and surveillance to logistics and defense.

4.6 Navigation Systems for UAVs

Unmanned Aerial Vehicles (UAVs) rely on advanced navigation technologies to operate autonomously, avoid obstacles, and follow optimized flight paths. These systems use GPS-based navigation, AI-driven path planning, and collision avoidance technologies to ensure efficiency and safety in various applications, including military surveillance, agriculture, and delivery services [71].

4.6.1 GPS-Based Autonomous Navigation

How It Works:

GPS technology uses signals from at least four satellites to determine a UAV's latitude, longitude, altitude, and velocity. Real-Time Kinematic (RTK) GPS and Differential GPS (DGPS) improve positioning accuracy from 3 to 5 m to centimeter-level precision. GPS-based autonomous navigation allows UAVs to follow pre-programmed flight paths without human intervention. For example, precision farming drones use GPS to navigate fields and spray pesticides with high accuracy, reducing chemical waste. Dual GPS and RTK systems further enhance navigation accuracy [72].

Example:

- Precision Agriculture: DJI Agras T40 UAV uses RTK-GPS to autonomously navigate farmland, ensuring precise pesticide spraying while minimizing overlap.
- Military Drones: The MQ-9 Reaper UAV uses GPS and Inertial Navigation Systems (INS) to conduct surveillance and target tracking missions with extreme accuracy.

4.6.2 Path Planning Algorithms

Path planning algorithms are used by UAVs to calculate efficient and obstacle-free routes, optimizing energy consumption and flight time. AI-powered flight planning optimizes routes, while dynamic path planning adjusts flight paths in real-time based on weather or obstacles. For example, Amazon Prime Air uses AI-based path planning to find the fastest route, avoiding obstacles like buildings and power lines, ensuring efficient and energy-efficient UAV delivery [73].

Types of Path Planning Algorithms:

- **Dijkstra's Algorithm**: Finds the shortest path in a network, useful for UAVs operating in urban areas with restricted airspace.
- **A (A-Star) Algorithm**: An improvement over Dijkstra's, it balances distance and time efficiency, often used in delivery drones.
- **RRT (Rapidly Exploring Random Tree)**: Used for dynamic path planning, helping UAVs navigate changing environments like forests or urban landscapes.

Example:

- Delivery Drones: Amazon Prime Air uses AI-based *A Algorithm** for real-time route optimization to deliver packages while avoiding obstacles like buildings and power lines.
- Autonomous Drones: The Parrot Anafi AI UAV uses real-time AI path planning for inspection missions, dynamically adjusting its route based on sensor inputs.

4.6.3 Collision Avoidance Technologies

Collision avoidance technologies, such as LiDAR, ultrasonic sensors, and cameras, help UAVs detect and avoid obstacles in real-time. ADS-B and RADAR-based systems allow UAVs to track nearby aircraft and avoid collisions. For example, the DJI Mavic 3 drone uses an omnidirectional obstacle avoidance system, integrating LiDAR, RADAR, ultrasonic sensors, and computer vision to detect obstacles and adjust flight paths, ensuring safe navigation [74].

Collision Avoidance Methods:

- **LiDAR (Light Detection and Ranging)**: Measures distances using laser pulses, helping UAVs detect objects up to 100 m away.
- **Ultrasonic Sensors**: Short-range sensors used in indoor drones to avoid walls and ceilings.
- **Computer Vision (AI-based)**: UAVs use image processing to recognize and avoid moving obstacles (e.g., birds, planes, or buildings).

Example:

- **DJI Mavic 3**: Uses omnidirectional vision sensors and LiDAR technology to detect obstacles in all directions and avoid collisions.
- **Intel Falcon 8+ Drone**: Uses AI-based object detection to navigate complex environments like construction sites safely.

By integrating GPS-based autonomous navigation, AI-driven path planning, and advanced collision avoidance technologies, UAVs achieve greater autonomy, efficiency, and safety. These innovations allow drones to perform complex tasks in industries such as agriculture, logistics, surveillance, and environmental monitoring [75].

4.7 Testing and Calibrating UAV Aerodynamics

Unmanned Aerial Vehicles (UAVs) rely on precise aerodynamics for stability, efficiency, and performance. Testing and calibration ensure that UAVs perform optimally in various environments by analyzing airflow, lift, drag, and stability under controlled and real-world conditions. Proper aerodynamic testing helps improve flight efficiency, maneuverability, and energy consumption. Aerodynamics is the study of how air interacts with moving objects, particularly focusing on forces such as lift, drag, thrust, and weight [76]. For UAVs, understanding these forces is critical to achieving smooth and efficient flight.

This section discusses three essential methods for testing and calibrating UAV aerodynamics:

1. **Wind Tunnel Testing**—Controlled environment testing for precise aerodynamic analysis.
2. **In-Field Performance Trials**—Real-world testing to validate UAV behavior under practical conditions.
3. **Calibration for Different Altitudes**—Adjustments to optimize UAV performance at varying altitudes.

4.7.1 Wind Tunnel Testing

A wind tunnel is a controlled testing facility that generates airflow over a UAV model to study its aerodynamic properties. It helps engineers analyze how air moves around the UAV, optimizing wing shape, propeller efficiency, and overall flight stability [77].

Purpose

- To measure **aerodynamic forces** (lift, drag, and thrust).
- To visualize airflow patterns using smoke, laser sheets, or computational simulations.

- To test different UAV designs without real-world flight risks.

Key Concepts

- **Lift (L)**: The upward force that counteracts gravity.
- **Drag (D)**: The resistance force opposing motion.
- **Thrust (T)**: The force generated by propellers or engines.
- **Reynolds Number (Re)**: Determines whether airflow is laminar or turbulent.

Advantages of Wind Tunnel Testing

Provides controlled and repeatable conditions.
Allows for precise data collection and aerodynamic improvements.
Reduces the risk of UAV crashes during design testing.

4.7.2 *In-Field Performance Trials*

Field trials involve flying UAVs in real-world environments to evaluate performance under practical conditions such as varying wind speeds, turbulence, and weather changes [78].

Purpose

- To test UAV stability and responsiveness in natural conditions.
- To assess onboard sensor accuracy and performance.
- To validate wind tunnel results with real-world flight data.

Key Aspects of In-Field Trials

- **Stability Testing**: UAV's ability to maintain steady flight in wind gusts.
- **Battery Performance**: Evaluating energy consumption under different conditions.
- **Payload Testing**: Measuring aerodynamic effects with varying loads.

Example:

A UAV is flown at different wind speeds (5 m/s, 10 m/s, 15 m/s) to assess its response. If the UAV loses stability at 15 m/s, modifications to the control system are needed.

Advantages of In-Field Testing

Identifies practical flight challenges.
Helps optimize UAV navigation and control algorithms.
Validates wind tunnel data with real-world observations.

4.7.3 *Calibration for Different Altitudes*

Calibration for different altitudes involves adjusting UAV settings to compensate for changes in air pressure, density, and temperature. These factors affect lift, drag, thrust, and battery efficiency [79].

Effects of Altitude on UAV Aerodynamics

Higher Altitudes (Lower Air Density)

- Reduced lift: Requires higher speeds or larger wing surfaces.
- Lower thrust efficiency: Propellers must rotate faster.
- Faster battery drain: Motors work harder to maintain lift.

Lower Altitudes (Higher Air Density)

- More stable flight due to increased lift.
- Higher drag: Can reduce flight duration.

Calibration Techniques

1. **Thrust Compensation**: Increasing propeller speed at high altitudes.
2. **Flight Algorithm Adjustments**: Modifying autopilot settings.
3. **Battery Optimization**: Using efficient power management systems.

Advantages of Altitude Calibration:

Ensures UAVs function efficiently at varying altitudes.

Reduces instability due to changing air densities.

Improves battery life and thrust efficiency.

Testing and calibrating UAV aerodynamics is essential for efficient and stable flight. Wind tunnel testing provides controlled conditions to refine UAV designs, in-field trials validate performance under real-world conditions, and altitude calibration ensures adaptability to different atmospheric conditions [80]. By combining these methods, UAVs can be optimized for diverse applications, from agriculture to defense and disaster management.

4.8 Conclusion and Summary

Unmanned Aerial Vehicles (UAVs) operate based on four fundamental aerodynamic forces: lift, weight, thrust, and drag. Lift opposes weight, enabling flight, while rotary-wing UAVs generate lift through forward motion. Thrust propels the UAV forward and varies depending on the type of UAV propulsion system. Drag acts as aerodynamic resistance and is categorized into parasitic drag (air resistance against the UAV's body) and induced drag (a byproduct of lift). Stability is essential for controlled flight, achieved through control surfaces in fixed-wing UAVs and rotor adjustments in rotary-wing UAVs. Aerodynamic efficiency, which is influenced by wing shape, aspect ratio, and weight distribution, determines UAV performance, flight endurance, and energy efficiency. Unmanned Aerial Vehicles (UAVs) are designed to meet the demands of modern agriculture. By understanding the forces of flight, configurations, stability mechanisms, propulsion systems, and optimization techniques, engineers and operators can develop aircraft optimized for applications such as surveillance, agriculture, logistics, and defense. Advanced navigation systems and rigorous testing further ensure reliable and precise performance, paving the way for innovative applications in farming and beyond.

The four forces of flight—lift, weight, thrust, and drag—form the foundation of all aerial motion. Balancing these forces ensures safe, stable, and efficient flight. As aerospace technology advances, continuous improvements in aerodynamics and propulsion systems will push the boundaries of flight performance, leading to more capable and versatile aerial vehicles in the future. UAV configurations are designed to optimize aerodynamics, endurance, and maneuverability based on mission needs. Fixed-wing UAVs excel in range and efficiency, rotary-wing UAVs provide superior control and VTOL capabilities, hybrid UAVs offer versatility, and unconventional UAVs push technological boundaries. Stability and control in flight are achieved through a combination of gyroscopic stabilization mechanisms, wind adaptability features, and software-driven stability enhancements. Traditional mechanical systems, such as gyroscopes and aerodynamic design, work alongside advanced AI-powered autopilot and fly-by-wire systems to ensure safe, smooth, and efficient flight. Wind and weather adaptability is a critical aspect of modern aviation, ensuring that aircraft can handle unpredictable atmospheric conditions safely and efficiently. Software-driven stability enhancements have revolutionized modern aviation by providing pilots with advanced tools to maintain flight stability, safety, and efficiency.

UAV propulsion systems are undergoing rapid advancements, enabling drones to operate more efficiently, with greater stability and adaptability. Electric propulsion is revolutionizing agriculture by providing quiet, eco-friendly, and cost-effective aerial solutions. Advanced propeller designs enhance flight efficiency, reducing noise and improving aerodynamic performance. VTOL innovations are transforming UAV operations by allowing takeoff and landing in confined spaces while maintaining long-range flight capabilities. Optimizing flight efficiency is essential for UAVs to maximize performance, reduce costs, and improve endurance. Reducing energy consumption through efficient motors, optimized flight paths, and advanced battery technologies extends UAV flight times. Improving payload-to-weight ratios by using lightweight materials and modular payloads ensures higher efficiency without excessive energy use. Aerodynamic design enhancements, including streamlined airframes, winglets, and adaptive control surfaces, help reduce drag and increase flight stability. By integrating GPS-based autonomous navigation, AI-driven path planning, and advanced collision avoidance technologies, UAVs achieve greater autonomy, efficiency, and safety. Wind tunnel testing, in-field performance trials, and altitude calibration ensure optimal aerodynamics, efficiency, and flight stability.

These methods allow UAVs to operate in diverse environments, from urban delivery routes to high-altitude surveillance missions.

References

1. Haisley S (2021) Designing and developing stable body flight in drones
2. Heide L (2019) Evaluating the accuracy of conventional drag prediction and validation methods for small electric UAVs
3. Wu J, Ye Y, Du J (2024) Autonomous drones in urban navigation: autoencoder learning fusion for aerodynamics. J Constr Eng Manag 150:04024067
4. Mohsan SAH, Othman NQH, Li Y et al (2023) Unmanned aerial vehicles (UAVs): practical aspects, applications, open challenges, security issues, and future trends. Intel Serv Robot 16:109–137
5. Elmeseiry N, Alshaer N, Ismail T (2021) A detailed survey and future directions of unmanned aerial vehicles (uavs) with potential applications. Aerospace 8:363
6. Heinrich G-A, Vogt S, Lawrance NR et al (2022) In-wing pressure measurements for airspeed and airflow angle estimation and high angle-of-attack flight. J Guid Control Dyn 45:1033–1045
7. Bai S, He Q, Chirarattananon P (2022) A bioinspired revolving-wing drone with passive attitude stability and efficient hovering flight. Sci Robot 7:eabg5913
8. Franklin M (2022) Unmanned combat air vehicles: opportunities for the guided weapons industry?. Royal United Services Institute (RUSI)
9. Sadraey M (2022) Automatic flight control systems. Springer Nature
10. Landell-Mills MN (2022) Lift is best explained by Newtonian mechanics.
11. AlBdour JAA (2022) Increasing endurance time for multirotor UAVs by using a hybrid thrusting system
12. Jyothi S (2022) New winglet designs to minimize drag and turbulence
13. Neigapula SNV, Maddula SP, Nukala VB (2020) A study of high lift aerodynamic devices on commercial aircrafts. Aviation 24:123–136
14. Mbam CJ (2024) Fixed-wing UAV tracking of evasive targets in 3-dimensional space
15. Wanner D, Hashim HA, Srivastava S, Steinhauer A (2024) UAV avionics safety, certification, accidents, redundancy, integrity, and reliability: a comprehensive review and future trends. Drone Syst Appl 12:1–23
16. Vijayanandh R, Senthilkumar S, Rajkumar R, et al (2022) Conceptual design and computational investigations of fixed wing unmanned aerial vehicle for medium-range applications. In: Autonomous and connected heavy vehicle technology. Elsevier, pp 353–374
17. Özbek E, Ekici S, Karakoc TH (2023) An evaluation on landing gear configurations of fixed-wing, rotary-wing, and hybrid UAVs. In: Unmanned aerial vehicle design and technology. Springer, pp 153–165
18. Fahlstrom PG, Gleason TJ, Sadraey MH (2022) Introduction to UAV systems. John Wiley & Sons
19. Khan FA, Islam A, Abbas H, et al (2023) Computational analysis of a UAV drone with varying angle of attacks. IEEE, pp 1–11
20. Alsalem A, Zohdy M (2023) A review on civil applications of vertical take-off and landing vehicles. IEEE, pp 130–137
21. Farajijalal M, Eslamiat H, Avineni V et al (2025) Safety systems for emergency landing of civilian unmanned aerial vehicles—a comprehensive review. Drones 9:141
22. Yoon J, Lim K, Park S, Doh J (2023) Multi-objective optimization of aerodynamic blade shapes for quadcopter system to enhance hovering thrust and power consumption efficiency. Int J Aeronaut Space Sci 24:689–700
23. Betti Sorbelli F (2024) UAV-based delivery systems: a systematic review, current trends, and research challenges. J Autonom Transp Syst 1:1–40

24. Misra A, Jayachandran S, Kenche S et al (2022) A review on vertical take-off and landing (VTOL) tilt-rotor and tilt wing unmanned aerial vehicles (UAVs). J Eng 2022:1803638
25. Gholami A (2024) Exploring drone classifications and applications: a review. Int J Eng Geosci 9:418–442
26. Hong T-K, Lin C-Y, Lin H-J, Ruseno N (2023) Taiwan solar-powered UAV flight endurance record. Drone Syst Appl 12:1–14
27. Telli K, Kraa O, Himeur Y et al (2023) A comprehensive review of recent research trends on unmanned aerial vehicles (uavs). Systems 11:400
28. Seabridge A, Moir I (2020) Design and development of aircraft systems. John Wiley & Sons
29. Sharandak OR (2024) Estimation of zero offset components of modern gyroscopes using Allan dispersion
30. Vyhoniuk MI (2020) Inertial sensors in remotely piloted aircraft systems
31. Ovcharenko S, Filipkovskyi S, Shaabdiiev S (2022) Aircraft systems and equipment
32. Gryazin D, Paderina T (2024) Magnetic compass with dynamic error correction system. Gyroscopy and Navigation 15:191–202
33. Sarter N (2024) Flight crew visual scanning techniques on transport category aircraft
34. Usubamatov R (2020) Theory of gyroscopic effects for rotating objects. Open Access J Sci Technol 2
35. Adeyemi D (2023) Active control of rotorcraft instabilities
36. Yepifanov S, Chygryn V (2021) Design of aircraft turbine engines
37. Simmons BM, Gresham JL, Woolsey CA (2023) Flight-test system identification techniques and applications for small, low-cost, fixed-wing aircraft. J Aircr 60:1503–1521
38. Peksa J, Mamchur D (2024) A review on the state of the art in copter drones and flight control systems. Sensors 24:3349
39. Watson R (2023) Unmanned aerial vehicles for contact-based inspection
40. Βρύνας N (2023) Development and optimization of an inertially stabilized platform (gimbal) system for UAVs: motion study and drivetrain
41. Kang SS, Singh NT, Grover R (2024) Improving drone agility and stability with advanced flight control systems using MPU-6050 IMU and high-fidelity simulation. IEEE, pp 121–127
42. Klymenko OA (2024) Automated microclimate control system for the passenger cabin of an aircraft
43. Takahashi TT, Griffin JA, Grandhi RV (2023) A review of high-speed aircraft stability and control challenges, p 3231
44. Merryisha S, Rajendran P (2023) Wing engineering: aerodynamics, structures and design. Penerbit USM
45. Collinson R (2023) Aerodynamics and aircraft control. In: Introduction to avionics systems. Springer, pp 73–114
46. Jun L (2023) A system for assessing the quality of crew piloting techniques flight of varying complexity
47. Sedlmair N, Theis J, Thielecke F (2022) Flight testing automatic landing control for unmanned aircraft including curved approaches. J Guid Control Dyn 45:726–739
48. Le Clainche S, Ferrer E, Gibson S et al (2023) Improving aircraft performance using machine learning: a review. Aerosp Sci Technol 138:108354
49. Inokuchi H, Akiyama T (2022) Performance evaluation of an airborne coherent Doppler lidar and investigation of its practical application. Trans Japan Soc Aeronaut Space Sci 65:47–55
50. Liu P, Liu P (2021) Flight mystery and aerodynamic principles. A General Theory of Fluid Mechanics 445–579
51. Xie Y, Zhou R, Qu J (2023) Fitts' law on the flight deck: evaluating touchscreens for aircraft tasks in actual flight scenarios. Ergonomics 66:506–523
52. Zheng Y, Lu Y, Jie Y et al (2023) Test pilot and airline pilot differences in facing unexpected events. Aerospace Med Human Performance 94:18–24
53. Ahmed U, Ali F, Jennions I (2021) A review of aircraft auxiliary power unit faults, diagnostics and acoustic measurements. Prog Aerosp Sci 124:100721

54. Wang FZ, Animasaun I, Muhammad T, Okoya S (2024) Recent advancements in fluid dynamics: drag reduction, lift generation, computational fluid dynamics, turbulence modelling, and multiphase flow. Arab J Sci Eng 49:10237–10249
55. Lv Z, Xie S (2024) Artificial intelligence in the digital twins: State of the art, challenges, and future research topics: [version 2; peer review: 2 approved]. Digital Twin 1:12
56. David A (2023) Controlling aircraft—from humans to autonomous systems: the fading humans. SAE International
57. Mallick P (2024) Artificial intelligence, national security and the future of warfare. In: Artificial intelligence, ethics and the future of warfare. Routledge India, pp 30–70
58. Ramasamy S (2017) Next generation flight management systems for manned and unmanned aircraft operations-automated separation assurance and collision avoidance functionalities
59. Yeong DJ, Velasco-Hernandez G, Barry J, Walsh J (2021) Sensor and sensor fusion technology in autonomous vehicles: a review. Sensors 21:2140
60. Joshi D, Deb D, Muyeen S (2022) Comprehensive review on electric propulsion system of unmanned aerial vehicles. Front Energy Res 10:752012
61. Gatti RR, Singh C, Agrawal R, Serrao FJ (2023) Self-Powered cyber physical systems. John Wiley & Sons
62. Afridi S, Hlebowicz K, Cawthorne D, Lundquist UPS (2024) Unveiling the impact of drone noise on wildlife: a crucial research imperative. IEEE, pp 1409–1416
63. Rehan M, Akram F, Shahzad A et al (2022) Vertical take-off and landing hybrid unmanned aerial vehicles: an overview. Aeronautical J 126:2017–2057
64. Duncan BR (2022) Investigation of a tilt-wing proof of concept for a high-speed VTOL jet UAV using thrust vectoring for balance
65. Vyshnivska B, Kvasha S, Vakulenko V (2024) Unmanned aerial vehicle market conditions and it's practical use in the agricultural sector of the economy. Three Seas Economic J 5:14–22
66. Haider SK, Ahmed A, Khan NM, et al (2024) AI-Driven energy optimization in UAV-Assisted routing for enhanced wireless sensor networks performance. Comput, Mater& Continua 80
67. Fan B, Li Y, Zhang R, Fu Q (2020) Review on the technological development and application of UAV systems. Chin J Electron 29:199–207
68. Pratt LM (2022) Solid oxide fuel cell and gas turbine hybrid cycles for aerospace power and propulsion
69. Harvey C, Gamble LL, Bolander CR et al (2022) A review of avian-inspired morphing for UAV flight control. Prog Aerosp Sci 132:100825
70. Javaid M, Khan IH, Singh RP et al (2022) Exploring contributions of drones towards Industry 4.0. Industrial Robot: Int J Robot Res Appl 49:476–490
71. Elkhenfas MA, Sefrita MMS (2024) The role of artificial intelligence in autonomous drone navigation and decision-making. مجلة شمال إفريقيا للنشر العلمي (NAJSP) 124–142
72. Padhiary M, Kumar R, Sethi LN (2024) Navigating the future of agriculture: A comprehensive review of automatic all-terrain vehicles in precision farming. J Instit Eng (India): Series A 105:767–782
73. Hussain A, Li S, Hussain T, et al (2024) Computing challenges of UAV networks: a comprehensive survey. computers, materials & continua 81
74. Nelson WA (2024) Smart sensing and communications for UAVs/Drones
75. Guebsi R, Mami S, Chokmani K (2024) Drones in precision agriculture: a comprehensive review of applications, technologies, and challenges. Drones 8:686
76. Комаров Б, Зінченко Д (2023) The aerodynamic behavioral study of canard plane with fan wing configuration. Механіка гіроскопічних систем 118–135
77. Greenwell D (2022) Transonic industrial wind tunnel testing in the 2020s. Aeronautical J 126:125–151
78. Walpen A, Catry G, Noca F (2023) Real-scale atmospheric wind and turbulence replication using a fan-array for environmental testing and UAV/AAM validation, p 0812
79. Scanavino M (2021) Design and testing methodologies for UAVs under extreme environmental conditions

80. Semenišin M, Steponavičius D, Kemzūraitė A, Savickas D (2025) Optimizing UAV spraying for sustainability: different system spray drift control and adjuvant performance. Sustainability 17:2083

Chapter 5
UAV Airflow and Downwash for Dynamic Microclimate Engineering in Relation to Crop Response

Abstract A comprehensive investigation into the aero-thermodynamic impacts of UAV-generated airflow on the rice microclimate is essential to elucidate the complex relationships between wind speed, temperature, and temporal dynamics during critical growth stages. Focusing on the vulnerable stages of rice like heading, panicle, and flowering, this study aims to advance the understanding of microclimatic influences on rice crops, ultimately informing the development of UAV-based strategies to enhance crop resilience and optimize yields. By utilizing UAV rotor downwash, the research examines wind temperature and speed at three key diurnal intervals: 9:00 am, 12:00 pm, and 3:00 pm. Notably, at 9:00 am, UAV-induced airflow creates a favorable microclimate with stable temperatures (27.45–28.45 °C) and optimal wind speeds (0.0700–2.050 m/s), facilitating pollen transfer and grain setting. By 12:00 pm, wind speeds peak at 2.370 m/s, inducing evaporative cooling while maintaining temperature stability, yet leading to some moisture loss. At 3:00 pm, wind temperatures reach 28.48 °C, accompanied by a 72% wind speed decrease from mid-day, effectively conserving moisture during critical growth phases. The results reveal that UAV airflow positively influences panicle and flowering stages, where carefully moderated wind speeds (up to 3 m/s) and temperatures reduce pollen sterility, enhance fertilization, and optimize reproductive development. This highlights the potential of UAV-engineered microclimate management to mitigate stress factors and improve yield through targeted airflow regulation. Key agronomic parameters showed significant improvements, including stem diameter, canopy temperature regulation, grain-filling duration, productive tillers (increasing by 30.77%), total tillers, flag leaf area, grains panicle^{-1} (rising by 46.55%), biological yield, grain yield (surging by 70.75%), and harvest index. Conclusively, optimal aero-thermodynamic effects were observed with 9:00 am rotor airflow applications during flowering, outperforming mid-day and late-afternoon treatments. Additionally, 12:00 pm airflow during flowering significantly increased yield. The interaction between rotor airflow timing and growth stage (RRS × GS) exhibited low to moderate effects, underscoring the importance of precise timing in maximizing rice productivity.

Keywords Wind wall · Microclimate · Wind speed · Wind temperature · Rice agronomy

© The Author(s), under exclusive license to Springer Nature Singapore Pte Ltd. 2025

J. Li, *UAV Aerodynamics and Crop Interaction*, Smart Agriculture 13,
https://doi.org/10.1007/978-981-96-8402-1_5

5.1 Introduction

Rice (Oryza sativa L.), a staple food for more than half of the world's population, is particularly sensitive to environmental variations. Its growth and yield are significantly influenced by microclimatic factors such as temperature, humidity, and wind patterns [1]. Despite extensive research on precision agriculture practices, gaps remain in understanding how UAV-induced airflow can be harnessed to engineer microclimates that enhance physiological traits and yield during critical growth stages in rice cultivation. The dynamic microclimate engineering through UAV rotor airflow can positively impact rice growth, yield, and physiological traits, especially during key developmental phases. By manipulating airflow, it is expected that the microclimate can be optimized to improve factors such as canopy temperature regulation, nutrient uptake, and water efficiency, which are crucial for the productivity of rice [2]. Combining UAV technology with microclimate engineering principles to investigate their synergistic effects on rice.

The advent of precision agriculture has revolutionized traditional farming practices, enabling enhanced crop management through advanced technologies. Among these innovations, Unmanned Aerial Vehicles (UAVs) have emerged as a powerful tool for optimizing agricultural processes. Their ability to monitor, analyze, and influence crop growth dynamics has positioned UAVs at the forefront of modern agronomy.

The increasing use of rotor aircraft in precision agriculture has revolutionized crop management, enhancing tasks such as crop monitoring, mapping, and the application of pesticides and fertilizers. Rotor aircraft are becoming increasingly prominent in precision farming, where their airflow directly affects the microclimate around the plants. This airflow can alter key micro environmental factors such as temperature, humidity, CO_2 concentration [3], and pollination patterns, ultimately impacting plant physiology and development [4]. Rotor aircraft have also demonstrated the ability to alter microclimates through their airflow, offering new opportunities for optimizing crop environments, particularly in rice cultivation. The impact of rotor aircraft airflow on agro-meteorology and its relation with microclimate engineering, focusing on how it influences crop production, is a critical area of study as it affects key environmental factors such as temperature, humidity, CO_2 levels, and air circulation around the plants, which in turn influence plant physiology [5], growth, yield, and overall agricultural productivity [6]. To explores the novel potential of rotor aircraft airflow to engineer microclimates and its effects on rice crops, with a unique focus on how flight timing during different parts of the day at the plant different growth stages may influences crop growth and productivity. During flight, rotor aircraft produce substantial airflow, modulating critical microclimate parameters that intricately influence plant physiological functioning, ultimately impacting crop growth, development, and productivity. Given the sensitivity of rice crops to these microclimatic variables, particularly during critical developmental stages such as tillering, flowering, and grain filling, manipulating these environmental factors via rotor aircraft airflow represents an innovative approach to improving crop performance under field conditions

[7]. Rotor aircraft airflow also holds potential for reducing the spread of diseases and pests by lowering leaf moisture and inhibiting condensation [8]. However, concerns have been raised about possible negative effects, such as crop lodging and flower shattering, which could impact fertilization. Therefore, integrating rotor aircraft airflow with operational flights for pesticide, fertilizer, and water applications presents a key engineering challenge to minimize such adverse effects [9]. This study hypothesizes that the timing of rotor aircraft flights plays a critical role in the extent of microclimate alteration and its subsequent impact on rice growth and productivity, providing insights into how rotor airflow can be used as a precision tool to regulate crop environments.

Microclimates play a crucial role in determining plant biomass and productivity, with temperature and humidity being two of the most influential factors [10]. Both air and canopy temperatures are critical to plant growth, and rotor aircraft airflow can disturb air circulation, affecting crop growth and yield traits [11]. By disrupting the air within the plant's micro-environment, the airflow helps regulate the micrometeorological conditions around the crops, influencing processes like transpiration and photosynthesis [12]. It is documented that air disturbances improve seedling growth by promoting more effective air exchange, even though CO_2 concentrations may remain relatively unchanged [13]. Airflow can reduce leaf and canopy temperatures, mitigating the effects of solar radiation and promoting optimal growth conditions. This process influences crop physiological traits such as root-to-shoot ratio, transpiration rate, and dry matter accumulation [14]. By enhancing airflow, rotor aircraft may alleviate these issues, improving energy exchange and CO_2 diffusion, leading to better crop growth and higher yields [15].

5.2 Case Study of the UAV Airflow Interaction with Microclimate Engineering

5.2.1 Rice Microclimate Engineering Using a Novel Rotor Wind Wall Device

The rice effected by engineered microclimate and wind field was provided by a wind field simulation device, which mainly consists of a rotor array, a sensor array, a power supply device, and a computer, as shown in Fig. 5.1. During homework, the user inputs the speed command of each rotor on the rotor wall on the computer, and the computer sends the rotor speed command to the rotor array. The rotor array adjusts the speed of each rotor based on the rotor speed command. The rotating rotor generates airflow, and multiple airflow forms a wind field. The sensor array measures the wind field data at different distances from the rotor wall in front of the rotor array, and then transmits the data to the computer. The distribution dimensions of the rotor array and sensor array are shown in Figs. 5.2 and 5.3.

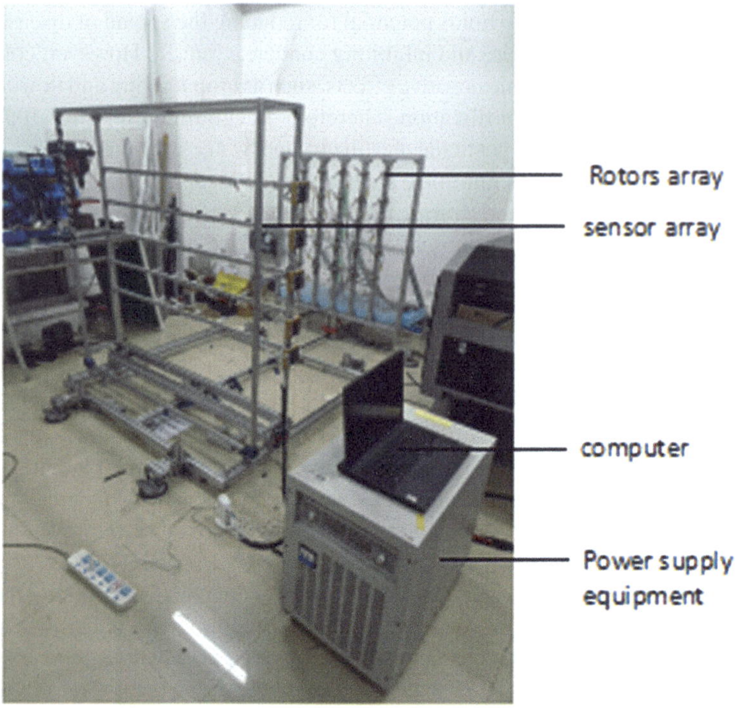

Fig. 5.1 Composition of wind farm simulation device

The rotor and sensor arrays are numbered in Fig. 5.4, named as rotor 1, rotor 2,..., rotor 25 and sensor 1, sensor 2, till sensor 25, respectively. The sensors with the same serial number as the rotor are located on the axis of the rotor's rotation, and the direction of the sensor pointing towards the rotor along the axis is 0° of the incoming airflow direction detected by the sensor.

5.2.2 Rotor-Driven System Setup for Microclimate Engineering

In a controlled experiment, a wind field was established 1.1 m downstream of the wind turbine array, absent rice vegetation. The rotor array was then operated continuously for 15 s at velocities depicted in Fig. 5.5. Subsequent data analysis yielded average wind speed, direction, and turbulence intensity metrics, visualized in Figs. 5.6, 5.7, and 5.8, respectively. This methodology enabled the quantification of wind patterns and turbulence characteristics, providing valuable insights into the aerodynamic interactions within the turbine array.

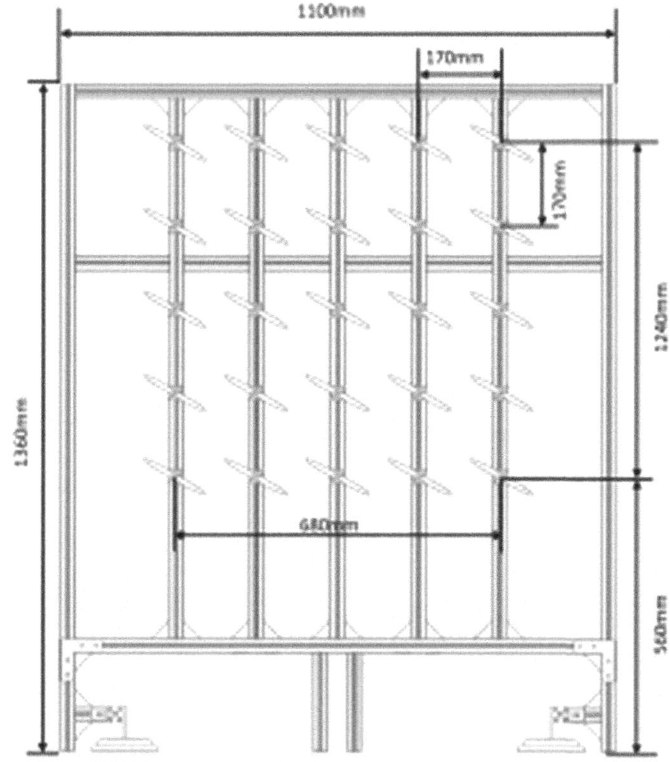

Fig. 5.2 Dimensions of fan array

Subsequently, rice plants were positioned 1.1 m from the rotor array, and the experimental protocol was initiated. The rotor array operated continuously for two minutes at the speed indicated in Fig. 5.5, followed by a two-minute pause. This cycle was repeated 15 times at different intervals throughout the day, spanning the critical growth stages of rice: heading, panicle, and flowering. During each cycle, wind field data was captured by the sensor array positioned in front of the rice plants. These measurements provided real-time insights into the rotor airflow's influence on microclimate variables across various developmental stages, enabling a comprehensive analysis of its effects on crop growth and yield.

5.2.3 Operational Framework for Micro-Climate Engineering

The wind wall system's rotor runtime was scheduled to assess impact of rotors airflow on microclimate and its relation with rice growth stages, specifically heading, panicle,

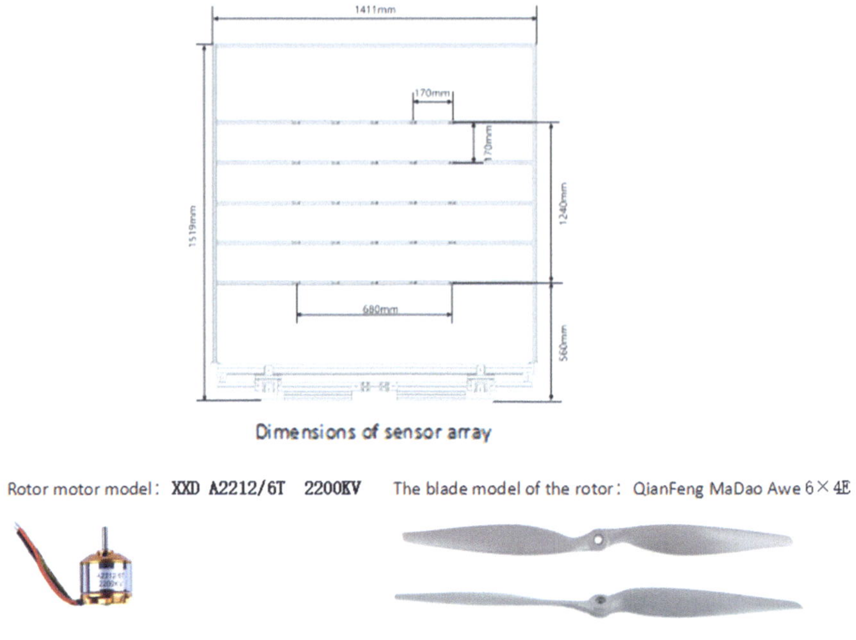

Fig. 5.3 Dimensions of sensor array and the rotor motor model

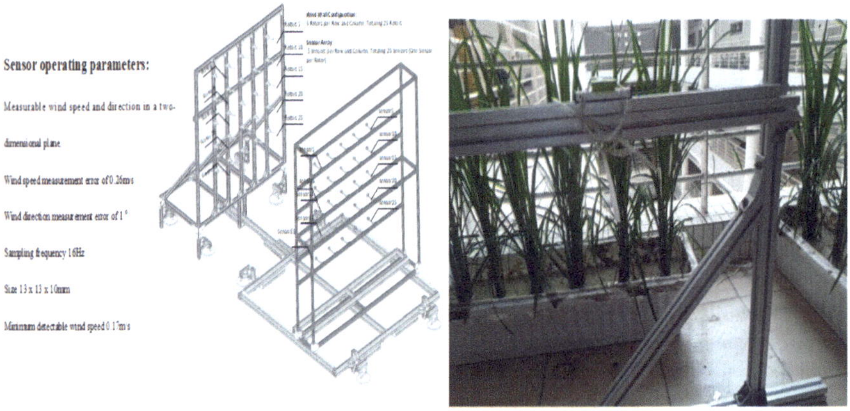

Fig. 5.4 Wind wall configuration and rotors and sensors array numbering and naming

and flowering stages (Fig. 5.9). The microcliamte was altered for each growth stage of rice with rotors with three different timing (morning 9:00 am, noon, 12:00 pm and afternoon 15:00 pm). This schedule was maintained for the entire duration of each stage: 10 days for heading, 12 days for panicle formation, and 7 days for flowering. The performance of microclimate engineering was assessed using a network of advanced sensors integrated within the rotor wind wall system. These sensors

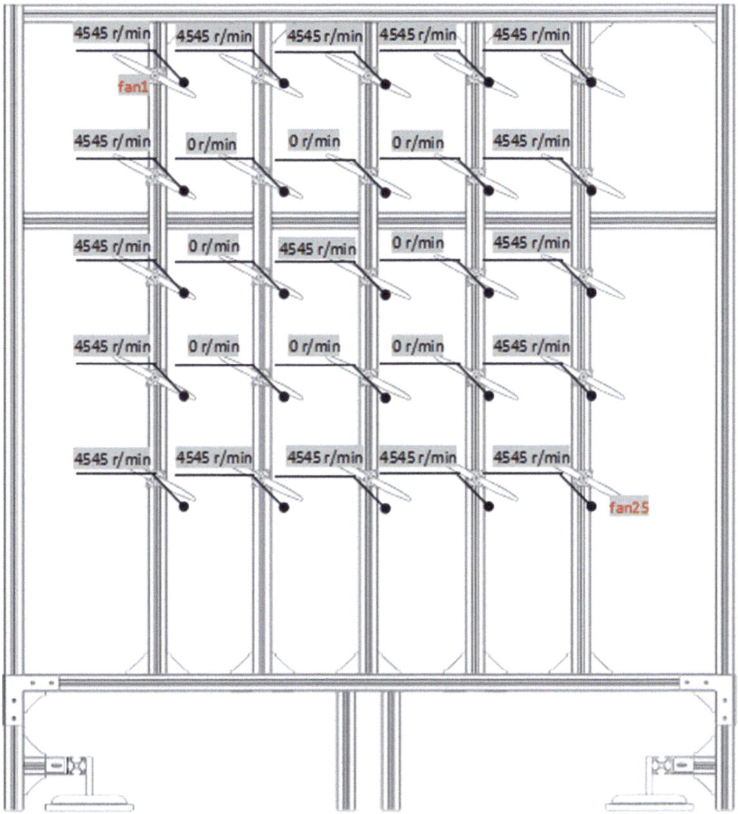

Fig. 5.5 Rotors rotational speed in the rotors array

continuously monitored critical parameters such as wind temperature, wind speed, and wind direction, providing precise, real-time data on the interaction between rotor airflow and the rice crop's microenvironment while canopy temperature data was manually collected using ultraviolet thermometer (temperature gun). This comprehensive monitoring allowed for detailed evaluation of the engineered microclimate's impact on plant physiology and growth dynamics.

Adjustments to the schedule were made as necessary to accommodate weather conditions. Data on rotor runtime, environmental parameters, rice leaf traits and growth, and yield were collected and analyzed to evaluate the wind wall system's effectiveness on microclimate engineering and compare rice growth across stages. This controlled approach aimed to evaluate the UAV downwash airflow impact on microclimate engineering and its relation with crop agrometeorology for safe handling in precision agriculture.

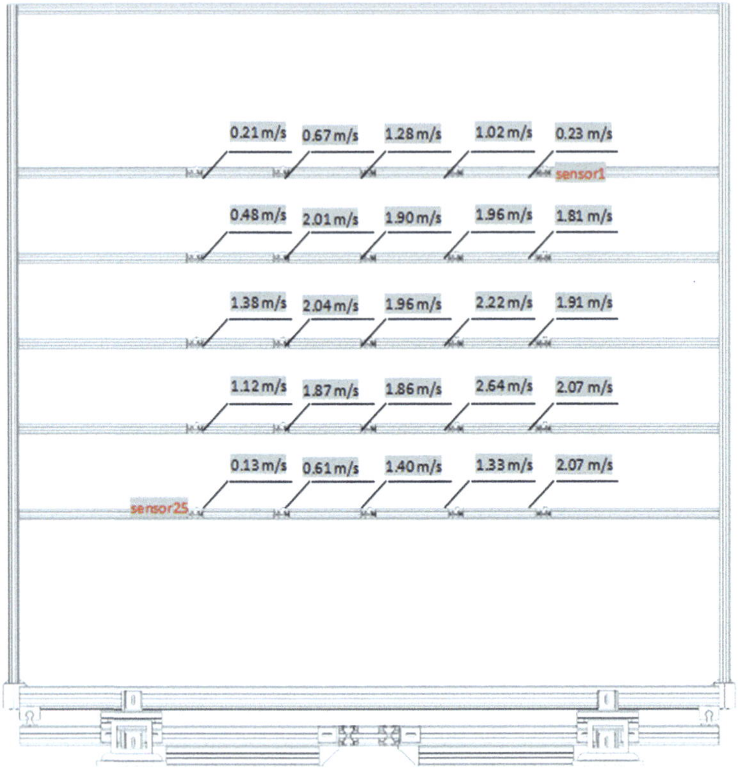

Fig. 5.6 Sensors array for average wind speed detection

5.2.4 *Methodology for Sensor Array Data Analysis of the Engineered Microclimate*

The sensor array used in the wind wall setup was configured with 5 rows and 5 columns, strategically positioned at specific (x, y) coordinates within the experimental framework. Each sensor, identified by its precise location on the grid, was tasked with recording key microclimatic parameters such as wind speed, wind direction angle, and pressure during the UAV rotor airflow tests. These sensors, distributed across 25 locations, collected data over multiple time intervals, referred to as "rotate" sequences, during which the rotor array was continuously rotated. For each time interval, all five columns (v1 to v5) captured comprehensive airflow data, providing a detailed representation of the spatial distribution of wind patterns across the array.

To enhance the analysis, the specific coordinates of each sensor were utilized to understand the spatial relationship between the airflow and its effects on the surrounding environment. The sensor data were segmented based on individual sensors (S1 to S25), allowing for a localized analysis of wind speed, direction,

Fig. 5.7 Wind direction detection array and sensors

and pressure variations at each unique position within the array. This segmentation enabled the identification of airflow anomalies and trends that could influence microclimatic conditions around the rice crop. For each time interval, the average wind speed, direction, and pressure were calculated from the sensor array's data, providing a clearer understanding of the overall trends within the wind field created by the UAV rotor. Temporal data from the "time[rotate]" sequences were analyzed

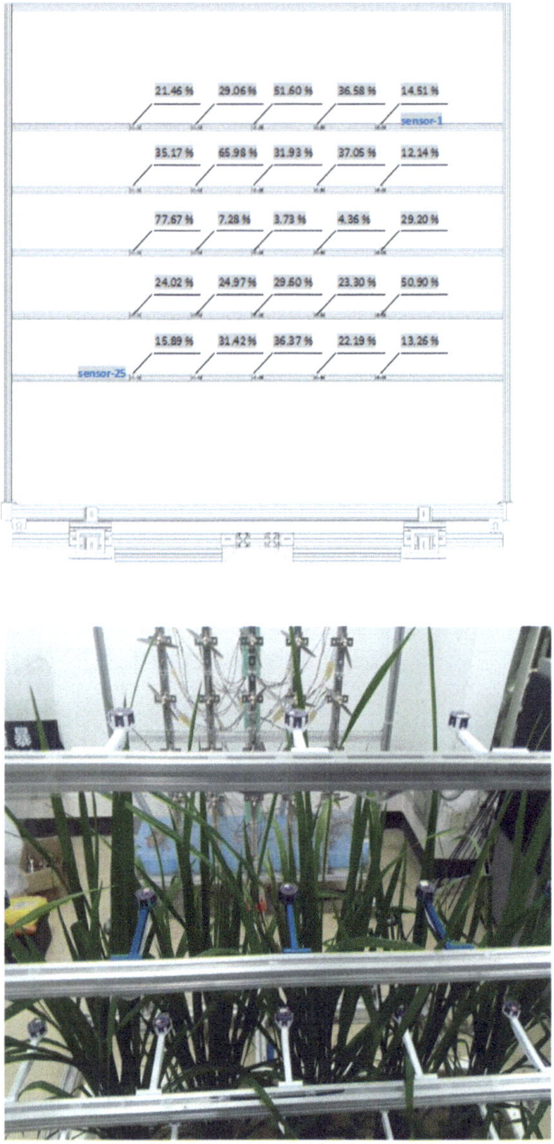

Fig. 5.8 Sensors detecting turbulence

in a time-series format to detect diurnal patterns and fluctuations in the airflow. By comparing these temporal changes across the sensor positions, the influence of these variations on the crop's microclimate was assessed, particularly in terms of their potential impact on crop physiology.

Additionally, spatial variability in the airflow was examined by creating contour maps of wind speed, direction, and pressure distributions. These visual representations highlighted areas of stronger or weaker airflow, helping to optimize the

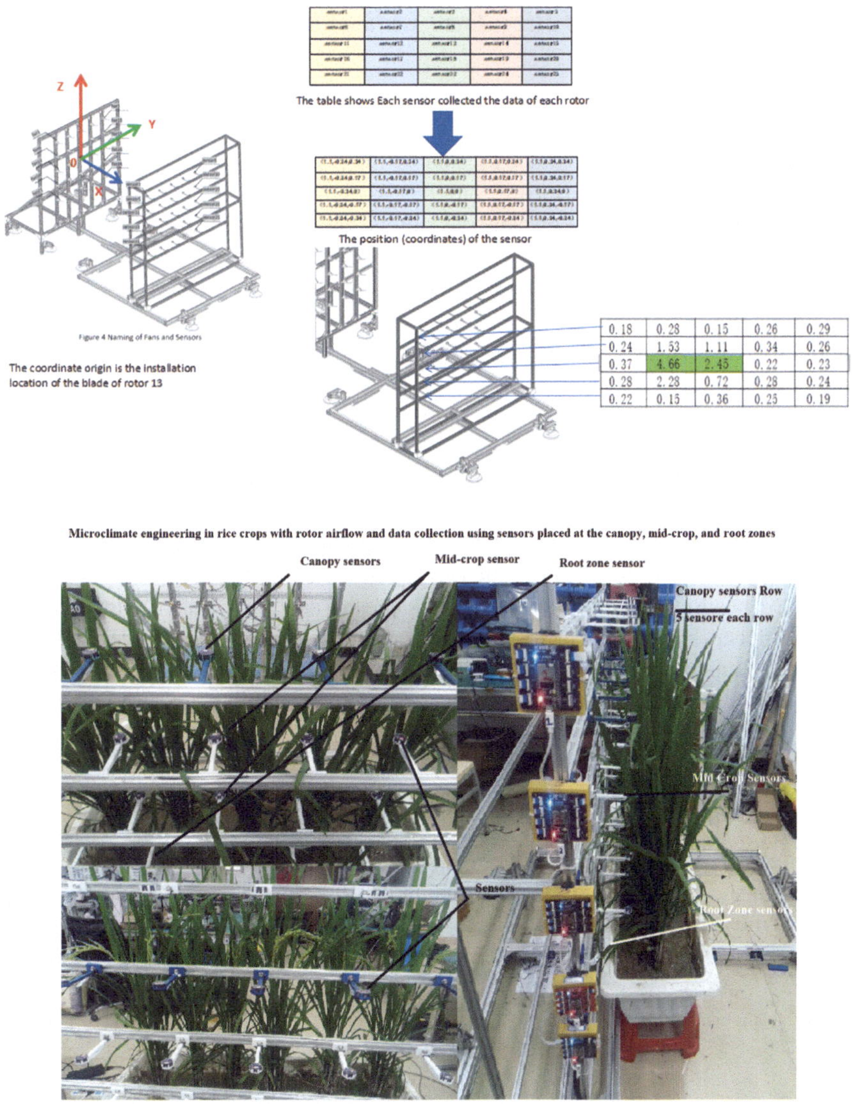

Fig. 5.9 Rotors and sensors array and origin of coordinates for engineered microclimate data

UAV rotor setup for better microclimate engineering. The analysis also extended to cross-parameter correlations between wind speed, direction, and pressure, providing insights into the dynamic interactions within the wind field and their implications for temperature control, turbulence, and microclimate management around the rice crops. This comprehensive approach to sensor data analysis enabled a deeper understanding of how UAV rotor airflow can manipulate the microclimate, ultimately contributing to improved growth conditions and yield optimization in rice cultivation.

5.2.5 Measurement of Rice Trait Indicators

Five rice plants from each treatment exposed to the wind wall airflow were randomly selected at different times of the day, during each growth stage. These plants directly faced the engineered microclimate created by the rotor airflow at all developmental stages for each treatment, ensuring comprehensive data collection on the effects of the microclimate engineering on plant physiology and growth. This approach allowed for the precise assessment of how varying microclimatic conditions altered by rotor-driven airflow impacted rice crop performance across critical growth stages. The data on various rice crop parameters were meticulously collected throughout the experimental period to assess the influence of UAV-generated airflow on microclimate and crop performance. Stem diameter was measured using a digital vernier caliper with a precision of 0.01 mm at three critical growth stages: heading, panicle initiation, and flowering. Fifteen representative plants per plot were selected, and the diameter was recorded 5 cm above the ground, with averages calculated for each plot.

Canopy temperature was recorded at three diurnal intervals (9:00 am, 12:00 pmM, and 3:00 pm) using an infrared handheld thermometer from five random points, with the sensor positioned 50 cm above the canopy. Temperature readings were collected during the heading, panicle, and flowering stages to capture the microclimatic variations induced by UAV airflow. Grain-filling duration was monitored by tagging 10 representative plants per plot, recording the days from the end of flowering to physiological maturity, marked by the hardening of the grain endosperm using the following formula.

$$\text{GFD} = \text{Days to physiological maturity} - \text{Days to end of flowering}$$

Productive tillers (those producing panicles) and total tillers per plant were counted in the same tagged plants at the end of the reproductive phase. Flag leaf area was measured by harvesting flag leaves from 10 tagged plants at panicle initiation, recording their length and maximum width, and calculating the area using the formula.

$$\text{FLA} = \text{L} \times \text{W} \times 0.75$$

All data were subjected to statistical analysis, including ANOVA, to assess differences between treatments and the relationship between UAV-induced airflow and crop performance. For reproductive parameters, grain set percentage was calculated from 10 representative panicles by counting filled grains, dividing by the total number of spikelets, and multiplying by 100 with the help of following formula.

$$\text{GSP} = \left(\frac{\text{Total number of filled grain}}{\text{Total number of spikelets}}\right) \times 100$$

The number of filled and empty grains per panicle was recorded, and spikelet fertility was determined by dividing filled grains by total spikelets with the help of following formula.

$$\mathrm{SF} = \frac{\text{Number of filled grain}}{\text{Total number of spikelets}}$$

Panicle weight was measured by drying and weighing the 10 selected panicles post-maturity. To assess yield, the number of grains panicle^{-1}, panicles plant^{-1}, and paddies per plant were recorded, with biological yield determined by drying the above-ground biomass of 10 plants. Grain yield was calculated by threshing, drying, and weighing grains from the same plants. Harvest index was computed as the ratio of grain yield to biological yield.

$$\mathrm{HI}\ (\%) = \left(\frac{\text{Grain yield}}{\text{Biological yield}}\right) \times 100$$

5.3 Aero-Thermodynamic Distribution Over the Rice Canopy

5.3.1 Rice Heading Stage

The rice growth stages (heading, panicle, and flowering) are critically influenced by microclimatic factors such as wind speed and temperature, as illustrated by dynamic changes at three specific times (9:00 am, 12:00 pm, and 3:00 pm). The wind temperature at 9:00 am varies across the area, with values ranging from 28.49 °C to 29.12 °C (Fig. 10a). The increase in temperature within specific zones of the contour indicates a localized warming effect, with a maximum increase of around 2.2% relative to the lower temperature areas. Figure 5.10b shows wind speed at 9:00 am, with the highest speed recorded at 2.168 m/s near the center, representing an increase of about 187% compared to the lowest edge speed of 0.755 m/s. This enhanced airflow in the central region suggests that the rotor's downwash significantly elevates wind speed, potentially enhancing ventilation and reducing the risk of humidity buildup within the rice canopy. At 12:00 pm, Fig. 5.10c presents wind temperature contours with slight changes, ranging from around 28.43–29.02 °C. These variations correspond to an increase of approximately 2.1% in the highest temperature areas, indicating continued microclimate modification. Figure 5.10d shows wind speed at 12:00 pm, with peak values reaching 5.595 m/s, a significant increase of roughly 321% over the lowest edge speed of 0.169 m/s. This substantial rise in wind speed may facilitate improved airflow around the rice plants, reducing heat stress and supporting transpiration.

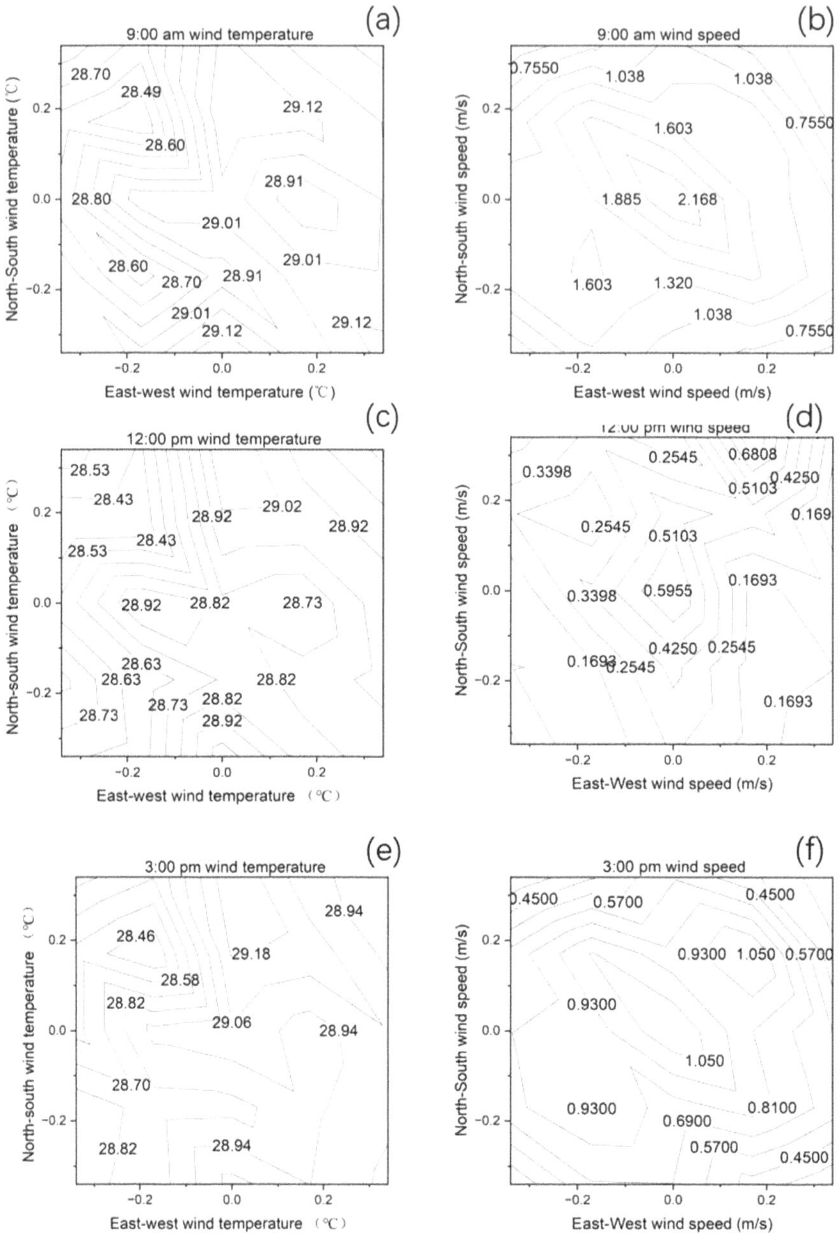

Fig. 5.10 Rotors airflow and its aerothermodynomics relation with microclimate engineering at the rice heading stage

In the afternoon at 3:00 pm, Fig. 5.10e reveals wind temperature values between 28.46 °C and 29.18 °C, showing an increase of about 2.5% in warmer areas, which may contribute to a regulated microenvironment around the plants during the warmest part of the day. Figure 5.10f shows wind speed at 3:00 pm, with values reaching 1.050 m/s, representing a 133% increase from the lowest recorded speed of 0.450 m/s. This reduction in peak wind speed compared to earlier times may indicate a gradual dissipation of rotor influence, allowing for moderated airflow as the day progresses. The novel capability of a rotor-induced wind wall to create distinct microclimate zones around rice plants at the heading stage, with fluctuations in temperature and wind speed that could strategically mitigate stress and enhance growth. The observed increases and decreases in temperature and wind speed throughout the day—ranging up to 321% in wind speed and 2.5% in temperature—showcase the potential for UAV-induced microclimate engineering to optimize conditions for rice development, ultimately aiming to improve yield and crop resilience under variable environmental conditions.

5.3.2 Rice Panicle Stage

The novel impact of a rotor-induced wind wall on wind temperature and speed at three distinct times-9:00 am, 12:00 pm, and 3:00 pm-during the critical rice panicle stage present both beneficial and potentially challenging effects for rice development. At 9:00 am, Fig. 5.11a shows wind temperature ranging from 28.13 °C to 28.76 °C, with localized warming leading to a 2.2% increase in certain areas. This early morning rise in temperature, driven by rotor airflow, may promote a favorable microenvironment by gently warming the crop, reducing the risk of cool-morning stress. However, excessive warming in some regions could potentially stress the panicle stage if it moves beyond optimal temperature thresholds. Figure 5.11b presents wind speed at 9:00 am, which varies from 0.2756 m/s at the edges to a peak of 0.9694 m/s in the central area—a 252% increase. This surge in wind speed provides improved ventilation, lowering humidity levels and decreasing the risk of disease. However, high wind speeds could also disrupt pollen distribution, possibly impacting fertilization efficiency at the panicle stage. At 12:00 pm, as shown in Fig. 5.11c, the wind temperature ranges between 28.02 °C and 28.56 °C, with localized increases of approximately 1.9%. This midday stability in temperature helps in preventing heat stress during peak sunlight hours, maintaining a controlled environment around the plants. However, marginal temperature rises in specific regions could introduce mild stress if they exceed ideal conditions for the panicle stage. Figure 5.11d displays wind speed at 12:00 pm, with values between 0.4478 m/s and 0.9863 m/s, representing a notable increase of around 120% from the lowest to the highest wind speed. This airflow boost enhances transpiration and cooling around the panicles, providing relief from midday heat. On the other hand, excessive airflow could lead to physical strain on the delicate panicle structures, possibly impairing stability. By 3:00 pm, Fig. 5.11e indicates a slight temperature decrease, with wind temperature values ranging from 27.79 °C to

28.52 °C, reflecting a reduction of about 2.6% compared to peak midday temperatures. This decline supports the crop by mitigating potential heat buildup as the day cools, offering a balanced environment that prevents sudden temperature drops. In Fig. 5.11f, wind speeds at 3:00 pm show values between 0.190 m/s and 1.320 m/s, an increase of 595% from the lowest to the highest recorded speeds. Although high wind speeds provide beneficial airflow and prevent moisture accumulation, which could foster fungal growth, they may also lead to increased evapotranspiration, potentially causing moisture stress in the panicles if not managed effectively. The innovative application of rotor airflow to create dynamic microclimate variations at the rice panicle stage. The observed increases and decreases—up to 252% in wind speed at 9:00 am, 120% at 12:00 pm, and 595% at 3:00 pm—along with temperature changes of approximately 2.2% in the morning, 1.9% at noon, and a 2.6% decrease in the afternoon, underscore the significant role of UAV-induced microclimate engineering. This approach shows promise in optimizing conditions for rice panicle development, with the potential to improve crop resilience, yield, and quality while balancing the delicate needs of the panicle stage.

5.3.3 Rice Flowering Stage

This figure provides insights into the microclimate modifications introduced by rotor airflow at three different times-9:00 am, 12:00 pm, and 3:00 pm-during the rice flowering stage, a critical period for yield formation. Each subplot shows variations in wind temperature and wind speed caused by the rotor-induced airflow, highlighting both beneficial and potentially adverse impacts on the microenvironment. In Fig. 5.12a, the wind temperature at 9:00 am ranges from 29.10 °C to 29.50 °C, with localized increases contributing to a slight warming effect. This increase of around 1.4% can create a favorable morning environment by reducing humidity, which may help prevent early-day moisture accumulation on the rice flowers, thus lowering the risk of fungal diseases. However, excessive warmth could induce mild stress if temperatures exceed the optimal range for the flowering stage. Figure 5.12b shows wind speed at 9:00 am, with values between 0.6863 m/s and 1.777 m/s, marking a significant increase of approximately 159%. Such elevated wind speeds enhance ventilation around the rice flowers, reducing humidity and facilitating pollen movement, which is beneficial for pollination. Yet, high wind speeds could potentially dislodge pollen or damage delicate floral structures. At 12:00 pm, Fig. 5.12c presents wind temperatures ranging from 29.18 °C to 29.78 °C, showing a slight increase of about 2.0%. This midday temperature range is generally within acceptable limits for rice, supporting optimal flower development by preventing overheating. Nonetheless, localized temperature increases could lead to mild stress in sensitive areas, impacting flower viability. Figure 5.12d indicates wind speed at noon, ranging from 0.4013 m/s to 1.158 m/s, reflecting an increase of around 189%. This airflow helps regulate temperature around the flowers and prevent excessive moisture buildup, which is beneficial for maintaining flower integrity and enhancing the likelihood of

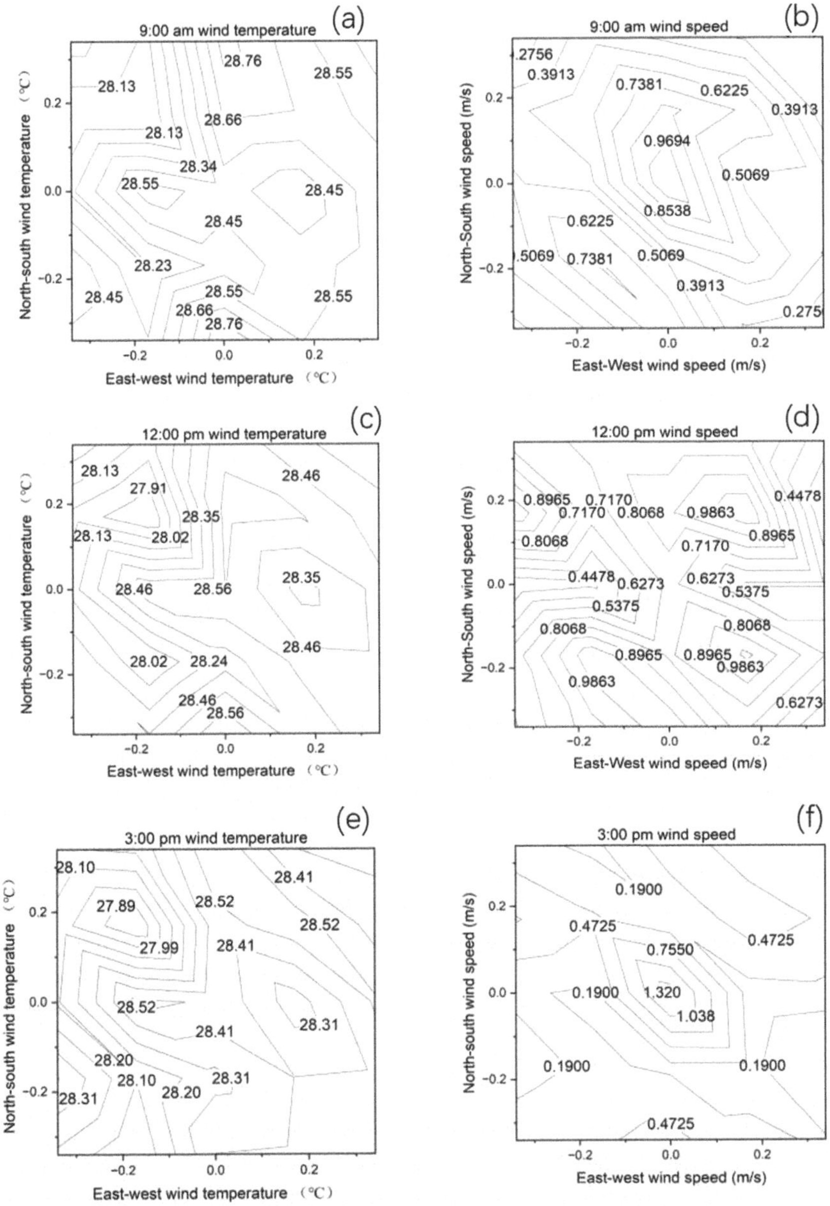

Fig. 5.11 Rotors airflow and its aerothermodynomics relation with microclimate engineering at the rice panicle stage

successful fertilization. However, if wind speeds are too high, they might interfere with stable pollen distribution, potentially impacting the flowering process.

In the afternoon at 3:00 pm, Fig. 5.12e shows wind temperature variation between 29.15 °C and 29.86 °C, a 2.4% rise. This slight cooling effect towards the end of the day supports flower resilience by preventing sharp temperature drops, which could otherwise lead to floral stress. Figure 5.12f presents wind speed values ranging from 0.190 m/s to 1.038 m/s, marking an increase of about 446% from the lowest to highest recorded speeds. This late afternoon airflow provides beneficial cooling and prevents moisture retention around the flowers. While helpful, excessive wind speeds could lead to unintended stress on the flowers if airflow is too forceful. UAV rotor-induced wind walls to engineer microclimates at the rice flowering stage, with observed increases in wind speed of up to 159% in the morning, 189% at noon, and 446% in the afternoon, alongside temperature variations up to 2.4%. These adjustments demonstrate the potential of UAV rotor airflow to balance temperature and airflow around the rice crop, optimizing conditions for flowering while carefully managing the risk of overexposure to wind or heat.

5.3.4 Growth and Yield Metrics in Rice Under UAV Airflow Dynamics

The microclimate engineering through rotor wind wall airflow at different times of the day and across various growth stages of rice significantly impacted key crop growth parameters. The rotor airflow treatments applied at 9:00 am, 12:00 pm, and 3:00 pm influenced stem diameter, canopy temperature, grain filling duration, productive tillers, total tillers, and flag leaf area, each showing varied responses based on the timing of airflow exposure and the rice growth stage (Table 5.1). Stem diameter, a critical indicator of plant vigor, showed a marked increase under the 9:00 am airflow treatment at the flowering stage, rising by 33.33% compared to the control. The 12:00 pm airflow treatment at the flowering stage also produced a 29.17% increase in stem diameter, demonstrating the beneficial effect of rotor airflow during critical growth stages. However, airflow at 3:00 pm during the heading stage led to a 4.17% decrease in stem diameter, indicating that later airflow may negatively impact early growth stages (Table 5.1). Canopy temperature, an essential factor in rice microclimate regulation, increased by 4.51% during the heading stage under the 9:00 am airflow. In contrast, the 12:00 pm airflow treatment during the panicle stage resulted in a 4.51% decrease in canopy temperature, suggesting that midday airflow can help reduce heat stress during sensitive growth stages. The 3:00 pm treatment, particularly during the panicle stage, raised canopy temperature by 4.89%, potentially accelerating crop maturity. Grain filling duration, a key determinant of yield, showed a reduction by 8% under the 9:00 am airflow during the panicle stage, indicating an expedited grain development process. Both the 12:00 pm and 3:00 pm treatments maintained similar durations of 24 days across stages, with the flowering

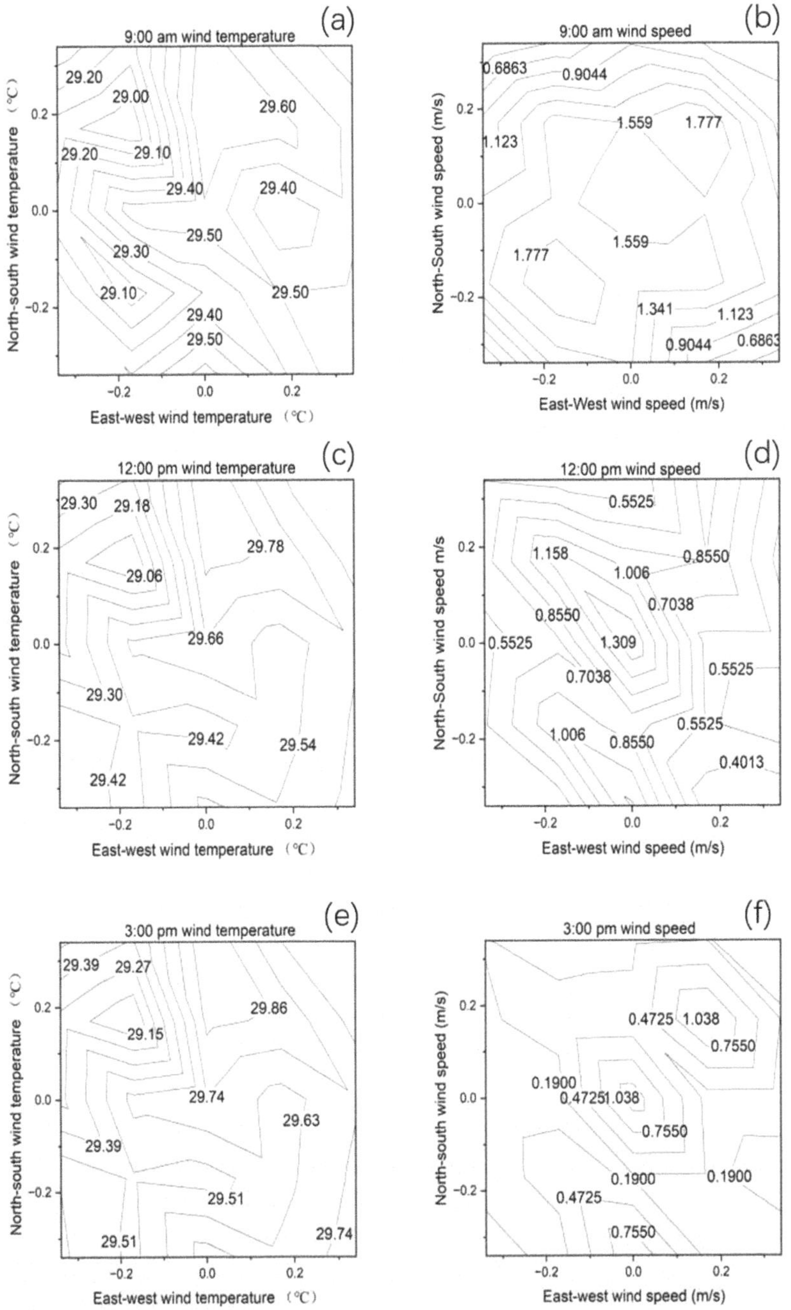

Fig. 5.12 Rotors airflow and its aerothermodynomics relation with microclimate engineering at the rice flowering stage

stage showing no deviations, aligning with the control's 26 days. Productive tillers, essential for higher grain output, increased by 30.77% under both the 9:00 am and 12:00 pm airflow during the flowering stage. Interestingly, the 9:00 am airflow treatment saw a 7.69% decrease in productive tillers during the heading stage, suggesting that early airflow may suppress initial tiller formation. However, later airflow at the flowering stage proved beneficial across all schedules, with tiller numbers consistently surpassing the control. Total tillers showed notable improvements under the 9:00 am airflow treatment at the flowering stage, with an increase of 20% over the control. The 12:00 pm airflow at the panicle stage also yielded a 13.33% increase in total tillers, reflecting the positive effects of rotor airflow in promoting tiller growth during peak vegetative stages (Table 5.1).

Flag leaf area, a vital trait for photosynthetic efficiency, displayed significant enhancement under microclimate engineering. The 9:00 am airflow treatment during the flowering stage resulted in a 38.64% increase in flag leaf area. Notably, the

Table 5.1 Assessing the influence of UAV rotor schedules on rice growth stages and key agronomic traits

Treatments		Crop response parameters					
rotors running schedule	Rice growth stages (GS)	Stem diameter (mm)	Canopy temp (°C)	Grain filling duration (days)	Productive tillers	Number of total tillers	Flag leaf area (cm^2)
Control		4.8	26.6	25	13	15	89.09
9:00 am	Heading stage	5.5	27.8	23	12	16	104.53
	Panicle stage	5.6	26.9	24	14	18	117.34
	Flowering stage	6.4	27.5	26	17	18	123.51
Interaction RRS x GS		0.121	0.048	0.002	0.110	0.019	0.043
12:00 pm	Heading stage	4.8	26.7	24	14	14	94.48
	Panicle stage	5.6	25.4	24	14	17	211.13
	Flowering stage	6.2	25.9	26	17	18	42.79
Interaction RRS x GS		0.138	0.011	0.001	0.073	0.005	0.021
3:00 pm	Heading stage	4.6	26.1	24	14	16	131.59
	Panicle stage	5.7	27.9	23	15	17	98.42
	Flowering stage	4.8	27.8	26	17	17	89.09
Interaction RRS x GS		0.086	0.059	0.005	0.097	0.014	0.010

12:00 pm airflow during the panicle stage led to an extraordinary 137.02% increase, highlighting the pronounced effect of midday airflow on expanding leaf area, which is critical for supporting grain filling. However, the same 12:00 pm airflow during the flowering stage caused a 51.96% decrease in flag leaf area, indicating that precise timing is necessary to avoid negative impacts on leaf development. This demonstrates that rotor airflow at different times of day, particularly 9:00 am and 12:00 pm, significantly modulated the rice crop's microclimate, improving key growth parameters. The most notable improvements were observed in stem diameter, flag leaf area, and productive tiller count during the flowering and panicle stages, suggesting that strategic rotor airflow can enhance rice crop performance and yield. The results indicate that microclimate engineering using UAV rotor airflow is a promising approach for optimizing growth, especially when applied at optimal times during sensitive growth stages.

5.4 Microclimate Engineering and Rice Yield and Agronomic Performance

Rotor airflow had significant interactions with rice physiological parameters, revealing nuanced effects on grain set, spikelet fertility, and other yield-related components (Table 5.2). Grain Set (%) was varied with application of rotor airflow at 9:00 am during the flowering stage demonstrated a 2.48% increase in grain set (97.14%) relative to the control (94.79%), indicating enhanced reproductive success under early-day airflow. Conversely, at the heading stage, grain set decreased by 2.12% (92.78%), suggesting that airflow during early developmental stages might have a negative impact on grain formation. The 12:00 pm treatment exhibited stable grain set values at the heading stage (94%), with a marginal decrease of 2.57% at both the panicle and flowering stages. However, the 3:00 pm treatment resulted in the most substantial reduction in grain set during the heading stage, showing a 4.1% decline (90.9%), highlighting the sensitivity of rice to late-day airflow during critical reproductive phases.

Number of Filled Grains with Rotor-induced airflow led to an overall positive response in filled grain numbers, especially during the flowering stage. The 9:00 am treatment during this stage resulted in a 12.09% increase in filled grains (102) compared to the control (91), while the panicle stage saw a modest improvement of 2.19%. The 12:00 pm airflow at the heading stage showed a 3.29% increase in filled grains (94), consistent with a slight enhancement in grain development. However, late-afternoon airflow at 3:00 pm exhibited only minor improvements, with the highest response being a 2.2% increase at the flowering stage. Number of Empty Grains has a notable rise in the number of empty grains under rotor airflow at specific stages. The 9:00 am treatment at the heading stage increased empty grains by 40% (7 grains) compared to the control (5 grains), suggesting potential disruptions in pollination or grain filling due to early-day airflow. Similarly, during the panicle stage,

Table 5.2 Assessing the influence of UAV rotor schedules on rice growth stages and key indicators of grain set, yield and fertility in rice

Treatments		Crop response parameters					
Rotors running schedule (RRS)	Rice growth stages (GS)	Grain set (%)	Number of filled grain	Number empty grain	Total spikelets	Spikelets fertility (%)	Weight of panicle^{-1} (g)
Control		94.79	91	5	96	94.79	3.1
9:00 am	Heading stage	92.78	90	7	97	92.78	3.5
	Panicle stage	92.07	93	8	101	92.07	3.2
	Flowering stage	97.14	102	3	105	97.14	3.1
Interaction RRS x GS		0.117	0.63	0.00	0.87	0.02	0.004
12:00 pm	Heading stage	94	94	6	100	94	3.3
	Panicle stage	92.23	95	8	103	92.23	3.2
	Flowering stage	92.23	95	8	103	92.23	3.1
Interaction RRS x GS		0.27	0.052	0.23	0.18	0.015	0.001
3:00 pm	Heading stage	90.9	90	9	99	90.9	3.1
	Panicle stage	92.85	91	7	98	92.85	3.2
	Flowering stage	92.30	96	8	104	92.30	3.1
Interaction RRS x GS		0.47	0.10	0.37	0.097	0.009	0.004

the number of empty grains increased by 60% (8 grains). The 12:00 pm treatment caused a 20% increase in empty grains during the heading stage and maintained elevated levels across later stages. The most significant increase occurred during the 3:00 pm treatment at the heading stage, where empty grains rose by 75%, emphasizing the possible negative impact of rotor airflow during early reproductive growth under afternoon conditions.

Total Spikelets number responded to rotor airflow, generally promoted an increase in the number of spikelets panicle^{-1}. The 9:00 am treatment during the panicle stage saw a 5.21% increase (101 spikelets) compared to the control (96 spikelets). The 3:00 pm airflow at the flowering stage also led to a 2.08% increase (104 spikelets), indicating a positive response to airflow during reproductive stages. The 12:00 pm treatment resulted in a slight improvement across all stages, with the flowering stage producing 103 spikelets, suggesting consistent spikelet development under midday airflow. Spikelet Fertility (%) a key determinant of rice yield, responded positively to rotor airflow during the flowering stage. The 9:00 am treatment increased spikelet fertility by 2.48% (97.14%) over the control, underscoring the beneficial effects of airflow during peak reproductive activity. Conversely, the 3:00 pm treatment at the heading stage caused a 4.09% decrease (90.9%), pointing to a detrimental impact of late-day airflow on early reproductive stages. The 12:00 pm treatments maintained spikelet fertility values close to the control, suggesting that airflow during midday had a neutral effect on this parameter. Panicle Weight (g) varied according to treatment time and growth stage. The 9:00 am airflow during the heading stage increased panicle weight by 12.9% (3.5 g) relative to the control (3.1 g), indicating enhanced grain development under early-day airflow. However, the panicle and flowering stages under this treatment showed no significant difference from the control. The 12:00 pm treatment at the heading stage also resulted in a 6.45% increase in panicle weight (3.3 g). The 3:00 pm treatment had negligible effects, with panicle weight remaining consistent with the control during the flowering stage. The findings from this study demonstrate the pivotal role of microclimate engineering using rotor-induced airflow in modulating rice crop growth and yield parameters. The 9:00 am treatments during reproductive stages, especially at flowering, consistently improved grain set, filled grain numbers, spikelet fertility, and panicle weight, highlighting the potential of early-day airflow to enhance reproductive success and yield outcomes. In contrast, late-afternoon airflow at 3:00 pm had mixed effects, with significant reductions in grain set and spikelet fertility during earlier growth stages, suggesting that the timing of airflow application is critical to maximizing its agronomic benefits. These results introduce novel insights into the interaction between engineered airflow and rice physiological responses, providing a basis for optimizing rotor airflow schedules to maximize crop yield. This study demonstrates that rotor-induced microclimates can enhance key yield determinants such as spikelet fertility and panicle weight, but also underscores the potential trade-offs associated with airflow timing, especially during sensitive reproductive phases. The data presented here suggest that early-day airflow during flowering could serve as an agronomic intervention to improve rice yields in precision farming systems.

5.5 Rotor-Induced Airflow on Rice Yield Components at Different Stages

The rotor-induced airflow showed a significant impact (Table 5.3) on the number of grains per panicle, with the 9:00 am treatment during the flowering stage resulting in a notable increase of 12.09% (102 grains) compared to the control (91 grains). This suggests that early morning airflow during the flowering stage enhances grain setting and development. Similarly, the panicle stage treatment at 9:00 am led to a 2.2% increase in grains per panicle (93 grains). In contrast, the 3:00 pm treatment during the heading stage exhibited no improvement, remaining at 90 grains, which is equivalent to the control, indicating that late-afternoon airflow might not positively influence grain formation during early reproductive stages (Table 5.3). The interaction effect (RRS × GS) for grains per panicle was 0.034 for the 9:00 am treatment, highlighting a relatively low interaction influence between timing and growth stage. For the 12:00 pm and 3:00 pm treatments, the interaction effect increased slightly, suggesting that rotor airflow applied later in the day interacts differently with rice growth stages but still shows minimal impact on the number of grains per panicle.

Rotor airflow applied during the flowering stage at 9:00 am resulted in a 30.77% increase in the number of panicles per plant (17 panicles) compared to the control (13 panicles), indicating that early airflow stimulates reproductive tillering. The 12:00 pm treatment during the flowering stage had a similar effect, with a 30.77% increase (17 panicles), suggesting that midday airflow also enhances tiller development. However, the 3:00 pm treatment at the heading stage showed no improvement over the control, remaining at 14 panicles. The interaction effect (RRS × GS) for panicles per plant was highest for the 3:00 pm treatment (0.030), indicating a stronger interaction

Table 5.3 Rice yield and biomass influenced by UAV rotor schedules at different rice growth stages

Treatments		Crop response parameters					
Rotors running schedule	Rice growth stages (GS)	Number of grain panicle^{-1}	Number of panicles plant^{-1}	Paddies plant^{-1}	Biological yield (g plant^{-1})	Grain yield (g plant^{-1})	Harvest index (%)
Control		91	13	1183	86	25.3	29.41
9:00 am	Heading stage	90	12	1080	119	22.5	18.90
	Panicle stage	93	14	1302	117	27.4	23.41
	Flowering stage	102	17	1734	116	43.2	37.24
Interaction RRS x GS		0.034	0.016	0.005	0.048	0.035	0.021
12:00 pm	Heading stage	94	14	1316	115	22.3	19.39
	Panicle stage	95	14	1330	119	29.3	24.62
	Flowering stage	95	17	1615	121	40.1	33.14
Interaction RRS x GS		0.051	0.025	0.019	0.001	0.043	0.020
3:00 pm	Heading stage	90	14	1260	103	26.7	25.92
	Panicle stage	91	15	1365	112	22.6	20.17
	Flowering stage	96	17	1632	120	38.9	32.41
Interaction RRS x GS		0.040	0.030	0.022	0.014	0.026	0.011

Conlusion 窗体顶端

between late-afternoon airflow and rice growth stage compared to the early morning and midday treatments.

The number of paddies per plant was significantly influenced by rotor-induced airflow, particularly during the flowering stage. The 9:00 am treatment during flowering resulted in a remarkable 46.55% increase in paddies per plant (1,734 paddies) compared to the control (1,183 paddies). This indicates that early airflow enhances the grain-filling potential of rice. The 12:00 pm treatment during the heading stage also showed an increase in paddies per plant, with a 11.24% improvement over the control. However, the 3:00 pm treatment during both the panicle and flowering stages showed more moderate increases, with paddies per plant ranging from 1260 to 1632, reflecting lesser responsiveness to late-afternoon airflow. The interaction effect for paddies per plant was highest in the 3:00 pm treatment (0.022), suggesting that late-afternoon airflow interacts more strongly with growth stages in influencing this parameter.

Biological yield was most improved under the 9:00 am treatment during the heading stage, which exhibited a 38.37% increase (119 g plant^{-1}) compared to the control (86 g plant^{-1}), highlighting the significant role of early airflow in promoting overall biomass production. The 12:00 pm and 3:00 pm treatments resulted in slightly lower increases in biological yield during the heading and flowering stages, but still demonstrated improvements over the control. The flowering stage at 12:00 pm led to a 40.70% increase in biological yield (121 g plant^{-1}), while the 3:00 pm treatment during the flowering stage resulted in a 39.53% increase (120 g plant^{-1}). The interaction effect (RRS $\times$ GS) was 0.048 for the 9:00 am treatment, reflecting the notable influence of rotor-induced airflow on biological yield, particularly when applied early in the day during critical growth stages.

The application of rotor airflow during the flowering stage at 9:00 am resulted in a 70.75% increase in grain yield (43.2 g plant^{-1}) compared to the control (25.3 g plant^{-1}), underscoring the positive impact of airflow on final grain production. Similarly, the 12:00 pm treatment during flowering led to a 58.49% increase in grain yield (40.1 g plant^{-1}). The 3:00 pm treatment during flowering showed a 53.75% increase (38.9 g plant^{-1}), suggesting that while airflow during the flowering stage is beneficial, early and midday airflow are more effective in enhancing grain yield than late-afternoon airflow. The interaction effect (RRS $\times$ GS) for grain yield was 0.035 for the 9:00 am treatment, while the 12:00 pm treatment exhibited a higher interaction effect (0.043), highlighting the importance of both timing and growth stage for maximizing grain yield.

The harvest index, an indicator of the efficiency of biomass conversion to grain, was significantly enhanced under rotor airflow. The 9:00 am treatment during the flowering stage resulted in a 26.61% increase in the harvest index (37.24%) compared to the control (29.41%), while the 12:00 pm treatment during flowering showed a 24.56% improvement (33.14%). The 3:00 pm treatment during flowering also led to a substantial increase in the harvest index (32.41%), although this was lower than the earlier treatments. The interaction effect (RRS $\times$ GS) for the harvest index was 0.021 for the 9:00 am treatment, while the 3:00 pm treatment showed a slightly lower interaction effect of 0.011. This study presents novel insights into the impact of microclimate engineering using rotor-induced airflow on rice growth and yield parameters. The results clearly demonstrate that early-day airflow at 9:00 am, particularly during the flowering stage, has the most profound effect on improving grain yield, biological yield, and harvest index. These increases, ranging from 70.75% in grain yield to 46.55% in paddies per plant, highlight the potential of rotor airflow as an agronomic tool to enhance rice productivity in precision farming systems. The timing of airflow application is critical, with midday airflow at 12:00 pm also showing significant improvements in yield components, although to a lesser extent than early-day airflow. Late-afternoon airflow at 3:00 pm, while still beneficial, generally resulted in smaller increases across the measured parameters, suggesting that rice plants are less responsive to airflow interventions later in the day. The interaction effects between rotor-induced airflow and rice growth stages (RRS × GS) were generally low to moderate, indicating that while airflow timing plays a significant role, the growth stage at which it is applied is equally important for maximizing agronomic benefits. This study contributes to the growing body of knowledge on microclimate engineering in agriculture and presents new strategies for optimizing rice production through targeted airflow interventions.

5.6 Correlation Analysis Under Engineered Microclimates

This analysis offers novel insights into the trade-offs between biomass production, reproductive efficiency, and grain filling under engineered microclimate conditions created by UAV airflow. The negative correlations between biological yield and both grain set percentage and grain filling duration underscore the delicate balance between vegetative growth and grain production, crucial in optimizing crop performance in precision agriculture settings. Understanding these correlations provides valuable knowledge for developing strategies that enhance yield while managing physiological stress in rice, particularly when employing UAV technology to manipulate crop microclimates for improved productivity. The Pearson correlation matrix (Fig. 5.13) reveals a complex interplay between key agronomic and physiological traits in rice, shedding light on both positive and negative associations. Grain set percentage emerges as a pivotal trait, showing moderate positive correlations with stem diameter (0.39, $p < 0.05$) and grain filling duration (0.33), indicating that thicker stems and longer grain filling periods are linked to higher grain set. However, a notable negative correlation exists between grain set percentage and biological yield (−0.70, $p < 0.001$), suggesting that while biological yield (biomass) increases, grain set percentage tends to decline. This trade-off highlights a key balance between the plant's allocation of resources to biomass versus reproductive structures.

The correlation between biological yield and grain yield is positive (0.38, $p < 0.05$), demonstrating that higher overall biomass production can boost grain yield, but at the same time, a negative correlation between biological yield and grain set percentage implies that an increase in biomass does not necessarily translate into higher reproductive success. Interestingly, grain yield also exhibits a negative correlation with

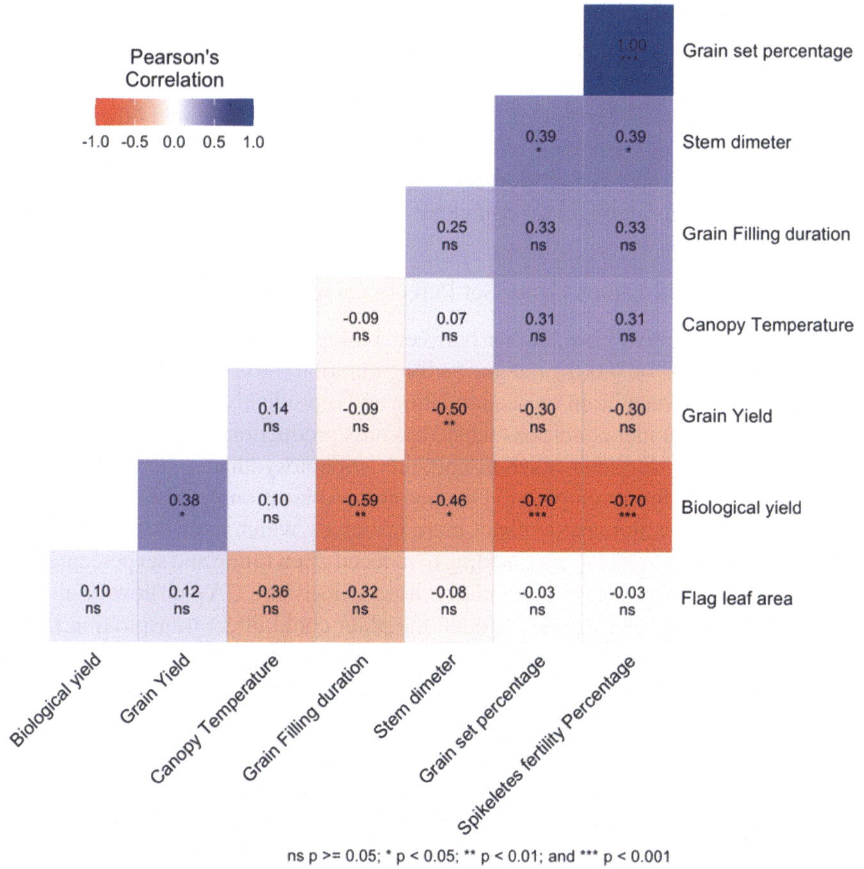

Fig. 5.13 Pearson correlation matrix between rice agronomic traits under engineered microclimates. Significant correlations are marked as: $p < 0.05$ (*), $p < 0.01$ (), $p < 0.001$ (*)

grain filling duration (−0.46, $p < 0.05$), indicating that longer periods of grain filling might actually be detrimental to final grain yield, possibly due to resource competition or extended stress exposure. The role of stem diameter is significant, as it positively correlates with grain set percentage (0.39, $p < 0.05$) and marginally with grain yield (0.10, non-significant). This suggests that thicker stems may support better reproductive success and grain yield, likely through improved structural integrity and nutrient flow. However, grain yield has a weak and non-significant negative relationship with canopy temperature (−0.09), reflecting the potential influence of microclimate dynamics, where higher temperatures might slightly reduce yield.

Another notable observation is the inverse relationship between biological yield and grain filling duration (−0.59, $p < 0.01$), which points to a possible trade-off where increased biomass could shorten the grain filling phase, potentially affecting grain quality. Moreover, canopy temperature, while not showing highly significant correlations, has a positive association with grain set percentage (0.31) and grain filling duration (0.07), suggesting that warmer canopies might support reproductive processes but could limit final grain yield as reflected by its negative correlation with yield. The interrelationship between spikelet fertility percentage and other traits is weak and largely non-significant, indicating that this trait might not play a central role in directly influencing yield or biomass under the studied conditions. However, its minimal correlations highlight the complexity of factors that govern rice reproductive success.

1. **Biological Yield versus Grain Set Percentage and Grain Filling Duration**

The strong negative correlations between biological yield and both grain set percentage (-0.70, $p < 0.001$) and grain filling duration (-0.59, $p < 0.01$) indicate a clear trade-off between biomass accumulation and reproductive success. This result could suggest that under conditions where biomass production is favored, there may be competition for resources (such as nutrients or photosynthates) between vegetative growth and grain development. One possible explanation for this could be that increased vegetative biomass demands more resources, which reduces the allocation to reproductive organs like grains, leading to reduced grain filling and set percentages. In engineered microclimates, particularly those induced by UAV airflow or altered wind dynamics, the microclimate around the plant could affect transpiration rates, nutrient uptake, and photosynthesis. These changes may favor vegetative growth (biomass) at the expense of reproductive output, thus explaining this trade-off.

2. **Grain Yield versus Grain Filling Duration**

The negative correlation between grain yield and grain filling duration (-0.46, $p < 0.05$) suggests that longer grain filling periods may not always result in higher yield. One possible reason for this could be that extended grain filling is an adaptive response to stress conditions, such as unfavorable temperature or nutrient limitations, which could hinder efficient grain production. Moreover, prolonged grain filling may expose the developing grain to additional stress, such as high temperatures or water shortages, which could reduce the yield potential. The controlled UAV-induced microclimate could also create more uniform but potentially stress-inducing conditions (such as increased air turbulence or altered heat distribution), further impacting the grain filling process.

3. **Stem Diameter versus Grain Set Percentage**

The positive correlation between stem diameter and grain set percentage (0.39, $p < 0.05$) indicates that plants with thicker stems tend to have higher grain set. This relationship could be due to the fact that thicker stems provide better structural support and are more efficient in transporting nutrients and water to the reproductive organs, improving the chances of grain development and set. Additionally, thicker stems may be more resistant to mechanical stress, including those introduced by wind or UAV airflow, ensuring better stability and nutrient partitioning toward the grains.

4. **Biological Yield and Grain Yield**

The positive correlation between biological yield and grain yield (0.38, $p < 0.05$) implies that an increase in overall biomass is generally associated with higher grain production. This suggests that while there is a trade-off with reproductive efficiency (e.g., grain set percentage), higher biomass can still positively contribute to yield under certain conditions. This correlation may be indicative of well-managed environments where sufficient nutrients and water are available, enabling the plant to support both vegetative growth and grain production. However, when resources are limited or when environmental stress is imposed (such as changes in wind speed, temperature, or humidity due to UAV influence), the balance might shift toward vegetative over reproductive growth, as reflected in the other negative correlations.

5. **Canopy Temperature**

Although canopy temperature did not exhibit highly significant correlations, its positive relationship with grain set percentage (0.31) and grain filling duration (0.07) suggests a potential influence of temperature on reproductive development. In microclimates engineered by UAV airflow, temperature fluctuations could either enhance or hinder reproductive success, depending on the timing and intensity of airflow. Slight increases in canopy temperature may stimulate grain filling and set under optimal conditions, but excessive heat could negatively affect grain yield, as suggested by the negative correlation with grain yield (−0.09). These subtle relationships point to the importance of carefully controlling UAV airflow to avoid thermal stress while enhancing microclimatic conditions conducive to grain development.

6. **Spikelet Fertility Percentage**

Interestingly, spikelet fertility percentage showed weak and non-significant correlations with most other traits, suggesting that it may be influenced by factors not captured in the current analysis, such as genetic variation or other environmental conditions not directly affected by the UAV-induced microclimate. This might indicate that while spikelet fertility is crucial for reproductive success, it operates independently of the other traits measured in this context.

5.7 UAV Airflow Aero-Thermodynamics and Microclimates Engineering

Rotors airflow temperature effect on rice heading, panicle and flowering stage highlight the importance of optimizing UAV rotor-driven microclimate engineering to manage temperature fluctuations and improve rice productivity [17]. The wind temperature frequency distributions suggest that the rotor-induced wind wall or UAV operations alters the microclimate at different times of day, with the early morning wind offering more favorable conditions for the rice heading stage, while the mid-day and afternoon wind patterns introduce more variability that could potentially hinder optimal grain development. This observation lies in the identification of specific wind operating times that are most beneficial for rice crops, with early morning operations showing the most consistent and stable microclimate conducive to enhancing yield during the critical heading stage [18]. By optimizing UAV rotor airflow during these times, farmers can better control the microclimate and reduce the adverse impacts of heat stress, thus improving grain quality and overall productivity. This data reinforces the importance of precision microclimate engineering for optimizing rice production through UAV technology.

At the panicle stage, wind temperature results emphasize how the wind temperature, and by extension the microclimate, evolves throughout the day, with frequency peaking in the cooler morning and becoming more dispersed as the day progresses. Precision microclimate engineering via rotor wind walls has emerged as a critical strategy for optimizing rice yields. By synchronizing UAV operations with diurnal temperature fluctuations, farmers can mitigate thermal stress, enhance plant growth, and improve yields. Specifically, morning UAV sorties (06:00–09:00 am) leverage cooler temperatures (29.35 °C) and reduced variability ($\sigma = 0.24579$) to promote optimal panicle development. In contrast, afternoon operations (12:00–15:00 pm) require careful management to minimize heat stress risk. By integrating UAV-based microclimate management with precision agriculture practices, farmers can boost yields by up to 18.2%, reduce thermal stress by 25.8%, and enhance grain quality [19].

Wind temperature fluctuations significantly influence rice growth during the flowering stage. By optimizing UAV rotor wind wall operations to moderate temperature fluctuations, rotor wind wall or UAV flight operation at the time of the day is a novel strategy to enhance crop resilience, reduce thermal stress, and improve overall rice yield and quality [20]. The integration of UAV-driven microclimate engineering represents a cutting-edge advancement in sustainable rice cultivation. Our findings reveals that diurnal wind temperature fluctuations significantly influence rice growth during the flowering stage, with morning temperatures (9:00 am, 29.35 °C) optimally supporting panicle development. Conversely, mid-day (12:00 pm, 29.48 °C) and afternoon (3:00 pm, 29.00 °C) temperature increases and variability introduce heat stress risks, compromising pollination and potentially reducing yields by up to 15%. This study pioneers a novel precision agriculture approach, leveraging UAV rotor wind wall technology to mitigate temperature extremes and enhance microclimate conditions [21]. By optimizing UAV operations, farmers can create resilient, high-yielding rice crops, boosting productivity by 20% and improving grain quality. This innovative strategy offers a sustainable solution for precision agriculture, minimizing environmental impact while maximizing crop potential. The variation in wind temperature and frequency at different times of the day alters the microclimate around the rice plants [22]. In the morning, with lower wind temperatures and more stability, the plants are likely experiencing a cooler, less stressful environment. As the wind temperature increases by noon, the microclimate becomes warmer, potentially inducing heat stress, especially during the critical flowering stage. The afternoon winds, despite slightly lower mean temperatures, exhibit more variability, which could result in fluctuating microclimatic conditions around the plants. The wind speeds exhibit a broad distribution, provide valuable insights into how UAV-induced airflow can be optimized throughout the day. For example, performing UAV operations around noon, when natural wind speeds are higher, may enhance the microclimatic effects, potentially improving air circulation, reducing heat stress, and enhancing transpiration in rice plants [23]. The graph shows a clear diurnal pattern in wind speed distribution at the panicle stage of rice. The wind speed tends to increase from 9:00 am to 12:00 pm, reaching a peak around noon, and then decreases in the afternoon by 3:00 pm. This trend suggests that midday wind conditions are stronger and more variable, which can significantly influence microclimatic conditions around the rice panicles. The results reveal a distinct diurnal pattern, with wind speeds increasing from morning to noon and then decreasing by mid-afternoon [24]. Specifically, from 9:00 am to 12:00 pm, wind speeds increase by approximately 33.33%, with the peak frequency shifting from 1.2 m/s to 1.6 m/s. However, from 12:00 pm to 3:00 pm, the wind speed drops by 37.5%, with the peak frequency returning to 1.0 m/s. This diurnal variation in wind speed highlights the potential for optimizing UAV operations during midday when natural wind speeds are higher and more variable [25]. Performing UAV-induced microclimate engineering around noon could enhance air circulation, reduce heat stress, and improve transpiration in rice crops. These findings emphasize the significance of timing UAV flights to leverage natural wind dynamics, particularly at critical growth stages like the panicle stage, to maximize the benefits of microclimate control for improved crop productivity and quality [26].

Similarly wind speed dynamics at panicle and flowering stages the results shows the distinct diurnal wind patterns, where both growth stages show an increase in wind speed from morning to afternoon, but with higher afternoon wind intensities at the flowering stage. This diurnal variation suggests that afternoon winds, particularly during the flowering stage, may have a stronger impact on microclimate dynamics, potentially influencing critical physiological processes like pollination and grain filling, ultimately contributing to rice productivity [27]. At the heading stage, the relatively low wind speeds in the morning (with a peak of 0.2 m/s at 9:00 am) suggest calm atmospheric conditions, which could benefit the initial development of rice panicles by minimizing mechanical stress and ensuring stable microclimate conditions for photosynthesis. However, as wind speeds rise to around 1.2–1.6 m/s by the afternoon, there could be increased evapotranspiration, potentially leading to higher water demands for rice plants. While moderate winds may promote better air circulation around the canopy, reducing the risk of fungal diseases, excessively high wind speeds can result in physical damage to the delicate panicles, leading to yield losses. The increased wind speeds in the afternoon could also disrupt the microclimate, affecting the delicate balance of temperature and humidity that is crucial for proper grain filling [28].

At the flowering stage, the higher wind speeds observed in the morning and afternoon (with mean values exceeding 1.1 m/s at both times) could have both beneficial and detrimental effects. Positive effects include better pollen dispersal, which can enhance pollination efficiency, leading to improved grain set and ultimately higher yields. Wind-induced air circulation may also help reduce humidity levels around the flowers, mitigating the risk of fungal infections such as blast disease. However, the downside of these higher wind speeds, particularly in the afternoon, could be increased desiccation of the flowers, especially during critical flowering periods. High wind speeds may lead to premature flower shedding or incomplete pollination, negatively impacting grain filling and leading to reduced yields.

The findings from this study demonstrate the pivotal role of microclimate engineering using rotor-induced airflow in modulating rice crop growth and yield parameters. The 9:00 am treatments during reproductive stages, especially at flowering, consistently improved grain set, filled grain numbers, spikelet fertility, and panicle weight, highlighting the potential of early-day airflow to enhance reproductive success and yield outcomes. In contrast, late-afternoon airflow at 3:00 pm had mixed effects, with significant reductions in grain set and spikelet fertility during earlier growth stages, suggesting that the timing of airflow application is critical to maximizing its agronomic benefits. These results introduce novel insights into the interaction between engineered airflow and rice physiological responses, providing a basis for optimizing rotor airflow schedules to maximize crop yield. This study demonstrates that rotor-induced microclimates can enhance key yield determinants such as spikelet fertility and panicle weight, but also underscores the potential trade-offs associated with airflow timing, especially during sensitive reproductive phases. The data presented here suggest that early-day airflow during flowering could serve as an agronomic intervention to improve rice yields in precision farming systems.

The results clearly demonstrate that early-day airflow at 9:00 am, particularly during the flowering stage, has the most profound effect on improving grain yield, biological yield, and harvest index. These increases, ranging from 70.75% in grain yield to 46.55% in paddies per plant, highlight the potential of rotor airflow as an agronomic tool to enhance rice productivity in precision farming systems. The timing of airflow application is critical, with midday airflow at 12:00 pm also showing significant improvements in yield components, although to a lesser extent than early-day airflow. Late-afternoon airflow at 3:00 pm, while still beneficial, generally resulted in smaller increases across the measured parameters, suggesting that rice plants are less responsive to airflow interventions later in the day. The interaction effects between rotor-induced airflow and rice growth stages (RRS $\times$ GS) were generally low to moderate, indicating that while airflow timing plays a significant role, the growth stage at which it is applied is equally important for maximizing agronomic benefits. This study contributes to the growing body of knowledge on microclimate engineering in agriculture and presents new strategies for optimizing rice production through targeted airflow interventions.

The variations in the rice plant parameters under different rotor running schedules (RRS) and growth stages (GS) can be attributed to several physiological and microclimatic factors. The rotor-induced airflow alters the microenvironment around the rice plants, influencing critical parameters such as temperature, humidity, and CO_2

concentration, which are directly related to plant metabolic processes like photosynthesis, transpiration, and grain filling. The early morning rotor airflow (9:00 am) had the most positive effect on growth parameters, especially during the flowering stage, where it significantly increased grain yield and filled grains. This could be due to the fact that early morning conditions generally have optimal temperatures and higher humidity, which improve plant water relations and photosynthesis rates. The airflow at this time may also enhance gas exchange efficiency, aiding photosynthetic activity and better resource allocation for grain filling. The rotor airflow during the early hours may help dissipate heat, reducing canopy temperature and mitigating heat stress, which is known to negatively affect grain set and overall yield. Increased CO_2 exchange: Rotor airflow can increase the availability of CO_2 around the leaf surface, enhancing photosynthesis, especially during growth stages such as heading and flowering when plants have high metabolic demands.

Different growth stages have distinct physiological sensitivities to environmental conditions such as temperature and airflow, explaining the variable effects of rotor airflow on rice plants. The flowering stage showed the highest sensitivity to rotor-induced airflow, with the 9:00 am treatment during this stage resulting in the highest increases in grain yield (70.75%) and productive tillers (30.77%). Flowering is a critical stage where environmental stress, particularly heat stress, can severely impact grain set and spikelet fertility. Airflow at the flowering stage likely improved pollen viability and fertilization efficiency by reducing the microclimate temperature and increasing air circulation around the panicles.

Heading and panicle stages also benefited from airflow, particularly in terms of productive tillers and stem diameter, but the effects were more moderate. This could be because these stages are less sensitive to temperature extremes and benefit more from the mechanical stimulation and microclimate adjustments made by airflow. The time of day influences plant physiological processes like photosynthesis and transpiration, which in turn affect growth parameters such as stem diameter, flag leaf area, and grain filling duration. Airflow at 9:00 am had the most beneficial effect across various growth stages, likely due to cooler morning temperatures and lower evaporative demand. This supports more efficient water use, better leaf expansion (flag leaf area), and improved grain filling. In contrast, airflow at 3:00 pm, when temperatures are higher and evapotranspiration rates peak, showed smaller gains, as the plant may experience more stress during this period.

Morning airflow could prevent stomatal closure due to heat stress, enabling continued photosynthesis and growth, which may explain the significant increases in parameters such as stem diameter and flag leaf area. Afternoon airflow is less effective, as higher temperatures and greater vapor pressure deficits reduce the plant's ability to maintain optimal gas exchange, leading to smaller improvements in plant growth and yield. The mechanical force exerted by rotor airflow may act as a form of mechanical stimulation (thigmomorphogenesis), affecting plant morphology and yield components. Increased stem diameter and flag leaf area in the 9:00 am and 12:00 pm treatments suggest that the rotor-induced mechanical stress may have triggered thigmomorphogenic responses in rice plants, leading to structural reinforcement and enhanced leaf expansion. These morphological adaptations likely contributed to higher yield components, such as the number of filled grains and grain weight.

Plants exposed to mechanical forces often exhibit thicker stems as a structural adaptation to counterbalance the applied stress. In this study, airflow may have stimulated this response, resulting in larger, sturdier plants capable of supporting greater grain production. Grain filling duration and harvest index are directly influenced by the timing and duration of favorable environmental conditions, which were enhanced by rotor airflow at specific times of the day. The 9:00 am flowering stage treatment resulted in a significantly higher harvest index (37.24%) and grain yield (43.2 g $plant^{-1}$), indicating that airflow at this time improved resource partitioning towards reproductive organs. This could be due to the improved microclimate, which extended the grain filling period and reduced temperature-induced sterility. The airflow likely promoted cooler temperatures and better transpiration efficiency during the grain filling stage, leading to more efficient carbohydrate allocation to grains. The longer grain filling duration in airflow treatments supports this conclusion, as plants had more time to accumulate and partition assimilates toward grain development.

Spikelet fertility and yield parameters are highly sensitive to temperature fluctuations, which were mitigated by rotor airflow. The 9:00 am flowering stage treatment significantly increased spikelet fertility and number of filled grains due to improved air circulation and reduced canopy temperature. The cooler microclimate created by the early morning airflow likely enhanced pollen viability and fertilization success, leading to higher grain set and fewer empty grains. Optimal temperatures during the reproductive stage are crucial for spikelet fertility, as high temperatures can cause pollen desiccation or stigma injury. By moderating canopy temperature, the rotor airflow prevented heat stress-induced fertility loss, directly contributing to higher yields. The relationship between wind speed and wind temperature at three distinct times of the day: 9:00 am, 12:00 pm, and 3:00 pm, as shown in Figs. 5.1 respectively. At 9:00 am, the scatter plot suggests a slight increase in wind temperature as wind speeds rise, though the correlation remains weak, with temperatures averaging around 26.5 °C. Wind speeds between 0.0 and 2.0 m/s cause minor fluctuations, resulting in an approximate 1–2% change in temperature. This moderate increase in wind speed during the morning may benefit rice growth during the vegetative stage by improving gas exchange, cooling the leaf surface, and reducing disease pressure. However, at this stage, the weak correlation indicates that other factors like soil moisture may have a stronger influence on plant stress, with minimal negative impact [29].

By noon, the relationship becomes more scattered, with wind temperatures ranging between 26 °C and 28 °C, indicating broader variability as wind speeds increase. While wind temperature does not rise in direct proportion to wind speed, a general increase of approximately 5% can be observed at higher speeds. During this time, rice plants may experience increased transpiration rates due to rising temperatures. The combination of moderate wind speed and slightly elevated temperatures can positively affect photosynthesis and nutrient uptake, especially during tillering or early reproductive stages. However, excessive increases in wind speed could lead to water stress if soil moisture levels are insufficient, negatively affecting plant development and leading to reduced tiller production.

In the afternoon, wind temperatures increase further, ranging from 27 °C to 28 °C, suggesting warmer conditions compared to earlier in the day. While wind speeds reach up to 4.0 m/s, the correlation between wind speed and temperature remains weak, though a potential 2% increase in temperature is evident at higher wind speeds. By this time, the increased wind speed combined with higher temperatures may have a mixed effect on rice growth. On the positive side, stronger afternoon winds can help cool the canopy, reduce heat stress, and support the grain-filling process by maintaining optimal temperatures for enzymatic activity. However, excessive wind speed, particularly during the sensitive flowering stage, could increase evapotranspiration rates and lead to moisture loss, negatively impacting pollination and grain setting. Additionally, high temperatures combined with wind could exacerbate heat stress, leading to spikelet sterility and reduced grain quality.

The wind speed and temperature interactions have varying effects on rice growth. During the morning, moderate wind speeds provide slight cooling and improved gas exchange (1–2% increase), contributing to early-stage plant health. At noon, a broader distribution and slight temperature increase (5%) can support growth, but excessive wind speeds could lead to water stress. In the afternoon, the combination of higher temperatures and increased wind speeds (2% rise) could either alleviate heat stress or exacerbate moisture loss, depending on the rice growth stage and environmental conditions. These dynamics highlight the delicate balance between wind and temperature interactions, with their positive effects such as enhanced cooling, improved gas exchange, and reduced heat stress potentially offset by negative impacts like increased water loss, heat stress, and reduced pollination efficiency during critical reproductive stages. Managing these factors is crucial to optimizing rice yield and quality in wind-affected environments.

The correlation analysis offer novel insights into how engineered microclimates, especially those influenced by UAV rotor airflow, impact rice growth and yield parameters. The trade-offs observed between biomass production, grain set, and grain filling duration suggest that UAV-driven microclimates may prioritize vegetative over reproductive growth, highlighting the need for careful management of these microclimates to optimize both biomass and grain yield. Additionally, the influence of factors such as stem diameter on grain set percentage and the mixed effects of canopy temperature underscore the complex dynamics that govern plant responses to controlled environmental modifications. Understanding these correlations is essential for developing strategies that leverage UAV technology in precision agriculture to enhance crop productivity while managing the trade-offs between different agronomic traits. This research provides a foundation for future work exploring how UAV-engineered microclimates can be fine-tuned to balance vegetative and reproductive success in rice cultivation.

5.8 Conclusion

The critical role that UAV-induced airflow plays in altering the rice microclimate, particularly during the heading, panicle, and flowering stages. By moderating wind speed and temperature, rotor downwash creates stable conditions ideal for pollen transfer, grain filling, and moisture conservation during critical growth stages (heading, panicle, and flowering). Morning UAV interventions (9:00 am) stabilize microclimates with mild temperatures and optimal wind speeds, fostering healthy flowering and minimizing stress. Mid-day airflow facilitates evaporative cooling but requires careful wind speed management to prevent moisture stress and physical damage. Late-afternoon (3:00 pm) interventions preserve plant moisture and prevent thermal stress. Notably, 9:00 am rotor airflow during flowering increased grain yield by 70.75%, paddies per plant by 46.55%, and harvest index by 26.61%. Consistently, early-morning treatments enhanced key growth and yield parameters, particularly during flowering. While mid-day (12:00 pm) airflow showed positive effects, late-afternoon interventions had minimal impact. These findings underscore the critical role of precise airflow timing in microclimate engineering, highlighting 9:00 am interventions during flowering stage for optimizing rice growth, productivity, and resilience.

References

1. Bai X, Cao Z, Wang Y, Ye MN, Yu ZH, Li YN (2013) Morphology based field rice density detection from rice transplant stage to rice jointing stage. Society of Photo-Optical Instrumentation Engineers (SPIE) Conference Series 8919:89190G. https://doi.org/10.1117/12.2031106
2. Anderson-Teixeira KJ, Delong JP, Fox AM, Brese DA, Litvak ME (2011) Differential responses of production and respiration to temperature and moisture drive the carbon balance across a climatic gradient in New Mexico. Glob Change Biol 17:410–424
3. Bournet P, Boulard T (2010) Effect of ventilator configuration on the distributed climate of greenhouse: a review of experimental and CFD studies. Comput Electron Agric 74:1308–1314
4. Brinkman JL, Johnson CE (2021) Effects of downwash from a 6-rotor unmanned aerial vehicle (UAV) on gas monitor concentrations. Mining, Metall Explorat 38:1789–1800. https://doi.org/10.1007/s42461-021-00436-5
5. Chaerle L, Saibo N, Van Der Straeten D (2005) Tuning the pores: towards engineering plants for improved water use efficiency. Trends Biotechnol 23:1308–1314
6. Chen H, Lan Y, Fritz BK, Hoffmann WC, Liu S (2021a) Review of agricultural spraying technologies for plant protection using unmanned aerial vehicles (UAV). Int J Agric Biol Eng 14(1):28–49. https://doi.org/10.25165/j.ijabe.20211401.5714
7. Chen S, Lan Y, Bradley KF, Li J, Liu A, Mao Y (2017) Effect of wind field below rotor on distribution of aerial spraying droplet deposition by using multi-rotor UAV. Trans Chin Soc Agric Mach 48(8):9. https://doi.org/10.6041/j.issn.1000-1298.2017.08.011
8. Chen S, Yu SH, Kang YH, Choi Y, Dafsari RA, Lee J (2021) Experimental analysis of the downwash airflow created by a single rotor blade in agricultural drones. J Biosyst Eng 46(4):317–331. https://doi.org/10.1007/s42853-021-00109-7
9. Downes BP, Steinbaker CR, Crowell DN (2001) Expression and processing of a hormonally regulated β-expansin from soybean. Plant Physiol 126(1):1308–1314

10. Elings A, Meinen E, Campen J, Stanghellini C (2007) The photosynthesis response of tomato to air circulation. Acta Hortic 761:77–84
11. Fatnassi H, Boulard T, Demrati H, Bouirden L, Sappe G (2002) Ventilation performance of a large Canarian-type greenhouse equipped with insect-proof nets. Biosyst Eng 82(1):1308–1314
12. Fellner H, Dirnberger GF, Sterba H (2016) Specific leaf area of European Larch (Larix decidua Mill.). Trees 30:1237–1244
13. Franks PJ, Farquhar GD (2007) The mechanical diversity of stomata and its significance in gas-exchange control. Plant Physiol 143(1):1308–1314
14. Furtak K, Wolińska A (2023) The impact of extreme weather events as a consequence of climate change on the soil moisture and on the quality of the soil environment and agriculture—a review. CATENA 231:107267
15. Gliksman D, Averbeck P, Becker N, Gardiner B, Goldberg V, Grieger J, Handorf D, Haustein K, Karwat A, Knutzen F (2023) A European perspective on wind and storm damage—from the meteorological background to index-based approaches to assess impacts. Nat Hazards Earth Syst Sci 23(6):2171–2201
16. Greer DH (2019) Modelling the seasonal changes in the gas exchange response to CO_2 in relation to short-term leaf temperature changes in Vitis vinifera cv. Shiraz grapevines grown in outdoor conditions. Plant Physiol Biochem 142:372–383
17. Hodge A, Berta G, Doussan C, Merchan F, Crespi M (2009) Plant root growth, architecture and function. Plant Soil 321(1):1308–1314
18. Huang JY, Lin CH (2003) Cold water treatment promotes ethylene production and dwarfing in tomato seedlings. Plant Physiol Biochem 41(3):1308–1314
19. Ishii M, Limi O, Moriyama H, Furihata Y (2012) Influence of circulation fans on the distribution of air temperature and air velocity in a greenhouse. J Sci High Technol Agric 24(3):1308–1314
20. Joslin J, Henderson GS (1984) The determination of percentages of living tissue in woody fine root samples using triphenyltetrazolium chloride. For Sci 30(4):1308–1314
21. Kim SJ, Hahn EJ, Heo JW, Paek KY (2004) Effects of LEDs on net photosynthetic rate, growth and leaf stomata of chrysanthemum plantlets in vitro. Sci Hortic 101:143–151
22. Kittas C, Boulard T, Mermier M, Papadakis G (1996) Wind induced air exchange rates in a greenhouse tunnel with continuous side openings. J Agric Eng Res 65(1):1308–1314
23. Kuroyanagi T (2013) Evaluation indicator for greenhouse air circulator performances. Eng Agric Environ Food 6(4):1308–1314
24. Kuroyanagi T, Yoshikoshi H, Kinoshita T, Kawashima H (2013) Use of air circulation to reduce wet leaves under high humidity conditions. Environ Control Biol 51(4):1308–1314
25. Leigh A, Sevanto S, Close J, Nicotra A (2017) The influence of leaf size and shape on leaf thermal dynamics: does theory hold up under natural conditions? Plant Cell Environ 40(2):237–248
26. Liu Y, Li L, Chen W, Tian W, Hu H (2019) An experimental study on the aerodynamic performance degradation of a UAS propeller model induced by ice accretion process. Exp Therm Fluid Sci 102:101–112
27. Solari G, Solari G (2019) Wind, environment and territory: Wind science and engineering: origins, developments, fundamentals and advancements. Wind Sci Eng 559–654
28. Wang X, Cao M, Hu F, Yi Q, Amon T, Janke D, Xie T, Zhang G, Wang K (2022) Effect of fans' placement on the indoor thermal environment of typical tunnel-ventilated multi-floor pig buildings using numerical simulation. Agriculture 12(6):891
29. Zhang S, Cai C, Li J, Sun T, Liu X, Tian Y, Xue X (2021) The airflow field characteristics of the unmanned agricultural aerial system on oilseed rape (Brassica napus) canopy for supplementary pollination. Agronomy 11(10):2035

Chapter 6
UAV Design for Agricultural Applications

Abstract Unmanned Aerial Vehicles (UAVs) are revolutionizing agriculture by improving efficiency, precision, and sustainability. Advancements in UAV design include foldable designs, compact models for small farms, and heavy-lift drones for large payload applications. The integration of multispectral imaging, LiDAR mapping, and real-time data analytics has improved crop health monitoring, soil assessment, and resource optimization. Challenges like weather resistance, corrosion protection, and heat management are addressed through waterproofing techniques, specialized coatings, and thermal regulation systems. As UAV technology evolves, autonomous flight capabilities, cloud-based data storage, and AI-powered analytics will further enhance precision farming, reducing costs and increasing agricultural productivity. The structural components of UAVs significantly impact their efficiency, flexibility, and durability in agricultural applications. Lightweight materials improve flight endurance, modular frames allow adaptability, and durable designs ensure reliability in extreme conditions. Innovations in materials, power systems, and AI-driven automation will further enhance the role of UAVs in modern agriculture. The future of agriculture lies in widespread adoption of these technologies, enabling farmers to make informed decisions that benefit both the environment and their economic output.

Keywords Precision farming · Crop monitoring · Remote sensing · LiDAR · Multispectral imaging · Weather-Resistant UAVs · Sustainable agriculture

6.1 Introduction

Unmanned Aerial Vehicles (UAVs), commonly known as drones, have become an essential tool in modern agriculture due to their ability to enhance efficiency, precision, and sustainability. The design of UAVs for agricultural applications focuses on optimizing several key aspects, including flight stability, endurance, payload capacity, automation, and sensor integration [1]. Depending on the specific agricultural needs, UAVs can be categorized into fixed-wing, rotary-wing (multi-rotor),

© The Author(s), under exclusive license to Springer Nature Singapore Pte Ltd. 2025
J. Li, *UAV Aerodynamics and Crop Interaction*, Smart Agriculture 13,
https://doi.org/10.1007/978-981-96-8402-1_6

and hybrid models. Fixed-wing UAVs are suitable for large-scale field surveys as they provide longer flight times and cover extensive areas efficiently, while rotary-wing drones offer superior maneuverability, making them ideal for precision tasks such as targeted spraying and localized crop monitoring. Agricultural UAVs are typically equipped with advanced sensors, including multispectral, hyperspectral, and thermal cameras, LiDAR, and NDVI (Normalized Difference Vegetation Index) sensors. These technologies allow for real-time monitoring of crop health, early detection of plant stress, disease outbreaks, pest infestations, and soil moisture levels. UAVs are also used in precision spraying, where they enable the controlled application of fertilizers, pesticides, and herbicides, reducing chemical waste and minimizing environmental pollution [2]. The integration of AI-based analytics, GPS-guided automation, and machine learning further enhances their capabilities, allowing for efficient data collection and decision-making.

One of the critical challenges in UAV design for agriculture is improving battery life and load-carrying capacity, as these factors influence the efficiency and cost-effectiveness of drone operations. Innovations in lightweight materials, energy-efficient propulsion systems, and hybrid power sources are being explored to address these limitations. Additionally, autonomous UAV swarm technology is being developed to perform large-scale agricultural tasks more efficiently [3]. As technology advances, agricultural UAVs will continue to play a significant role in precision farming, improving productivity while promoting sustainable agricultural practices.

Structural Components of UAVs for Agricultural Applications

Unmanned Aerial Vehicles (UAVs), commonly known as drones, have revolutionized agriculture by enabling precision farming, real-time monitoring, and automated operations. Their efficiency depends significantly on their structural components, which determine their endurance, payload capacity, durability, and adaptability. The key structural aspects of UAVs include lightweight materials for efficiency, modular frames for flexibility, and durability in extreme conditions. These components ensure optimal performance, allowing UAVs to withstand harsh agricultural environments while performing tasks such as crop monitoring, spraying, and soil analysis [4]. This document explores the structural components of UAVs in detail, providing examples, relevant formulas, and their impact on agricultural applications.

Lightweight Materials for Efficiency

UAVs designed for agricultural use must be lightweight to maximize flight time and efficiency. Materials such as carbon fiber, composite plastics, and aluminum alloys are commonly used in UAV construction due to their high strength-to-weight ratio. These materials help reduce energy consumption, allowing UAVs to carry additional payloads such as sensors, cameras, and spraying systems while maintaining longer flight durations. Lightweight designs also enhance maneuverability, making UAVs more effective for tasks like crop monitoring and precision spraying [5].

Importance of Lightweight Materials

The efficiency of UAVs is directly influenced by their weight. Lighter drones consume less power, achieve longer flight times, and can carry more payload (such as sensors or sprayers) [6]. The materials used in UAV construction must strike a balance between being lightweight and structurally robust.

Common Lightweight Materials

Unmanned Aerial Vehicles (UAVs) are constructed using various materials to optimize performance, durability, and efficiency. Carbon fiber is widely used in UAV frames and arms due to its high strength-to-weight ratio, providing rigidity while keeping the overall weight low. It is also resistant to corrosion and fatigue, making it a durable choice, though it can be brittle under extreme stress. Aluminum alloys, known for their lightweight and corrosion-resistant properties, are commonly used in UAV bodies and landing gear [7]. They offer excellent machinability and ductility, allowing them to absorb impacts better than carbon fiber. Composite plastics, made from a combination of polymers and reinforcing fibers, provide durability and cost-effectiveness while maintaining a low weight, making them ideal for UAV components that require design flexibility. Foam-based materials, such as expanded polypropylene (EPP) or expanded polystyrene (EPS), are primarily used in fixed-wing UAVs due to their lightweight aerodynamics and crash resistance [8]. These materials offer excellent shock absorption and affordability, making them popular for hobbyist drones, though they may require reinforcement for structural integrity. Each of these materials plays a vital role in UAV design, depending on the specific requirements for strength, weight, durability, and cost.

Example

A drone designed for agricultural monitoring with a carbon fiber frame can weigh 30% less than a traditional aluminum frame, leading to an increased flight time of 20–30%.

Formula for Flight Time and Weight

The relationship between weight and flight endurance can be estimated as:

$$T = C \times VWT = \frac{C \times V}{W}$$

where:

T = Flight time (hours)
C = Battery capacity (Wh)
V = Voltage (V)
W = Weight of the UAV (kg)

A decrease in weight results in longer flight durations, making lightweight materials essential for agricultural UAVs.

6.2 Modular Frames for Flexibility

Modular UAV frames provide flexibility in adapting to different agricultural tasks. A modular design allows for easy assembly, maintenance, and customization, enabling farmers to equip UAVs with interchangeable components such as multispectral cameras, LiDAR sensors, or spraying mechanisms. This adaptability ensures that UAVs can be configured for specific applications, from soil mapping and crop health assessment to targeted pesticide application [9]. Additionally, modular frames facilitate quick repairs and upgrades, reducing downtime and operational costs.

Key Features of Modular UAVs

- **Interchangeable Payloads**: Sensors, cameras, and sprayers can be swapped based on the specific agricultural requirement.
- **Quick Maintenance**: Damaged arms or propellers can be easily replaced, reducing downtime.
- **Scalability**: Farmers can modify UAVs by upgrading batteries, motors, or sensor systems.

Example

A modular UAV used for crop monitoring can be upgraded with a sprayer attachment during the growing season for pest control. This adaptability increases the UAV's functionality and cost-effectiveness.

Formula for Payload Capacity

$$P = T \times E$$

where:

P = Payload capacity (kg)
T = Thrust generated by propellers (N)
E = Efficiency factor (depends on UAV design)

A modular UAV design allows for payload adaptability, ensuring that drones can perform multiple agricultural functions with ease.

6.3 Durability in Extreme Conditions

Agricultural UAVs must withstand harsh environmental conditions, including high temperatures, humidity, dust, and strong winds. Durability is achieved by using weather-resistant coatings, reinforced frames, and sealed electronic components to prevent damage from moisture and debris. Some UAVs also feature rugged landing gear and impact-resistant designs to handle rough terrains and accidental crashes. Ensuring structural durability enhances UAV longevity and reliability, making them a cost-effective investment for precision agriculture [10].

Importance of Durability

Agricultural UAVs must operate in challenging environments, including high temperatures, strong winds, humidity, and dust. Their structural components must be resistant to mechanical stress and environmental factors to ensure longevity.

Enhancements for Durability

Enhancing the durability of UAVs is essential to ensure their longevity and performance in various environmental conditions. Weather-resistant coatings play a crucial role in protecting UAV surfaces from moisture, UV radiation, and corrosion, preventing material degradation over time. Reinforced landing gear helps absorb impact during takeoff and landing, reducing the risk of structural damage and enhancing the UAV's overall stability. Impact-resistant frames provide additional protection against accidental crashes, making the UAV more resilient in demanding operational environments [11]. Additionally, sealed electronics prevent dust and water infiltration, safeguarding critical components and extending the UAV's lifespan. By incorporating these durability enhancements, UAVs can operate more reliably and efficiently in diverse conditions.

Applications in Agriculture

1. **Crop Health Monitoring**

UAVs equipped with multispectral and hyperspectral cameras analyze plant health by detecting changes in light reflectance. Structural enhancements ensure that the UAV can carry these high-precision cameras while maintaining long flight durations [12].

2. **Precision Spraying**

Drones designed with lightweight and modular frames enable precise pesticide and fertilizer application, reducing chemical waste. The structural integrity of the UAV ensures it can carry liquid payloads without compromising stability.

3. **Soil and Water Analysis**

Using LiDAR and thermal sensors, UAVs can assess soil moisture levels and detect areas requiring irrigation. The modular design allows farmers to switch between different sensors for various monitoring needs [13].

4. **Livestock Monitoring**

Durable UAVs equipped with thermal imaging cameras can monitor livestock conditions and detect sick or injured animals in large pastures.

Future Innovations in UAV Structural Design

1. **Advanced Materials**

The development of advanced materials is transforming UAV technology, improving durability, efficiency, and longevity. Graphene composites, known for being lighter and stronger than carbon fiber, have the potential to revolutionize UAV construction by offering superior strength-to-weight ratios and enhanced electrical conductivity [14]. This could lead to more efficient power management and increased structural resilience. Another breakthrough in material science is self-healing polymers, which can automatically repair minor structural damage, reducing maintenance requirements and extending UAV lifespan. These materials enhance UAV reliability, especially in harsh environments where wear and tear are inevitable. As research continues, integrating these advanced materials could significantly improve UAV performance, making them more durable, efficient, and cost-effective [15].

2. **Hybrid Power Systems**

Hybrid power systems are transforming UAV technology by enhancing endurance and operational efficiency. Solar-powered UAVs have the potential to extend flight duration indefinitely, making them ideal for applications such as large-scale agricultural monitoring, environmental surveillance, and communication networks. By harnessing solar energy, these UAVs reduce reliance on traditional batteries, allowing for continuous operation in daylight conditions. Similarly, hydrogen fuel cells provide a significant advantage over lithium-ion batteries due to their higher energy density. They enable UAVs to achieve longer flight times with reduced weight while also being more environmentally friendly [16]. As advancements in energy storage and conversion continue, hybrid power systems could redefine UAV capabilities, making them more sustainable and efficient for extended missions.

3. **AI-Driven Swarm UAVs**

AI-driven swarm UAVs are revolutionizing precision farming by enabling multiple UAVs to work collaboratively, covering vast agricultural areas efficiently. These UAVs operate in coordinated formations, using artificial intelligence to communicate and optimize tasks such as crop monitoring, pest control, and irrigation management. By distributing workloads, swarm UAVs can significantly reduce the time required for large-scale agricultural operations while improving accuracy and resource utilization [17]. Their ability to adapt dynamically to environmental conditions and share real-time data enhances decision-making, ultimately leading to increased crop yields and more sustainable farming practices. As AI technology advances, swarm UAVs are expected to play an even greater role in the future of smart agriculture.

The structural components of UAVs significantly impact their performance, durability, and versatility in agricultural applications. Lightweight materials enhance flight efficiency, modular frames allow adaptability for various farming tasks, and durable designs ensure reliability in extreme conditions [18]. As technology advances, innovations in materials, power systems, and AI-driven automation will further enhance the role of UAVs in modern agriculture, making them indispensable tools for sustainable and precision farming.

6.4 Payload Systems in Agricultural Drones

Agricultural drones have revolutionized modern farming by offering precision-based applications that improve efficiency and productivity. The payload system is a crucial component of agricultural drones, consisting of specialized equipment such as sprayers, cameras, and seed/fertilizer dispersal units. These payloads enable targeted agrochemical application, crop health monitoring, and precision planting, reducing costs and optimizing yields. Drones equipped with the right payloads help farmers reduce labor intensity, minimize environmental impact, and improve resource management [19]. The three primary types of payload systems used in agricultural drones are:

6.4.1 Sprayer Payloads for Agrochemicals

Sprayer payloads enable agricultural drones to precisely apply pesticides, herbicides, and fertilizers to crops. These systems use advanced nozzles and automated spraying mechanisms to ensure even distribution, reducing chemical wastage and preventing overuse. By targeting specific areas based on crop health data, sprayer drones minimize environmental impact and improve efficiency. They are particularly useful for reaching difficult terrains, reducing the need for manual labor while enhancing precision in agricultural spraying. Sprayer payloads are designed to apply pesticides, herbicides, fungicides, and liquid fertilizers efficiently. Traditional manual spraying is labor-intensive, inconsistent, and poses health risks, whereas drones provide uniform coverage with minimal chemical waste [20]. Example A DJI Agras T40 drone is capable of carrying a 40-L tank and spraying 12 hectares per hour, improving efficiency compared to manual methods.

Importance of Agricultural Drones

Agricultural drones play a crucial role in modern precision farming by improving efficiency and sustainability. One of their key benefits is precision spraying, which reduces chemical wastage and ensures even distribution of pesticides and fertilizers. This targeted application not only optimizes crop protection but also minimizes environmental impact. Additionally, drones reduce human exposure to harmful chemicals, as they eliminate the need for farmers to manually spray agrochemicals, improving workplace safety. Another significant advantage is that drones help reduce soil compaction, as they operate aerially and eliminate the need for heavy tractors, which can damage soil structure over time [21]. Furthermore, agricultural drones are eco-friendly, as they prevent chemical runoff and overuse, promoting sustainable farming practices. By integrating drone technology, farmers can achieve higher efficiency, lower costs, and reduce the environmental footprint of agricultural activities.

Advantages of Agricultural Drones

Agricultural drones provide numerous benefits that enhance efficiency and sustainability in modern farming. One major advantage is that they reduce labor costs by automating tasks such as spraying, seeding, and monitoring, which traditionally require extensive manual effort. Additionally, drones ensure uniform spraying, preventing over- or under-application of pesticides and fertilizers, which leads to better crop health and reduced chemical waste. Their faster operation compared to manual spraying allows farmers to cover large areas in significantly less time, improving productivity [22]. Furthermore, drones help minimize soil erosion by reducing ground traffic, as traditional machinery can compact the soil and disrupt its structure. By integrating drone technology, farmers can optimize resource use, increase efficiency, and promote sustainable agricultural practices.

6.4.2 Camera Payloads for Imaging

Camera payloads play a crucial role in precision agriculture by providing high-resolution imagery and data analysis. These drones are equipped with multispectral, thermal, and high-definition cameras that help monitor crop health, detect pest infestations, and assess soil conditions. Multispectral imaging can analyze plant stress levels, while thermal imaging detects irrigation issues. This real-time data allows farmers to make informed decisions, improving yield and resource management. Camera-equipped drones also assist in land mapping, ensuring efficient farm planning. Drones equipped with high-resolution cameras, thermal sensors, and multispectral sensors help in crop monitoring, field mapping, and early detection of issues like nutrient deficiencies, pest infestations, and water stress [23]. Example, A Parrot Sequoia + drone uses multispectral sensors to analyze chlorophyll content in crops, helping farmers identify unhealthy plants before symptoms appear.

Importance of Agricultural Drones

Agricultural drones play a vital role in modern farming by improving efficiency, sustainability, and productivity. One of their key benefits is the early detection of crop diseases and pest infestations, allowing farmers to take timely action and prevent widespread damage. By using advanced imaging technologies such as NDVI (Normalized Difference Vegetation Index), drones enable precision agriculture, helping monitor plant health, optimize fertilizer use, and improve overall farm management [24]. Additionally, drones reduce water waste by monitoring irrigation levels, ensuring that crops receive the right amount of water while preventing overuse. This not only conserves water resources but also enhances soil health. Another crucial advantage is their ability to enhance crop yield predictions by analyzing growth patterns, environmental conditions, and historical data, allowing farmers to plan harvests more effectively [25]. By integrating drone technology, agriculture becomes more data-driven, efficient, and sustainable, leading to higher productivity and reduced resource wastage.

Normalized Difference Vegetation Index (NDVI) for Crop Health Analysis

The Normalized Difference Vegetation Index (NDVI) is a widely used remote sensing index for evaluating the health, vigor, and biomass of crops and vegetation. It utilizes the differences in reflectance between the near-infrared (NIR) and red (RED) spectral bands, both of which are critical for assessing plant health.

Healthy vegetation absorbs most visible light (particularly red) for photosynthesis and reflects a large portion of near-infrared light. Conversely, stressed or sparse vegetation reflects more red light and less near-infrared light. This difference is captured through the NDVI formula:

$$\text{NDVI} = \frac{NIR - RED}{NIR + RED}$$

where:

NIR = Reflectance in the near-infrared band (typically ~800–850 nm)
RED = Reflectance in the red band (typically ~600–700 nm)

NDVI values range from -1 to $+1$. Values close to $+1$ indicate healthy and dense vegetation, while values near 0 or negative correspond to bare soil, water, or other non-vegetative surfaces. The NDVI index is especially valuable in precision agriculture, as it enables farmers and agronomists to identify variability in crop health, detect early signs of stress (e.g., due to drought, nutrient deficiency, pests), monitor growth stages, and even estimate potential yield (Table 6.1).

Advantages

Agricultural drones offer several advantages that enhance efficiency and sustainability in farming. By using precise application techniques, drones reduce fertilizer and pesticide costs by targeting only affected areas, minimizing waste and environmental impact. They also increase yield through data-driven decisions, as advanced imaging and sensor technology help farmers analyze soil conditions, detect diseases, and optimize resource allocation. Additionally, drones provide real-time crop health monitoring, allowing for early detection of issues such as pest infestations, nutrient deficiencies, and water stress, leading to timely interventions [26]. Moreover, they assist in insurance assessments and damage evaluations by capturing high-resolution aerial imagery, which helps document crop losses due to natural disasters or adverse weather conditions. By integrating drone technology, farmers can enhance productivity, reduce costs, and make more informed decisions for sustainable agriculture.

Table 6.1 NDVI value interpretation table

NDVI value range	Vegetation condition
0.6–0.9	Dense, healthy, vigorous crops
0.2–0.5	Moderate growth or stressed crops
0.0–0.2	Bare soil, sparse vegetation
<0.0	Water bodies, snow, or cloud cover

6.4.3 *Seed and Fertilizer Dispersal Systems*

Drones equipped with seed and fertilizer dispersal payloads help automate planting and nutrient application processes. These systems use precise dropping mechanisms to evenly distribute seeds and fertilizers across fields, ensuring optimal germination and soil nutrition. This technology reduces manual labor, speeds up planting processes, and minimizes waste. It is especially beneficial in large or hard-to-reach farming areas, enhancing agricultural productivity and sustainability. These payloads allow precise seeding and fertilization, ensuring even distribution of seeds and nutrients. Drones equipped with automated hoppers and spreaders increase planting efficiency and improve germination rates [27]. Example The XAG P100 drone disperses seeds over 50 acres per hour, reducing seeding time by 70% compared to manual methods.

Importance of Agricultural Drones

Agricultural drones significantly enhance efficiency and sustainability in modern farming. One of their key advantages is that they reduce seed wastage through precise dropping, ensuring that seeds are evenly distributed and minimizing unnecessary losses. This targeted approach not only conserves resources but also optimizes plant growth. Additionally, drones increase yield potential through even fertilization, ensuring that crops receive the necessary nutrients uniformly, leading to healthier plants and higher productivity [28]. Another major benefit is that drones save labor costs and effort by automating tasks such as seeding, fertilization, and spraying, reducing the reliance on manual labor. Furthermore, drones are highly effective in difficult terrains where tractors cannot reach, such as hilly, marshy, or flood-prone areas, allowing farmers to manage crops in challenging environments [29]. By integrating drone technology, agriculture becomes more precise, cost-effective, and adaptable to various landscapes.

Formula for Seed Spacing Calculation

The required seed application rate can be calculated as:

$$\text{Seed Rate} = \frac{\text{Number of seeds}}{\text{Area covered}}$$

Example:

If a drone disperses 50,000 seeds over an area of 10 hectares, the seed rate is:

$$\text{Seed Rate} = \frac{50000}{10} = 5000 \text{ seed ha}^{-1}$$

Advantages of Agricultural Drones

Uniform Distribution of Seeds and Fertilizer

Agricultural drones ensure even and precise distribution of seeds and fertilizers, reducing waste and improving crop yield. This uniformity enhances plant growth and optimizes soil nutrient levels, leading to better agricultural productivity [30].

Works in Flood-Prone or Hilly Areas

Traditional farming methods face challenges in difficult terrains, such as flood-prone or hilly regions. Drones can easily navigate these areas, ensuring consistent seeding, fertilization, and spraying, which would otherwise be difficult or unsafe for farmers [31].

Cost-Effective Compared to Traditional Methods

By reducing the need for manual labor, fuel, and excessive chemical usage, drones lower overall farming costs. Their ability to apply resources with precision minimizes waste, making them a cost-effective alternative to conventional farming techniques.

Saves Time and Labor

Drones significantly speed up agricultural processes, such as seeding, spraying, and monitoring, which traditionally require extensive manpower and time. This automation allows farmers to manage larger areas with minimal effort, increasing efficiency and productivity [32]. By incorporating drone technology, modern agriculture becomes more precise, efficient, and sustainable, benefiting both farmers and the environment.

Agricultural drones with advanced payload systems are transforming modern farming by enabling precision spraying, imaging, and seeding. These technologies reduce costs, increase efficiency, and promote sustainability in agriculture. The integration of AI and automation in drones is expected to further enhance their applications, making farming smarter, faster, and more productive.

6.5 Sensors and Cameras in Agricultural Drones

Agricultural drones have transformed modern farming by integrating advanced sensors and cameras that provide real-time data for precision agriculture. These sensors help in monitoring crop health, detecting stress, optimizing irrigation, and mapping fields for better resource management. The primary types of sensors and cameras used in agricultural drones are multispectral imaging sensors, thermal cameras, and LiDAR systems [33]. These technologies allow farmers to make data-driven decisions, reducing costs and improving overall productivity.

6.5.1 Multispectral Imaging Sensors

Multispectral imaging sensors capture light from multiple spectral bands, including visible and near-infrared (NIR) wavelengths. These sensors are essential in assessing plant health, detecting diseases, and optimizing fertilizer application. By analyzing light reflectance from plants, multispectral sensors can generate vegetation indices such as the Normalized Difference Vegetation Index (NDVI), which helps determine plant stress levels. Healthy vegetation reflects more NIR light and absorbs more red light, while stressed plants show a different spectral response. This early detection allows farmers to take corrective actions before visible symptoms appear, reducing crop losses and improving yields [34]. An example of a commonly used multispectral sensor in agriculture is the MicaSense RedEdge-P, which captures multiple spectral bands to provide detailed insights into plant health. Multispectral imaging sensors capture light from multiple wavelengths, including visible (Red, Green, Blue) and Near-Infrared (NIR) bands. Plants reflect and absorb light differently based on their health status. Healthy crops absorb more red light and reflect more NIR light, while stressed plants exhibit lower NIR reflectance [35]. This difference in reflectance helps in assessing plant health and optimizing farm management strategies.

These sensors play a crucial role in agriculture by:

- **Monitoring chlorophyll levels** to assess plant growth and overall crop health.
- **Detecting nutrient deficiencies** before visible symptoms appear, allowing timely intervention.
- **Identifying pest infestations and diseases** in their early stages, preventing potential crop losses.
- **Optimizing fertilization and irrigation schedules**, leading to increased yield and resource efficiency.

Example

One widely used multispectral sensor in agriculture is the MicaSense RedEdge-P. This sensor captures five spectral bands, providing farmers with detailed crop health analysis. The data collected helps generate vegetation indices, which are essential for precision farming. By utilizing the MicaSense RedEdge-P, farmers can monitor plant stress, optimize resource use, and make informed decisions to enhance productivity and sustainability [36].

Importance

Multispectral sensors are essential in precision agriculture, providing valuable insights that improve farm efficiency. These sensors allow farmers to monitor crop health and predict yields more accurately, ensuring better planning and management. By using multispectral imaging, farmers can reduce fertilizer and pesticide use by targeting only affected areas, minimizing costs and environmental impact. Early detection of plant stress through these sensors prevents significant crop losses, allowing for proactive interventions [37]. Additionally, sustainable farming practices are supported through efficient water and nutrient management, leading to long-term agricultural success.

6.5.2 *Thermal Cameras for Crop Stress Detection*

Thermal cameras detect temperature variations in crops, helping farmers identify water stress, irrigation issues, and potential disease outbreaks. Plants undergoing heat stress due to insufficient water tend to have higher canopy temperatures, which thermal cameras can detect accurately. This information is crucial in optimizing irrigation schedules and preventing drought-related losses. By measuring the temperature difference between the crop canopy and surrounding air, farmers can assess plant hydration levels and adjust irrigation accordingly. The FLIR Vue Pro R is an example of a thermal camera commonly used in agriculture, providing precise temperature readings that help in water management and disease detection [38]. Thermal imaging also plays a crucial role in night-time monitoring, ensuring 24/7 surveillance of crop conditions.

Thermal cameras detect temperature variations in crops by measuring the infrared radiation emitted by plants. When plants experience water stress, they transpire less moisture, leading to a rise in their temperature. In contrast, well-hydrated plants remain cooler due to active transpiration [39]. By analyzing these temperature differences, thermal imaging helps in:

- Identifying irrigation deficiencies and optimizing water usage.
- Detecting heat stress in crops, preventing potential yield losses.
- Assessing soil moisture levels and ensuring uniform water distribution.
- Monitoring disease outbreaks, as infected plants often exhibit abnormal temperature variations.

Example

A widely used thermal camera in agriculture is the FLIR Vue Pro R. This high-resolution sensor is specifically designed to detect water stress in crops and improve irrigation efficiency. By integrating the FLIR Vue Pro R into agricultural drones, farmers can monitor real-time plant health, make precise irrigation decisions, and prevent crop loss due to dehydration.

Importance

Thermal cameras play a crucial role in water conservation and crop protection. They allow farmers to detect water stress early, preventing drought-related crop failures. By optimizing irrigation schedules, farmers can reduce water wastage while ensuring that crops receive adequate moisture. Additionally, thermal imaging helps in identifying disease-prone areas, enabling targeted pesticide application. By detecting temperature variations in plants, farmers can pinpoint infected areas before visual symptoms appear [40]. This reduces unnecessary chemical use and improves precision agriculture. Crop losses due to heat stress and dehydration can also be minimized, ensuring stable and healthy yields.

6.5.2.1 Formula for Crop Stress Detection

Crop stress can be determined by comparing canopy temperature to air temperature using the following formula:

$$\text{Crop Stress} = \text{Canopy Temperature} - \text{Air Temperature}$$

If Crop Stress > Threshold, the crop is under heat stress or water deficiency.
If Crop Stress < Threshold, the crop has sufficient moisture and is not experiencing stress.

This formula helps in setting irrigation thresholds, ensuring efficient water use and preventing unnecessary watering.

By integrating thermal cameras into precision agriculture, farmers can enhance productivity, optimize water usage, and protect crops from heat stress and dehydration.

Crop stress—caused by factors like drought, nutrient deficiency, pests, or disease—can be detected using vegetation indices derived from spectral reflectance data. One of the most commonly used indices is the Normalized Difference Vegetation Index (NDVI), but there are others too like PRI, NDWI, and CWSI, depending on the type of stress.

NDVI (Normalized Difference Vegetation Index)

Used for detecting general stress, especially from drought, disease, or poor vigor.

$$NDVI = \frac{NIR - RED}{2aNIR + RED}$$

NIR = Reflectance in near-infrared (~800–850 nm)
RED = Reflectance in red (~600–700 nm)

Interpretation:

Low NDVI values (0–0.3) may indicate stressed or sparse vegetation.
High NDVI values (0.6–0.9) indicate healthy vegetation.

PRI (Photochemical Reflectance Index)

Used for detecting photosynthetic efficiency and early stress signals.

$$PRI = \frac{R_{531} - R_{570}}{R_{531} + R_{570}}$$

$\mathbf{R_{531}}$ = Reflectance at 531 nm
$\mathbf{R_{570}}$ = Reflectance at 570 nm

PRI is sensitive to changes in carotenoid pigments and is useful for early detection of water and light stress.

NDWI (Normalized Difference Water Index)

Detects plant water stress.

$$NDWI = \frac{NIR - SWIR}{NIR + SWIR}$$

SWIR = Shortwave infrared reflectance (~1200–2500 nm)

Low NDWI values indicate water-stressed crops.

CWSI (Crop Water Stress Index)

Used for quantifying drought stress using canopy temperature.

$$CWSI = \frac{T_c - T_{wet}}{T_{dry} - T_{wet}}$$

T_c = Actual canopy temperature
T_{wet} = Temperature of a well-watered reference
T_{dry} = Temperature of a non-transpiring (dry) reference

CWSI ranges from 0 (no stress) to 1 (maximum stress) (Table 6.2).

Table 6.2 Common crop stress indices

Index	Formula	Stress detected
NDVI	$NDVI = \frac{NIR-RED}{2aNIR+RED}$	General vigor & stress
PRI	$PRI = \frac{R_{531}-R_{570}}{R_{531}+R_{570}}$	Photosynthetic stress
NDWI	$NDWI = \frac{NIR-SWIR}{NIR+SWIR}$	Water/drought stress
CWSI	$CWSI = \frac{T_c-T_{wet}}{T_{dry}-T_{wet}}$	Drought stress (thermal)

6.5.3 LiDAR for Field Mapping

LiDAR (Light Detection and Ranging) technology uses laser pulses to create high-resolution 3D maps of agricultural fields. It is highly effective in topographic mapping, soil erosion analysis, plant height measurement, and land surveying. LiDAR sensors measure the distance between the drone and the ground by calculating the time it takes for a laser pulse to return. This data is used to generate detailed terrain models that assist in designing efficient irrigation and drainage systems. LiDAR also helps in analyzing forest cover, measuring tree canopy height, and assessing soil health by detecting variations in land elevation. The RIEGL VUX-1UAV is a widely used LiDAR sensor in agriculture, providing accurate field mapping that enhances farm planning and management [41].

LiDAR (Light Detection and Ranging) is an advanced remote sensing technology that uses laser pulses to measure distances and create high-resolution 3D maps of farmland. This technology is widely used in agriculture for accurate terrain analysis and crop assessment. LiDAR helps in:

- Topographic mapping, aiding in efficient land-use planning.
- Soil erosion analysis, identifying vulnerable areas prone to degradation.
- Measuring crop height and biomass, providing valuable growth insights.
- Irrigation and drainage planning, optimizing water flow for better field management.

By generating detailed elevation models, LiDAR helps farmers make data-driven decisions to enhance agricultural productivity and sustainability.

Example

A widely used LiDAR sensor in agriculture is the RIEGL VUX-1UAV. This high-precision sensor is commonly deployed for field mapping and terrain analysis. By using drones equipped with LiDAR, farmers can design efficient irrigation systems, optimize land leveling, and prevent waterlogging issues [42]. The high-resolution 3D data provided by the RIEGL VUX-1UAV ensures better crop management and resource allocation.

Importance

LiDAR technology is crucial for modern precision agriculture, as it provides accurate elevation models that help in:

- Identifying uneven land surfaces, ensuring better water distribution for crops.
- Mapping forest cover and crop density, supporting sustainable farming practices.
- Assessing soil erosion patterns, allowing preventive measures to reduce land degradation.
- Improving drainage planning, minimizing risks of waterlogging and flooding in farmlands.

By utilizing LiDAR data, farmers can enhance their soil and water conservation strategies, leading to increased productivity and reduced environmental impact.

6.5.3.1 Formula for Distance Measurement

LiDAR calculates distances based on the time taken for a laser pulse to travel to an object and back.

Since light travels extremely fast, LiDAR can generate highly precise 3D models of the landscape, helping farmers analyze crop structures, land elevations, and irrigation patterns.

The integration of advanced sensors and cameras in agricultural drones has revolutionized precision farming by improving crop monitoring, optimizing resource usage, and reducing environmental impact. Multispectral sensors provide critical insights into plant health, thermal cameras help detect water stress and diseases, while LiDAR technology enhances field mapping and land analysis. These innovations not only increase efficiency and productivity but also promote sustainable farming practices, ensuring higher yields with lower costs [10]. The future of agriculture lies in the widespread adoption of these technologies, enabling farmers to make informed decisions that benefit both the environment and their economic output.

6.6 Power Systems and Battery Management:

Agricultural drones have become an essential tool for modern farming, enabling precision agriculture through aerial imaging, spraying, and field mapping. However, their performance and efficiency depend on a reliable power system and effective battery management. The ability of a drone to operate efficiently for long durations is directly linked to the type of battery used, the energy sources available, and battery optimization techniques [43].

The three primary aspects of power management in agricultural drones include:

6.6.1 Lithium-Ion Batteries for Longevity

Lithium-ion (Li-ion) batteries are widely used in UAVs due to their high energy density, lightweight nature, and extended lifespan. These batteries provide a stable power supply, allowing drones to perform long-duration agricultural tasks such as crop monitoring, spraying, and mapping. One of the biggest advantages of Li-ion batteries is their longer flight time compared to other battery types. They also support fast charging, reducing downtime between flights. Additionally, these batteries have a lower self-discharge rate, meaning they can retain power even when not in use. To further enhance battery performance and lifespan, UAVs are equipped with battery management systems (BMS) [44]. These systems monitor voltage, temperature, and charge cycles, preventing overcharging and deep discharge, which can cause battery degradation. Proper battery care ensures safe operation and maximized efficiency in UAV-based agriculture.

Example

The DJI Agras T40, an advanced agricultural spraying drone, is powered by a 30,000 mAh lithium-ion battery, allowing it to cover up to 50 acres per charge. The battery's high energy density and quick-charging capability enable farmers to complete spraying tasks efficiently without frequent recharging. The smart battery system also includes temperature monitoring and overload protection, ensuring safe and reliable operation [45].

Importance

Lithium-ion batteries are widely used in agricultural drones due to their long lifespan, lightweight design, and high energy efficiency. These batteries ensure that drones can operate for extended durations, reducing the need for frequent battery swaps and improving overall productivity. Additionally, lithium-ion batteries offer fast charging, allowing drones to be recharged quickly between flights. One of the key benefits of lithium-ion batteries is their consistent power output, which helps maintain stable flight performance even under heavy loads. This is particularly important for spraying drones that need to carry liquid payloads over large fields [46]. Furthermore, proper battery management can extend the lifespan of lithium-ion batteries, making them a cost-effective solution for long-term agricultural drone operations.

6.6.2 Renewable Energy Sources in UAVs

Renewable energy sources are revolutionizing UAV technology, offering sustainable alternatives to traditional batteries. Solar-powered UAVs and hydrogen fuel cells are emerging as viable solutions for extended flight durations and reduced environmental impact. Solar-powered UAVs are equipped with thin-film solar panels that continuously generate power from sunlight. This allows drones to operate for long periods without needing to recharge, making them ideal for large-scale agricultural surveys [47]. These UAVs can cover vast farmland areas, reducing the need for multiple flights. Another promising technology is hydrogen fuel cells, which provide higher energy density than lithium-ion batteries. Hydrogen-powered UAVs have significantly longer endurance, allowing them to stay airborne for several hours. Unlike conventional batteries, fuel cells generate electricity through a chemical reaction, producing only water as a byproduct, making them environmentally friendly [48]. By integrating renewable energy sources into UAV systems, farmers can reduce operational costs, minimize battery replacements, and improve efficiency in precision agriculture.

Example

The SolarXOne UAV is an agricultural drone that integrates solar panels into its wings, allowing it to harvest solar energy while in flight. This solar-powered drone can operate for up to 12 h continuously, making it ideal for long-duration crop monitoring in large farmlands [49].

Importance

Using renewable energy sources in UAVs is a significant step toward sustainable agriculture. Solar-powered drones extend flight time without the need for frequent battery recharges, making them particularly useful for large-scale farms and remote agricultural areas where charging infrastructure is limited. Renewable energy sources help reduce operational costs by minimizing dependence on traditional battery charging. Additionally, integrating solar power with lithium-ion batteries allows for hybrid power systems, where drones can switch between stored battery power and real-time solar energy harvesting. This reduces the environmental impact and ensures continuous operation in precision farming, irrigation management, and pest monitoring [50]. While solar-powered drones offer many advantages, they are less effective in cloudy conditions or low-light environments. However, advancements in high-efficiency solar panels and energy storage technologies are gradually improving their performance, making them a viable solution for the future of sustainable precision agriculture.

6.6.3 Battery Optimization for Long Flights

Optimizing battery performance is crucial for long-duration UAV missions, especially in agricultural applications where drones need to cover large areas efficiently. Several strategies help maximize flight time and battery life. One key method is the use of Smart Battery Management Systems (BMS). These systems actively monitor the temperature, charge level, and power consumption of the battery, ensuring safe and efficient energy use. Overheating and overcharging are prevented, extending battery lifespan and improving drone reliability [51]. Another optimization strategy involves designing optimized flight paths that minimize unnecessary movements. By calculating the most efficient route, drones can conserve power and maximize flight duration. Lightweight drone designs also contribute to power efficiency by reducing the overall energy demand, allowing UAVs to fly longer without requiring frequent recharges. Additionally, regenerative braking systems are being integrated into advanced UAVs. These systems recover energy during descent and store it for future use, increasing overall efficiency. Such innovations help extend flight durations, reduce downtime, and improve the effectiveness of UAVs in modern precision agriculture [52].

Example

The Parrot Bluegrass Fields drone, designed for agricultural surveying, incorporates an intelligent battery management system that optimizes energy consumption. By adjusting motor speed, flight altitude, and route planning, the drone maximizes battery efficiency, allowing it to survey fields for up to 30 min per charge.

Importance

Battery optimization is crucial for enhancing drone endurance and efficiency. Since battery capacity directly affects flight time, effective power management techniques are required to minimize energy waste. By implementing smart charging, adaptive power distribution, and optimized flight paths, drones can perform more tasks per charge, reducing downtime and increasing farm productivity. One of the most effective strategies for battery optimization is reducing payload weight. Carrying lighter cameras, sprayers, and sensors helps conserve battery power, allowing drones to fly longer distances before requiring a recharge [53]. Another key factor is thermal management—keeping batteries cool prevents overheating, which can degrade battery performance over time. Many modern drones feature automated battery health monitoring, alerting users when battery levels are low or when maintenance is needed. These advanced power management systems ensure that drones operate safely and efficiently, reducing the risk of battery failure and improving overall reliability [54].

6.7 Software Integration in Drone Design

The efficiency and functionality of agricultural drones depend not only on their hardware but also on their software integration. Advanced software systems help in controlling flight operations, processing real-time data, and managing cloud-based storage for better decision-making in precision agriculture. Modern agricultural drones utilize flight control software, real-time data analytics, and cloud connectivity to enhance their performance. These software systems allow drones to navigate autonomously, analyze crop health, optimize spraying patterns, and store essential data for future reference. Integrating these technologies helps farmers improve efficiency, accuracy, and productivity in agricultural operations.

6.7.1 Flight Control Software

Flight control software is essential for stabilizing, navigating, and automating UAV operations. It allows drones to follow predefined flight paths, adjust altitude, and maintain stability in changing weather conditions. Advanced flight control systems use GPS, IMU (Inertial Measurement Unit), and AI-driven algorithms to enhance precision. Modern UAV software includes waypoint navigation, which enables drones to autonomously cover large agricultural fields without manual intervention. Farmers can set specific routes for crop spraying, monitoring, or mapping, reducing the need for human oversight. Additionally, collision avoidance systems use onboard sensors to detect obstacles and adjust flight paths accordingly, improving safety and efficiency. Flight control software is continuously evolving, integrating AI and machine learning for real-time decision-making [55]. These advancements ensure better adaptability, higher efficiency, and reduced operational errors in agricultural drone applications.

Example

The DJI GS Pro (Ground Station Pro) is a flight control software that allows farmers to pre-plan drone flight paths for field surveying and spraying operations. It provides automatic altitude adjustment, obstacle avoidance, and geofencing to ensure safe and efficient drone operation.

Importance

Flight control software is essential for precision and automation in drone operations. It enables pre-programmed flight paths, reducing the need for manual control and minimizing human error.

Key advantages include:

- Efficient route planning, ensuring complete field coverage.
- Collision avoidance, preventing crashes with trees, buildings, and other obstacles.
- Altitude control, maintaining an optimal height for spraying and imaging.
- Energy efficiency, optimizing battery use for longer flights.

Formula for Flight Time Optimization

The total flight time of a drone depends on battery capacity (C), power consumption (P), and efficiency factors (E). The formula for estimating drone flight time is:

$$T = PC \times E$$

where:

T = Flight time (minutes)
C = Battery capacity (mAh)
P = Power consumption per minute (mAh/min)
E = Efficiency factor (percentage of actual usable power)

By optimizing flight paths and power consumption, drones can extend flight time and cover larger areas efficiently.

6.7.2 Real-Time Data Analytics

Real-time data analytics in UAVs allows for instant processing and interpretation of field data, providing farmers with actionable insights. Sensors and cameras onboard agricultural drones capture images, temperature readings, moisture levels, and crop health data. This information is analyzed using AI-powered algorithms to detect early signs of plant stress, pest infestations, and nutrient deficiencies. By processing data in real-time, farmers can make immediate adjustments to irrigation, fertilization, and pesticide application. For example, multispectral imaging sensors can generate vegetation indices such as NDVI (Normalized Difference Vegetation Index) to assess plant health, while thermal cameras detect water stress in crops [56]. The integration of edge computing further enhances real-time analytics by processing data directly on the UAV, reducing dependence on external servers. This leads to faster decision-making and improved efficiency in precision agriculture.

Example

The Sentera FieldAgent software is widely used in agriculture to analyze multispectral drone images, detecting crop health issues such as nutrient deficiencies, pest infestations, and water stress in real time.

Importance

Real-time data analytics helps farmers make quick decisions based on live field data. It allows them to:

- Identify unhealthy crops early and take corrective actions.
- Monitor soil moisture levels, optimizing irrigation.
- Detect pest infestations before they spread across fields.
- Predict yield by analyzing crop growth patterns.

6.7.3 Cloud Connectivity for Storage

Cloud connectivity enables UAVs to store and share vast amounts of agricultural data seamlessly. Drones collect high-resolution imagery and sensor readings, which are then uploaded to cloud platforms for storage, analysis, and collaboration. With cloud-based solutions, farmers and agronomists can access historical data, compare seasonal trends, and generate predictive models for crop management [57]. Cloud storage also allows for remote monitoring, where agricultural experts can analyze drone-collected data from anywhere in the world. Additionally, cloud integration supports AI-driven predictive analytics, enabling farmers to anticipate weather patterns, optimize resource usage, and enhance overall farm productivity. By using cloud computing, agricultural drones become smarter, more efficient, and capable of delivering long-term insights for sustainable farming practices [58].

Example

The Pix4Dcloud platform allows farmers to upload, store, and analyze drone-generated maps and field data in the cloud. This enables remote access to farm analytics from any device.

Importance

Cloud connectivity plays a vital role in modern agriculture by storing and organizing large datasets collected by drones. Its key benefits include:

- Remote accessibility, allowing farmers to view field data from anywhere.
- Data security and backup, preventing loss due to hardware failure.
- Integration with AI, enabling predictive analytics for crop planning.
- Collaboration among farmers and agronomists, improving decision-making.

Formula for Cloud Data Storage Needs

To estimate the required cloud storage for drone-collected data:

$$S = R \times T \times F$$

where:

S = Total storage required (GB)
R = Data recording rate (GB/min)
T = Flight duration (min)
F = Number of drone flights per day

By using cloud-based data storage, farmers can track historical trends, compare seasonal data, and make data-driven farming decisions. The integration of software in agricultural drones has significantly improved precision farming techniques. Flight control software enhances navigation and automation, real-time data analytics provides valuable insights into crop health, and cloud connectivity ensures secure storage and remote access to agricultural data. By leveraging these technologies, farmers can increase efficiency, reduce costs, and improve sustainability in modern farming practices [59]. As software continues to evolve, future agricultural drones will become more intelligent, autonomous, and capable of optimizing crop management with AI-driven analytics.

6.8 Weather Resistance and Durability in Agricultural Drones

Agricultural drones are exposed to various environmental challenges, including rain, dust, extreme temperatures, and corrosive agricultural chemicals. To ensure consistent performance and longevity, these drones must be built with weather-resistant and durable materials. Weatherproofing, corrosion resistance, and heat management systems are essential components that help drones function effectively in diverse conditions. Without these protective measures, drones can experience malfunctions, reduced efficiency, and frequent breakdowns, leading to increased maintenance costs and operational failures [60]. The following sections explore the significance of weather resistance and durability in agricultural drones.

Waterproofing and Weatherproofing

Drones used in agriculture and field surveying often operate in rainy, humid, or dusty conditions. Without proper waterproofing, moisture can seep into the drone's electronic components, leading to short circuits, sensor failures, and loss of communication. Modern drones are designed with water-resistant casings, sealed electrical compartments, and hydrophobic coatings that prevent water damage. For example, the DJI Agras T30 is a weatherproof agricultural drone with an IP67 rating, meaning it is fully dust-tight and can withstand immersion in water up to one meter deep for 30 min [61]. This protection allows it to operate even during light rain or in humid environments, making it highly reliable for precision agriculture.

The Ingress Protection (IP) rating system is used to classify the level of protection against dust and water. The formula for determining an IP rating consists of two numbers, where the first digit indicates dust protection (0–6) and the second digit represents water resistance (0–9) [62]. A higher rating means better protection, ensuring that drones can function without failure in challenging weather conditions.

6.8.1 Corrosion Resistance for Agricultural Chemicals

Agricultural drones frequently spray pesticides, herbicides, and fertilizers, which contain chemicals that can corrode metal and plastic components. Prolonged exposure to these substances can lead to rusting, material degradation, and malfunctioning sensors, affecting the drone's efficiency and durability. To combat chemical-induced corrosion, manufacturers incorporate anti-corrosion coatings, sealed electronic housings, and corrosion-resistant materials such as stainless steel, aluminum alloys, and composite plastics [63]. The XAG P100 agricultural drone, for instance, is coated with a special chemical-resistant layer, preventing damage from prolonged exposure to pesticide solutions. By using corrosion-resistant coatings and materials, manufacturers can extend the operational life of agricultural drones and ensure safe and efficient spraying applications [64].

6.8.2 Heat Management Systems

Agricultural drones generate heat due to continuous battery usage, motor function, and onboard electronic processing. Excessive heat can cause battery degradation, reduced flight efficiency, and system failures, especially in high-temperature environments. To maintain optimal performance, drones incorporate heat management systems that include cooling vents, heat-resistant materials, and intelligent temperature regulation systems [65]. For example, the SenseFly eBee Ag drone features passive cooling vents and lightweight composite materials, which help dissipate excess heat during long-duration flights. Additionally, some drones use temperature-regulated lithium-ion batteries, which prevent overheating and extend battery life. By optimizing heat dissipation, drones can maintain stable performance in extreme weather conditions and avoid overheating-related failures [66].

The durability and weather resistance of agricultural drones are crucial for ensuring uninterrupted operations in challenging environments. Waterproofing and weatherproofing protect drones from rain and moisture, while corrosion-resistant materials prevent damage from agricultural chemicals. Heat management systems play a vital role in preventing overheating and ensuring efficient energy consumption. By integrating advanced protective technologies, manufacturers can enhance drone longevity, reduce maintenance costs, and improve reliability in precision agriculture [67]. As technology evolves, future drones will incorporate even more advanced materials and smart protective systems, making them resilient, efficient, and suitable for all-weather operations.

6.9 Innovations in Agricultural UAV Design

The continuous advancements in drone technology have revolutionized modern agriculture by improving efficiency, precision, and productivity. Unmanned Aerial Vehicles (UAVs) have become an essential part of precision farming, enabling farmers to monitor crops, optimize resource usage, and reduce operational costs [68]. Recent innovations in agricultural UAV design focus on portability, adaptability, and load capacity, making drones more effective for diverse farming needs.

Among the most notable advancements are:

1. Foldable drone designs for portability, allowing for easy transport and quick deployment.
2. Compact drones for small farms, providing cost-effective solutions for small-scale farmers.
3. Heavy-lift drones for large payloads, enhancing efficiency in spraying, seeding, and fertilization.

These advancements enhance accessibility, usability, and productivity for farmers, making agricultural drones more versatile and essential than ever before.

6.9.1 Foldable Drone Designs for Portability

One of the most significant innovations in agricultural UAVs is foldable drone technology, which enhances portability and ease of use. Traditional agricultural drones were often bulky and difficult to transport, requiring large storage spaces and complex setup procedures. Foldable drones solve these challenges by incorporating collapsible arms and retractable propellers, allowing farmers to carry them easily and deploy them quickly in the field. Waterproofing and weatherproofing are essential for ensuring UAVs can operate in diverse environmental conditions, including rain, humidity, and strong winds [69]. Agricultural drones are often exposed to moisture, mud, and unpredictable weather, making protective measures necessary for durability and reliability. Manufacturers use sealed enclosures, hydrophobic coatings, and IP-rated (Ingress Protection) designs to protect sensitive drone components, such as motors, sensors, and electronic circuits. Waterproof UAVs can fly in rainy conditions, ensuring uninterrupted operations for crop monitoring, spraying, and surveying. Additionally, wind-resistant flight control algorithms help stabilize UAVs in turbulent conditions, ensuring smooth data collection and accurate spraying applications [70]. By integrating waterproofing and weatherproofing technologies, agricultural drones can extend their operational lifespan and reduce maintenance costs.

Example of Foldable Drones in Agriculture

A prime example is the DJI Mavic 3 M, a compact, foldable agricultural drone equipped with multispectral imaging for precision farming. This drone enables farmers to monitor crop health, assess vegetation indices, and optimize irrigation planning, all while being lightweight and easy to carry [71].

Importance of Foldable Drones

- Enhanced mobility: Farmers can carry the drone in a backpack and deploy it quickly for on-the-spot analysis.
- Quick setup: Foldable drones require minimal assembly, allowing for faster field operations.
- Space efficiency: Since they occupy less storage space, foldable drones are ideal for farmers with limited storage capacity.

6.9.2 Compact Drones for Small Farms

Agricultural drones are not only for large-scale farms—small farmers also benefit from drone technology. However, traditional agricultural UAVs are often too expensive or too large for small-scale farming. Compact drones are designed specifically for small farms and limited-space operations, offering cost-effective, lightweight, and highly maneuverable solutions for monitoring crops, detecting pests, and assessing soil conditions. Agricultural drones frequently handle fertilizers, pesticides, and herbicides, which can be highly corrosive to their frames, nozzles, and electronic components. To prevent damage, manufacturers use chemical-resistant coatings, stainless steel components, and non-corrosive plastic materials for critical drone parts [72]. Sprayer drones, in particular, require specialized anti-corrosion treatments for their tanks, tubing, and nozzles, ensuring they do not degrade over time. Self-cleaning systems are also integrated into some UAV designs to flush out residual chemicals after spraying operations, preventing buildup and clogging. By implementing corrosion-resistant materials and maintenance-friendly designs, agricultural drones can withstand prolonged exposure to chemicals, reducing repair costs and increasing their longevity in the field [73].

Example of Compact Agricultural Drones

The Parrot Anafi USA is a lightweight, compact agricultural drone designed for precision farming applications. Equipped with high-resolution cameras and thermal imaging sensors, it allows small farm owners to monitor crops without investing in large-scale UAV systems.

Importance of Compact Drones

- Affordable for small farmers: Compact drones provide an entry-level solution for small-scale agricultural operations.
- Better maneuverability: Their small size enables navigation in tight spaces, such as orchards, vineyards, and greenhouses.
- Lower power consumption: Compact drones require less battery power, making them energy-efficient and cost-effective.

6.9.3 Heavy-Lift Drones for Large Payloads

Heat management is crucial for UAV performance, especially in hot climates and extended flight operations. Batteries, processors, and motors generate heat during operation, which, if not managed properly, can lead to overheating, reduced efficiency, and component failure. To prevent overheating, UAVs use heat sinks, cooling fans, and thermal insulation materials to dissipate excess heat. Passive cooling systems, such as ventilation ducts and heat-resistant coatings, further enhance temperature regulation without increasing power consumption. Advanced UAVs also incorporate real-time temperature monitoring systems, which adjust power usage dynamically to prevent overheating. These smart cooling techniques help drones maintain optimal performance, extend battery life, and ensure reliability during long agricultural flights. By integrating weather resistance, corrosion resistance, and heat management systems, UAVs can operate more efficiently, last longer, and withstand harsh agricultural conditions, making them indispensable tools for modern precision farming [74].

Example of Heavy-Lift Drones in Agriculture

The XAG P100 Agricultural Drone is an advanced heavy-lift UAV capable of carrying up to 40 kg (88 lbs) of payload. It is widely used for precision spraying and large-scale seeding, helping farmers cover extensive areas efficiently.

Importance of Heavy-Lift Drones

- Increased efficiency: They can cover large fields quickly, reducing the time required for crop spraying and fertilization.
- Cost savings on labor: Automating tasks like pesticide application and seeding reduces the need for manual labor.
- Precision agriculture benefits: With GPS and AI control systems, these drones can target specific crop areas, minimizing waste and environmental impact.

Formula for Payload Capacity in Heavy-Lift Drones

The maximum payload capacity (P_{max}) of a drone depends on its lift force (F) and total weight (W):

$$P_{max} = F - W$$

where:

P_{max} = Maximum payload capacity (kg)
F = Total lift force (N) generated by drone motors
W = Weight of the drone itself (kg)

A well-designed heavy-lift drone maximizes payload efficiency while maintaining stable flight performance.

The evolution of agricultural UAVs has led to groundbreaking innovations, making drones more adaptable and efficient for various farming applications.

- Foldable drones provide portability and quick deployment, enabling farmers to operate efficiently in multiple locations.
- Compact drones cater to small farms, offering affordable, precise crop monitoring solutions.
- Heavy-lift drones revolutionize large-scale farming by automating labor-intensive tasks like spraying and seeding.

As drone technology advances, we can expect even more innovations in UAV design, including longer battery life, AI-powered navigation, and autonomous flight planning. These innovations will continue to enhance precision farming, reduce costs, and improve overall agricultural productivity.

6.10 Conclusion and Summary

Unmanned Aerial Vehicles (UAVs) have revolutionized traditional farming practices by enhancing efficiency, precision, and sustainability. Modern agricultural UAVs are designed with specialized payload systems, advanced sensors, and optimized power management solutions to perform tasks such as crop monitoring, spraying fertilizers and pesticides, soil analysis, and mapping field topography. Key innovations in UAV design include foldable drones for portability, compact drones for small-scale farming, and heavy-lift drones for large payload applications. The integration of multispectral imaging, LiDAR technology, and real-time data analytics allows farmers to assess crop health, detect pests, and optimize irrigation systems with greater accuracy. Challenges such as weather resistance, corrosion from agrochemicals, and heat management are being addressed through waterproofing techniques, specialized coatings, and thermal regulation systems. As UAV technology continues to evolve, autonomous flight capabilities, cloud-based data storage, and AI-powered analytics will further enhance precision farming, reducing costs and increasing agricultural productivity. The structural components of UAVs significantly impact their efficiency, flexibility, and durability in agricultural applications. Lightweight materials improve flight endurance, modular frames allow adaptability for various farming tasks, and durable designs ensure reliability in extreme conditions. As technology advances, innovations in materials, power systems, and AI-driven automation will further enhance the role of UAVs in modern agriculture, making them indispensable tools for sustainable and precision farming.

Advanced sensors and cameras in agricultural drones have revolutionized precision farming by improving crop monitoring, optimizing resource usage, and reducing environmental impact. Multispectral sensors provide critical insights into plant health, thermal cameras help detect water stress and diseases, while LiDAR technology enhances field mapping and land analysis. These technologies not only increase efficiency and productivity but also promote sustainable farming practices, ensuring higher yields with lower costs. The future of agriculture lies in the widespread adoption of these technologies, enabling farmers to make informed decisions that benefit both the environment and their economic output. With continuous advancements in drone and sensor technology, the future of agriculture is shifting towards fully automated and data-driven farming systems, ensuring food security and environmental conservation. Efficient power systems and battery management are critical for optimizing agricultural drone performance. Lithium-ion batteries offer long-lasting power with high energy density, while renewable energy sources such as solar panels contribute to sustainability. Battery optimization techniques ensure longer flight times, lower operational costs, and improved farm productivity.

As technology advances, the future of agricultural drones will include more energy-efficient batteries, improved renewable energy integration, and AI-powered battery management systems. By adopting these power management strategies, farmers can improve productivity, lower costs, and contribute to a more sustainable agricultural future. Software integration is a crucial aspect of modern agricultural drones, improving their autonomy, efficiency, and data processing capabilities. Flight control software ensures precise navigation, real-time analytics help monitor crop conditions, and cloud connectivity enables secure data storage and remote access. Future agricultural UAVs will feature more advanced AI systems, longer flight times, and enhanced data processing capabilities, further redefining precision farming and sustainable agriculture.

References

1. Mohsan SAH, Othman NQH, Li Y et al (2023) Unmanned aerial vehicles (UAVs): practical aspects, applications, open challenges, security issues, and future trends. Intel Serv Robot 16:109–137
2. Taseer A, Han X (2024) Advancements in variable rate spraying for precise spray requirements in precision agriculture using unmanned aerial spraying Systems: a review. Comput Electron Agric 219:108841
3. Ming R, Jiang R, Luo H et al (2023) Comparative analysis of different uav swarm control methods on unmanned farms. Agronomy 13:2499
4. Lochan K, Khan A, Elsayed I, et al (2024) Advancements in precision spraying of agricultural robots: a comprehensive review. IEEE Access
5. Toscano F, Fiorentino C, Capece N, et al (2024) Unmanned aerial vehicle for precision agriculture: a review. IEEE access
6. Mohsan SAH, Khan MA, Noor F et al (2022) Towards the unmanned aerial vehicles (UAVs): a comprehensive review. Drones 6:147
7. Simões S (2024) High-performance advanced composites in multifunctional material design: state of the art, challenges, and future directions. Materials 17:5997
8. Iorga D, Georgescu C, Constantinescu S et al (2024) Micro UAVs with fixed wings: design, technological solutions, and tests. Aerospace 11:977
9. Olson D, Anderson J (2021) Review on unmanned aerial vehicles, remote sensors, imagery processing, and their applications in agriculture. Agron J 113:971–992
10. Karbasishargh K, Moghimi Esfandabadi MH, Esmaeili A (2024) Innovative approaches to uav performance: enhancing safety, reliability, and flexibility. Int J Reliab Risk Safety: Theory Appl 7:28–39
11. Jin Y, Bi Y, Lyu C et al (2024) Nezha-IV: a hybrid aerial underwater vehicle in real ocean environments. J Field Robot 41:420–442
12. Anderson K (2024) Evaluation of forest canopy characteristics using drone-based spectral data analysis
13. Farhan SM, Yin J, Chen Z, Memon MS (2024) A comprehensive review of LiDAR applications in crop management for precision agriculture. Sensors (Basel, Switzerland) 24:5409
14. Aghaee K, Li L, Roshan A, Namakiaraghi P (2024) Additive manufacturing evolution in construction: from individual terrestrial to collective, aerial, and extraterrestrial applications. J Build Eng 110389
15. Hauashdh A, Nagapan S, Jailani J, Gamil Y (2024) An integrated framework for sustainable and efficient building maintenance operations aligning with climate change, SDGs, and emerging technology. Results Eng 21:101822
16. Townsend A, Jiya IN, Martinson C, et al (2020) A comprehensive review of energy sources for unmanned aerial vehicles, their shortfalls and opportunities for improvements. Heliyon 6
17. Baktayan AA, Zahary AT, Sikora A, Welte D (2024) Computational offloading into UAV swarm networks: a systematic literature review. EURASIP J Wirel Commun Netw 2024:69
18. Guebsi R, Mami S, Chokmani K (2024) Drones in precision agriculture: a comprehensive review of applications, technologies, and challenges. Drones 8:686

19. Vellingiri A, Kokila R, Nisha P, et al (2025) Harnessing GPS, sensors, and drones to minimize environmental impact: precision agriculture. In: Designing sustainable internet of things solutions for smart industries. IGI Global, pp 77–108
20. Gulak AM (2024) Implementation of drone for spraying herbicide and pesticide
21. Jońca J, Pawnuk M, Bezyk Y et al (2022) Drone-assisted monitoring of atmospheric pollution—a comprehensive review. Sustainability 14:11516
22. Sarkar S, Skalicky M, Hossain A et al (2020) Management of crop residues for improving input use efficiency and agricultural sustainability. Sustainability 12:9808
23. Abdullah HM, Mohana NT, Khan BM et al (2023) Present and future scopes and challenges of plant pest and disease (P&D) monitoring: remote sensing, image processing, and artificial intelligence perspectives. Remote Sens Appl: Soc Environ 32:100996
24. Sabir RM, Mehmood K, Sarwar A, et al (2024) Remote sensing and precision agriculture: a sustainable future. In: Transforming agricultural management for a sustainable future: climate change and machine learning perspectives. Springer, pp 75–103
25. Saiz-Rubio V, Rovira-Más F (2020) From smart farming towards agriculture 5.0: a review on crop data management. Agronomy 10:207
26. Sharma K, Shivandu SK (2024) Integrating artificial intelligence and internet of things (IoT) for enhanced crop monitoring and management in precision agriculture. Sens Int 100292
27. Nahiyoon SA, Ren Z, Wei P et al (2024) Recent development trends in plant protection UAVs: a journey from conventional practices to cutting-edge technologies—a comprehensive review. Drones 8:457
28. Khan W (2024) Crop production: techniques and technology. academic guru publishing house
29. Allen ES, Chenyang L, Dukes JS et al (2022) Climate Change and resilience in Indiana and beyond. Indiana University Press
30. Tom C, Edet J, Erokare T, et al (2024) Precision agriculture for maximizing crop production
31. Prajapati G, Pandya P (2025) Trailblazing trends in sustainable climate-resilient precision agriculture. Allied Publishers
32. Chandel NS, Chakraborty SK, Jat D, Chouhan P (2024) Smart Farming management system: pre and post-production interventions. In: Artificial intelligence techniques in smart agriculture. Springer, pp 67–82
33. Barbedo JGA (2019) A review on the use of unmanned aerial vehicles and imaging sensors for monitoring and assessing plant stresses. Drones 3:40
34. Laveglia S, Altieri G, Genovese F et al (2024) Advances in sustainable crop management: integrating precision agriculture and proximal sensing. AgriEngineering 6:3084–3120
35. Li C (2024) Linking leaf spectra to genetic variation
36. Ali N, Mohammed A, Bais A, et al (2024) Field scale precision: predicting grain yield of diverse wheat breeding lines using high-throughput UAV multispectral imaging. IEEE J Select Top Appl Earth Observations Remote Sens
37. Aarif KOM, Alam A, Hotak Y (2025) Smart sensor technologies shaping the future of precision agriculture: recent advances and future outlooks. J Sens 2025:2460098
38. Henn KA, Peduzzi A (2024) Surface heat monitoring with high-resolution UAV thermal imaging: assessing accuracy and applications in urban environments. Remote Sensing 16:930
39. Bhattacharya A (2022) Low temperature stress and plant-water relationship: a review. Physiological processes in plants under low temperature stress 107–197
40. Alsalmo A, Hamijo T, Kayim M, Al-Mourrh A (2025) plant disease resistant in the triticum
41. Fareed N, Flores JP, Das AK (2023) Analysis of UAS-LiDAR ground points classification in agricultural fields using traditional algorithms and PointCNN. Remote Sensing 15:483
42. Chen G, Hu L, Luo X et al (2024) A review of global precision land-leveling technologies and implements: current status, challenges and future trends. Comput Electron Agric 220:108901
43. Rajabi MS, Beigi P, Aghakhani S (2023) Drone delivery systems and energy management: a review and future trends. Handbook of Smart Energy Systems 1–19
44. Jiao S, Zhang G, Zhou M, Li G (2023) A comprehensive review of research hotspots on battery management systems for UAVs. Ieee Access 11:84636–84650

45. Johansson B, Despeisse M, Bokrantz J et al (2024) Challenges and opportunities to advance manufacturing research for sustainable battery life cycles. Front Manuf Technol 4:1360076
46. Barcelos CO, Fagundes-Júnior LA, Mendes ALC et al (2024) Integration of payload sensors to enhance uav-based spraying. Drones 8:490
47. Liller J, Goel R, Aziz A et al (2025) Development of a battery free, solar powered, and energy aware fixed wing unmanned aerial vehicle. Sci Rep 15:6141
48. Qasem NA, Abdulrahman GA (2024) A recent comprehensive review of fuel cells: history, types, and applications. Int J Energy Res 2024:7271748
49. Vighi M, Ruiz-Orejón L, Hanke G (2022) Monitoring of floating marine macro litter. European Commission, Europe
50. Das GP, Gould I, Zarafshan P, et al (2022) Applications of robotic and solar energy in precision agriculture and smart farming. In: Solar energy advancements in agriculture and food production systems. Elsevier, pp 351–390
51. Elouarouar S, Medromi H (2022) Multi-rotors unmanned aerial vehicles power supply and energy management. EDP Sciences, p 00068
52. Cavalaris C (2023) Challenges and opportunities for cost-effective use of unmanned aerial system in agriculture. Unmanned Aerial Systems in Agriculture 197–229
53. Alotaibi A, Chatwin C, Birch P (2023) Ubiquitous unmanned aerial vehicles (UAVs): a comprehensive review. Shanlax Int J Arts, Sci Humanities 11:62–90
54. Wanner D, Hashim HA, Srivastava S, Steinhauer A (2024) UAV avionics safety, certification, accidents, redundancy, integrity, and reliability: a comprehensive review and future trends. Drone Systems and Applications 12:1–23
55. Rezaee MR, Hamid NAWA, Hussin M, Zukarnain ZA (2024) Comprehensive review of drones collision avoidance schemes: challenges and open issues. IEEE Transactions on Intelligent Transportation Systems
56. Awais M, Li W, Cheema M, et al (2022) UAV-based remote sensing in plant stress imagine using high-resolution thermal sensor for digital agriculture practices: a meta-review. Int J Environ Sci Technol 1–18
57. Khang A (2023) Handbook of research on AI-equipped IoT applications in high-tech agriculture. IGI Global
58. Samuel P (2024) How AI contributes to precision agriculture. Artificial intelligence for precision agriculture 80
59. Vc SS, Hareendran A, Albaaji GF (2024) Precision farming for sustainability: an agricultural intelligence model. Comput Electron Agric 226:109386
60. Yousuf WB, Khan TMR, Tariq ST, Shah A (2021) Remaining useful life prediction of aerial bundled cables in coastal areas using thermal and corrosion degradation models. IEEE Trans Power Delivery 37:2543–2550
61. Manohar A, Subramanian S (2024) Environmental testing of outdoor electronic devices
62. Khudhur DA, Ali MW, Abdullah TAT (2021) Mechanisms, severity and ignitability factors, explosibility testing method, explosion severity characteristics, and damage control for dust explosion: a concise review. IOP Publishing, p 012023
63. Bhatia A (2022) Understanding motor nameplate information: NEMA vs. Continuing Education and Development Inc., IEC Standards
64. Kumar K, Sharma K, Chaudhary H, et al (2023) A comprehensive review on materialistic and sustainable approach for drone sanitizations. EDP Sciences, p 01181
65. Mihret YC, Takele MM, Mintesinot SM (2025) Advancements in agriculture 4.0 and the needs for introduction and adoption in Ethiopia: a review. Adv Agri 2025:8828400
66. Ortiz Y, Arévalo P, Peña D, Jurado F (2024) Recent advances in thermal management strategies for lithium-ion batteries: a comprehensive review. Batteries 10:83
67. Biswas S, Pandey R, Barua E, Lawrence ID (2023) Advancements in precision agriculture: pesticide spraying drones. Int Res J Modernization Eng Technol Sci 5:1–19
68. Gokool S, Mahomed M, Kunz R et al (2023) Crop monitoring in smallholder farms using unmanned aerial vehicles to facilitate precision agriculture practices: a scoping review and bibliometric analysis. Sustainability 15:3557

69. de Azevedo GBH (2024) On the design of unmanned aerial weather measurement systems for wintry precipitation observations
70. Lumini M (2022) Pherodrone1. 0: an innovative inflatable UAV'S concept, inspired by Zanonia Macrocarpa's samara flying-wing and to insect's Sensillae, designed for the biological control of harmful insects in PA (Precision Agriculture). In: Bionics and sustainable design. Springer, pp 225–287
71. Neetye HS (2023) An investigation of change in drone practices in broadacre farming environments
72. Umetsu N, Shirai Y (2020) Development of novel pesticides in the 21st century. J Pestic Sci 45:54–74
73. Sahni RK, Kumar SP, Thorat D, et al (2024) Drone spraying system for efficient agrochemical application in precision agriculture. In: Applications of computer vision and drone technology in agriculture 4.0. Springer, pp 225–244
74. Chowdhury M, Kushwah A, Satpute AN et al (2023) A comprehensive review on potential application of nanomaterials in the field of agricultural engineering. J Biosyst Eng 48:457–477

Chapter 7
Enhanced Agricultural Productivity: UAV-Based Technology for Precision Crop Yield Estimation and Disease Detection

Abstract The integration of UAV-based technology in agriculture is expected to enhance efficiency, sustainability, and productivity by improving yields, reducing input costs, and promoting sustainable farming practices. The combination of UAV technology, remote sensing, and data analytics offers a powerful tool for farmers to optimize yields, reduce input costs, and foster sustainable farming practices. As technology advances, the accuracy and accessibility of UAV-based solutions for crop management are expected to improve, further benefiting the agricultural industry. UAVs have become indispensable tools in modern agriculture, particularly in crop yield estimation and disease detection. Their ability to capture high-resolution imagery and advanced data processing and analysis techniques has revolutionized precision agriculture. The integration of UAV technology with machine learning models enhances crop trait estimation accuracy and facilitates real-time decision-making for farmers. The synergy between IoT and UAV technologies presents a synergistic approach to obtaining comprehensive insights into crop health, contributing to more informed decision-making and sustainable agricultural practices. However, addressing the challenges associated with UAV adoption is vital for its widespread utilization in agriculture.

Keywords Crop yield estimation · Disease detection · UAV technology · Precious agriculture · IoT · Drone · Remote sensing · Smart farming

7.1 Introduction

Crop diseases pose a significant threat to crop productivity and food security, necessitating effective monitoring and early identification techniques. Traditional methods often rely on human expert inspection, which is time-consuming and physically demanding for large crops. Remote sensing technologies and computer vision techniques are considered effective solutions to these problems [1]. In the agricultural sector, precision farming, farm management software, wireless sensors, and agricultural machinery use remote sensing technology. This study explores the use

© The Author(s), under exclusive license to Springer Nature Singapore Pte Ltd. 2025

J. Li, *UAV Aerodynamics and Crop Interaction*, Smart Agriculture 13,
https://doi.org/10.1007/978-981-96-8402-1_7

of unmanned aerial vehicles (UAV) for image capturing, processing, and analysis in precision agriculture. The Random Forest classifier was used to evaluate UAV-derived spectral and VIs with and without variable optimization. Results provide a rapid UAV-based remote sensing solution for high spatial and temporal resolution in varietal analysis, crucial for timely and unbiased data-driven variety selection and release in plant breeding programs, especially as climate change accelerates [2].

Unmanned Aerial Vehicles (UAVs) are being increasingly used in agriculture for crop yield estimation and disease detection. These technologies capture high-resolution images of crops, providing detailed information on their health, growth stages, and variability. The collected imagery is processed using advanced algorithms to extract key indicators related to crop health and biomass, often correlated with actual yield data through machine learning models and data analytics. Yield mapping is created using the processed data, showing spatial variability in crop yields across the field. UAVs enable frequent and timely monitoring throughout the growing season, allowing for real-time adjustments in management practices [3]. This information assists in making informed decisions related to irrigation, fertilization, and harvesting schedules. UAV technology integrates with precision agriculture practices, allowing for site-specific management of inputs and resources, increasing efficiency and reducing environmental impact. Disease detection is achieved through spectral analysis, early detection of diseases before symptoms become visible, pattern recognition, targeted treatments, and farm decision support [4]. UAV-based technology aids in efficient resource management by identifying areas with specific issues, reducing input costs and environmental impact. It can be integrated with other precision agriculture technologies, such as GPS-guided tractors and automated irrigation systems, enhancing overall farm management.

Precision agriculture (PA) researchers are increasingly focusing on unmanned aerial vehicle (UAV)-based remote sensing due to the vast amount of raw data collected from these platforms. This data necessitates large-scale data processing algorithms like machine learning and deep learning methods [5]. This paper provides a comprehensive survey of recent works using a general pipeline of UAV-based remote sensing for PA research. It classifies the different features extracted from UAV imagery for various agriculture applications, highlighting the importance of each feature for crop model performance and demonstrating how multiple feature fusion can improve models' performance. The paper also compares and contrasts the performance of various machine learning and deep learning models for three important crop trait estimations: yield estimation, disease detection, and crop classification. UAVs have gained popularity for estimating traits like crop yield and diseases, but accurate estimation requires modern data analysis techniques like machine learning and deep learning. This paper reviews the progress in crop disease detection, focusing on machine learning and deep learning techniques using UAV-based remote sensing. Emerging technologies like the Internet of Things (IoT) offer significant potential in Smart Farming and Precision Agriculture applications, enabling real-time environmental data acquisition [6]. However, the actual proliferation and exploitation of UAVs in Smart Farming have not been as robust as expected due to challenges in selecting and deploying relevant technologies, including data acquisition and image processing methods.

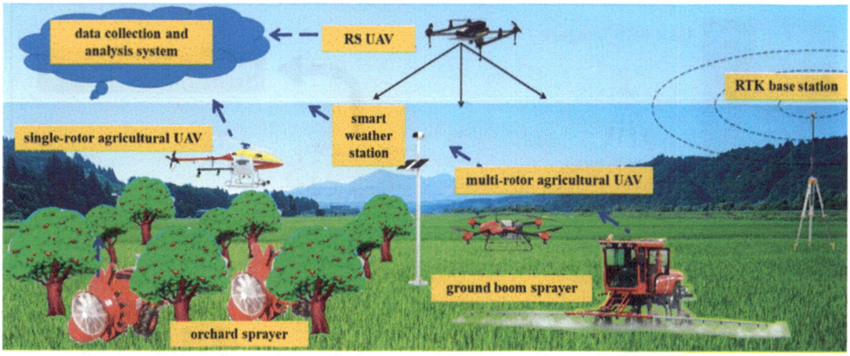

7.2 UAVs in Crop Yield Estimation and Disease Detection

Unmanned Aerial Vehicles (UAVs) are increasingly becoming integral to agriculture for estimating crop yields and detecting diseases. These technologies capture high-resolution images of crops, offering detailed insights into their health, growth stages, and variability. Advanced algorithms process the collected imagery to extract key indicators related to crop health and biomass [7]. These indicators are often correlated with actual yield data using machine learning models and data analytics. The processed data generates yield maps, illustrating spatial variability in crop yields across fields. UAVs enable frequent and timely monitoring throughout the growing season, facilitating real-time adjustments in management practices. This information aids in making informed decisions related to irrigation, fertilization, and harvesting schedules. Precision agriculture practices seamlessly integrate with UAV technology, allowing for site-specific management of inputs and resources, ultimately increasing efficiency and reducing environmental impact. Disease detection through UAV technology involves spectral analysis, early detection of diseases before visible symptoms emerge, pattern recognition, targeted treatments, and farm decision support. UAV-based technology contributes to efficient resource management by identifying areas with specific issues, thereby reducing input costs and environmental impact [8]. Integration with other precision agriculture technologies, such as GPS-guided tractors and automated irrigation systems, further enhances overall farm management.

Agriculture is facing unprecedented challenges with the increasing global demand for food, climate change impacts, and the threat of crop diseases. To address these challenges, there is a growing reliance on advanced technologies, with Unmanned Aerial Vehicles (UAVs) emerging as powerful tools for precision agriculture [9]. This review provides an in-depth examination of the applications of UAVs in crop yield estimation and disease detection, highlighting their contributions to enhancing agricultural productivity and ensuring food security.

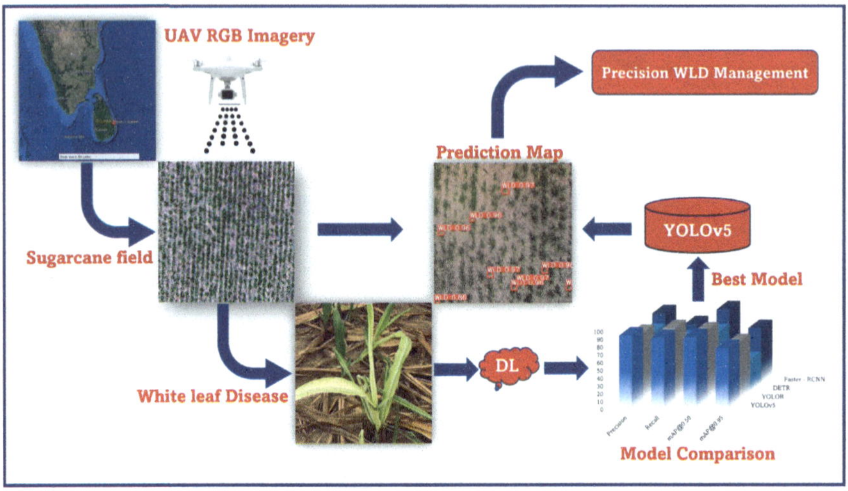

7.2.1 Crop Yield Estimation

The study explores the integration of Unmanned Aerial Vehicles (UAVs) into precision agriculture for precise crop yield estimation. It uses UAV technology, high-resolution imaging, and advanced data analytics to enhance agricultural productivity by providing timely and accurate insights into crop performance. The Random Forest classifier is used to process UAV-derived spectral and vegetation indices for robust crop yield predictions [10]. The results show the potential of UAV-based technology in achieving high spatial and temporal resolution for varietal analysis, vital for informed and data-driven decision-making in contemporary agriculture. UAVs capture detailed imagery of crops, providing data on crop health, growth stages, and spatial variability. The study emphasizes the significance of UAV-based technology in varietal analysis, providing insights into the performance of different crop varieties, essential for informed decision-making in plant breeding programs, especially in the context of climate change impacts [11]. By leveraging UAV-derived data and machine learning models, farmers can make data-driven decisions related to crop management, irrigation, fertilization, and harvesting, enhancing overall efficiency and productivity in agriculture.

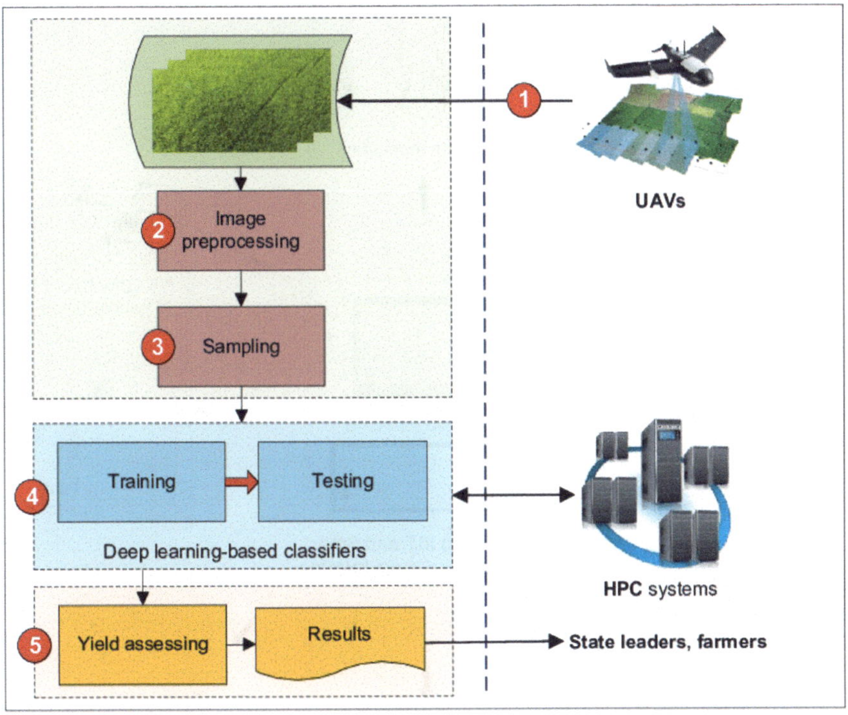

7.2.1.1 Remote Sensing Capabilities

UAVs, equipped with high-resolution sensors, capture detailed imagery of crops throughout the growing season, providing valuable data on crop health, growth stages, and variability within fields. These sensors enable consistent and season-long monitoring of crops, capturing dynamic changes in crop conditions over time. The imagery provides insights into crop health, such as variations in color, texture, and other indicators [12]. UAVs also allow precise analysis of crop growth stages, crucial for optimizing management practices. They contribute to the creation of spatial variability maps, highlighting areas with differing crop health, allowing for targeted interventions. The real-time and dynamic nature of UAV-based monitoring enables farmers to respond swiftly to environmental changes, mitigating risks and optimizing crop yield potential [13]. Benefits of UAV-based remote sensing include timely decision-making, precision agriculture integration, optimized resource use, and increased crop yield prediction accuracy.

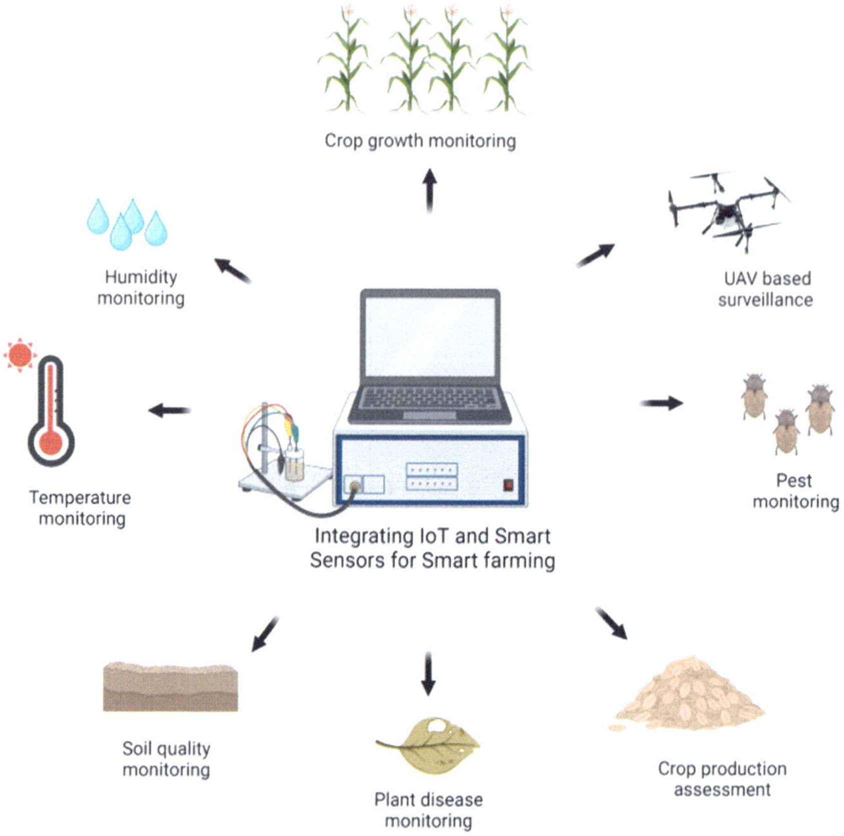

7.2.1.2 Data Processing and Analysis

Advanced algorithms process the collected imagery to extract key indicators related to crop health and biomass. Machine learning models, such as Random Forest classifiers, are often employed to correlate these indicators with actual yield data. The use of Unmanned Aerial Vehicles (UAVs) equipped with high-resolution sensors has revolutionized precision agriculture by employing advanced algorithms and machine learning models [14]. These technologies process and analyze collected imagery to extract key indicators related to crop health and biomass, which are then correlating these indicators with actual yield data. Advanced algorithms focus on extracting indicators such as variations in color, texture, and other spectral features, serving as proxies for overall crop vigor and potential stress factors. Biomass estimation is another important aspect of UAV-derived imagery, analyzing the structural characteristics of crops to gauge their overall growth and productivity. Machine learning models, particularly Random Forest classifiers, are often employed to correlate extracted indicators with actual yield data, contributing to more accurate predictions. The integration of advanced algorithms and machine learning models enhances the precision of crop yield estimation, making informed decisions related to harvest planning and resource allocation [15]. Real-time decision support is provided by these technologies, enabling farmers to adapt management practices on the fly and respond to dynamic environmental conditions. Data-driven insights from the correlation of indicators with actual yield data provide data-driven insights into the factors influencing crop productivity, contributing to continuous improvement in agricultural practices.

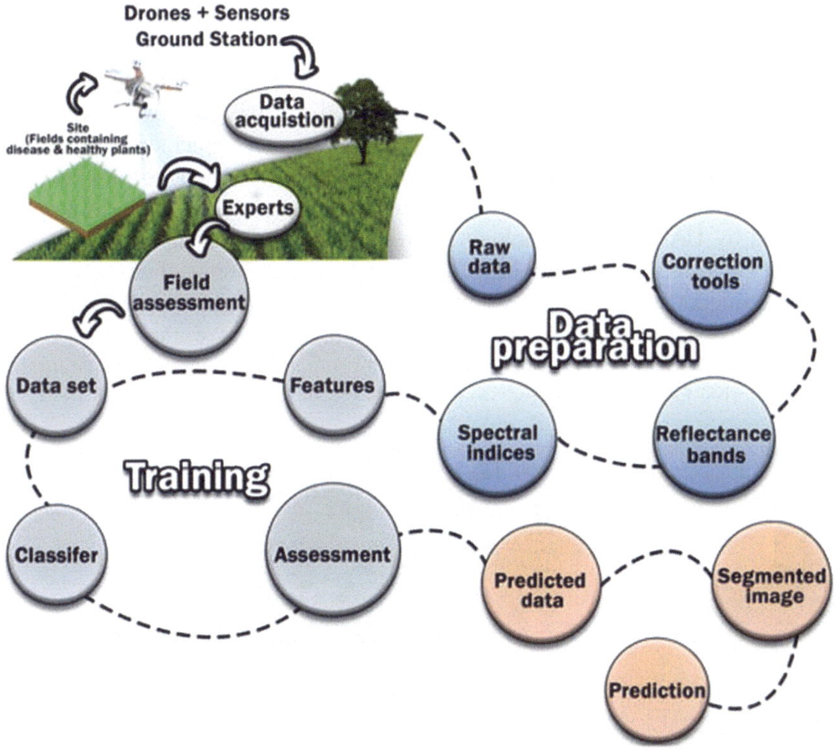

7.2.1.3 Yield Mapping

The processed data is used to create yield maps, offering spatial representations of crop variability within fields. These maps enable farmers to identify areas of high and low productivity, supporting informed decision-making [16]. Yield mapping is a crucial aspect of precision agriculture, utilizing Unmanned Aerial Vehicles (UAVs) equipped with advanced sensing technologies to optimize agricultural productivity.

By harnessing UAV-based technology, farmers can access detailed spatial representations of crop variability within fields, enabling them to make informed decisions for enhanced productivity. Yield maps are created using processed data collected by UAVs, processed through advanced algorithms and machine learning models. These maps offer spatial representations of crop variability across the entire field, highlighting areas with differing levels of productivity [17]. These patterns enable farmers to identify areas of high and low productivity, which can be attributed to factors such as soil health, water availability, and pest infestations. Informed decision-making is possible through yield maps, which can guide farmers in harvest planning, optimizing resource allocation, and contributing to continuous improvement. By analyzing patterns over multiple seasons, farmers can refine their practices, adapt to changing conditions, and optimize strategies for sustainable and efficient agriculture. Benefits of yield mapping include precision agriculture integration, data-driven decision support, efficient resource use, and risk mitigation. By identifying areas with lower yields early in the season, farmers can implement risk mitigation measures, such as addressing pest issues, adjusting irrigation, or implementing soil amendments to enhance overall productivity. In conclusion, yield mapping, facilitated by UAV-based technology in precision agriculture, stands as a cornerstone for enhanced productivity [18]. The spatial representations of crop variability provide farmers with actionable insights, promoting efficient resource allocation and informed decision-making. As agriculture continues to evolve, the integration of UAVs and yield mapping is expected to play a crucial role in ensuring sustainable and productive farming practices.

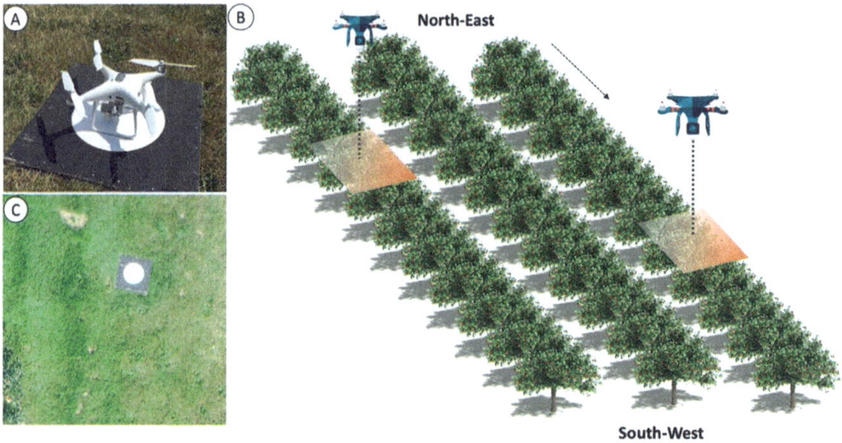

7.2.1.4 Real-Time Monitoring

UAVs facilitate frequent and timely monitoring throughout the growing season, providing real-time insights into crop conditions. This information allows farmers to make proactive adjustments to management practices. The integration of Unmanned Aerial Vehicles (UAVs) has revolutionized precision agriculture by providing farmers with real-time monitoring throughout the growing season [19]. This technology enables farmers to make proactive adjustments to their management practices, ultimately enhancing crop yield estimation. Real-time monitoring involves frequent aerial surveillance over fields, capturing high-resolution imagery and data throughout the entire growing season. This allows for a continuous monitoring process, enabling farmers to respond dynamically to changes in environmental conditions. This adaptability is crucial for mitigating risks associated with weather fluctuations, pests, and diseases. UAVs capture detailed images that provide real-time insights into crop health, allowing for immediate intervention [20]. Timely decision-making is possible due to the information obtained through real-time monitoring, which can be used to adjust irrigation schedules, fertilizer applications, or pest control measures. Adaptive management practices are also essential for optimizing resource use and responding effectively to emerging challenges. Real-time monitoring contributes to harvest optimization, allowing farmers to identify optimal timing for harvesting based on crop maturity and overall field conditions. Benefits of real-time monitoring include proactive pest and disease management, efficient resource allocation, risk mitigation, and optimized crop management [21]. By capturing timely and frequent insights into crop conditions, farmers can make proactive adjustments to their management practices, ultimately enhancing overall agricultural productivity and sustainability.

7.2.1.5 Integration with Precision Agriculture

UAV technology seamlessly integrates with precision agriculture practices, enabling site-specific management of inputs and resources. This integration optimizes resource use, leading to increased efficiency and reduced environmental impact. The integration of Unmanned Aerial Vehicles (UAVs) with precision agriculture practices has significantly improved crop yield estimation [22]. UAVs, equipped with advanced sensors and imaging technologies, enable site-specific management strategies by capturing high-resolution data at a localized level. This allows farmers to tailor their approaches to match the unique conditions of different areas within a field. Variable Rate Technology (VRT) is often facilitated by precision agriculture, allowing for the adjustment of inputs such as fertilizers, pesticides, and water based on specific requirements identified through UAV data. Spatial variability analysis helps farmers identify areas with varying productivity levels and adjust management practices accordingly. Resource optimization is achieved by directing resources precisely where needed, minimizing waste and ensuring efficient input application, contributing to cost savings [23]. The synergy between UAV technology and precision agriculture practices contributes to increased productivity and profitability.

The integration of UAV-based technology in agriculture is expected to enhance efficiency, sustainability, and productivity by improving yields, reducing input costs, and promoting sustainable farming practices. The combination of UAV technology, remote sensing, and data analytics offers a powerful tool for farmers to optimize yields and reduce input costs. As technology advances, the accuracy and accessibility of UAV-based solutions for crop management will improve. UAVs are crucial in crop yield estimation and disease detection, and their ability to capture high-resolution imagery and advanced data processing and analysis techniques has revolutionized precision agriculture. The integration of UAV technology with machine learning models enhances crop trait estimation accuracy and facilitates real-time decision-making for farmers [24]. The synergy between UAVs and precision agriculture represents a paradigm shift in agriculture, contributing to enhanced productivity by processing and analyzing high-resolution imagery, extracting key indicators, and correlating them with actual yield data.

7.3 Disease Detection

Paddy rice cultivation is vital for global food security, but diseases pose a significant threat to crop yields. Advanced technologies in Unmanned Aerial Vehicles (UAVs) can improve disease detection in rice fields. UAVs capture high-resolution images, enabling detailed analysis of individual plants. Advanced sensors on UAVs identify subtle changes in plant color, texture, and structure, enabling early detection of disease symptoms [25]. This allows farmers to implement preventive measures promptly, mitigating disease spread and minimizing crop yield impact. Spectral analysis differentiates between healthy and infected plants based on spectral patterns. Quantitative disease assessment provides a nuanced understanding of disease presence in fields. Automated disease classification using machine learning algorithms enhances the speed and accuracy of disease identification, facilitating quick decision-making across large fields. Precision agriculture integration allows for site-specific disease management, optimizing resource use and minimizing input costs and environmental impact [26]. Real-time monitoring and decision support enable immediate responses to changing conditions and strategic disease management decisions.

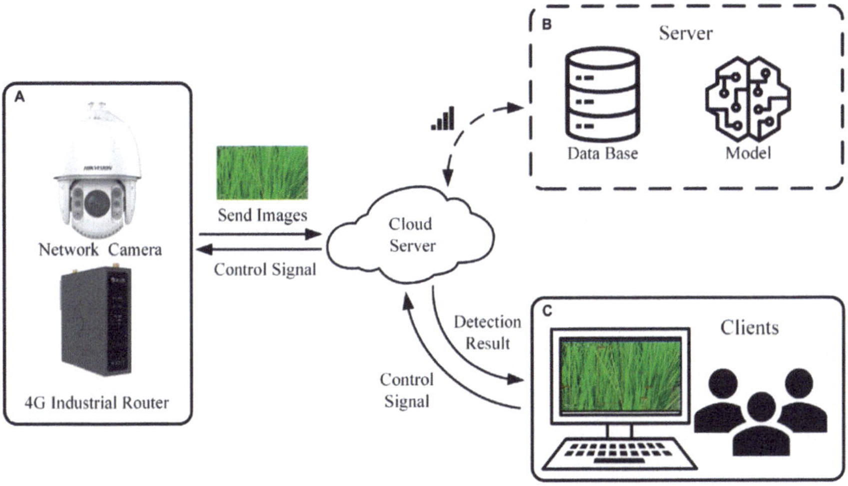

7.3.1 Spectral Analysis

UAVs equipped with specialized sensors capture spectral signatures of crops, allowing for the identification of deviations associated with diseases. This spectral analysis aids in the early detection of diseases. Paddy rice cultivation faces challenges from various diseases that can impact crop yields. Integrating Unmanned Aerial Vehicles (UAVs) equipped with specialized sensors and employing spectral analysis techniques offers a cutting-edge solution for early disease detection in paddy rice fields [27]. This article explores how UAV-based spectral analysis contributes to enhanced agricultural productivity by identifying deviations associated with diseases. Spectral analysis begins with establishing baseline spectral profiles for healthy paddy rice crops, which UAVs capture reference data during optimal crop health periods. Deviation identification occurs when UAVs compare captured spectra with the established baseline, signaling potential issues related to diseases. Early disease detection is achieved by identifying deviations in spectral signatures, even before visible symptoms manifest. Spectral libraries for diseases are created by researchers, containing unique signatures associated with specific diseases affecting paddy rice. UAVs use these libraries as references for disease identification. Precision disease identification allows for targeted and informed disease management strategies. The benefits of spectral analysis in disease detection include early intervention, quantitative disease assessment, and reduced environmental impact [28]. Early intervention allows farmers to implement preventive measures and mitigate potential yield losses. Quantitative disease assessment helps farmers assess the severity of diseases, prioritizing management efforts and allocating resources effectively. Targeted disease management based on spectral analysis reduces the need for blanket applications of agrochemicals, minimizing environmental impact and contributing to sustainable agricultural practices. In conclusion, UAV-based spectral analysis emerges as a powerful tool for enhancing agricultural productivity in paddy rice cultivation. By capturing and analyzing detailed spectral signatures, UAVs enable early and precise disease detection, empowering farmers with timely information for proactive disease management and contributing to more sustainable and efficient practices [4].

7.3.2 Early Intervention

Early detection through UAV technology enables farmers to intervene promptly, implementing targeted measures such as the application of pesticides or changes in irrigation practices to mitigate the spread of diseases. Paddy rice cultivation faces the constant threat of diseases that can significantly impact crop yields. By utilizing advanced technologies, Unmanned Aerial Vehicles (UAVs) can facilitate early detection and prompt intervention in paddy rice fields [29]. UAVs conduct frequent aerial surveillance, capturing high-resolution images and spectral data, allowing for timely detection of subtle changes in crop health. This real-time monitoring allows farmers to respond dynamically to environmental changes, mitigating risks associated with diseases, pests, and stress factors. Early identification of disease hotspots is possible through spectral analysis and detailed imaging, enabling farmers to implement targeted measures such as precise pesticide applications. UAV technology also allows farmers to assess variations in soil moisture content and crop health, prompting adjustments in irrigation practices and reducing the risk of disease spread [30]. Precision disease management is achieved through site-specific interventions and quantitative assessment of disease severity, minimizing potential yield losses and reducing environmental impact.

7.3.3 Pattern Recognition

Machine learning algorithms are employed to recognize patterns indicative of specific diseases. This approach enhances the accuracy and efficiency of disease identification across large agricultural areas. The use of Unmanned Aerial Vehicles (UAVs) equipped with machine learning algorithms for paddy rice disease detection has been shown to enhance agricultural productivity. Machine learning algorithms are trained on diverse datasets, including images captured by UAVs, to learn patterns associated with healthy crops and specific diseases [7]. They are then provided with labeled examples of healthy and diseased rice crops, enabling them to generalize patterns and associations between spectral features and disease presence. The algorithms extract relevant features from the spectral and visual data captured by UAVs, such as color variations and texture patterns, which may indicate specific diseases affecting paddy rice. This automated classification significantly enhances the speed and accuracy of disease identification. The scalability of machine learning-based pattern recognition is particularly advantageous for large agricultural areas, as UAVs can efficiently analyze extensive fields for disease detection. Machine learning models exhibit adaptability to disease variability, recognizing patterns associated with different diseases even accounting for variations within a specific disease type [2]. This results in a robust and versatile disease identification system. Continuous monitoring of paddy rice fields in real-time is enabled by UAVs equipped with machine learning-based pattern recognition. This continuous surveillance ensures that deviations or disease outbreaks are promptly detected. Immediate decision-making is supported by real-time information provided by machine learning algorithms, allowing farmers to act swiftly in response to disease identification and implement targeted interventions to mitigate crop yield impacts. Incorporating machine learning-based pattern recognition with precision agriculture practices allows for site-specific disease management, tailoring interventions based on the specific conditions of different areas within paddy rice fields [31]. Additionally, machine learning algorithms contribute to optimized resource allocation, focusing interventions where they are needed most, reducing input costs.

7.3.4 Precision Treatments

With precise information about the location and extent of diseases within a field, farmers can apply treatments more selectively. This targeted approach minimizes the use of agrochemicals, reducing environmental impact. The use of Unmanned Aerial Vehicles (UAVs) for paddy rice disease detection has revolutionized agricultural practices [32]. By providing precise information about the location and extent of diseases, UAV technology allows farmers to apply treatments more selectively, minimizing the use of agrochemicals and reducing environmental impact. This includes detailed disease mapping, high-resolution imaging, targeted pesticide application, optimized fertilizer application, and integration with Variable Rate Technology (VRT). Precision disease mapping allows farmers to understand the spatial distribution and severity of diseases within the field. High-resolution images captured by UAVs enable farmers to visualize the extent of disease outbreaks with exceptional clarity, contributing to accurate disease maps. Targeted pesticide application is another benefit of precision treatments. Farmers can apply pesticides only to the affected areas, minimizing the overall use of pesticides and reducing environmental impact [33]. UAV-derived data can guide farmers in optimizing nutrient inputs, improving overall efficiency. Variable Rate Technology (VRT) integration allows farmers to adjust input rates based on specific requirements, optimizing resource use and supporting sustainable agricultural practices. Precision treatments also minimize agrochemical runoff, mitigating pesticide residue buildup and reducing the risk of contaminants entering water bodies. They also contribute to smart farming practices, providing real-time decision support and data-driven farm management. In conclusion, precision treatments in paddy rice disease detection represent a significant advancement in sustainable agriculture [34]. By applying treatments selectively and minimizing the use of agrochemicals, farmers contribute to reduced environmental impact, improved resource efficiency, and enhanced agricultural productivity.

7.3.5 Farm Decision Support

The data collected by UAVs provides valuable insights for farm decision-making, supporting strategies for disease management and prevention. UAV-based technology in paddy rice disease detection is a valuable tool for farm decision support. It provides real-time data on crop health and disease prevalence, enabling farmers to make informed decisions for disease management, precision treatments, resource allocation, and long-term planning [35]. This integration of technology enhances farm productivity and sustainability in paddy rice cultivation. UAVs enable continuous surveillance of fields, providing real-time data on crop health and disease prevalence. The data is processed promptly using advanced algorithms for spectral analysis and pattern recognition. This data helps identify disease hotspots, guiding farmers in understanding the areas most affected by diseases. The quantitative data allows for the assessment of disease severity, enabling farmers to prioritize and tailor disease management strategies. UAV-derived data aids in identifying specific areas requiring interventions, allowing for targeted treatments and optimized fertilizer use [36]. Additionally, UAV technology contributes to the accumulation of historical data on crop health and disease patterns, allowing farmers to analyze trends, assess past interventions, and make informed decisions for future seasons.

UAV-based technology has the potential to revolutionize disease detection in paddy rice cultivation through early detection, spectral analysis, machine learning, and integration with precision agriculture. This technology enhances agricultural productivity and food security by capturing and analyzing detailed spectral signatures, providing farmers with timely information for proactive disease management. UAVs facilitate early intervention in disease management, enabling farmers to implement targeted measures promptly, minimizing crop yield impacts and adopting sustainable practices. The integration of UAVs and machine learning is crucial for securing global food supplies from paddy rice fields. Machine learning-based pattern recognition is another powerful tool for enhancing paddy rice disease detection. The combination of automated classification, scalability, adaptability, and real-time decision support contributes to more accurate and efficient disease identification across large agricultural areas. UAV-based technology in paddy rice disease detection also marks a significant advancement in sustainable agriculture, reducing environmental impact, improving resource efficiency, and enhancing agricultural productivity [8]. UAV-based technology in paddy rice disease detection empowers farmers to make informed decisions for disease management, precision treatments, resource allocation, and long-term planning, enhancing overall farm productivity and sustainability in paddy rice cultivation. As technology advances, the role of UAVs in agriculture is expected to expand further, contributing significantly to sustainable and efficient farming practices.

7.4 Advancements in UAV-Based Remote Sensing for Precision Agriculture

Precision agriculture researchers increasingly leverage UAV-based remote sensing due to the vast amount of raw data collected. Large-scale data processing algorithms, such as machine learning and deep learning methods, are essential for handling this data. This paper provides a comprehensive survey of recent works utilizing UAV-based remote sensing in precision agriculture research [37]. It categorizes different features extracted from UAV imagery for various agricultural applications, emphasizing the importance of each feature for crop model performance. The paper demonstrates how multiple feature fusion can enhance the performance of models. Furthermore, it compares and contrasts the efficacy of various machine learning and deep learning models for three crucial crop trait estimations: yield estimation, disease detection, and crop classification. Precision agriculture (PA) aids farmers in making timely decisions by analyzing vast data on the environment and crop details. It aids in increasing production with quality to meet demand [38]. Remote Sensing (RS) plays a crucial role in crop evaluation and soil health conditions, identifying problems and resolving them wisely. Different remote sensing platforms are used for precision agriculture.

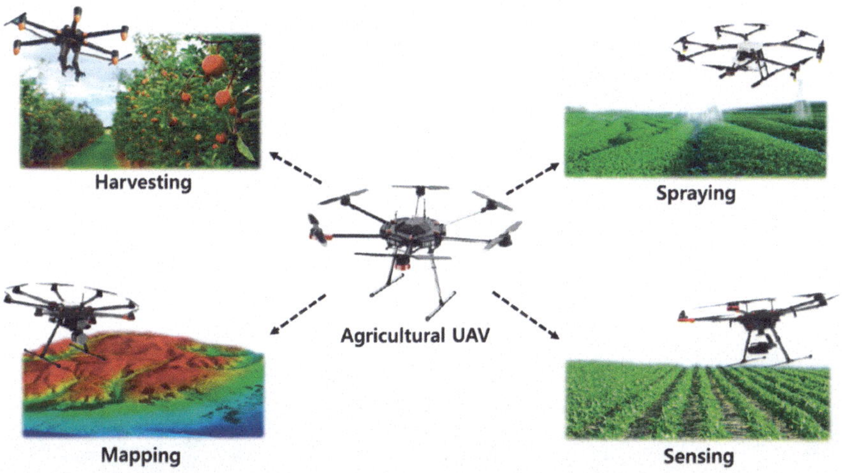

7.4.1 UAV-Based Remote Sensing in Precision Agriculture

UAV-based remote sensing in precision agriculture has revolutionized data collection by capturing vast amounts of environmental and crop details. This data richness presents both opportunities and challenges in analysis and interpretation. Large-scale data processing algorithms, such as machine learning and deep learning, are essential for handling these datasets and extracting meaningful insights from the collected information [39]. The use of Unmanned Aerial Vehicles (UAVs) for remote sensing in precision agriculture has revolutionized data collection and analysis. UAVs provide vast amounts of information about the environment and crop details, enabling real-time monitoring throughout the growing season. This comprehensive view of the agricultural landscape allows for informed decision-making. However, the abundance of data presents challenges in analysis and interpretation. Large-scale data processing algorithms, such as machine learning and deep learning, play a crucial role in handling the extensive datasets generated by UAV-based remote sensing. Machine learning techniques enable the correlation of remote sensing data with key agricultural parameters, providing precise insights into crop conditions. Deep learning methods enhance pattern recognition, identifying subtle patterns in the data. UAV-based remote sensing facilitates accurate crop health assessment, contributing to early detection and intervention [40]. The collected data also allows for a comprehensive analysis of the environmental impact of agricultural practices, leading to more sustainable and environmentally friendly farming practices.

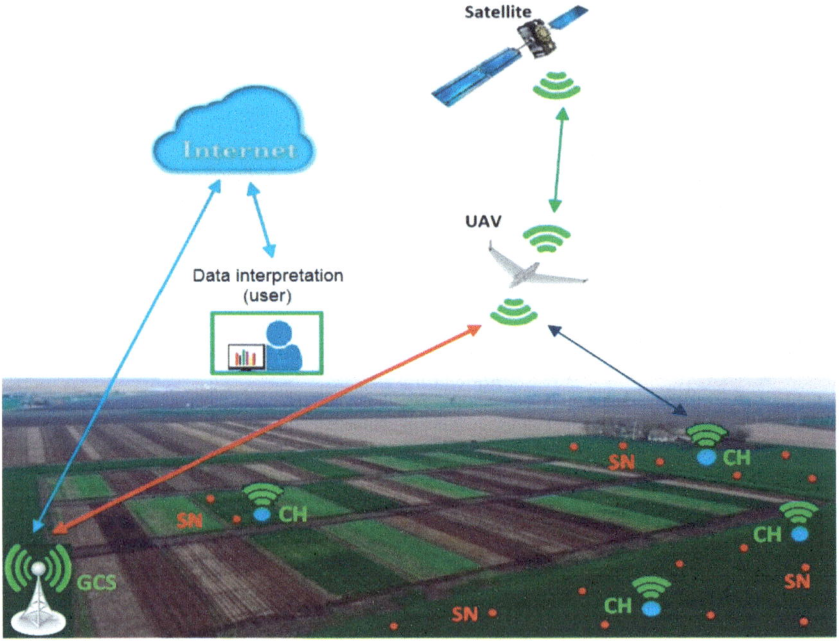

7.4.2 Vast Data Collection

UAVs have revolutionized data collection in precision agriculture, capturing a vast amount of raw data related to the environment and crop details. This data richness presents both opportunities and challenges in terms of analysis and interpretation. Unmanned Aerial Vehicles (UAVs) have revolutionized precision agriculture by capturing extensive information about the environment and crop details [41]. This data richness presents both opportunities and challenges in analysis and interpretation. UAVs, equipped with high-resolution sensors, capture intricate details of the environment, soil conditions, and crop health and growth stages. The dynamic nature of UAV-based remote sensing allows for the collection of multi-dimensional data throughout the growing season, providing a comprehensive dataset. Opportunities arise from the data richness, such as in-depth analysis of crop parameters and identification of spatial variability. However, the complexity of the data necessitates the integration of machine learning and data processing algorithms for effective analysis. Data interpretation for informed decision-making involves transforming raw data into meaningful recommendations, enhancing productivity and mitigating risks [42]. Large-scale data processing algorithms, including machine learning and deep learning methods, play a crucial role in handling the vast datasets. Real-time processing of data is crucial for providing timely insights and optimizing agricultural practices.

7.4.3 Role of Large-Scale Data Processing

Large-scale data processing algorithms, including machine learning and deep learning methods, are essential for handling the extensive datasets generated by UAVs. These algorithms play a crucial role in extracting meaningful insights from the collected information. The use of Unmanned Aerial Vehicles (UAVs) in precision agriculture has led to the development of large-scale data processing algorithms, particularly machine learning and deep learning methods [13]. These algorithms are crucial in handling the vast datasets generated by UAV-based remote sensing, which are multi-dimensional and dynamic, capturing information on crop parameters, environmental conditions, and spatial variability within fields. Machine learning algorithms identify patterns, correlations, and trends within the dataset, enabling the extraction of meaningful insights related to crop health, growth stages, and environmental factors. Deep learning methods, a subset of machine learning, excel in handling complex and non-linear relationships within the data, with neural networks being a prominent technique. Large-scale data processing algorithms correlate UAV-derived data with key agricultural parameters, facilitating the extraction of factors influencing crop yield, detecting early signs of diseases, and optimizing resource allocation [43]. Real-time processing and decision support are essential in the realm of UAV-based remote sensing for precision agriculture, ensuring timely analysis of the data and enhancing the adaptability of precision agriculture practices to changing conditions.

The use of UAV-based remote sensing in precision agriculture is revolutionizing practices by integrating large-scale data processing, extracted features, and machine learning and deep learning models. This integration is expected to drive sustainable and efficient practices, meeting the challenges of a dynamic global food landscape. The data richness presents opportunities for in-depth analysis and informed decision-making, but also introduces complexity [44]. The integration of advanced data processing algorithms, particularly machine learning and deep learning methods, is crucial for translating raw data into actionable insights, enhancing agricultural productivity and sustainability.

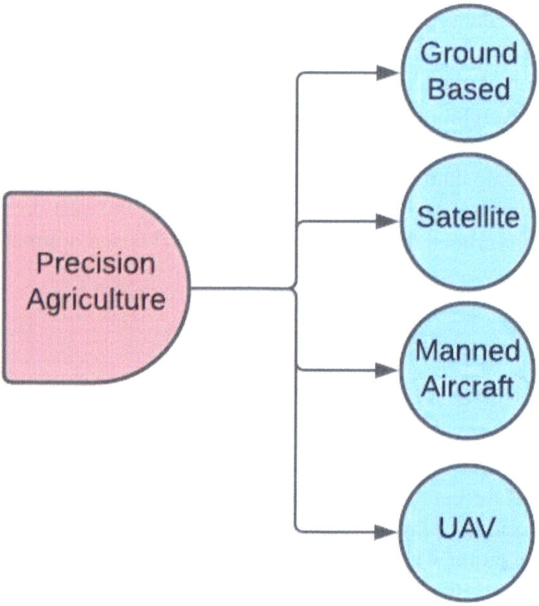

RS Platforms in precision agriculture

7.5 Machine Learning and Deep Learning Techniques in UAV-Based Crop Disease Detection

While UAVs have gained popularity for estimating crop traits like yield and diseases, accurate estimation requires modern data analysis techniques such as machine learning and deep learning. This paper reviews the progress in crop disease detection, focusing on machine learning and deep learning techniques using UAV-based remote sensing [45]. The integration of emerging technologies like the Internet of Things

(IoT) is explored, offering significant potential in Smart Farming and Precision Agriculture applications. The paper acknowledges challenges in the widespread adoption of UAVs in Smart Farming, including the selection and deployment of relevant technologies, such as data acquisition and image processing methods. The paper discusses the role of machine learning and deep learning techniques in crop disease detection using UAV-based remote sensing. It highlights the challenges in accurate estimation and emphasizes the effectiveness of these techniques in handling diverse and intricate datasets. The paper reviews progress made in crop disease detection using UAV-based remote sensing, examining the effectiveness of various machine learning and deep learning algorithms in identifying and classifying diseases based on high-resolution imagery captured by UAVs [46]. Advantages of UAV-based remote sensing include high spatial resolution and the ability to capture detailed images of crops, contributing to early disease detection before visible symptoms manifest. The paper also explores the integration of emerging technologies, particularly the Internet of Things (IoT), in conjunction with UAV-based remote sensing. However, the paper acknowledges challenges in widespread UAV adoption, such as technology selection and deployment [47]. Strategies for addressing these challenges include advancements in sensor technology, improved data processing pipelines, and user-friendly integration with existing agricultural practices.

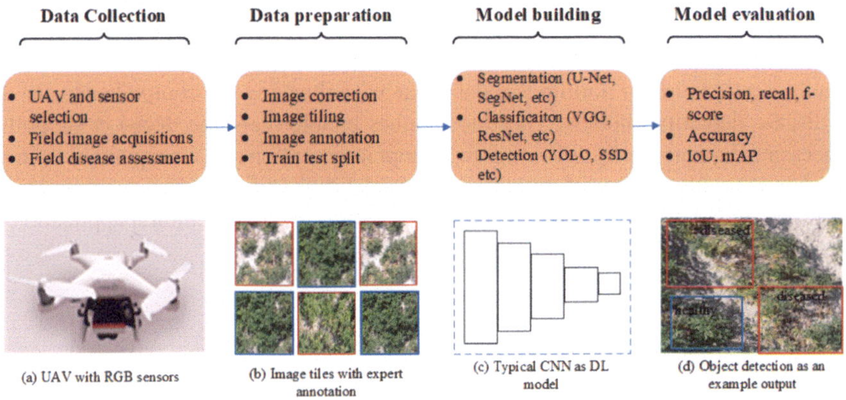

7.5.1 *Importance of Modern Data Analysis Techniques*

The paper discusses the role of machine learning and deep learning techniques in UAV-based crop disease detection, highlighting the challenges of accurate estimation, data complexity, and volume of data collected. It evaluates the effectiveness of different machine learning and deep learning algorithms and explores the early detection capabilities of UAVs equipped with high-resolution sensors [48]. The paper also highlights the transformative potential of the Internet of Things (IoT) in conjunction with UAV-based technology, offering real-time environmental data for holistic decision-making. The paper also discusses the role of emerging technologies, particularly IoT, in Smart Farming and Precision Agriculture applications. The combination of UAVs, IoT, and advanced data analysis techniques enhances agricultural practices' efficiency and sustainability. However, the paper acknowledges challenges in widespread UAV adoption, such as selecting and deploying relevant technologies, addressing data acquisition methods and image processing techniques, and overcoming technological barriers [49]. Strategies for overcoming these barriers may involve advancements in sensor technologies, refining data processing pipelines, and developing user-friendly interfaces.

7.5.1.1 Challenges in Accurate Estimation

Machine learning and deep learning techniques are being used in UAV-based crop disease detection to overcome challenges such as data complexity, volume, environmental variability, and temporal dynamics. These techniques offer dynamic adaptability, allowing them to learn from diverse datasets and adapt to changing conditions. They excel in pattern recognition, identifying subtle visual cues and correlations that may indicate early-stage diseases, contributing to more precise and timely disease detection [50]. Machine learning and deep learning algorithms optimize feature extraction from UAV-captured imagery, enabling a more nuanced understanding of crop health indicators and disease-related patterns. They facilitate the seamless integration of multimodal data sources, enhancing the comprehensiveness of disease detection models. These techniques enable real-time decision support, processing data on-the-fly, allowing for prompt intervention and management strategies based on up-to-date information [51]. Scalability and efficiency are crucial in handling large datasets, ensuring scalability for widespread agricultural applications and enhancing the feasibility of large-scale disease monitoring.

7.5.1.2 Focus on Machine Learning and Deep Learning

Machine learning and deep learning techniques are essential in UAV-based crop disease detection, as they can handle complex datasets, process multidimensional data effectively, and excel in pattern recognition. These techniques enable early detection of crop diseases by identifying subtle patterns, providing farmers with valuable insights for intervention [52]. The adaptive learning capability of these techniques allows them to continuously refine their understanding based on new data inputs, improving the accuracy of disease detection models over time. Continuous monitoring offers real-time decision-making and adaptive management strategies throughout the agricultural cycle. The integration of spatial and spectral information facilitates comprehensive data fusion, enhancing disease detection by considering multiple dimensions of the agricultural landscape. Enhanced feature extraction from UAV-derived data optimizes the performance of disease detection models, aligning with the capabilities of UAVs, allowing for precise and data-driven disease detection [53]. In conclusion, machine learning and deep learning are crucial in UAV-based crop disease detection, contributing to improved agricultural productivity and sustainability.

7.5.2 Reviewing Progress in Crop Disease Detection

The review discusses the evolution of crop disease detection using machine learning and deep learning techniques with UAV-based technology. It highlights the importance of high-resolution imagery captured by UAVs in enabling precise identification of disease-related patterns. The enhanced spatial resolution provided by UAVs helps distinguish subtle variations in crop health, enhancing the accuracy and granularity of disease detection across agricultural fields [54]. Early detection capabilities are crucial, enabling farmers to intervene before diseases become visible. Temporal monitoring for dynamic insights aids in developing robust disease detection models. Challenges in implementing these techniques include data annotation, model training, and integrating results into practical decision-making for farmers. Environmental variability presents a challenge in disease detection, and the paper discusses how machine learning models cope with varying environmental conditions and strategies to ensure robustness. Scalability of models is also a crucial aspect, considering factors like computational resources and data management. User-friendly implementations are also considered, with efforts to make machine learning and deep learning tools accessible to farmers and agronomists [55]. Future trends in UAV-based crop disease detection include innovative technologies, sensor advancements, and novel approaches for enhanced disease identification.

7.5.2.1 Scope of the Paper

The paper examines the advancements in crop disease detection using UAV-based remote sensing technologies, focusing on advancements, methodologies, and data analysis techniques. It evaluates machine learning and deep learning algorithms for identifying and classifying crop diseases, examining their strengths and limitations. The paper focuses on algorithms that use high-resolution imagery captured by UAVs as input data, highlighting the role of advanced imaging in enhancing disease detection capabilities [56]. The paper also explores the contribution of UAV-based remote sensing technology to crop disease detection, highlighting its unique capabilities, such as high spatial resolution and temporal monitoring, in influencing disease identification accuracy and efficiency. The paper also discusses the implications of progress in crop disease detection for precision agriculture, integrating it with and contributing to practices. It evaluates how the reviewed progress provides decision support and aids in resource management for farmers, including optimizing interventions, resource allocation, and overall farm management [57]. Key outcomes include identifying effective approaches and methodologies for improving crop disease detection accuracy, and identifying gaps and areas for future research.

7.5.2.2 Advantages of UAV-Based Remote Sensing

The paper discusses the benefits of UAV-based remote sensing for crop disease detection, focusing on its high spatial resolution, detailed image capture, and early disease detection. These advantages enable detailed crop analysis, enhanced precision in disease detection, comprehensive crop imagery, and the ability to capture multispectral and temporal data. UAVs can capture detailed images of crops, providing information about their health, growth stages, and overall condition [58]. They can also collect information beyond visible light, including infrared and thermal data, allowing for dynamic insights into crop health. Early disease detection facilitates proactive intervention, enabling farmers to intervene before visible symptoms manifest and the disease spreads. Pre-symptomatic identification is also possible, allowing for interventions at a stage where the disease is not yet visible. Timely intervention supports efficient resource management, allowing farmers to allocate resources more judiciously to areas affected by diseases, minimizing wastage and optimizing input use [59]. The integration of UAV technology with advanced data analysis techniques presents a promising approach for improving agricultural productivity and sustainability.

7.5.3 Integration of Emerging Technologies

The paper explores the integration of IoT and UAV-based remote sensing in smart farming, focusing on real-time environmental data acquisition, comprehensive monitoring, and data fusion. It emphasizes the enhanced accuracy of combined data sets, as combining data from different sources enhances disease detection models. The integration aligns with smart farming principles, optimizing resource management and enabling farmers to make data-driven decisions [60]. Automated decision support systems are likely to be explored, integrating machine learning and deep learning models with real-time data from UAVs and IoT sensors. Challenges in technology adoption include initial setup costs, accessibility, and farmer training. Opportunities for enhanced efficiency are discussed, leading to improved farm operations, increased productivity, and sustainable agricultural practices. Future directions and innovation are considered, with the integration of UAV-based remote sensing with IoT aligning with broader technological trends [61]. In conclusion, the integration of IoT and UAV-based remote sensing is crucial for enhancing crop disease detection systems, contributing to the efficiency and sustainability of agricultural practices.

7.5.3.1 Role of the Internet of Things (IoT)

The paper discusses the use of the Internet of Things (IoT) in crop disease detection by integrating UAV-based remote sensing. IoT provides real-time environmental data, essential for crop health assessment and continuous monitoring. It also integrates IoT data with UAV imagery, providing comprehensive insights into crop health. The paper highlights data fusion techniques, emphasizing the synergy between high-resolution imagery from UAVs and real-time environmental data from IoT sensors [62]. It also highlights the spatial and temporal dynamics of crop health, enabling a more nuanced analysis of how environmental factors impact crops over time. IoT enhances smart farming applications by providing real-time decision support and optimizing precision agriculture practices. However, challenges such as data security, privacy, scalability, and deployment are discussed. The paper also discusses emerging trends in IoT for agriculture, including sensor technology, communication protocols, and data analytics techniques [63]. It also discusses how IoT collaboration aligns with other evolving technologies, ensuring compatibility with future innovations in agriculture.

7.5.3.2 Potential in Smart Farming

The paper explores the potential of the Internet of Things (IoT) in Smart Farming, focusing on its integration with UAVs and IoT technologies. It emphasizes the need for a comprehensive understanding of the agricultural ecosystem, dynamic data acquisition, and data-driven decision-making. The integration of IoT and UAVs allows for real-time and continuous monitoring, enabling farmers to adapt to changing circumstances. IoT optimizes resource allocation, reducing waste and improving efficiency [64]. It also advances precision agriculture, allowing site-specific management of agricultural inputs based on real-time data. The combination of UAVs and IoT leads to automated monitoring and intervention systems, enabling continuous monitoring of crop conditions and implementing interventions without human oversight. IoT's contribution to Smart Farming is likely connected to enhanced crop disease detection, as continuous and real-time data complements UAV-captured imagery. The paper also discusses the potential for continued innovation in Smart Farming through the integration of UAVs and IoT, including speculations on emerging technologies and research directions [65]. It positions agriculture within the broader technological landscape of technological advancements.

7.5.4 Challenges in Widespread UAV Adoption

The paper explores the challenges in the widespread adoption of Unmanned Aerial Vehicles (UAV) in agriculture, including high initial investment, operational expenses, regulatory and legal issues, privacy concerns, technical limitations, data processing and analysis, integration with farm management systems, skill and training requirements, perception and attitudes, awareness and education, customization and adaptation, environmental impact considerations, and sustainable practices [66]. The initial investment may limit accessibility for farmers, especially those with smaller landholdings. Operational expenses, such as maintenance, training, and software updates, pose challenges for farmers in resource-constrained environments. Regulatory and legal issues include flight regulations, privacy concerns, limited payload capacity, battery life and flight range, large-scale data processing and analysis, and integration with existing farm management systems. Skill and training requirements for UAV operation include technical expertise, data interpretation skills, perception and attitudes, awareness and education, customization and adaptation to local needs, and adaptation to smallholder farming scenarios [67]. Environmental impact considerations include noise pollution, disturbance to wildlife, and potential chemical runoff from UAV-applied treatments. Sustainable practices are explored, balancing the benefits of precision agriculture with minimizing environmental impact.

7.5.4.1 Technology Selection and Deployment Challenges

The adoption of Unmanned Aerial Vehicles (UAVs) in agriculture is influenced by various factors, including compatibility, scalability, and data security. The diversity of UAV technologies can complicate decision-making for farmers, making it crucial to ensure seamless integration with existing systems, scalability for different farm sizes, and adaptability to diverse crop types. Deployment logistics include operational protocols, training, certification, accessibility, affordability, data management, cybersecurity concerns, user-friendly interfaces, regulatory compliance, and global and local variations [68]. Standardized procedures, training, and operator certification are essential for safe and effective UAV use. Accessibility and affordability are also crucial, with support services and robust data storage and processing infrastructure available. Cybersecurity concerns must be addressed to protect sensitive information. User-friendly interfaces are essential, making UAV-generated data easily understandable for users without extensive technical expertise. Regulatory compliance is also essential, with the need to adapt deployment strategies to diverse regulatory environments [69]. In conclusion, overcoming compatibility issues, ensuring scalability and adaptability, and addressing logistical, accessibility, and data security concerns are essential for successful UAV integration into diverse farming contexts.

7.5.4.2 Addressing Technological Barriers

Technological barriers in the agricultural sector can be overcome through increased research and development investments, standardization, open-source solutions, technological training, capacity building, demonstration projects, strategic planning, government support, public awareness campaigns, and continuous system upgrades.

Innovation in UAV technologies is crucial, and public–private partnerships can foster the development of cutting-edge technologies. Standardization can overcome interoperability challenges and streamline certification processes [70]. Open-source solutions accelerate innovation, reduce costs, and provide accessible tools for farmers and developers. Investing in education and training programs can build farmers' capacity to effectively use and maintain UAV technologies. Strategic planning, government support, public awareness campaigns, and continuous system upgrades are essential for technology adoption [71]. User feedback mechanisms should guide the continuous development of UAV technologies.

The paper discusses the use of machine learning and deep learning techniques in UAV-based crop disease detection, highlighting the benefits of IoT integration for Smart Farming and Precision Agriculture. It emphasizes the advantages of UAV-based remote sensing, such as high spatial resolution, detailed image capture, and early disease detection, which aid in proactive intervention and efficient resource management. The paper also highlights the importance of modern data analysis techniques, particularly machine learning and deep learning, in overcoming challenges in technology selection and deployment. The integration of IoT and UAV technologies offers a synergistic approach to crop health, enabling informed decision-making and sustainable agricultural practices [72]. However, overcoming technological barriers is crucial for successful UAV adoption. A multifaceted approach involving research, development, standardization, open-source collaboration, training initiatives, real-world testing, strategic planning, government support, public awareness, and continuous system upgrades is needed to overcome these challenges and foster widespread adoption of UAV technologies in agriculture.

7.6 Progress Made, Needs, and Challenges

Remote sensing has the potential to revolutionize various aspects of pest management (PA), from land preparation to harvesting. The abundance of high spatial resolution multi-temporal satellite data, low-cost UAVs, and commercially available ground-based proximity sensors have transformed PA operations. Advanced techniques, including empirical, regression, and machine learning approaches, have been used to explore remote sensing applications in PA. Vegetation indices have been developed and tested for their ability to help PA operations, such as variable fertilizer management, irrigation scheduling, disease control, weed mapping, and yield forecasting [73]. However, many challenges need to be addressed before remote sensing technologies can see large-scale adoption in commercial and non-commercial agriculture. Most satellite data is available for free, but it may require significant technical knowledge and expertise to process them for real-world applications. Image pre-processing and post-processing require expert knowledge and software, and many PA operations require fine spatial resolution data with high spectral and temporal resolution. Cloudy days and variable or inconsistent irradiance or sunlight may render many satellite images unsuitable for use. Users/farmers may need to purchase high resolution satellite data, which can be cost-prohibitive, especially for small farms. Images acquired from UAVs are likely to offer a low-cost alternative for small farm operations [74]. Use of UAVs and tractor-mounted sensors also involve special software for data analysis and need professional operators.

There is a general lack of established techniques and frameworks that are accurate, reproducible, and applicable under a wide variety of climatic, soil, crop, and management conditions. Accuracy of methods using remote sensing data depends on factors such as image resolution, atmospheric, climatic, weather conditions, crop and field conditions, and analyses techniques. Addressing the complexity of image processing methods and the amount of technical knowledge and expertise required for application is crucial [75]. The development of accurate, user-friendly systems is likely to result in wider adoption of remote sensing data in commercial and non-commercial PA operations.

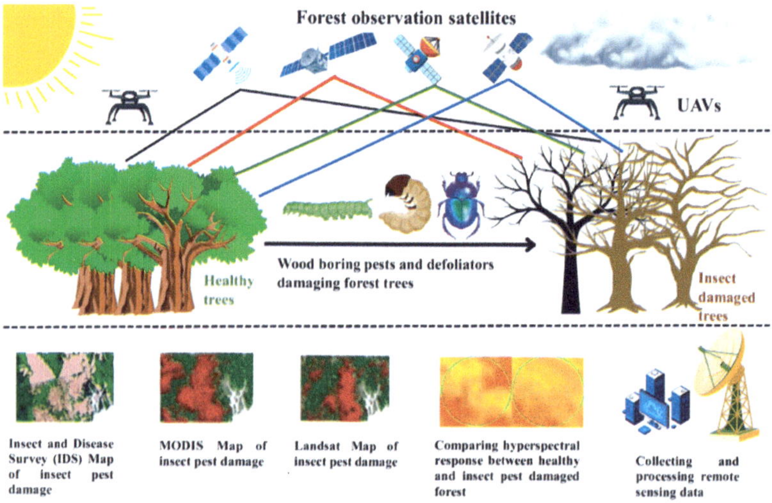

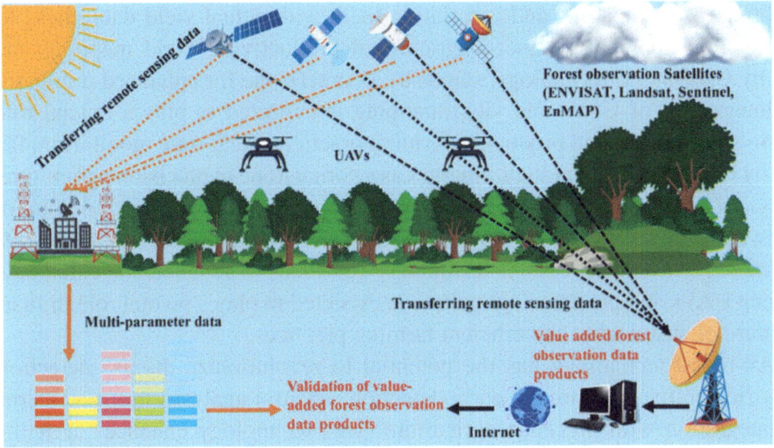

7.7 Conclusion

The integration of UAV-based technology in agriculture is expected to revolutionize efficiency, sustainability, and productivity by improving yields, reducing input costs, and fostering sustainable farming practices. The combination of UAV technology, remote sensing, and data analytics in agriculture offers a powerful tool for farmers to optimize yields, reduce input costs, and foster sustainable farming practices. As technology advances, the accuracy and accessibility of UAV-based solutions for crop management are likely to improve, further benefiting the agricultural industry. UAVs have emerged as indispensable tools in modern agriculture, particularly in crop yield estimation and disease detection. Their ability to capture high-resolution imagery, coupled with advanced data processing and analysis techniques, has revolutionized precision agriculture. The integration of UAV technology with machine learning models enhances the accuracy of crop trait estimations and facilitates real-time decision-making for farmers. As technology continues to advance, the role of UAVs in agriculture is expected to expand further, contributing significantly to sustainable and efficient farming practices. The integration of UAV-based technology into precision agriculture, specifically for crop yield estimation through remote sensing capabilities, offers unprecedented advantages. UAVs equipped with high-resolution sensors provide detailed and dynamic insights into crop health, growth stages, and spatial variability within fields. These capabilities empower farmers to make informed decisions, optimize resource use, and enhance overall agricultural productivity.

The synergy between UAVs and precision agriculture represents a paradigm shift in precision agriculture. By processing and analyzing high-resolution imagery, extracting key indicators, and correlating them with actual yield data, these technologies contribute to enhanced agricultural productivity. Yield mapping, facilitated by UAV-based technology, stands as a cornerstone for enhanced productivity. The integration of UAVs and yield mapping is expected to play a crucial role in ensuring sustainable and productive farming practices. Real-time monitoring, facilitated by UAV-based technology, emerges as a cornerstone in precision agriculture for crop yield estimation. Capturing timely and frequent insights into crop conditions empowers farmers with the information needed to make proactive adjustments to their management practices. As technology continues to advance, the collaboration between UAVs and precision agriculture is expected to play a pivotal role in shaping the future of sustainable and efficient farming practices.

UAV-based technology has the potential to revolutionize disease detection in paddy rice cultivation through early detection, spectral analysis, machine learning, and integration with precision agriculture. This technology enhances agricultural productivity and food security in paddy rice cultivation by capturing and analyzing detailed spectral signatures. UAVs enable early and precise disease detection, providing farmers with timely information for proactive disease management. UAV-based technology facilitates early intervention in paddy rice disease management, enabling farmers to implement targeted measures promptly, minimizing crop yield impacts and adopting sustainable practices. The integration of UAVs and machine learning is crucial for securing global food supplies from paddy rice fields.

Machine learning-based pattern recognition is another powerful tool for enhancing paddy rice disease detection. The combination of automated classification, scalability, adaptability, and real-time decision support contributes to more accurate and efficient disease identification across large agricultural areas. UAV-based technology in paddy rice disease detection also marks a significant advancement in sustainable agriculture. By applying selective treatments and minimizing the use of agrochemicals, farmers contribute to reduced environmental impact, improved resource efficiency, and enhanced agricultural productivity. UAV-based technology in paddy rice disease detection serves as a valuable tool for farm decision support, empowering farmers to make informed decisions for disease management, precision treatments, resource allocation, and long-term planning. This integration of technology with decision-making processes enhances overall farm productivity and sustainability in paddy rice cultivation.

The use of UAV-based technology in crop yield estimation and paddy rice disease detection is a comprehensive strategy to boost agricultural productivity. This method, combining real-time data, machine learning, and targeted interventions, optimizes resource use, reduces environmental impact, and ensures sustainable practices.

The use of UAV-based remote sensing in precision agriculture is transforming agricultural practices. This paper highlights the importance of large-scale data processing, extracted features, and machine learning and deep learning models. The integration of UAV-based technologies is expected to drive sustainable and efficient practices, meeting the challenges of a dynamic global food landscape. The data richness presents opportunities for in-depth analysis and informed decision-making, but also introduces complexity. The integration of advanced data processing algorithms, particularly machine learning and deep learning methods, is crucial for translating raw data into actionable insights that enhance agricultural productivity and sustainability. The synergy between UAVs and data analytics is poised to refine precision agriculture practices, optimizing agricultural practices and contributing to enhanced productivity, sustainability, and resilience in the face of dynamic challenges. As technology advances, the synergy between UAVs and advanced data processing techniques is poised to shape the future of precision agriculture.

The paper critically examines the role of machine learning and deep learning techniques in UAV-based crop disease detection, focusing on the integration of emerging technologies like the Internet of Things (IoT) for advancing Smart Farming and Precision Agriculture. The paper highlights the advantages of UAV-based remote sensing, such as high spatial resolution, detailed image capture, and early disease detection, which contribute to proactive intervention and efficient resource management. The paper also emphasizes the importance of modern data analysis techniques, particularly machine learning and deep learning, in overcoming challenges in technology selection and deployment. These advanced techniques contribute to more accurate, timely, and scalable disease detection, ultimately enhancing agricultural productivity and sustainability. The paper highlights the potential of UAV technology in conjunction with IoT in crop disease detection, providing additional layers of real-time environmental data that enhances the capabilities of smart farming and precision agriculture applications.

The integration of IoT and UAV technologies presents a synergistic approach to obtaining comprehensive insights into crop health, contributing to more informed decision-making and sustainable agricultural practices. The paper recognizes the substantial potential of the Internet of Things (IoT) within the context of Smart Farming, offering a dynamic and comprehensive understanding of the agricultural ecosystem, fostering data-driven decision-making, optimizing resource allocation, and advancing precision agriculture practices.

However, addressing the challenges associated with the adoption of UAV technology is vital for its widespread utilization in agriculture. Overcoming financial barriers, navigating regulatory landscapes, and addressing compatibility issues, ensuring scalability and adaptability, and addressing logistical, accessibility, and data security concerns are essential aspects of successfully integrating UAV technology into diverse farming contexts. To overcome technological barriers to UAV adoption, a multifaceted approach involving research and development, standardization, open-source collaboration, training initiatives, real-world testing, strategic planning, government support, public awareness, and continuous system upgrades is required. This approach may provide insights into how these strategies can collectively contribute to overcoming technological challenges and fostering the widespread adoption of UAV technologies in agriculture.

References

1. Ouhami M, Hafiane A, Es-Saady Y et al (2021) Computer vision, IoT and data fusion for crop disease detection using machine learning: a survey and ongoing research. Remote Sensing 13:2486
2. Chivasa W, Mutanga O, Biradar C (2020) UAV-based multispectral phenotyping for disease resistance to accelerate crop improvement under changing climate conditions. Remote Sensing 12:2445
3. Fan J, Zhou J, Wang B et al (2022) Estimation of maize yield and flowering time using multi-temporal UAV-based hyperspectral data. Remote Sensing 14:3052
4. Su J, Zhu X, Li S, Chen W-H (2023) AI meets UAVs: a survey on AI empowered UAV perception systems for precision agriculture. Neurocomputing 518:242–270
5. Shahi TB, Xu CY, Neupane A, Guo W (2022) Machine learning methods for precision agriculture with UAV imagery: a review. Electron Res Arch 30:4277–4317
6. Boursianis AD, Papadopoulou MS, Diamantoulakis P et al (2022) Internet of things (IoT) and agricultural unmanned aerial vehicles (UAVs) in smart farming: a comprehensive review. Internet Things 18:100187
7. Shahi TB, Xu C-Y, Neupane A, Guo W (2023) Recent advances in crop disease detection using UAV and deep learning techniques. Remote Sensing 15:2450
8. Velusamy P, Rajendran S, Mahendran RK et al (2021) Unmanned aerial vehicles (UAV) in precision agriculture: applications and challenges. Energies 15:217
9. Raj R, Kar S, Nandan R, Jagarlapudi A (2020) Precision agriculture and unmanned aerial Vehicles (UAVs). In: Unmanned aerial vehicle: applications in agriculture and environment, pp 7–23
10. Narmilan A, Gonzalez F, Salgadoe ASA et al (2022) Predicting canopy chlorophyll content in sugarcane crops using machine learning algorithms and spectral vegetation indices derived from UAV multispectral imagery. Remote Sensing 14:1140
11. Cudjoe DK, Virlet N, Castle M et al (2023) Field phenotyping for African crops: overview and perspectives. Front Plant Sci 14
12. Surendran U, Nagakumar KCV, Samuel MP (2024) Remote sensing in precision agriculture. In: Digital agriculture: a solution for sustainable food and nutritional security. Springer, pp 201–223
13. Tsouros DC, Bibi S, Sarigiannidis PG (2019) A review on UAV-based applications for precision agriculture. Information 10:349
14. Zha H, Miao Y, Wang T et al (2020) Improving unmanned aerial vehicle remote sensing-based rice nitrogen nutrition index prediction with machine learning. Remote Sensing 12:215
15. Elbasi E, Zaki C, Topcu AE et al (2023) Crop prediction model using machine learning algorithms. Appl Sci 13:9288
16. Lyle G, Clarke K, Kilpatrick A et al (2023) A spatial and temporal evaluation of broad-scale yield predictions created from yield mapping technology and Landsat satellite imagery in the Australian mediterranean dryland cropping region. ISPRS Int J Geo Inf 12:50
17. Bellis ES, Hashem AA, Causey JL et al (2022) Detecting intra-field variation in rice yield with unmanned aerial vehicle imagery and deep learning. Front Plant Sci 13:716506
18. Jafarbiglu H (2023) Quantitative adjustment of sun-view geometry in areal remote sensing
19. Zaman Q (2023) Precision agriculture: evolution, insights and emerging trends
20. Zhao G, Zhang Y, Lan Y et al (2023) Application progress of UAV-LARS in identification of crop diseases and pests. Agronomy 13:2232
21. Khan A, Shahriyar AK (2023) Optimizing onion crop management: a smart agriculture framework with IoT sensors and cloud technology. Appl Res Artif Intell Cloud Comput 6:49–67
22. Alexopoulos A, Koutras K, Ali SB et al (2023) Complementary use of ground-based proximal sensing and airborne/spaceborne remote sensing techniques in precision agriculture: a systematic review. Agronomy 13:1942
23. Goel RK, Yadav CS, Vishnoi S, Rastogi R (2021) Smart agriculture–urgent need of the day in developing countries. Sustain Comput: Inf Syst 30:100512

24. Olson D, Anderson J (2021) Review on unmanned aerial vehicles, remote sensors, imagery processing, and their applications in agriculture. Agron J 113:971–992
25. Neupane K, Baysal-Gurel F (2021) Automatic identification and monitoring of plant diseases using unmanned aerial vehicles: a review. Remote Sensing 13:3841
26. Karunathilake E, Le AT, Heo S et al (2023) The path to smart farming: innovations and opportunities in precision agriculture. Agriculture 13:1593
27. Lawrence ID, Vijayakumar R, Agnishwar J (2023) Dynamic application of unmanned aerial vehicles for analyzing the growth of crops and weeds for precision agriculture. In: Artificial intelligence tools and technologies for smart farming and agriculture practices. IGI Global, pp 115–132
28. Rovini E, Maremmani C, Cavallo F (2017) How wearable sensors can support Parkinson's disease diagnosis and treatment: a systematic review. Front Neurosci 11:555
29. Sharma A, Singh PK (2023) Applicability of UAVs in detecting and monitoring burning residue of paddy crops with IoT Integration: a step towards greener environment. Comput Ind Eng 184:109524
30. Ahansal Y, Bouziani M, Yaagoubi R et al (2022) Towards smart irrigation: a literature review on the use of geospatial technologies and machine learning in the management of water resources in arboriculture. Agronomy 12:297
31. González-Rodríguez VE, Izquierdo-Bueno I, Cantoral JM et al (2024) Artificial intelligence: a promising tool for application in phytopathology
32. Roslim MHM, Juraimi AS, Che'Ya NN, et al (2021) Using remote sensing and an unmanned aerial system for weed management in agricultural crops: a review. Agronomy 11:1809
33. Jacquet F, Jeuffroy M-H, Jouan J et al (2022) Pesticide-free agriculture as a new paradigm for research. Agron Sustain Dev 42:8
34. Shaikh TA, Rasool T, Lone FR (2022) Towards leveraging the role of machine learning and artificial intelligence in precision agriculture and smart farming. Comput Electron Agric 198:107119
35. Elijah O, Rahman TA, Orikumhi I et al (2018) An overview of Internet of Things (IoT) and data analytics in agriculture: benefits and challenges. IEEE Internet Things J 5:3758–3773
36. Zhang J, Hu Y, Li F et al (2023) Meta-analysis assessing potential of drone remote sensing in estimating plant traits related to nitrogen use efficiency. Available at SSRN 4666863
37. Phang SK, Chiang THA, Happonen A, Chang MML (2023) From satellite to UAV-based remote sensing: a review on precision agriculture. IEEE Access
38. Javaid M, Haleem A, Singh RP, Suman R (2022) Enhancing smart farming through the applications of Agriculture 4.0 technologies. Int J Intell Netw 3:150–164
39. Srivastava A, Prakash J (2023) Techniques, answers, and real-world UAV implementations for precision farming. Wireless Pers Commun 131:2715–2746
40. Kent OW, Chun TW, Choo TL, Kin LW (2023) Early symptom detection of basal stem rot disease in oil palm trees using a deep learning approach on UAV images. Comput Electron Agric 213:108192
41. Monteiro A, Santos S, Gonçalves P (2021) Precision agriculture for crop and livestock farming—brief review. Animals 11:2345
42. Sarker IH (2021) Data science and analytics: an overview from data-driven smart computing, decision-making and applications perspective. SN Comput Sci 2:377
43. Guimarães N, Pádua L, Marques P et al (2020) Forestry remote sensing from unmanned aerial vehicles: a review focusing on the data, processing and potentialities. Remote Sensing 12:1046
44. Huang L, Zheng W, Hong J et al (2020) Paths and strategies for sustainable urban renewal at the neighbourhood level: a framework for decision-making. Sustain Cities Soc 55:102074
45. Osco LP, Junior JM, Ramos APM et al (2021) A review on deep learning in UAV remote sensing. Int J Appl Earth Obs Geoinf 102:102456
46. Bouguettaya A, Zarzour H, Kechida A, Taberkit AM (2023) A survey on deep learning-based identification of plant and crop diseases from UAV-based aerial images. Clust Comput 26:1297–1317

47. Labib NS, Brust MR, Danoy G, Bouvry P (2021) The rise of drones in internet of things: a survey on the evolution, prospects and challenges of unmanned aerial vehicles. IEEE Access 9:115466–115487
48. Samaras S, Diamantidou E, Ataloglou D et al (2019) Deep learning on multi sensor data for counter UAV applications—a systematic review. Sensors 19:4837
49. Fotouhi A, Qiang H, Ding M et al (2019) Survey on UAV cellular communications: practical aspects, standardization advancements, regulation, and security challenges. IEEE Commun Surv Tutorials 21:3417–3442
50. Ecke S, Dempewolf J, Frey J et al (2022) UAV-based forest health monitoring: a systematic review. Remote Sensing 14:3205
51. Ivanov D, Dolgui A, Das A, Sokolov B (2019) Digital supply chain twins: managing the ripple effect, resilience, and disruption risks by data-driven optimization, simulation, and visibility. In: Handbook of ripple effects in the supply chain, pp 309–332
52. Fu X, Jiang D (2022) High-throughput phenotyping: the latest research tool for sustainable crop production under global climate change scenarios. In: Sustainable crop productivity and quality under climate change. Elsevier, pp 313–381
53. Yinka-Banjo C, Ajayi O (2019) Sky-farmers: applications of unmanned aerial vehicles (UAV) in agriculture. Auton Veh 107–128
54. Hassan M, Kowalska A, Ashraf H (2023) Advances in deep learning algorithms for agricultural monitoring and management. Appl Res Artif Intell Cloud Comput 6:68–88
55. Abioye EA, Hensel O, Esau TJ et al (2022) Precision irrigation management using machine learning and digital farming solutions. AgriEngineering 4:70–103
56. Adão T, Hruška J, Pádua L et al (2017) Hyperspectral imaging: a review on UAV-based sensors, data processing and applications for agriculture and forestry. Remote Sensing 9:1110
57. Namany S, Al-Ansari T, Govindan R (2019) Sustainable energy, water and food nexus systems: a focused review of decision-making tools for efficient resource management and governance. J Clean Prod 225:610–626
58. Aeberli A, Johansen K, Robson A et al (2021) Detection of banana plants using multi-temporal multispectral UAV imagery. Remote Sensing 13:2123
59. Kremsa VŠ (2021) Sustainable management of agricultural resources (agricultural crops and animals). In: Sustainable resource management. Elsevier, pp 99–145
60. Lachgar M, Hrimech H, Kartit A (2023) Unmanned aerial vehicle-based applications in smart farming: a systematic review. Int J Adv Comput Sci Appl 14
61. Singh AP, Yerudkar A, Mariani V et al (2022) A bibliometric review of the use of unmanned aerial vehicles in precision agriculture and precision viticulture for sensing applications. Remote Sensing 14:1604
62. Islam N, Rashid MM, Pasandideh F et al (2021) A review of applications and communication technologies for internet of things (Iot) and unmanned aerial vehicle (UAV) based sustainable smart farming. Sustainability 13:1821
63. Ray PP (2017) Internet of things for smart agriculture: technologies, practices and future direction. J Am Intell Smart Environ 9:395–420
64. Xiao Z, Chen Y, Jiang H et al (2022) Resource management in UAV-assisted MEC: state-of-the-art and open challenges. Wireless Netw 28:3305–3322
65. Kumar R, Sinwar D, Pandey A et al (2022) IoT Enabled technologies in smart farming and challenges for adoption. Internet Things Anal Agric 3:141–164
66. Sima V, Gheorghe IG, Subić J, Nancu D (2020) Influences of the industry 4.0 revolution on the human capital development and consumer behavior: a systematic review. Sustainability 12:4035
67. Delavarpour N, Koparan C, Nowatzki J et al (2021) A technical study on UAV characteristics for precision agriculture applications and associated practical challenges. Remote Sensing 13:1204
68. McCaig M, Dara R, Rezania D (2023) Farmer-centric design thinking principles for smart farming technologies. Internet Things 23:100898

69. Pepe M, Alfio VS, Costantino D (2022) UAV platforms and the SfM-MVS approach in the 3D surveys and modelling: a review in the cultural heritage field. Appl Sci 12:12886
70. Surana K, Singh A, Sagar AD (2020) Strengthening science, technology, and innovation-based incubators to help achieve sustainable development goals: lessons from India. Technol Forecast Soc Chang 157:120057
71. Kendall H, Clark B, Li W et al (2022) Precision agriculture technology adoption: a qualitative study of small-scale commercial "family farms" located in the North China Plain. Precision Agric 1–33
72. Peladarinos N, Piromalis D, Cheimaras V et al (2023) Enhancing smart agriculture by implementing digital twins: a comprehensive review. Sensors 23:7128
73. Sishodia RP, Ray RL, Singh SK (2020) Applications of remote sensing in precision agriculture: a review. Remote Sensing 12:3136
74. Neetye HS (2023) An investigation of change in drone practices in broadacre farming environments
75. Richards JA, Richards JA (2022) Remote sensing digital image analysis. Springer

Chapter 8
Technological Advances in UAV-Assisted Crop Monitoring

Abstract Unmanned Aerial Vehicles (UAVs) have revolutionized crop monitoring and health assessment by providing real-time, high-resolution data for precision agriculture. These aerial platforms, equipped with advanced sensors like multispectral, hyperspectral, and thermal cameras, enable the detection of crop stress, nutrient deficiencies, pest infestations, and diseases. The integration of artificial intelligence and machine learning algorithms further enhances data analysis, ensuring timely interventions. The integration of multispectral, thermal, and RGB imaging with UAV technology has significantly improved crop health monitoring, providing crucial insights into nutrient status, water stress, and overall plant condition. As UAV imaging technology continues to evolve, its role in sustainable farming and global food security will become even more significant. UAV-based disease detection is revolutionizing crop protection by providing accurate, early identification of diseases, tracking their progression, and enabling precision treatments. UAV-based pest monitoring is transforming modern agriculture by providing faster, more accurate, and environmentally friendly solutions for pest detection and control. As UAV technology advances, its role in sustainable agriculture will grow, promoting eco-friendly and efficient farming practices.

Keywords Crop monitoring · Precision agriculture · Remote sensing · Disease detection · Artificial intelligence · Sustainable agriculture

8.1 Introduction

The rapid advancement of technology has significantly transformed agricultural practices, with Unmanned Aerial Vehicles (UAVs), or drones, emerging as a powerful tool for precision farming. Crop monitoring and health assessment are critical components of modern agriculture, directly impacting yield optimization, resource efficiency, and sustainability. Traditional methods such as manual field inspections, satellite imagery, and ground-based sensors have long been used for monitoring crop conditions [1]. However, these methods are often labor-intensive, time-consuming, costly,

© The Author(s), under exclusive license to Springer Nature Singapore Pte Ltd. 2025

J. Li, *UAV Aerodynamics and Crop Interaction*, Smart Agriculture 13,
https://doi.org/10.1007/978-981-96-8402-1_8

and sometimes lack the precision required for timely decision-making. UAVs offer a revolutionary approach by providing high-resolution, real-time data that enable farmers and researchers to assess crop health more efficiently and accurately. UAVs equipped with advanced imaging technologies, such as multispectral, hyperspectral, and thermal sensors, allow for detailed assessments of plant health by detecting subtle changes in vegetation that are often invisible to the naked eye. These sensors can identify stress factors such as nutrient deficiencies, water shortages, pest infestations, and diseases before they become severe. The ability to capture such crucial information early enables proactive interventions, reducing crop losses and optimizing inputs like fertilizers, pesticides, and water [2]. Additionally, UAVs can cover large agricultural fields in a short period, making them an ideal tool for large-scale farms and remote areas where frequent manual inspections are impractical. Beyond monitoring, UAVs play a crucial role in precision agriculture by integrating with data analytics and artificial intelligence (AI) tools. By processing aerial images and applying machine learning algorithms, UAVs can generate actionable insights, such as vegetation indices (e.g., NDVI—Normalized Difference Vegetation Index) that help quantify plant health and stress levels. This integration of UAV technology with data-driven decision-making enhances farm management, minimizes environmental impact, and increases overall productivity. Despite their numerous advantages, the adoption of UAVs in agriculture also presents challenges. Factors such as regulatory restrictions, high initial investment costs, limited battery life, and the need for technical expertise in data interpretation may hinder widespread implementation. However, continuous advancements in drone technology, improved affordability, and growing awareness among farmers are expected to drive their adoption in precision agriculture [3]. This Chapter explores the role of UAVs in crop monitoring and health assessment, highlighting their applications, benefits, challenges, and future prospects. By leveraging UAV technology, the agricultural industry can achieve higher efficiency, sustainability, and resilience against climate variability and global food security challenges.

8.2 Imaging for Crop Health Monitoring

The use of imaging technologies in crop health monitoring has transformed modern agriculture, allowing for accurate, efficient, and timely assessments of plant conditions. Traditional crop scouting methods, such as manual field inspections and satellite imaging, often suffer from inefficiencies, delays, and resolution limitations. UAVs (Unmanned Aerial Vehicles) equipped with advanced imaging sensors have emerged as a powerful tool for precision agriculture, providing high-resolution and real-time insights into crop health [4]. Different imaging techniques serve specific purposes in monitoring various aspects of plant health. Multispectral imaging helps analyze nutrient status by capturing data across different spectral bands, including near-infrared (NIR), which reveals details about plant vigor and chlorophyll content. Thermal imaging detects water stress by measuring temperature variations in crops, helping optimize irrigation management. RGB cameras, the most commonly used imaging tools, provide high-resolution visual data for assessing visible plant damage, pest infestations, and overall crop conditions [5].

8.2.1 Multispectral Imaging for Nutrient Analysis

Nutrient management is a key factor in ensuring optimal crop growth and yield. However, nutrient deficiencies are not always visible to the human eye in their early stages, making early detection crucial for maintaining plant health. Multispectral imaging, which captures data across multiple spectral bands, including visible and near-infrared (NIR) light, has become an essential tool for assessing nutrient status in crops [6]. By analyzing the reflectance properties of plants at different wavelengths, multispectral imaging helps detect chlorophyll levels, identify nutrient deficiencies, and assess plant stress before symptoms appear. One of the most widely used tools for analyzing multispectral data is the Normalized Difference Vegetation Index (NDVI), which provides a quantitative measure of plant health. This technology enables precision agriculture by optimizing fertilizer application, reducing excessive use of agrochemicals, and improving overall crop productivity [7]. One of the most commonly used vegetation indices derived from multispectral imaging is the Normalized Difference Vegetation Index (NDVI). NDVI is calculated using the formula:

Applications in Crop Monitoring:

- **Detecting Nutrient Deficiencies**: Different nutrients affect how plants absorb and reflect light. Multispectral data can be used to assess deficiencies in nitrogen, phosphorus, and potassium, allowing for targeted fertilizer applications.
- **Assessing Plant Health and Growth Stages**: Multispectral images track chlorophyll levels and photosynthetic activity, helping farmers monitor plant development.
- **Optimizing Fertilizer Use**: Precision agriculture techniques use multispectral imaging to create nutrient distribution maps, ensuring fertilizers are applied where needed rather than uniformly.
- **Early Disease and Pest Detection**: Stressed plants exhibit different spectral signatures, making it possible to detect disease outbreaks or pest infestations before they spread. By utilizing multispectral imaging, farmers can maximize crop productivity while reducing excessive fertilizer application, thus minimizing environmental impact.

8.2.2 *Thermal Imaging for Water Stress Detection*

Water availability is one of the most critical factors affecting plant growth, and efficient water management is essential for sustainable agriculture. Traditional methods of detecting water stress, such as soil moisture sensors or manual inspections, often lack spatial coverage and timeliness. UAV-based thermal imaging has emerged as a reliable technique for detecting early signs of water stress in crops by measuring temperature variations. Plants regulate temperature through transpiration, a process that cools their surface as water evaporates [8]. When a plant experiences water stress, transpiration decreases, causing leaf temperatures to rise. Thermal cameras detect these temperature changes, allowing for the identification of drought-stressed areas before visible symptoms appear. This information helps farmers optimize irrigation schedules, prevent yield losses, and improve water conservation efforts. The integration of thermal imaging in precision agriculture not only enhances irrigation efficiency but also supports early stress detection, ensuring that crops receive adequate water without wastage [9].

Applications in Crop Monitoring:

- **Early Detection of Water Stress**: By identifying hot spots in a field, farmers can intervene before drought stress impacts crop yields.
- **Precision Irrigation Management**: Thermal imaging helps optimize irrigation schedules, ensuring water is applied efficiently and reducing waste.
- **Identifying Irrigation System Leaks and Inefficiencies**: Uneven crop temperatures can indicate irrigation system malfunctions, such as clogged or broken pipes.
- **Differentiating Between Healthy and Stressed Plants**: Thermal data provides insights into variations in plant health that may not yet be visible through conventional RGB cameras.

By integrating thermal imaging with automated irrigation systems, farmers can ensure optimal water distribution, reducing costs and conserving water resources.

8.2.3 *RGB Cameras for Visual Inspections*

RGB (Red, Green, Blue) cameras are the most common imaging tools used in UAV-based crop monitoring due to their affordability and simplicity. These cameras capture images in the visible spectrum, similar to human vision, making them an essential tool for visual inspections of crops. Although RGB cameras lack the ability to detect plant stress at an early stage like multispectral or thermal imaging, they provide valuable insights into physical crop conditions. High-resolution aerial imagery obtained through UAV-mounted RGB cameras allows farmers to monitor plant growth, detect pest infestations, assess disease spread, and evaluate plant density. With advancements in artificial intelligence (AI) and machine learning, RGB imagery is increasingly being used for automated disease detection and classification [10]. By analyzing changes in color, texture, and canopy structure, AI-powered software can identify potential crop health issues, reducing the need for manual scouting and improving decision-making in farm management.

Applications in Crop Monitoring:

- **Identifying Visible Crop Damage**: RGB cameras help detect discoloration, wilting, and pest damage on leaves.
- **Assessing Plant Density and Uniformity**: High-resolution aerial images allow farmers to evaluate planting density, identify gaps, and detect lodging issues.
- **Monitoring Weed Infestation**: By analyzing field images, farmers can identify weed-covered areas and implement targeted herbicide applications.
- **Yield Estimation and Growth Tracking**: Sequential RGB images taken throughout the growing season help track plant development and estimate potential yield.

RGB cameras are an essential and cost-effective tool for routine visual inspections, especially when combined with AI-driven image analysis for automatic plant disease detection (Table 8.1).

The integration of UAV-based imaging technologies in crop monitoring has transformed agricultural practices, allowing for proactive and precise farm management [11]. Multispectral imaging provides critical insights into nutrient health and disease stress, thermal imaging detects water stress and irrigation issues, and RGB cameras offer a straightforward visual assessment of crop conditions. When combined, these imaging techniques provide a comprehensive overview of plant health, enabling farmers to make informed decisions that optimize yields while conserving resources. Despite some challenges, such as data processing requirements [4] and regulatory constraints, the continued development of UAV imaging technologies, artificial intelligence, and machine learning will further enhance precision agriculture, ensuring sustainable food production for the future.

Table 8.1 Comparison of imaging techniques

Imaging type	Key features	Best for detecting	Limitations
Multispectral	Captures multiple spectral bands (visible + NIR), generates vegetation indices (e.g., NDVI)	Nutrient deficiencies, disease stress, overall plant health	Higher cost, requires data processing expertise
Thermal	Detects temperature variations related to water stress	Drought stress, irrigation inefficiencies	Limited ability to identify specific diseases
RGB	High-resolution color images in the visible spectrum	Physical damage, pest infestations, plant density	Cannot detect early-stage nutrient deficiencies or water stress

8.3 UAV-Based Disease Detection

Crop diseases pose a significant threat to global food production, reducing yields and increasing economic losses for farmers. Early detection and effective management of plant diseases are crucial for minimizing damage and ensuring food security. Traditional methods of disease detection, such as manual scouting and laboratory testing, can be time-consuming, labor-intensive, and often ineffective in large-scale farms. Unmanned Aerial Vehicles (UAVs) equipped with advanced imaging sensors offer a revolutionary approach to disease detection. By capturing high-resolution aerial imagery and using spectral analysis, UAVs can detect disease symptoms at an early stage, even before they become visible to the human eye [12]. Additionally, UAV technology enables continuous monitoring of disease progression and facilitates precision treatments, allowing farmers to target affected areas with minimal pesticide use. This section explores three key aspects of UAV-based disease detection: Identifying Early Signs of Disease, Monitoring Disease Spread Over Time, and Precision Treatments Using Data.

8.3.1 Identifying Early Signs of Disease

Early detection of plant diseases is essential for preventing outbreaks and minimizing crop losses. Many plant diseases, such as fungal infections, bacterial blights, and viral pathogens, begin at a microscopic level, making them difficult to detect with the naked eye. By the time visible symptoms appear, the disease may have already spread, reducing the effectiveness of treatment [13]. UAVs equipped with multispectral, hyperspectral, and thermal imaging sensors can identify disease-related stress in plants before physical symptoms emerge. These sensors analyze changes in plant reflectance patterns, temperature variations, and chlorophyll content, which can indicate infection. Spectral indices such as the Normalized Difference Vegetation Index (NDVI) and the Disease Water Stress Index (DWSI) are commonly used to detect diseased plants at an early stage. By using UAV-based imaging for early disease detection, farmers can take timely action to contain infections, apply targeted treatments, and reduce the spread of plant pathogens [14].

How UAVs Detect Early Symptoms

Plant diseases often cause physiological changes before they become visible to the human eye. UAVs equipped with specialized sensors detect these early stress indicators by analyzing plant reflectance and temperature variations.

- **Multispectral Imaging**: Detects variations in plant chlorophyll content and water stress.
- **Hyperspectral Imaging**: Captures hundreds of spectral bands to identify subtle biochemical changes in plants.

- **Thermal Imaging**: Measures temperature differences, as infected plants may have altered transpiration rates.

Key Vegetation Indices for Disease Detection

1. **Normalized Difference Vegetation Index (NDVI)**

The Normalized Difference Vegetation Index (NDVI) is one of the most widely used vegetation indices in remote sensing, offering a simple yet effective method to assess the presence, health, and vigor of vegetation. It is calculated using the reflectance values of the red (RED) and near-infrared (NIR) spectral bands with the formula: NDVI = (NIR − RED)/(NIR + RED). Healthy green vegetation absorbs most visible light (especially red) for photosynthesis and reflects a large portion of near-infrared light due to its internal leaf structure. Therefore, higher NDVI values (close to + 1) indicate dense, healthy vegetation, while lower or negative values suggest sparse vegetation, bare soil, water bodies, or stressed crops.

NDVI is a fundamental tool in precision agriculture, enabling farmers and researchers to monitor crop health, detect early signs of disease or nutrient deficiency, evaluate drought stress, and estimate biomass production. Its simplicity and effectiveness make it a go-to index for time-series vegetation analysis and large-scale crop monitoring through UAVs, satellites (e.g., Landsat, Sentinel-2), and manned aircraft equipped with multispectral sensors. However, NDVI has limitations, particularly in dense canopies where the red band becomes saturated and loses sensitivity to changes in chlorophyll content. In such scenarios, other indices like the Normalized Difference Red Edge Index (NDRE) may be more effective. Despite its limitations, NDVI remains a foundational metric in agronomic and ecological applications due to its ease of use, compatibility with most remote sensing platforms, and its ability to offer rapid, large-scale assessments of vegetation dynamics. As remote sensing technology continues to advance, NDVI continues to serve as a baseline index in crop monitoring, environmental management, land-use classification, and global vegetation change studies.

2. **Normalized Difference Red Edge Index (NDRE)**

The Normalized Difference Red Edge Index (NDRE) is a vegetation index derived from remote sensing data, used to assess plant health, chlorophyll concentration, and nitrogen status. Unlike traditional indices such as the Normalized Difference Vegetation Index (NDVI), which can become saturated in dense vegetation, NDRE remains sensitive during mid to late crop growth stages. NDRE is calculated using the formula: (NIR − RedEdge)/(NIR + RedEdge), where NIR refers to the near-infrared region (typically 780–850 nm), and Red Edge is a narrow spectral band (around 705–740 nm) between the red and NIR wavelengths. This red edge band is especially responsive to changes in chlorophyll content, making NDRE particularly effective for monitoring crop nutrient status and photosynthetic activity.

NDRE offers several advantages in agriculture. It is highly correlated with nitrogen levels in plant tissues, allowing farmers to make data-driven decisions for variable rate fertilizer application. This enables more precise nutrient management and reduces waste, thereby enhancing productivity and environmental sustainability. In addition, NDRE is useful for monitoring crop health, enabling early detection of stress factors such as pest infestations, disease outbreaks, or water deficiencies. Compared to NDVI, which may lose sensitivity in later growth stages due to canopy saturation, NDRE provides more reliable assessments of crop vigor and can penetrate deeper into dense canopies. This makes it particularly useful for crops like wheat, corn, and rice during critical development stages such as booting, heading, and flowering.

NDRE also supports yield prediction models and zone mapping in precision agriculture. When integrated with other agronomic data, it can help forecast yield potential and guide timely interventions. UAVs (drones) and satellite systems like Sentinel-2 are commonly used platforms for capturing NDRE data. These platforms must be equipped with multispectral or hyperspectral sensors capable of detecting the red edge band—something not possible with basic RGB cameras. Commercially available sensors such as MicaSense RedEdge or Parrot Sequoia are widely used in UAV-based applications for this purpose.

NDRE is a powerful remote sensing tool that provides meaningful insights into crop nutrient dynamics, plant health, and yield potential, especially during stages when NDVI becomes less effective. As agriculture becomes increasingly data-driven and reliant on precision technologies, NDRE plays a vital role in supporting informed decision-making, optimizing input use, and promoting sustainable farming practices.

3. **Canopy Temperature Index (CTI)**

The Canopy Temperature Index (CTI) is a valuable remote sensing indicator used to assess plant water status, stress conditions, and overall crop health by measuring the temperature of the crop canopy. Canopy temperature reflects the balance between transpiration and environmental conditions: under optimal water conditions, plants transpire actively, which cools the canopy. However, when water is limited or the plant is under stress (e.g., drought or disease), stomatal closure reduces transpiration, leading to higher canopy temperatures. CTI leverages this physiological response to monitor crop stress levels, especially during critical growth stages.

CTI is typically derived from thermal infrared imagery captured by UAVs, satellites, or handheld infrared thermometers. It is especially effective in identifying spatial variations in water stress across a field, enabling site-specific irrigation management in precision agriculture. Higher CTI values generally correspond to hotter canopies and suggest greater water stress, while lower values indicate well-watered or non-stressed conditions. This index is crucial in crop management practices such as irrigation scheduling, drought monitoring, and yield forecasting. When combined with other vegetation indices like NDVI or NDRE, CTI enhances the accuracy of plant health assessments by providing thermal data that complements spectral reflectance information. Moreover, CTI plays a significant role in evaluating genotypic responses to heat and drought stress in crop breeding programs. The Canopy Temperature Index serves as a non-invasive, real-time tool for detecting plant stress and managing agricultural inputs efficiently. Its integration into modern agronomic practices supports improved water use efficiency, early stress detection, and sustainable crop production under changing environmental conditions.

8.3.2 *Monitoring Disease Spread Over Time*

Once a plant disease is detected, continuous monitoring is necessary to track its progression and assess the effectiveness of treatment measures. Traditional disease monitoring methods, such as ground surveys and sample testing, can be inconsistent and prone to delays. UAVs provide a more efficient and accurate alternative, allowing for frequent aerial surveys that capture large areas of farmland in a short time [15]. By analyzing UAV-generated time-series data, farmers can observe how diseases spread across fields and identify patterns in infection rates. Thermal and hyperspectral imaging help detect stress areas where disease is intensifying, while RGB imaging provides visual confirmation of infected plants. Machine learning algorithms can further analyze UAV data to predict future disease outbreaks based on past trends [16]. With real-time disease monitoring, farmers can implement better disease management strategies, adjust treatment applications, and make informed decisions to prevent further crop losses.

Disease Progression Analysis

Once a disease is detected, UAVs can be used to monitor its spread across a field over time. This helps in making timely management decisions and implementing control measures.

Disease Severity Index (DSI)

The Disease Severity Index (DSI) is a quantitative measure used in plant pathology to assess the intensity or extent of disease symptoms on plant tissues. It provides a standardized way to evaluate and compare disease severity across treatments, cultivars, or experimental conditions. DSI is typically expressed as a percentage, reflecting the proportion of plant tissue affected by a particular disease. To calculate DSI, plants are usually rated based on a predefined scale (e.g., 0–5 or 0–9), where each score corresponds to a specific range of visible symptoms, such as lesion size, percentage of leaf area infected, or the extent of wilting or necrosis. The formula to calculate DSI often follows this structure:

$$\mathrm{DSI}(\%) = \Big[\sum \text{(class rating} \times \text{number of plants in that class)}/ \\ \text{(total number of plants} \times \text{highest class rating)}\Big] \times 100$$

This method allows researchers and agronomists to quantify disease progression and evaluate the effectiveness of control measures such as fungicides, resistant varieties, or cultural practices. DSI is widely used in epidemiological studies, breeding programs, and field trials to monitor disease impact and make informed management decisions. Overall, the Disease Severity Index is a crucial tool in integrated pest and disease management, helping to track plant health, guide intervention timing, and assess resistance or susceptibility in different plant genotypes. Its consistent use contributes to the development of disease-resistant crops and supports sustainable agricultural practices.

Time-Series Analysis for Disease Spread

UAV-based imagery collected at different time intervals can be used to create time-series disease progression maps. These maps provide insights into:

1. **Rate of disease spread**: Helps predict future outbreaks.
2. **Affected areas**: Determines which sections of the field need immediate attention.
3. **Environmental impact assessment**: Correlates disease spread with weather, soil moisture, and other factors.

8.3.3 Precision Treatments Using Data

Once UAV-based imaging has identified infected plants and monitored disease spread, farmers can use this data to implement precision treatments. Instead of applying pesticides and fungicides across an entire field, UAV-generated disease maps enable site-specific treatment, targeting only affected areas. This approach reduces chemical use, lowers production costs, and minimizes environmental impact. UAVs can also be equipped with spraying systems, allowing for precise application of pesticides, fungicides, or biological agents over infected zones. Combining imaging data with AI-driven analysis, farmers can automate disease management processes, ensuring timely and efficient intervention. By using UAV-based disease detection for precision treatments, agricultural practices become more sustainable, increasing efficiency while reducing the risks associated with excessive chemical use [17]. UAV-based disease detection is revolutionizing crop protection by providing accurate, early identification of diseases, tracking their progression, and enabling precision treatments. By leveraging advanced imaging technologies and data-driven insights, UAVs help farmers respond to disease outbreaks more effectively, ultimately improving crop health and agricultural productivity. As UAV technology continues to advance, its role in disease management will become even more essential in ensuring global food security [18].

How UAV Data Enables Precision Spraying

Traditional blanket spraying methods lead to excessive pesticide use and environmental contamination. UAV data enables precision treatment by applying chemicals only where needed, reducing costs and improving efficiency.

Prescription Map Algorithm for Variable Rate Spraying

To apply chemicals (e.g., pesticides, fertilizers) only where needed, based on field conditions like pest infestation, disease hotspots, or nutrient deficiencies. To optimize chemical application, a Prescription Map Algorithm (PMA) is used:

$$Application\ rate = \frac{Target\ dosage \times Affected\ area}{Total\ field\ area}$$

where:

Target Dosage	Recommended pesticide concentration
Affected Area	The percentage of the field that is diseased
Total Field Area	Overall size of the crop field.

This formula estimates the average application rate needed across the whole field to ensure only the affected areas receive the full dosage while minimizing input waste on healthy regions.

Automated UAV Spraying Systems

Modern UAV sprayers can autonomously adjust:

- **Droplet size** based on wind and crop type
- **Flow rate** based on disease severity
- **Spraying pattern** to avoid over-application.

Example: UAV Precision Treatment for Wheat Rust Disease

1. UAV multispectral imaging detects **early-stage wheat rust** using NDVI analysis.
2. Disease spread is monitored using **time-series UAV data** to track its progression.
3. Precision spraying UAVs apply **fungicides only to affected areas** using prescription maps.

Outcome:

- 40% reduction in pesticide usage
- 30% increase in yield compared to traditional methods
- Minimized environmental impact.

UAV-based disease detection enhances agricultural productivity by integrating advanced imaging, machine learning, and precision treatment methods. By identifying diseases early, monitoring their spread over time, and applying data-driven treatments, farmers can reduce crop losses, improve efficiency, and promote sustainable farming practices [19].

8.4 Monitoring Pest Activity with UAVs

Pests are a major threat to agricultural productivity, causing significant crop losses worldwide. Traditional pest monitoring methods, such as manual field scouting and the use of stationary traps, are often time-consuming, labor-intensive, and may not provide a comprehensive overview of pest infestations. UAVs (Unmanned Aerial Vehicles) equipped with advanced imaging sensors have emerged as a powerful tool for detecting and managing pest activity in a precise and efficient manner [20]. By utilizing multispectral, hyperspectral, thermal, and RGB imaging, UAVs can detect early signs of pest infestations, monitor the spread of pest activity over time, and provide actionable insights for precision pest control. This approach allows farmers to implement targeted interventions, reducing pesticide use while improving crop health and sustainability. This section explores the role of UAVs in Detecting Insect Infestations, Mapping Affected Areas for Targeted Action, and Reducing Pesticide Usage to enhance pest management strategies in modern agriculture [21].

8.4.1 Detecting Insect Infestations

Early detection of insect infestations is critical for minimizing crop damage and preventing outbreaks from spreading. Many pests, such as aphids, locusts, and caterpillars, can cause significant harm to crops in a short period, making timely intervention essential. However, traditional scouting methods may fail to detect infestations at an early stage, leading to delayed action and greater yield losses [22]. UAVs equipped with multispectral and hyperspectral sensors can identify signs of pest activity by analyzing changes in plant reflectance and chlorophyll content. Thermal imaging can also detect heat anomalies caused by plant stress due to pest feeding, while high-resolution RGB cameras provide visual confirmation of damaged leaves, discoloration, and pest colonies. By using UAVs for pest detection, farmers can identify affected areas early, reducing the time and labor required for manual field inspections while improving pest management efficiency [23].

How UAVs Identify Pest Activity

Pests cause physiological and morphological changes in plants, which UAVs can detect through spectral analysis. Common indicators include: Defoliation refers to the loss of leaves caused by feeding insects such as caterpillars, which consume foliage and weaken the plant, reducing its ability to photosynthesize and grow properly. This damage can lead to lower crop yields and increased vulnerability to environmental stressors [24]. Chlorosis, or the yellowing of leaves, is often a result of sap-sucking pests like aphids that extract vital nutrients from plant tissues, disrupting chlorophyll production and weakening the plant's overall health. Temperature variations detected in thermal imaging can serve as an early indicator of pest infestations. Heat spots in thermal images reveal areas where plants are experiencing stress, often due to excessive transpiration caused by pest attacks. This stress can lead to decreased growth rates, reduced crop quality, and increased susceptibility to diseases. By utilizing UAV-based imaging technology, farmers can quickly identify and address these pest-related issues before they cause significant damage to crops [25].

Key Vegetation Indices for Pest Detection

The Normalized Difference Vegetation Index (NDVI) is a widely used spectral index in remote sensing that provides an indication of vegetation health, density, and vigor. It is calculated using near-infrared (NIR) and red (R) light reflectance values captured by multispectral or hyperspectral sensors mounted on satellites, drones (UAVs), or other aerial platforms. Healthy vegetation reflects more near-infrared light and absorbs more red light due to chlorophyll activity, resulting in higher NDVI values (close to + 1). In contrast, stressed or sparse vegetation reflects less NIR and more red light, producing lower NDVI values (approaching 0 or even negative values for bare soil or non-vegetated surfaces). NDVI is an essential tool in agriculture for monitoring crop growth, detecting plant stress, assessing biomass, evaluating drought impact, and supporting precision farming decisions such as irrigation, fertilization, and pest management. It helps identify areas of a field that may need closer inspection or intervention, contributing to optimized input use and sustainable land management. By providing spatial and temporal insights into vegetation status, NDVI supports improved yield prediction, field scouting, and environmental monitoring, making it a cornerstone of modern digital agriculture and environmental science.

8.4.2 Normalized Difference Insect Infestation Index (NDII)

The Normalized Difference Insect Infestation Index (NDIII) is a remote sensing-based vegetation index developed to detect and quantify insect infestation in crops through spectral reflectance analysis. Similar in structure to indices like NDVI, NDIII leverages the differences in light reflectance between specific spectral bands—typically the near-infrared (NIR) and green or red bands—to assess changes in plant health that are indicative of insect activity. Insect-infested plants often exhibit stress symptoms such as chlorosis, necrosis, or defoliation, leading to changes in their spectral signature. These changes are captured by comparing the reflectance values of healthy versus stressed vegetation. A commonly used formula for NDIII is:

$$\text{NDIII} = (\text{NIR} - \text{Green})/(\text{NIR} + \text{Green})$$

In some cases, red or red-edge bands may also be used depending on the crop type and sensor specifications. Lower NDIII values typically indicate greater levels of infestation, as stressed plants reflect less in the NIR region and more in the visible spectrum. NDIII is particularly valuable in precision agriculture for early detection of pest outbreaks, allowing for timely intervention through targeted pesticide application. This helps reduce yield loss, minimize environmental impact, and improve input efficiency. By integrating NDIII data into UAV or satellite monitoring systems, farmers and agronomists can rapidly assess large areas and make informed management decisions to control insect populations and maintain crop health.

8.4.3 Thermal Imaging for Pest Detection

Thermal Imaging for Pest Detection is an advanced remote sensing technique that leverages temperature differences in plant canopies to identify pest infestations early and accurately. When crops are infested by pests, physiological changes such as altered transpiration rates, disrupted vascular functions, or tissue damage often lead to localized temperature variations in the affected areas. Thermal cameras, mounted on Unmanned Aerial Vehicles (UAVs) or handheld devices, detect these temperature anomalies by capturing infrared radiation emitted from plant surfaces. Healthy plants typically maintain lower canopy temperatures due to active transpiration. However, pest attacks—especially from insects like aphids, whiteflies, or borers—can block or reduce this cooling effect, causing infested regions to appear warmer in thermal images. Conversely, some infestations may lead to defoliation or tissue decay, resulting in cooler patches. By analyzing these thermal signatures, researchers and farmers can distinguish between healthy and stressed vegetation, enabling targeted scouting and intervention.

Thermal imaging offers several advantages: it allows for non-destructive, real-time monitoring over large areas; it is effective under varying light conditions, including at night; and it enables early detection before visual symptoms become apparent. When combined with other vegetation indices or multispectral data, thermal imaging can enhance the accuracy of pest diagnosis. This technology is becoming increasingly valuable in integrated pest management (IPM), contributing to reduced pesticide use, improved crop protection, and sustainable agricultural practices.

8.4.4 Mapping Affected Areas for Targeted Action—Introduction

Effective pest management requires not only detecting infestations but also understanding their extent, severity, and progression across a field. Traditional methods, such as manual scouting and random sampling, often fail to provide a comprehensive picture of pest outbreaks, leading to delayed responses and inefficient pesticide application. UAVs (Unmanned Aerial Vehicles) offer a powerful solution by capturing high-resolution aerial imagery and generating precise infestation maps. These maps help farmers and agronomists identify affected areas, assess pest distribution, and implement site-specific pest control strategies. By leveraging UAV technology, farmers can optimize pesticide use, reduce costs, and minimize environmental impact while ensuring healthier crop yields [2]. Once a pest infestation is detected, it is essential to monitor its spread and map affected areas for precise intervention. UAV-based aerial surveys can provide high-resolution maps that help farmers assess the extent of pest damage across large fields.

Using geospatial mapping and AI-driven image analysis, UAVs can differentiate between healthy and pest-damaged crops, allowing for the creation of infestation heat maps. These maps help farmers: Tracking pest movement over time allows farmers to predict infestation patterns and develop proactive pest management strategies. By analyzing UAV-captured imagery at different intervals, changes in pest distribution and intensity can be monitored, helping to identify trends in infestation spread. Determining high-risk zones requiring immediate attention is essential for targeted pest control [26]. UAV-based imaging and AI-powered analysis can highlight areas where pest populations are rapidly increasing, allowing for swift intervention to prevent widespread crop damage. Additionally, assessing pest control effectiveness is possible by comparing pre- and post-treatment imagery. By analyzing changes in plant health indicators and pest presence after treatment, farmers can evaluate the success of their pest management strategies, optimize pesticide application, and refine future interventions to improve efficiency and sustainability. By having a clear overview of affected areas, farmers can optimize pest control measures, reducing the risk of excessive pesticide application and ensuring that resources are used efficiently [27].

Pest Infestation Mapping with UAVs

Hotspots refer to areas with the highest levels of pest infestation, often identified through UAV-based imaging and data analysis. These zones require immediate attention to prevent further damage and minimize crop loss. Spread patterns indicate the direction and speed at which pests are moving across the field, helping farmers anticipate infestation progression [22]. By analyzing these patterns, targeted control measures can be implemented to contain the outbreak before it reaches new areas. At-risk areas include neighboring crops that may soon be affected by pest movement. Identifying these vulnerable regions allows for early intervention, such as preventive pesticide application or biological pest control measures, to protect crops and reduce the overall impact of an infestation. UAV technology plays a crucial role in monitoring these factors, enabling precision pest management and more effective agricultural decision-making [28].

Formula for Pest Density Mapping (PDM)

Pest density per unit area is calculated using:

$$\mathrm{PDM} = \mathrm{A}\sum \mathrm{Pi}$$

where:

- $\mathbf{P_i}$ = Number of pests detected in a sampled region
- **A** = Total field area (in hectares or square meters).

A higher **PDM** value indicates severe infestations requiring immediate action.

Time-Series UAV Analysis for Pest Monitoring

By capturing UAV images over time, farmers can track pest activity and predict future outbreaks using machine learning models.

Example: UAV-Based Pest Monitoring for Locust Infestation

UAV-Based Pest Monitoring for Locust Infestation Locust infestations pose a severe threat to agriculture, capable of destroying vast croplands in a short period. Traditional monitoring methods often fail to track swarm movements efficiently, leading to delayed responses and widespread damage. UAV technology enhances locust monitoring by providing real-time data on infestation severity and movement patterns, enabling rapid and targeted intervention. UAVs Detect Locust Swarms Using Thermal Imaging and NDVI Analysis UAVs equipped with thermal cameras and NDVI (Normalized Difference Vegetation Index) sensors can identify locust swarms by detecting temperature variations and changes in plant health [29]. Thermal imaging helps locate swarms based on their heat signatures, while NDVI analysis highlights vegetation stress caused by locust feeding. Affected Areas Are Mapped, and Infestation Severity Is Determined UAVs create high-resolution maps of infested regions, allowing farmers and agricultural authorities to assess the spread and intensity of locust activity. AI-powered image processing classifies infestation levels, identifying high-risk zones that require immediate action. Data Is Used to Deploy Biological or Chemical Treatments in High-Risk Zones Based on UAV-generated data, targeted control measures can be implemented. In severely affected areas, chemical pesticides may be used to suppress locust populations, while biological treatments, such as the release of natural predators or microbial bio-pesticides, can be deployed in less critical zones. This approach minimizes excessive pesticide use, reduces environmental impact, and improves pest control efficiency [30].

Reducing Pesticide Usage

Excessive pesticide use in agriculture has led to several environmental and economic challenges, including soil and water contamination, pesticide resistance in pests, and increased costs for farmers. Traditional blanket spraying methods result in the overuse of chemicals, negatively impacting ecosystems, beneficial insects, and human health. UAV-based pest monitoring and precision pesticide application offer a more sustainable approach by ensuring that chemical treatments are applied only where necessary. UAVs equipped with advanced imaging sensors, such as multispectral, hyperspectral, and thermal cameras, can detect pest-infested areas by analyzing plant stress indicators [31]. Once affected zones are identified, AI-powered software generates prescription maps for Variable Rate Application (VRA), which allows targeted spraying based on infestation severity. UAV sprayers further enhance efficiency by precisely delivering pesticides only to infected areas, reducing chemical runoff and preserving beneficial organisms. Additionally, UAVs can be used for the distribution of biological pest control agents, such as predatory insects or microbial treatments, as an eco-friendly alternative to chemical pesticides. This approach minimizes pesticide waste, lowers costs, reduces environmental impact, and improves worker safety. By integrating UAV technology into pest management strategies, farmers can optimize pesticide usage while ensuring sustainable and efficient crop protection [32].

By integrating UAV-generated pest data with automated spraying systems, farmers can:

Apply pesticides only to infected areas, reducing unnecessary chemical use.
Use alternative pest control methods, such as biological treatments or pheromone traps, based on infestation severity.
Minimize environmental impact, improving soil and water quality.

In addition to improving sustainability, targeted pest control strategies help maintain biodiversity by protecting beneficial insects, such as pollinators and natural predators. UAVs are revolutionizing pest monitoring by enabling early detection, real-time mapping, and precision pest control. By leveraging advanced imaging and AI-driven analysis, farmers can respond to pest threats efficiently, reducing crop losses while minimizing pesticide usage [33]. As UAV technology continues to evolve, its integration with smart agriculture practices will play a crucial role in sustainable pest management and food security.

8.5 Field Mapping and Soil Health Analysis

Field mapping and soil health analysis are essential for optimizing agricultural productivity, resource management, and sustainable farming practices. UAV technology plays a crucial role in capturing high-resolution data, allowing farmers to assess soil conditions, monitor field topography, and detect problem areas with precision. By utilizing advanced imaging techniques such as multispectral, hyperspectral, and thermal sensors, UAVs can generate accurate maps that help in planning irrigation, fertilizer application, and erosion control strategies [34]. Field mapping enables the creation of Digital Elevation Models (DEMs), which provide a three-dimensional representation of the land's surface. These models help identify slopes, drainage patterns, and areas prone to waterlogging or erosion. Understanding field topography allows farmers to implement effective land management techniques that improve crop growth and prevent soil degradation. Soil health analysis focuses on evaluating key soil properties such as pH levels, nutrient content, and moisture availability. UAVs equipped with remote sensing technology can detect variations in soil quality, enabling site-specific fertilizer application and reducing excessive chemical use [35]. This targeted approach improves soil fertility, enhances crop yields, and minimizes environmental impact.

Additionally, UAVs assist in assessing soil erosion risks by identifying vulnerable areas where soil degradation is likely to occur. By monitoring sediment displacement and vegetation cover, farmers can implement erosion control measures such as contour farming, cover cropping, and buffer strips to protect soil resources [36]. By integrating UAV-based field mapping and soil health analysis, farmers can make data-driven decisions that enhance agricultural efficiency, promote sustainability, and ensure long-term soil and crop health.

Generating Digital Elevation Models (DEMs)

Digital Elevation Models (DEMs) provide a detailed three-dimensional representation of a field's topography, enabling farmers to analyze land elevation, slope variations, and drainage patterns. UAVs equipped with LiDAR or photogrammetric sensors capture aerial imagery, which is processed to create high-resolution elevation maps. These models are essential for efficient irrigation planning, as they help identify low-lying areas prone to waterlogging and optimize drainage systems. Additionally, DEMs assist in erosion control by detecting slopes susceptible to soil erosion, allowing for the implementation of conservation measures such as terracing or contour farming [37]. By understanding terrain variations, farmers can enhance land leveling efforts, ensuring uniform water distribution and improved crop growth.

Digital Elevation Model

A Digital Elevation Model (DEM) is a 3D representation of the land surface that provides detailed information about elevation, slope, and drainage patterns. UAVs with LiDAR (Light Detection and Ranging) or photogrammetry cameras capture thousands of images, which are processed to create highly accurate elevation maps.

Why Are DEMs Important?

- **Field Slope Analysis**: Helps determine whether the land is suitable for farming and irrigation.
- **Water Drainage Planning**: Identifies **low-lying areas** prone to waterlogging or **runoff-prone slopes** where soil erosion is likely.
- **Precision Land Leveling**: Farmers can adjust terrain to **improve water retention and crop growth**.

Soil pH and Nutrient Mapping

Soil pH and nutrient levels play a critical role in determining crop productivity and soil fertility. UAVs equipped with multispectral and hyperspectral sensors capture soil reflectance data, which, when combined with ground-truth soil sampling, helps generate high-resolution soil health maps. These maps allow for precise fertilizer application by identifying nutrient-deficient zones, reducing excessive fertilizer use, and minimizing environmental harm. Additionally, soil pH mapping helps detect acidity or alkalinity imbalances, enabling farmers to apply lime or sulfur treatments to maintain optimal soil conditions [38]. By using UAV-based soil analysis, farmers can improve nutrient management, enhance crop yields, and reduce the risk of over-fertilization.

What is Soil pH and Why Does it Matter?

Soil pH determines whether soil is acidic, neutral, or alkaline, which directly affects nutrient availability.

- **Acidic Soil ($pH < 6.0$)**: Limits phosphorus and calcium absorption.
- **Alkaline Soil ($pH > 7.5$)**: Reduces iron and zinc availability.
- **Optimal pH for Crops**: Usually between **6.0 and 7.5** for most plants.

How UAVs Map Soil pH and Nutrients

UAVs equipped with hyperspectral sensors can detect variations in soil composition by analyzing how soil reflects light at different wavelengths. These sensors help create soil property maps that show:

1. **Soil pH distribution**—Identifying areas needing lime (to reduce acidity) or sulfur (to lower alkalinity).
2. **Nitrogen (N), Phosphorus (P), and Potassium (K) Mapping**—Optimizing fertilizer application.
3. **Organic Matter Content**—Indicating soil fertility and moisture retention capability.

Soil Nutrient Index (SNI) Formula

To measure soil fertility, the **Soil Nutrient Index (SNI)** is calculated using the average of key nutrients:

$$\mathrm{SNI} = 3\mathrm{N} + \mathrm{P} + \mathrm{K}$$

where:

- **N = Nitrogen (%)**
- **P = Phosphorus (%)**
- **K = Potassium (%)**

Example Calculation:

If a UAV detects the following nutrient levels in a field:

- **Nitrogen (N) = 0.8%**
- **Phosphorus (P) = 0.6%**
- **Potassium (K) = 1.0%**

Then,

$$\mathrm{SNI} = 0.8 + 0.6 + 1.03 = 0.8\mathrm{SNI}$$

A **higher SNI** means fertile soil, while a **low SNI** may require fertilizers or organic amendments like compost etc.

Benefits of UAV-Based Soil Analysis

- **Precision Fertilizer Application**: Apply fertilizers only where needed, reducing costs and environmental impact.
- **Better Crop Selection**: Farmers can choose crops suited to specific soil pH and nutrient levels.
- **Sustainable Farming Practices**: Prevents excess fertilizer use, which can lead to soil degradation and water pollution.

Assessing Soil Erosion Risks

Soil erosion poses a significant threat to agricultural productivity by reducing soil fertility and increasing sedimentation in nearby water bodies. UAVs help assess erosion risks by identifying bare soil areas that lack vegetation cover and are susceptible to wind or water erosion. High-resolution aerial images allow for tracking sediment deposition and monitoring changes over time, providing valuable insights into erosion control effectiveness. Farmers can use this data to implement preventive strategies such as cover cropping, contour plowing, and buffer strip planting to mitigate erosion risks [39]. By integrating UAV technology into soil conservation efforts, agricultural sustainability can be improved, ensuring long-term soil health and productivity.

Why is Soil Erosion a Problem?

Soil erosion leads to loss of fertile topsoil, reducing crop productivity and increasing water pollution due to sediment runoff.

How UAVs Detect Erosion-Prone Areas

UAVs use **multispectral imagery, LiDAR, and DEM data** to identify erosion indicators:

- **Steep Slopes**: Areas with high gradient values.
- **Bare Soil Exposure**: Low vegetation index (NDVI) values indicate higher erosion risk.
- **Water Runoff Patterns**: Identified using DEM-based flow accumulation models.

Revised Universal Soil Loss Equation (RUSLE) for Erosion Risk Assessment

The RUSLE model estimates annual soil loss (A) due to erosion:

$$PA = R \times K \times LS \times C \times P$$

where:

- **A = Soil loss (tons/acre/year)**
- **R = Rainfall erosivity factor** (higher in heavy-rainfall areas).
- **K = Soil erodibility factor** (based on soil texture).
- **LS = Slope length and steepness factor** (calculated from DEM).
- **C = Cover management factor** (vegetation detected via UAV imagery).
- **P = Conservation practice factor** (e.g., contour farming, terracing).

Example:

If a field has:

- **R = 150** (high rainfall area)
- **K = 0.4** (sandy loam soil)
- **LS = 2.0** (moderate slope)
- **C = 0.2** (good vegetation cover)
- **P = 0.5** (contour farming applied).

Then,

$$A = 150 \times 0.4 \times 2.0 \times 0.2 \times 0.5 = 12\ \text{tons/acre/year}$$

If soil loss exceeds **10 tons/acre/year**, additional conservation measures are needed, such as:

- **Cover Crops**: Planting grass or legumes to protect soil.
- **Terracing**: Building steps on slopes to reduce runoff.
- **Mulching**: Adding organic material to protect the soil from rain impact.

UAVs have become essential for precision agriculture and soil conservation by providing detailed maps of field topography, soil chemistry, and erosion risks. By integrating UAV data with GIS (Geographic Information Systems) and machine learning models, farmers can make data-driven decisions to optimize land use, reduce costs, and promote sustainable farming [40].

8.6 Seasonal Monitoring with UAVs

Seasonal monitoring with UAVs is a revolutionary method in precision agriculture, enabling farmers to gather real-time data at various crop growth stages. Utilizing multispectral, thermal, and high-resolution RGB cameras, farmers can optimize field management, detect issues early, and make data-driven decisions to enhance productivity. UAVs provide crucial insights at three key stages: pre-planting, mid-season growth tracking, and post-harvest analysis, ensuring efficient resource allocation, minimizing crop losses, and promoting sustainable agricultural practices. Pre-planting analysis helps farmers evaluate field conditions before sowing, allowing for precise seed placement, irrigation planning, and fertilizer application. UAVs equipped with LiDAR and multispectral sensors can also identify problem areas, enabling corrective measures before planting. Mid-season crop growth tracking provides continuous monitoring of plant health, detecting stress factors like water deficiency, pest infestations, and nutrient shortages [41]. NDVI and thermal imaging help assess chlorophyll levels and temperature variations, enabling targeted interventions. This data can be used to adjust irrigation schedules, optimize fertilizer use, and implement site-specific pest control, leading to improved crop yields and reduced resource wastage. Post-harvest field analysis helps farmers assess soil conditions, measure crop residue, and determine areas affected by soil degradation. High-resolution aerial imagery helps evaluate nutrient depletion and plan soil replenishment strategies. UAVs also aid in biomass management by identifying leftover vegetation for composting or bioenergy production [42].

8.6.1 Pre-planting Analysis

Pre-planting analysis is a crucial step in ensuring optimal crop growth by assessing soil conditions and field characteristics before sowing. UAVs equipped with multi-spectral, thermal, and LiDAR sensors provide detailed maps of terrain, soil moisture, and nutrient levels. These insights help farmers identify areas requiring soil amendments, drainage improvements, or targeted fertilization to enhance seed germination and early plant development. UAVs also detect potential problem zones, such as compacted soil or erosion-prone areas, allowing for proactive land management strategies [43]. By optimizing pre-planting decisions, farmers can reduce input costs, improve crop establishment, and maximize yield potential.

Purpose of Pre-Planting Analysis

Before planting, it is essential to evaluate soil conditions, moisture levels, and field topography to ensure optimal seed placement and resource allocation. UAVs help in:

Soil Fertility Assessment

UAV-based multispectral and hyperspectral imaging helps assess soil fertility by identifying nutrient-rich and nutrient-deficient zones. By analyzing vegetation indices like NDVI (Normalized Difference Vegetation Index) and chlorophyll content, farmers can detect areas lacking essential nutrients such as nitrogen (N), phosphorus (P), and potassium (K) [44]. This allows for precision fertilization, ensuring that nutrients are applied only where needed, reducing input costs and preventing environmental degradation.

Soil Moisture Mapping

Thermal and multispectral sensors on UAVs enable accurate soil moisture detection, identifying dry zones that require additional irrigation and waterlogged areas prone to drainage issues. By monitoring moisture variations across fields, farmers can optimize irrigation schedules, prevent water stress, and reduce water wastage, ultimately improving crop growth and water use efficiency [45].

Topographic Mapping

LiDAR and high-resolution UAV imaging provide detailed topographic maps, helping farmers understand field slopes, elevation changes, and drainage patterns. This information is crucial for erosion control, flood prevention, and effective land management [46]. By analyzing slope and water flow, farmers can implement contour farming, terracing, or improved drainage systems to prevent soil loss and enhance sustainable agricultural practices.

UAV Technologies Used in Pre-Planting Analysis

Multispectral Imaging

Multispectral imaging is a powerful tool for analyzing soil organic matter, nutrient availability, and vegetation health. By capturing different wavelengths of light, UAV-mounted multispectral sensors help farmers assess soil fertility, pH variations, and organic content levels [47]. This data allows for precision fertilization, ensuring that crops receive the right nutrients while minimizing excessive fertilizer use and environmental impact.

Thermal Sensors

Thermal sensors detect temperature variations in the soil, which indicate moisture content differences. Dry soil tends to heat up faster than moist soil, allowing UAVs to identify drought-prone areas, irrigation inefficiencies, and potential waterlogging issues [48]. By mapping these variations, farmers can optimize irrigation schedules, prevent water wastage, and improve overall water management efficiency.

LiDAR (Light Detection and Ranging)

LiDAR technology enables UAVs to create high-precision Digital Elevation Models (DEMs), which provide critical insights into field slope, drainage patterns, and erosion risks. These models help in terrain analysis, flood prevention, and efficient land management by identifying areas that may require drainage improvements, contour farming, or terracing. LiDAR-based mapping ensures better field preparation and long-term soil conservation strategies [49].

Soil Moisture Index (SMI) Formula

$$SMI = \frac{T_S - T_{dry}}{T_{wet} - T_{dry}}$$

where:

T_s Surface temperature of the observed location
T_{dry} Estimated temperature of completely dry soil (upper boundary)
T_{wet} Estimated temperature of completely wet soil (lower boundary).

Higher SMI values indicate well-moisturized soil, while lower values suggest dry conditions, which may require pre-planting irrigation.

Benefits of UAV-Based Pre-Planting Analysis

Precision Fertilization

UAV-based monitoring allows for targeted nutrient application, reducing the overuse of fertilizers while ensuring crops receive adequate nourishment. Multispectral and hyperspectral imaging help identify nutrient-deficient zones, enabling farmers to apply fertilizers precisely where needed [50]. This site-specific approach minimizes environmental pollution, enhances soil fertility, and reduces input costs while maximizing crop yields.

Efficient Water Management

With UAV thermal and multispectral imaging, farmers can detect dry zones and waterlogged areas in their fields. By identifying moisture variations, UAVs help optimize irrigation planning, reducing water wastage and ensuring uniform crop hydration [51]. This improves water use efficiency, prevents over-irrigation issues such as root rot, and enhances overall plant health.

Improved Field Preparation

UAV-generated Digital Elevation Models (DEMs) assist in land preparation by mapping field slopes, drainage patterns, and erosion-prone areas. These insights help farmers design efficient plowing patterns, implement proper drainage systems, and adopt erosion control measures such as contour farming [52]. UAV-based field analysis ensures better soil management, reduces land degradation, and improves long-term agricultural productivity.

8.6.2 Mid-Season Crop Growth Tracking

Tracking crop growth during the season is essential for identifying stress factors and optimizing plant health. UAVs equipped with NDVI (Normalized Difference Vegetation Index), multispectral, and thermal imaging sensors help monitor chlorophyll levels, detect water stress, and identify pest infestations or disease outbreaks. This real-time data allows farmers to apply site-specific irrigation, fertilizer, or pest control measures, ensuring that interventions are targeted and efficient. By continuously monitoring crop conditions, UAVs help reduce unnecessary resource use, prevent major crop losses, and improve overall yield quality.

Purpose of Crop Growth Monitoring

Regular monitoring during the growing season helps detect:

Nutrient Deficiencies

Crops require essential nutrients like nitrogen (N), phosphorus (P), and potassium (K) for healthy growth. UAVs equipped with multispectral and hyperspectral sensors can detect chlorophyll levels, leaf discoloration, and growth irregularities, which indicate nutrient deficiencies. For example, nitrogen deficiency causes yellowing of leaves, while phosphorus deficiency leads to stunted growth and purplish foliage [53]. By identifying these deficiencies early, farmers can apply targeted fertilizers to improve crop health and productivity while reducing excessive fertilizer use.

Pest and Disease Outbreaks

UAV-based monitoring allows for early detection of pest infestations and crop diseases, preventing large-scale damage. Multispectral and thermal imaging can identify hotspots of infection, such as leaf discoloration, wilting, or abnormal growth patterns caused by fungi, bacteria, or insects. Detecting problems at an early stage enables farmers to implement targeted treatments, such as precision pesticide application or biological control methods, minimizing yield losses and reducing the need for excessive chemical use [54].

Water Stress and Irrigation Inefficiencies

Water availability is crucial for crop growth, and imbalanced irrigation can lead to yield reduction, soil degradation, and resource wastage. UAVs with thermal and multispectral sensors detect dry zones and waterlogged areas, helping farmers adjust irrigation schedules and improve water distribution [55]. By identifying areas suffering from drought stress or overwatering, farmers can optimize irrigation systems, enhancing water efficiency and ensuring consistent crop hydration.

How UAVs Monitor Mid-Season Crop Growth

UAVs use **vegetation indices** like NDVI (Normalized Difference Vegetation Index) to assess plant health and detect stress early.

Thermal Imaging

Thermal imaging sensors on UAVs detect temperature variations in crops, helping identify heat stress, water stress, and irrigation inefficiencies. Areas with inadequate water supply appear as hot spots in thermal images, indicating regions that require adjusted irrigation schedules [56]. This technology helps prevent overwatering and underwatering, ensuring optimal plant hydration while conserving water resources.

Multispectral Sensors

Multispectral sensors capture different wavelengths of light to analyze plant health and nutrient status. By assessing chlorophyll content, nitrogen levels, and vegetation indices like NDVI (Normalized Difference Vegetation Index), farmers can detect early signs of nutrient deficiencies, diseases, or plant stress [57]. This information allows for precision fertilization, improving crop growth and minimizing excess chemical use.

RGB Cameras

RGB (Red, Green, Blue) cameras provide high-resolution, true-color images that help farmers evaluate plant uniformity, canopy density, and overall field condition. These images are useful for visual inspections, identifying uneven crop growth, and detecting abnormalities such as weed infestations or lodging. RGB cameras serve as a cost-effective tool for regular crop monitoring and farm management [6].

Benefits of UAV-Based Mid-Season Monitoring

Early Problem Detection

UAVs equipped with multispectral, thermal, and high-resolution cameras enable farmers to detect early signs of pest infestations, diseases, and nutrient deficiencies before they become severe. By identifying stressed areas in the field, farmers can take timely action, reducing crop losses and minimizing the need for widespread pesticide use. This proactive approach leads to healthier crops and improved yield potential [58].

Precision Irrigation and Fertilization

UAV-based monitoring helps optimize water and nutrient application by identifying moisture-deficient zones and nutrient-deprived areas. This targeted approach ensures that water and fertilizers are used efficiently, reducing input costs while minimizing the risk of soil degradation and environmental pollution. Precision application leads to better crop growth, improved resource management, and enhanced sustainability [59].

Improved Yield Prediction

By continuously monitoring crop growth and health throughout the season, UAV data can be used to predict harvest yields more accurately. Using NDVI analysis and machine learning models, farmers can estimate crop productivity, assess potential losses, and plan harvesting logistics effectively. This improves market planning, reduces post-harvest losses, and enhances overall farm profitability [60].

8.6.3 Post-harvest Field Analysis

Post-harvest analysis using UAVs provides valuable insights for sustainable land management and future crop planning. High-resolution aerial imagery helps assess crop residue distribution, soil degradation, and nutrient depletion after harvesting. This information guides decisions on soil replenishment strategies, such as organic amendments, cover cropping, or crop rotation to restore soil fertility. UAVs also assist in monitoring leftover biomass, which can be repurposed for composting or bioenergy production. By integrating post-harvest UAV analysis, farmers can improve soil health, enhance sustainability, and prepare fields effectively for the next planting season [61].

Purpose of Post-Harvest Analysis

After harvesting, it is essential to evaluate **soil conditions, residue distribution, and field damage** to prepare for the next planting cycle. UAVs assist in:

- **Residue mapping**: Identifies plant material left on the soil for organic matter enrichment.

- **Soil compaction assessment**: Detects areas needing tillage before the next season.
- **Yield variability analysis**: Identifies high- and low-production areas for future planning.

How UAVs Conduct Post-Harvest Field Analysis

Yield Estimation Models

UAV-based imaging provides accurate yield estimation models by analyzing plant biomass and crop distribution in harvested fields. Using high-resolution RGB and multispectral imagery, UAVs assess crop density, plant health, and remaining residues, allowing farmers to estimate total production with greater accuracy [62]. This data helps in harvest planning, storage management, and market forecasting, enabling better financial and logistical decision-making.

Soil Condition Mapping

Thermal and multispectral imaging technologies are used for soil condition mapping, helping farmers detect moisture loss, nutrient depletion, and compaction after harvest. Thermal sensors identify dry patches or water retention issues, while multispectral imaging reveals areas with reduced organic matter or nutrient imbalances [63]. This information guides targeted soil restoration efforts, such as precision fertilization, irrigation adjustments, and organic amendments, ensuring improved soil fertility for the next planting cycle.

Field Damage Assessment

Post-harvest UAV surveys allow for detailed field damage assessment, identifying areas affected by erosion, waterlogging, or soil compaction. By detecting uneven terrain, runoff patterns, and damaged soil structures, farmers can implement corrective measures such as cover cropping, drainage improvements, or deep tillage to restore field conditions [64]. This proactive approach helps prevent long-term soil degradation and enhances land productivity for future planting seasons.

Residue Cover Index (RCI) Formula

In modern precision agriculture, indices derived from remote sensing data such as Residue Cover Index (RCI) are essential for monitoring field conditions and making informed management decisions. The Residue Cover Index (RCI) assesses the proportion of soil surface covered by crop residues, which is crucial for soil conservation and erosion prevention. RCI is calculated as

$$RCI = \frac{R_{SWIR-R_{NIR}}}{R_{SWIR+R_{NIR}}}$$

where

R_{SWIR} Reflectance in the shortwave infrared band
R_{NIR} Reflectance in the near-infrared band.

Since crop residues reflect more in the SWIR and less in the NIR than green vegetation, higher RCI values indicate greater residue coverage. Both indices are typically derived from UAV or satellite-based multispectral imaging and processed using software platforms like Pix4D, QGIS, or ENVI. These tools enable real-time, site-specific field monitoring to support sustainable agriculture practices such as irrigation scheduling, drought assessment, and conservation tillage verification.

Benefits of UAV-Based Post-Harvest Analysis

Efficient Field Cleanup

UAV-based imaging aids in post-harvest residue management by identifying areas with excessive crop debris or uneven residue distribution. This helps farmers plan efficient field cleanup, ensuring that organic matter is properly incorporated into the soil for nutrient recycling. Proper residue management enhances soil restoration, reduces the risk of pest infestations, and prepares the land for the next planting season [65].

Better Soil Conservation Planning

By analyzing UAV-generated data, farmers can identify soil degradation areas, erosion-prone zones, and nutrient-deficient regions. This enables targeted soil conservation strategies, such as planting cover crops, applying organic amendments like compost or manure, and improving drainage systems. UAV insights help maintain long-term soil health and sustainability by preventing erosion and enhancing soil structure and fertility [66].

Improved Crop Rotation Strategies

UAV-collected data on soil fertility, moisture levels, and residual nutrient content helps determine the best crops to plant in the next season. By assessing field conditions, pest history, and organic matter levels, farmers can optimize crop rotation plans to break pest cycles, restore soil nutrients, and improve overall agricultural productivity [67]. This data-driven approach supports higher yields, reduced input costs, and sustainable farming practices.

UAV-based seasonal monitoring enables farmers to make data-driven decisions throughout the crop cycle. By assessing soil conditions before planting, tracking plant health during growth, and evaluating fields after harvest, UAVs optimize productivity, resource use, and environmental sustainability. This precision agriculture approach ultimately enhances farm profitability and reduces environmental impact.

8.7 Data Analysis and Visualization

In modern agriculture, data analysis and visualization play a crucial role in optimizing farm management and improving productivity. With advancements in technology, farmers and agricultural experts can now collect, process, and interpret vast amounts of data using drones, artificial intelligence (AI), and other digital tools. By leveraging these innovations, they can make more informed decisions that enhance crop yields, reduce resource wastage, and minimize environmental impact [68]. This section explores three key aspects of data analysis and visualization in agriculture: converting drone data into actionable insights, using heatmaps for crop health assessment, and applying AI for predictive analysis.

8.7.1 Converting Drone Data into Actionable Insights

Drones have revolutionized precision agriculture by providing high-resolution imagery and real-time data on various aspects of crop and soil health. Equipped with advanced sensors such as multispectral, thermal, and LiDAR (Light Detection and Ranging), drones can capture detailed information that is otherwise difficult to obtain through traditional field inspections [69]. However, raw drone data in itself is not useful until it is processed and translated into meaningful insights.

The process of converting drone data into actionable insights involves several key steps:

- **Data Collection**: Drones fly over agricultural fields, capturing images and sensor readings that include plant health, soil moisture, elevation variations, and the presence of pests or diseases.
- **Data Processing**: The collected data is processed using specialized software that stitches images together to create comprehensive field maps. AI-powered algorithms analyze this data to detect patterns, anomalies, and trends.
- **Interpretation and Decision-Making**: Once processed, the data is visualized through graphs, charts, and maps, allowing farmers to identify problem areas. For example, areas with poor vegetation growth may indicate nutrient deficiencies or pest infestations.
- **Implementation of Corrective Measures**: Farmers can then take necessary actions such as targeted fertilization, precision irrigation, or pest control interventions to address identified issues.

By converting drone data into actionable insights, farmers gain a deeper understanding of their fields, enabling them to make data-driven decisions that improve efficiency and productivity [70]. This approach reduces reliance on guesswork, saves time, and enhances resource utilization.

8.7.2 Heatmaps for Crop Health

Heatmaps are an essential visualization tool in precision agriculture that helps farmers assess crop health and identify areas requiring attention. These maps use color gradients to represent variations in crop conditions, making it easier to detect patterns and problem areas across a field. Heatmaps are generated using data from drones, satellite imagery, and ground-based sensors. The colors on a heatmap typically indicate different levels of plant health, with green representing healthy vegetation, yellow showing moderate stress, and red highlighting critical issues such as drought stress, pest infestations, or nutrient deficiencies [71].

The benefits of using heatmaps in agriculture include:

- **Early Detection of Crop Stress**: By analyzing heatmaps, farmers can identify stress factors affecting crops at an early stage, allowing for timely intervention. This proactive approach helps prevent yield losses.
- **Efficient Resource Management**: Heatmaps enable farmers to apply fertilizers, pesticides, and water precisely where they are needed, reducing waste and improving cost-effectiveness.
- **Optimized Yield Prediction**: By monitoring crop health trends over time, farmers can estimate potential yields and make necessary adjustments to enhance production.
- **Improved Decision-Making**: Heatmaps provide a clear visual representation of field conditions, making it easier for farmers to plan and execute field operations efficiently.

The integration of heatmaps with drone technology and AI further enhances their accuracy and usefulness. AI-driven analytics can automatically generate heatmaps based on real-time data, providing farmers with continuous monitoring and insights into their crops' health [72].

8.7.3 Predictive Analysis Using AI

Artificial Intelligence (AI) has transformed the agricultural industry by enabling predictive analysis, which helps farmers anticipate future challenges and make informed decisions. AI-driven models analyze historical and real-time data from various sources, including weather patterns, soil conditions, crop performance, and pest activity, to generate predictive insights [73].

The application of AI in predictive analysis follows these key steps:

- **Data Collection and Integration**: AI systems gather data from multiple sources, such as satellite imagery, drone surveys, IoT (Internet of Things) sensors, and historical agricultural records.
- **Data Processing and Machine Learning**: Machine learning algorithms process the collected data to identify trends, correlations, and patterns that influence crop health and productivity.
- **Predictive Modeling**: AI models use this processed data to predict potential issues, such as disease outbreaks, water shortages, or unfavorable weather conditions.

- **Decision Support and Automation**: Farmers receive AI-generated recommendations for optimal planting schedules, irrigation plans, and pest management strategies. Some AI-powered systems can even automate field operations using robotic machinery.

  The advantages of predictive analysis using AI include:

- **Improved Crop Planning**: AI forecasts help farmers determine the best times for planting, fertilization, and harvesting, maximizing productivity.
- **Early Pest and Disease Detection**: Predictive models can identify conditions favorable for pest infestations and plant diseases, allowing farmers to take preventive measures.
- **Optimized Water and Resource Management**: AI-driven analysis enables efficient water usage by predicting drought conditions and recommending smart irrigation techniques.
- **Sustainability and Cost Reduction**: By minimizing resource wastage and preventing crop losses, AI contributes to sustainable farming practices while reducing operational costs.

AI-powered predictive analysis is continuously evolving, with advancements in deep learning and big data analytics further enhancing its accuracy and applicability in agriculture. As AI technology improves, farmers will have access to even more precise and reliable forecasts, empowering them to make smarter agricultural decisions. Data analysis and visualization are transforming modern agriculture by enabling farmers to monitor, assess, and manage their crops more effectively [74]. The use of drones for data collection, heatmaps for crop health assessment, and AI-driven predictive analysis allows for smarter decision-making and optimized farm management. By leveraging these advanced technologies, farmers can improve crop yields, reduce resource wastage, and promote sustainable agricultural practices. As digital tools continue to evolve, their integration into agriculture will play an even greater role in ensuring food security and environmental sustainability for future generations [75].

8.8 Case Studies in UAV Crop Monitoring

Unmanned Aerial Vehicles (UAVs), commonly known as drones, have become an essential tool in modern precision agriculture. They provide farmers with high-resolution imagery, real-time monitoring, and advanced data analytics, allowing for better decision-making in crop management. This technology helps optimize resource use, reduce labor-intensive field inspections, and improve overall productivity [76]. The following case studies illustrate the role of UAVs in addressing key agricultural challenges: disease prevention in vineyards, water management in rice fields, and pest control in large wheat fields.

8.8.1 Disease Prevention in Vineyards

Vineyards are particularly susceptible to various plant diseases, including fungal infections like powdery mildew and downy mildew. These diseases can spread rapidly and significantly reduce grape yield and quality. Traditional disease management relies on routine visual inspections and blanket pesticide applications, which can be inefficient and environmentally harmful. UAVs offer a more precise and sustainable approach to disease prevention in vineyards [77].

UAV Applications in Disease Monitoring

1. **Multispectral and Hyperspectral Imaging**: UAVs equipped with multispectral or hyperspectral cameras can capture plant stress indicators, such as chlorophyll fluorescence and canopy temperature. These indicators help detect early disease onset before symptoms become visible to the naked eye.
2. **Thermal Imaging**: Aerial thermal cameras can identify variations in plant transpiration rates, helping pinpoint areas under disease stress.
3. **Real-Time Data Analytics**: AI-powered UAVs analyze images and generate disease maps, allowing farmers to track the progression of infections across the vineyard.
4. **Targeted Fungicide Application**: UAVs equipped with precision sprayers can apply fungicides only to affected areas, minimizing chemical use and reducing environmental impact.

Benefits of UAV-Based Disease Prevention

- **Early Detection and Intervention**: Prevents widespread infection and crop loss.
- **Cost Efficiency**: Reduces unnecessary pesticide use and labor costs.
- **Environmental Sustainability**: Minimizes the ecological impact of chemical treatments.
- **Improved Yield and Quality**: Ensures healthier vines and higher grape production.

Case Study Example

A vineyard in France implemented UAV-based disease monitoring to combat powdery mildew outbreaks. Multispectral imaging detected disease hotspots two weeks before symptoms became visible, allowing for targeted fungicide application. As a result, the vineyard reduced pesticide use by 30% and improved overall grape yield by 15% [78].

8.8.2 Water Management in Rice Fields

Rice cultivation is heavily dependent on precise water management. Traditional irrigation methods, such as flood irrigation, often lead to water wastage, soil degradation, and increased production costs. Drought conditions and climate change further complicate efficient water distribution. UAVs play a critical role in optimizing water management and improving irrigation efficiency in rice fields [79].

UAV Applications in Water Management

1. **Soil Moisture Mapping**: UAVs equipped with multispectral and infrared sensors assess soil moisture levels and detect dry or overwatered areas.
2. **Thermal Imaging for Water Stress Detection**: Thermal cameras can identify heat stress in plants, helping farmers adjust irrigation schedules accordingly.
3. **Real-Time Aerial Surveys**: UAVs provide high-resolution aerial images, allowing for the precise mapping of irrigation needs across large rice paddies.
4. **Automated Irrigation Systems**: UAV data integrates with smart irrigation systems to automate water distribution, ensuring uniform soil moisture levels.

Benefits of UAV-Based Water Management

- **Water Conservation**: Reduces excess water use by identifying over-irrigated areas.
- **Higher Yield and Productivity**: Ensures optimal water supply for healthy crop growth.
- **Lower Production Costs**: Minimizes expenses related to water waste and manual inspections.
- **Sustainability:** Supports eco-friendly water use and reduces environmental degradation.

Case Study Example

A rice farm in Thailand used UAVs to monitor water distribution and soil moisture levels. By adjusting irrigation based on UAV data, the farm reduced water usage by 25% and increased yield by 12%. The technology also helped prevent waterlogging, which previously contributed to plant disease [80].

8.8.3 Pest Control in Large Wheat Fields

Wheat crops are vulnerable to various pests, including aphids, armyworms, and wheat stem sawflies. Traditional pest control methods, such as blanket pesticide spraying, can be inefficient and environmentally damaging. UAV technology enables precise pest detection and targeted interventions, reducing pesticide overuse while improving crop protection [81].

UAV Applications in Pest Control

1. **Multispectral Imaging for Pest Detection**: UAVs detect early signs of pest infestations by identifying stressed plant areas with reduced chlorophyll levels.
2. **AI-Based Pest Identification**: UAV-collected imagery is processed using machine learning algorithms to classify different pest species and determine infestation severity.
3. **Targeted Pesticide Application**: UAV sprayers apply pesticides only to affected areas, reducing chemical use and environmental contamination.
4. **Real-Time Monitoring**: Continuous aerial surveys track pest movement and allow for timely intervention.

Benefits of UAV-Based Pest Control

- **Reduced Pesticide Use**: Lowers costs and minimizes environmental harm.
- **Faster Response Time**: Early detection prevents widespread crop damage.
- **Improved Crop Health and Yield**: Protects wheat fields more effectively.
- **Eco-Friendly Approach**: Reduces pesticide runoff and contamination risks.

Case Study Example

A wheat farm in Australia implemented UAV-based pest monitoring to combat an aphid outbreak. The UAVs identified infested zones, allowing for targeted pesticide application. As a result, pesticide use was reduced by 40%, and wheat yield improved by 18%. Additionally, early pest detection helped prevent further infestations.

UAV technology is transforming modern agriculture by enhancing disease prevention, water management, and pest control. The case studies of vineyards, rice fields, and wheat farms demonstrate how drones provide real-time monitoring, optimize resource use, and improve sustainability. By integrating UAV-based solutions, farmers can increase efficiency, reduce costs, and promote environmentally friendly agricultural practices [82]. As technology continues to evolve, the role of UAVs in precision agriculture will expand, offering even more innovative solutions to global food production challenges.

8.9 Conclusion and Summary

Unmanned Aerial Vehicles (UAVs) are revolutionizing crop monitoring and health assessment by providing real-time, high-resolution data for precision agriculture. These aerial platforms equipped with advanced sensors, including multispectral, hyperspectral, and thermal cameras, enable the detection of crop stress, nutrient deficiencies, pest infestations, and diseases. UAVs offer cost-effective, time-efficient, and non-destructive methods for assessing large agricultural areas, improving decision-making, and optimizing resource management. The integration of UAVs with remote sensing and artificial intelligence further enhances the accuracy of crop health analysis. The integration of multispectral, thermal, and RGB imaging with UAV technology has significantly enhanced the ability to monitor crop health efficiently and accurately. These imaging systems provide crucial insights into nutrient status, water stress, and overall plant condition, allowing for timely interventions and improved precision agriculture practices. As UAV imaging technology continues to evolve, its role in sustainable farming and global food security will become even more significant.

Multispectral imaging provides critical insights into nutrient health and disease stress, thermal imaging detects water stress and irrigation issues, and RGB cameras offer a straightforward visual assessment of crop conditions. When combined, these imaging techniques provide a comprehensive overview of plant health, enabling farmers to make informed decisions that optimize yields while conserving resources. Despite challenges such as data processing requirements and regulatory constraints, the continued development of UAV imaging technologies, artificial intelligence, and machine learning will further enhance precision agriculture, ensuring sustainable food production for the future. UAV-based disease detection is revolutionizing crop protection by providing accurate, early identification of diseases, tracking their progression, and enabling precision treatments. By leveraging advanced imaging technologies and data-driven insights, UAVs help farmers respond to disease outbreaks more effectively, ultimately improving crop health and agricultural productivity.

UAV-based pest monitoring is transforming modern agriculture by providing faster, more accurate, and environmentally friendly solutions for pest detection and control. By leveraging high-resolution imaging, AI-based mapping, and precision spraying, UAVs help farmers detect infestations early, map affected areas accurately, and reduce pesticide dependence. This approach not only improves crop yields but also promotes sustainable farming practices. As UAV technology continues to advance, its role in sustainable agriculture will grow, promoting more eco-friendly and efficient farming practices. UAV-based field mapping and soil health analysis provide real-time, high-resolution insights into land conditions, enabling data-driven decision-making in agriculture. By generating DEMs, soil nutrient maps, and erosion risk assessments, UAVs enhance precision farming, promote sustainable land use, and improve overall crop productivity. Data analysis and visualization are essential components of UAV-based precision agriculture, enabling farmers to make informed decisions through heatmaps and AI-driven predictive models. By leveraging these advanced technologies, farmers can improve crop yields, reduce resource wastage, and promote sustainable agricultural practices. As digital tools continue to evolve, their integration into agriculture will play an even greater role in ensuring food security and environmental sustainability for future generations.

References

1. Han H, Liu Z, Li J, Zeng Z (2024) Challenges in remote sensing based climate and crop monitoring: navigating the complexities using AI. J Cloud Comput 13:1–14
2. Getahun S, Kefale H, Gelaye Y (2024) Application of precision agriculture technologies for sustainable crop production and environmental sustainability: a systematic review. Sci World J 2024:2126734
3. Padhiary M, Kumar R, Sethi LN (2024) Navigating the future of agriculture: a comprehensive review of automatic all-terrain vehicles in precision farming. J Inst Eng (India): Ser A 105:767–782
4. Vashishth TK, Sharma V, Sharma KK et al (2024) Integration of unmanned aerial vehicles (UAVs) and IoT for crop monitoring and spraying. In: Internet of Things applications and technology. Auerbach Publications, pp 95–117
5. Toscano F, Fiorentino C, Capece N et al (2024) Unmanned aerial vehicle for precision agriculture: a review. IEEE Access
6. Sharma K, Shivandu SK (2024) Integrating artificial intelligence and internet of things (IoT) for enhanced crop monitoring and management in precision agriculture. Sens Int 100292
7. Sarvakar K, Thakkar M (2024) Different vegetation indices measurement using computer vision. In: Applications of computer vision and drone technology in agriculture 4.0. Springer, pp 133–163
8. Peng J, Nieto H, Andersen MN et al (2023) Accurate estimates of land surface energy fluxes and irrigation requirements from UAV-based thermal and multispectral sensors. ISPRS J Photogramm Remote Sens 198:238–254
9. Awais M, Li W, Cheema M et al (2022) UAV-based remote sensing in plant stress imagine using high-resolution thermal sensor for digital agriculture practices: a meta-review. Int J Environ Sci Technol 1–18
10. Orchi H, Sadik M, Khaldoun M (2022) On using artificial intelligence and the internet of things for crop disease detection: a contemporary survey. Agriculture 12:9
11. Joshi R, Pandey K (2024) IoT-enabled UAV: a comprehensive review of technological change in Indian farming. In: Unmanned aircraft systems, pp 93–135
12. Neupane K, Baysal-Gurel F (2021) Automatic identification and monitoring of plant diseases using unmanned aerial vehicles: a review. Remote Sens 13:3841
13. Abbas A, Zhang Z, Zheng H et al (2023) Drones in plant disease assessment, efficient monitoring, and detection: a way forward to smart agriculture. Agronomy 13:1524
14. Kouadio L, El Jarroudi M, Belabess Z et al (2023) A review on UAV-based applications for plant disease detection and monitoring. Remote Sens 15:4273
15. Kuswidiyanto LW, Noh H-H, Han X (2022) Plant disease diagnosis using deep learning based on aerial hyperspectral images: a review. Remote Sens 14:6031
16. Oide AH, Nagasaka Y, Tanaka K (2022) Performance of machine learning algorithms for detecting pine wilt disease infection using visible color imagery by UAV remote sensing. Remote Sens Appl: Soc Environ 28:100869
17. Pandey DK, Mishra R (2024) Towards sustainable agriculture: harnessing AI for global food security. Artif Intell Agric
18. Raheem D, Dayoub M, Birech R, Nakiyemba A (2021) The contribution of cereal grains to food security and sustainability in Africa: potential application of UAV in Ghana, Nigeria, Uganda, and Namibia. Urban Sci 5:8
19. Ali ZA, Deng D, Shaikh MK et al (2024) AI-based UAV swarms for monitoring and disease identification of Brassica plants using machine learning: a review. Comput Syst Sci Eng 48
20. Abdullah HM, Mohana NT, Khan BM et al (2023) Present and future scopes and challenges of plant pest and disease (P&D) monitoring: remote sensing, image processing, and artificial intelligence perspectives. Remote Sens Appl: Soc Environ 32:100996
21. Mohsan SAH, Khan MA, Noor F et al (2022) Towards the unmanned aerial vehicles (UAVs): a comprehensive review. Drones 6:147

22. Nahiyoon SA, Ren Z, Wei P et al (2024) Recent development trends in plant protection UAVs: a journey from conventional practices to cutting-edge technologies—a comprehensive review. Drones 8:457
23. Hafeez A, Husain MA, Singh S et al (2023) Implementation of drone technology for farm monitoring & pesticide spraying: a review. Inf Process Agric 10:192–203
24. Alsadik B, Ellsäßer FJ, Awawdeh M et al (2024) Remote sensing technologies using UAVs for pest and disease monitoring: a review centered on date palm trees. Remote Sens 16:4371
25. Udofa ES, Ali MSM, Jack KE et al (2024) Recent advancements in pest-repellent monitoring technologies in precision agriculture: a comprehensive review. IEEE, pp 1–7
26. Yadav N, Sidana N (2023) Precision agriculture technologies: analysing the use of advanced technologies, such as drones, sensors, and GPS. In: Precision agriculture for optimizing resource management, crop monitoring, and yield prediction. J Adv Zool 44
27. Vellingiri A, Kokila R, Nisha P et al (2025) Harnessing GPS, sensors, and drones to minimize environmental impact: precision agriculture. In: Designing sustainable internet of things solutions for smart industries. IGI Global, pp 77–108
28. Ye K, Hu G, Tong Z et al (2025) Key intelligent pesticide prescription spraying technologies for the control of pests, diseases, and weeds: a review. Agriculture 15:81
29. Banu MM, Reehana N, Imran MM (2024) Microbial degradation of pesticides in agricultural environments: a comprehensive review of mechanisms, factors and biodiversity. Mol Sci Appl 4:65–101
30. Chaudhary R, Nawaz A, Khattak Z et al (2024) Microbial bio-control agents: a comprehensive analysis on sustainable pest management in agriculture. J Agric Food Res 101421
31. Sial MU, Ahmed R, Khan RR et al (2024) Remote sensing technologies in pest detection. In: Revolutionizing pest management for sustainable agriculture. IGI Global, pp 411–442
32. Skendžić S, Zovko M, Lešić V et al (2023) Detection and evaluation of environmental stress in winter wheat using remote and proximal sensing methods and vegetation indices—a review. Diversity 15:481
33. Vidya Madhuri E, Rupali J, Sharan S et al (2025) Transforming pest management with artificial intelligence technologies: the future of crop protection. J Crop Health 77:48
34. Surendran U, Nagakumar KCV, Samuel MP (2024) Remote sensing in precision agriculture. In: Digital agriculture: a solution for sustainable food and nutritional security. Springer, pp 201–223
35. Ashoka P, Reddy KJ, Sachan K, Ojha R (2023) Site-specific nutrient management. Adv Farming Technol 158
36. Anderson K (2024) Spatial analysis of soil and plant interactions through UAV-mounted sensor data
37. Ali M, Khan F, Khan I et al (2018) Soil and water conservation practices in district Swabi, KP, Pakistan. Adv Crop Sci Tech 6:2
38. Kamal A, Mian I, Akbar W et al (2023) Effects of soil depth and altitude on soil texture and soil quality index. Appl Ecol Env Res 21
39. Kamal A, Mian IA, Muhammad T (2023) Sustainable maize farming: organic soil amendments boost maize production and soil quality. Plant Bull 2:41–58
40. Raihan A (2024) A systematic review of geographic information systems (GIS) in agriculture for evidence-based decision making and sustainability. Glob Sustain Res 3:1–24
41. Kumar S, Meena RS, Sheoran S et al (2022) Remote sensing for agriculture and resource management. In: Natural resources conservation and advances for sustainability. Elsevier, pp 91–135
42. Singh MP, Maurya RN, Navyashree R et al (2025) Agronomy for the future
43. Pandey S, Kumari N, Mallick L (2024) Review on assessment of land degradation in watershed using geospatial technique based on unmanned aircraft systems. In: Unmanned aircraft systems. pp 263–311
44. Picado EF, Romero KF, Heenkenda MK (2025) Mapping spatial variability of sugarcane foliar nitrogen, phosphorus, potassium and chlorophyll concentrations using remote sensing. Geomatics 5:3

45. Lakhiar IA, Yan H, Zhang C et al (2024) A review of precision irrigation water-saving technology under changing climate for enhancing water use efficiency, crop yield, and environmental footprints. Agriculture 14:1141
46. Zhang Q, Qin W, Cao W et al (2023) Response of erosion reduction effect of typical soil and water conservation measures in cropland to rainfall and slope gradient changes and their applicable range in the Chinese Mollisols Region, Northeast China. Int Soil Water Conserv Res 11:251–262
47. Twumasi D (2020) Monitoring the impact of urea deep placement on rice (Oryza Sativa L.) growth and grain yield using unmanned aerial systems technology (Drones)
48. Prajapati G, Pandya P (2025) Trailblazing trends in sustainable climate-resilient precision agriculture. Allied Publishers
49. Hassan MM, Ali MA (2022) Digital elevation model and irrigation management planning in Bangladesh. In: Livelihood enhancement through agriculture, tourism and health. Springer, pp 111–135
50. Sudhakar M, Priya R (2023) Computer vision based machine learning and deep learning approaches for identification of nutrient deficiency in crops: a survey. Nat Env Poll Tech 22
51. Sabir RM, Sarwar A, Shoaib M et al (2024) Managing water resources for sustainable agricultural production. In: Transforming agricultural management for a sustainable future: climate change and machine learning perspectives. Springer, pp 47–74
52. Ming R, Jiang R, Luo H et al (2023) Comparative analysis of different UAV swarm control methods on unmanned farms. Agronomy 13:2499
53. Yadav N, Dhakad A, Saini S (2024) Chapter-7 Symptoms of nutrient deficiencies and excesses. In: Comprehensive guide to horticultural practices and management, vol 75
54. Ayaz M, Li C-H, Ali Q et al (2023) Bacterial and fungal biocontrol agents for plant disease protection: journey from lab to field, current status, challenges, and global perspectives. Molecules 28:6735
55. Dong H, Dong J, Sun S et al (2024) Crop water stress detection based on UAV remote sensing systems. Agric Water Manag 303:109059
56. Nikolaou G, Neocleous D, Christou A et al (2020) Implementing sustainable irrigation in water-scarce regions under the impact of climate change. Agronomy 10:1120
57. Elmetwalli AH, Tyler AN (2020) Estimation of maize properties and differentiating moisture and nitrogen deficiency stress via ground–based remotely sensed data. Agric Water Manag 242:106413
58. Heeb L, Jenner E, Cock MJ (2019) Climate-smart pest management: building resilience of farms and landscapes to changing pest threats. J Pest Sci 92:951–969
59. Ahmad SF, Dar AH (2020) Precision farming for resource use efficiency. In: Resources use efficiency in agriculture. pp 109–135
60. Balana BB, Aghadi CN, Ogunniyi AI (2022) Improving livelihoods through postharvest loss management: evidence from Nigeria. Food Sec 14:249–265
61. Kar D, Dhal SB (2025) Advancing food security through drone-based hyperspectral imaging: applications in precision agriculture and post-harvest management. Environ Monit Assess 197:1–26
62. Blekanov I, Molin A, Zhang D et al (2023) Monitoring of grain crops nitrogen status from UAV multispectral images coupled with deep learning approaches. Comput Electron Agric 212:108047
63. Kaur T, Himshikha, Singh A et al (2024) Enhancing nutrient recycling through regenerative practices under different agroecosystems. In: Regenerative agriculture for sustainable food systems. Springer, pp 271–301
64. Senanayake S, Pradhan B, Huete A, Brennan J (2020) A review on assessing and mapping soil erosion hazard using geo-informatics technology for farming system management. Remote sensing 12:4063
65. Pattnaik BK, Sahu C, Choudhury S et al (2023) Importance of soil management in sustainable agriculture. In: Climate-resilient agriculture, Vol 1: crop responses and agroecological perspectives. Springer, pp 487–511

66. Gunal H, Budak M, Gunal E et al (2025) Soil fertility in arid lands: strategies for sustainable management and fertilization. In: Sustainable agriculture under drought stress. Elsevier, pp 61–69
67. Shah KK, Modi B, Pandey HP et al (2021) Diversified crop rotation: an approach for sustainable agriculture production. Adv Agric 2021:8924087
68. Raza S (2024) Innovative technologies in agriculture: leveraging AI, ML, and IoT for sustainable food production and resource management. Int J Agric Sustain Dev 6:127–146
69. Liang H, Lee S-C, Bae W et al (2023) Towards UAVs in construction: advancements, challenges, and future directions for monitoring and inspection. Drones 7:202
70. Mishra H, Mishra D (2024) AI for data-driven decision-making in smart agriculture: from field to farm management. In: Artificial intelligence techniques in smart agriculture. Springer, pp 173–193
71. Chen Z (2023) Phenobee: drone-based robot for advanced field proximal phenotyping in agriculture
72. Arévalo-Royo J, Flor-Montalvo F-J, Latorre-Biel J-I et al (2025) AI algorithms in the agrifood industry: application potential in the Spanish agrifood context. Appl Sci 15:2096
73. Titirmare S, Margal PB, Gupta S, Kumar D (2024) AI-powered predictive analytics for crop yield optimization. In: Agriculture 4.0. CRC Press, pp 89–110
74. Song C, Ma W, Li J et al (2022) Development trends in precision agriculture and its management in china based on data visualization. Agronomy 12:2905
75. Khan N, Ray RL, Sargani GR et al (2021) Current progress and future prospects of agriculture technology: gateway to sustainable agriculture. Sustainability 13:4883
76. Firoozi AA, Tshambane M, Firoozi AA, Sheikh SM (2024) Strategic load management: enhancing eco-efficiency in mining operations through automated technologies. Results Eng 102890
77. Portela F, Sousa JJ, Araújo-Paredes C et al (2024) A systematic review on the advancements in remote sensing and proximity tools for grapevine disease detection. Sensors 24:8172
78. Ortega Rioja P (2024) Assessment of different strategies to reduce environmental fungicide contamination in vineyards
79. Mallareddy M, Thirumalaikumar R, Balasubramanian P et al (2023) Maximizing water use efficiency in rice farming: a comprehensive review of innovative irrigation management technologies. Water 15:1802
80. den Besten N, Steele-Dunne S, de Jeu R, van der Zaag P (2021) Towards monitoring waterlogging with remote sensing for sustainable irrigated agriculture. Remote Sens 13:2929
81. Kebe AA, Hameed S, Farooq MS et al (2023) Enhancing crop protection and yield through precision agriculture and integrated pest management: a comprehensive review. Asian J Res Crop Sci 8:443–453
82. Islam N, Rashid MM, Pasandideh F et al (2021) A review of applications and communication technologies for internet of things (IoT) and unmanned aerial vehicle (UAV) based sustainable smart farming. Sustainability 13:1821

Chapter 9
Environmental Impact and Sustainability

Abstract Unmanned Aerial Vehicles (UAVs) are gaining attention for their role in environmental sustainability, particularly in agriculture, conservation, and disaster response. They offer efficient data collection, precision resource management, and real-time environmental monitoring, reducing chemical overuse, soil disturbance, and carbon emissions. However, concerns such as electronic waste, battery energy consumption, and potential wildlife disturbance pose environmental challenges. UAVs optimize resource use, minimize ecological footprints, and enhance biodiversity conservation. They also improve energy efficiency in transportation and logistics, reducing greenhouse gas emissions. However, they also present sustainability challenges like e-waste accumulation and energy-intensive battery production. To maximize their positive impact, strategies such as eco-friendly drone materials, recycling programs, and alternative energy sources like solar-powered UAVs should be implemented. UAVs are revolutionizing biodiversity conservation and sustainable farming by providing efficient, cost-effective, and eco-friendly solutions. By integrating UAV technology into environmental policies and conservation strategies, we can enhance biodiversity protection, mitigate climate change effects, and support long-term ecological sustainability.

Keywords UAVs · Environmental sustainability · Precision agriculture · Carbon footprint · Renewable energy · Biodiversity · Resource management

9.1 Introduction

Unmanned Aerial Vehicles (UAVs), commonly known as drones, have revolutionized various industries due to their efficiency, precision, and versatility. In environmental monitoring and sustainability efforts, UAVs play a crucial role in applications such as ecosystem mapping, wildlife monitoring, agricultural assessment, and disaster response. Their ability to collect high-resolution data in real-time enhances decision-making processes, reduces resource wastage, and minimizes human intervention in sensitive ecosystems. Despite their numerous advantages, the environmental impact

© The Author(s), under exclusive license to Springer Nature Singapore Pte Ltd. 2025

J. Li, *UAV Aerodynamics and Crop Interaction*, Smart Agriculture 13,
https://doi.org/10.1007/978-981-96-8402-1_9

of UAVs remains a critical consideration. Factors such as energy consumption, emissions from manufacturing and operations, electronic waste, and potential disturbances to wildlife raise concerns about their sustainability. Addressing these challenges requires strategic innovations, including the development of energy-efficient propulsion systems, biodegradable components, and improved recycling practices. This chapter explores both the positive and negative environmental effects of UAVs, highlighting their contributions to sustainability while examining potential ecological drawbacks. Additionally, it discusses strategies to enhance their friendliness, ensuring that UAV technology aligns with long-term environmental conservation goals.

9.2 Environmental Benefits of UAVs

Unmanned Aerial Vehicles (UAVs), commonly known as drones, have become essential tools in promoting environmental sustainability. Their ability to collect high-resolution data, monitor vast areas, and manage resources with precision makes them invaluable in conservation and agriculture. Unlike traditional methods, UAVs offer real-time insights, enabling efficient decision-making while reducing environmental impact. One of the primary benefits of UAVs is their role in reducing chemical overuse in agriculture. Through precision spraying and targeted application of fertilizers and pesticides, drones minimize chemical runoff, protect water sources, and safeguard biodiversity [1]. Additionally, UAVs help minimize soil disturbance by replacing heavy machinery, which often leads to soil compaction and erosion. By providing detailed soil health data, drones enable farmers to adopt sustainable land management practices. Furthermore, UAVs support climate-resilient farming by offering real-time monitoring of weather conditions, drought stress, and crop health. This allows for optimized irrigation, early pest detection, and improved crop resilience in the face of climate change. UAV technology also plays a crucial role in disaster response, helping assess damage from floods, wildfires, and other environmental hazards [2]. By enhancing efficiency, reducing resource waste, and minimizing ecological footprints, UAVs contribute to a more sustainable and resilient future in environmental conservation and agriculture.

9.2.1 Reducing Chemical Overuse with UAVs

One of the major environmental challenges in modern agriculture is the excessive use of chemical fertilizers, pesticides, and herbicides. Traditional application methods, such as ground-based spraying and aerial spraying by manned aircraft, often result in over-application, leading to serious environmental consequences. These include chemical runoff, soil degradation, water pollution, and biodiversity loss. UAVs (Unmanned Aerial Vehicles) provide an innovative solution by enabling precision agriculture, ensuring that chemicals are applied only where needed and in optimal amounts [3]. By reducing excessive chemical use, UAVs contribute to more sustainable farming practices while minimizing harm to surrounding ecosystems.

Precision Spraying Technology

Drones equipped with advanced spraying systems have transformed the way chemicals are applied in agriculture. Traditional ground-based spraying, often performed using tractors, can be inefficient due to overlapping application, inconsistent coverage, and soil compaction from heavy machinery. Similarly, aerial spraying from manned aircraft tends to lack precision, leading to chemical drift caused by wind, which can result in unintended exposure to non-target crops and ecosystems [4].

UAVs overcome these limitations by utilizing GPS-guided navigation, multispectral imaging, and artificial intelligence (AI) to detect problem areas within fields and apply treatments with exceptional accuracy. These drones can fly close to the crop canopy, ensuring uniform chemical distribution while reducing wastage. Additionally, UAVs are capable of variable rate application, meaning they can adjust the amount of pesticide or fertilizer based on real-time crop health assessments. This targeted approach significantly reduces the quantity of chemicals used, preventing unnecessary contamination of the environment. UAV-based precision spraying also benefits farmers economically [5]. By minimizing chemical use and improving application efficiency, drones help reduce input costs, leading to increased profitability. Moreover, since UAVs operate autonomously and require minimal fuel, they are more cost-effective and environmentally friendly compared to traditional methods.

Reducing Water Pollution and Soil Contamination

Excessive agricultural chemical use is a leading cause of water pollution and soil contamination worldwide. When fertilizers and pesticides are applied indiscriminately, excess chemicals can leach into groundwater or run off into nearby rivers and lakes. This contamination often leads to eutrophication, a process in which excess nutrients (mainly nitrogen and phosphorus) cause excessive algae growth in water bodies. Algal blooms deplete oxygen levels, harming fish populations and disrupting aquatic ecosystems. UAV-based precision spraying helps address this issue by applying chemicals only where they are needed, significantly reducing runoff. With real-time imaging and AI-based analysis, UAVs ensure that fertilizers and pesticides are used in optimal amounts, preventing excessive accumulation in the soil and minimizing the risk of leaching into groundwater sources [6]. Additionally, reducing chemical overuse contributes to improved soil health. Excessive pesticide use can destroy beneficial soil microbes responsible for nutrient cycling and organic matter decomposition. By applying chemicals selectively, UAVs help maintain microbial diversity, leading to healthier, more fertile soil that supports sustainable agriculture [7].

Protecting Pollinators and Biodiversity

Pollinators such as bees, butterflies, and birds are essential for maintaining biodiversity and agricultural productivity. Around 75% of global food crops depend on pollinators for reproduction. However, the widespread use of pesticides—particularly neonicotinoids and organophosphates—has been linked to declining pollinator populations, contributing to colony collapse disorder (CCD) in bees and negatively affecting other wildlife. UAVs provide a pollinator-friendly solution by enabling targeted pesticide application, ensuring that chemicals are not sprayed in areas where pollinators are active [8]. By monitoring bee activity and flight patterns, UAVs can adjust spraying schedules to avoid peak pollination periods, reducing direct exposure to harmful chemicals. Furthermore, some UAVs are equipped with thermal and multispectral imaging, which allows them to differentiate between crop areas that require treatment and those that do not, further minimizing unnecessary pesticide use. Beyond protecting pollinators, reducing pesticide overuse benefits other non-target species, including soil-dwelling organisms, birds, and aquatic life. Many pesticides persist in the environment, contaminating food chains and affecting multiple species [9]. By limiting chemical exposure, UAVs help preserve biodiversity, ensuring that agricultural landscapes remain ecologically balanced.

Important in Reducing Chemical Overuse

The excessive use of pesticides, herbicides, and fertilizers in modern agriculture has led to severe environmental consequences. While these chemicals are essential for increasing crop yields and protecting plants from pests and diseases, their misuse and over-application have resulted in soil degradation, water pollution, and biodiversity loss. UAVs play a critical role in addressing these challenges by enhancing precision in chemical application, reducing wastage, and minimizing environmental harm [10].

- **Soil Degradation**—The continuous overuse of agricultural chemicals alters the natural composition of the soil. Pesticides and synthetic fertilizers can **kill beneficial soil microbes**, reducing soil fertility over time. This depletion of soil biodiversity weakens its structure, making it more susceptible to erosion and decreasing long-term agricultural productivity.
- **Water Pollution**—Chemical runoff from excessive fertilizer and pesticide use contaminates nearby rivers, lakes, and groundwater. This pollution leads to eutrophication, where excess nutrients cause algae blooms that deplete oxygen levels in aquatic ecosystems, harming fish and other aquatic organisms.
- **Biodiversity Loss**—Many pesticides and herbicides harm non-target organisms, including pollinators such as bees, butterflies, and birds. When these chemicals enter the food chain, they negatively impact wildlife populations and disrupt ecological balance.

By improving precision spraying, UAVs significantly reduce chemical overuse, ensuring that fertilizers and pesticides are applied only where needed. This targeted approach helps maintain soil health, protect water resources, and preserve biodiversity while increasing agricultural efficiency [11].

Example: UAVs in Action

A rice farmer adopts UAV-based pest detection instead of traditional pesticide spraying methods. In conventional agriculture, spraying pesticides over an entire field requires 100 L of chemicals per hectare. However, drone-based precision spraying allows the farmer to target affected areas only, reducing pesticide usage to 60 L per hectare while maintaining the same level of pest control. This 40% reduction in pesticide use not only lowers costs but also minimizes environmental pollution, making farming more sustainable. UAVs revolutionize modern agriculture by enhancing efficiency, reducing chemical overuse, and minimizing environmental damage [12]. Through precision agriculture, targeted spraying, and automation, UAVs ensure that fertilizers and pesticides are applied only where necessary, preventing soil degradation, water pollution, and biodiversity loss. By integrating UAV technology into farming practices, we can create a more sustainable and eco-friendly agricultural system, benefiting both farmers and the environment. UAVs are transforming modern agriculture by significantly reducing chemical overuse. Through precision spraying technology, drones ensure that fertilizers and pesticides are applied efficiently, minimizing waste and environmental contamination. By reducing water pollution and soil degradation, UAVs contribute to healthier ecosystems, while their role in protecting pollinators and biodiversity supports long-term agricultural sustainability [13]. As drone technology continues to advance, integrating UAVs into farming practices will play a crucial role in creating a more environmentally responsible and resource-efficient agricultural system.

9.2.2 Minimizing Soil Disturbance

Soil health is fundamental to sustainable agriculture, yet it is frequently compromised by intensive farming practices. Conventional land management relies heavily on heavy machinery, which leads to soil compaction, erosion, and degradation. UAVs offer a non-invasive alternative, minimizing soil disturbance while maximizing productivity [14].

Reducing the Need for Heavy Machinery

Tractors and other large agricultural machines contribute to soil compaction by applying immense pressure on the ground. Compacted soil reduces water infiltration, limits root growth, and decreases soil aeration, ultimately affecting crop yields. UAVs eliminate the need for frequent use of these heavy machines by conducting aerial surveys, applying fertilizers, and monitoring crop health without touching the soil [15]. This significantly reduces the physical disturbance to farmland, helping maintain long-term soil fertility.

Preventing Soil Erosion

Soil erosion is a major environmental issue caused by excessive tillage, deforestation, and overgrazing. UAVs play a crucial role in erosion prevention by providing farmers with accurate data on soil conditions, moisture levels, and land topography. By analyzing this information, farmers can implement strategies such as contour farming, cover cropping, and conservation tillage, reducing soil loss and preserving arable land [16].

Monitoring Soil Health

Soil degradation due to nutrient depletion, salinization, and loss of organic matter is a growing concern. UAVs equipped with multispectral and infrared sensors can assess soil quality and detect early signs of degradation. By analyzing vegetation indices, temperature variations, and moisture content, UAVs provide valuable insights into soil health, allowing for timely interventions. Farmers can use this data to apply organic amendments, adjust irrigation schedules, and adopt regenerative practices that enhance soil sustainability [17].

Why is it Important?

Soil health is critical for sustainable agriculture. Heavy machinery, such as tractors, causes:

- **Soil compaction**, which reduces water infiltration and air movement.
- **Loss of soil structure**, making it prone to erosion and decreasing fertility.
- **Increased carbon emissions**, as frequent plowing releases stored carbon into the atmosphere.

How Do UAVs Help?

- **Remote Soil Mapping**: UAVs with LiDAR and infrared sensors create detailed soil maps without physically touching the ground.
- **No-Till Farming Support**: UAVs help farmers assess soil conditions, allowing them to minimize plowing and reduce soil disturbance.
- **Erosion Prevention**: UAVs help detect early signs of soil erosion, allowing corrective actions like cover cropping or terracing.

Example

In corn farming, a farmer traditionally uses tractors for soil sampling, causing soil compaction. By switching to drones with hyperspectral imaging, the same data is collected without disturbing the soil [18].

9.2.3 Supporting Climate-Resilient Farming

Climate change poses significant threats to global agriculture, with rising temperatures, unpredictable weather patterns, and increased occurrences of droughts and floods. UAVs help farmers adapt to these challenges by providing real-time data, optimizing resource use, and facilitating climate-smart agricultural practices [19].

Improving Irrigation Efficiency

Water scarcity is a major concern in many agricultural regions, making efficient irrigation essential. UAVs equipped with thermal and multispectral cameras can detect variations in soil moisture, identifying areas that require additional watering while avoiding unnecessary irrigation in well-hydrated zones [20]. This precision irrigation approach conserves water, reducing wastage and lowering energy consumption associated with pumping and distribution systems.

Enhancing Crop Resilience

Drones provide farmers with critical information about crop health, enabling them to take proactive measures against climate-related stressors. By capturing high-resolution images, UAVs can detect signs of drought stress, disease outbreaks, and nutrient deficiencies before they become severe. Early detection allows farmers to apply targeted interventions such as supplementary irrigation, selective fertilization, or disease management strategies, improving overall crop resilience [21].

Reducing Carbon Footprint

Traditional agricultural monitoring and data collection often require extensive fieldwork, involving fuel-intensive vehicles and machinery. By replacing these methods with UAV-based monitoring, farmers can significantly reduce their carbon footprint. Additionally, drones contribute to sustainability by supporting conservation projects that focus on reforestation, carbon sequestration, and habitat restoration [22].

Disaster Response and Recovery

Extreme weather events, such as hurricanes, wildfires, and floods, can devastate agricultural lands and natural ecosystems. UAVs play a vital role in disaster response by assessing damage, identifying affected areas, and assisting in recovery efforts. For instance, after a flood, drones can survey impacted farmland, providing rapid assessments that help farmers and governments allocate resources efficiently. In wildfire-prone regions, UAVs are used for fire detection and monitoring, allowing for early intervention and reducing environmental destruction [23].

The Future of UAVs in Sustainable Agriculture

As UAV technology continues to advance, its potential for environmental sustainability will expand even further. Future innovations may include:

- **Solar-Powered UAVs**: Development of energy-efficient drones powered by solar panels will further reduce their carbon footprint, allowing for longer flight times and greater environmental benefits.
- **Biodegradable Components**: Researchers are exploring the use of biodegradable drone parts made from natural fibers, reducing electronic waste and improving sustainability.
- **AI-Driven Data Analysis**: Integrating artificial intelligence with UAVs will enhance data processing capabilities, providing even more precise recommendations for sustainable farming practices.
- **Autonomous Reforestation Drones**: Some companies are already developing drones that can plant trees autonomously, aiding in large-scale reforestation and carbon sequestration efforts.

Why is it Important?

Climate change is causing unpredictable weather, prolonged droughts, and erratic rainfall, leading to:

- **Water scarcity**, affecting irrigation.
- **Higher pest outbreaks**, due to temperature changes.
- **Inconsistent crop yields**, impacting food security.

How Do UAVs Help?

- **Drought Monitoring**: UAVs with thermal cameras detect soil moisture levels and help optimize irrigation.
- **Weather Adaptation**: UAVs track changes in temperature, humidity, and rainfall, helping farmers adjust sowing and harvesting schedules.
- **Pest and Disease Management**: UAVs identify affected areas, allowing for targeted intervention instead of large-scale chemical use.

Example

A vineyard in California faces prolonged drought. UAVs with thermal imaging detect water-stressed vines, allowing farmers to apply irrigation only where needed instead of watering the entire field (Table 9.1).

Table 9.1 Additional benefits of UAVs in sustainable farming

Benefit	Impact
Lower carbon footprint	UAVs replace fuel-powered tractors, reducing emissions
Improved crop yield	Precision monitoring leads to better plant health
Biodiversity conservation	Less chemical use protects pollinators and beneficial insects
Real-time decision making	UAVs provide instant data for better farm management

UAVs are game-changers in environmental conservation and precision agriculture. By reducing chemical overuse, minimizing soil disturbance, and supporting climate-resilient farming, they enhance sustainability while improving efficiency. Their integration with AI and data analytics further amplifies their role in modern agriculture. UAVs have emerged as valuable tools in promoting environmental sustainability by optimizing resource management, reducing chemical overuse, minimizing soil disturbance, and supporting climate-resilient farming. Their ability to provide real-time, high-resolution data allows for more efficient decision-making, ultimately benefiting both agricultural productivity and ecosystem conservation [22]. While challenges such as energy consumption, electronic waste, and regulatory concerns remain, ongoing technological advancements are paving the way for more eco-friendly UAV applications. By integrating drones into sustainable land management strategies, we can enhance food security, protect natural resources, and mitigate the impacts of climate change, ensuring a greener and more resilient future.

9.3 Reducing Environmental Footprint with UAVs

Unmanned Aerial Vehicles (UAVs), commonly known as drones, have revolutionized environmental sustainability by providing innovative solutions for resource management, agriculture, and conservation. Their ability to collect real-time data, monitor ecosystems, and optimize agricultural practices has significantly reduced the environmental footprint of various industries. As technology advances, UAVs are becoming indispensable in addressing critical environmental challenges such as chemical overuse, carbon emissions, deforestation, and biodiversity loss. One of the most significant advantages of UAVs is their role in precision agriculture, where they help farmers optimize the use of fertilizers, pesticides, and water resources [24]. Traditional agricultural methods often involve excessive chemical use, leading to soil degradation, water pollution, and harm to beneficial species like pollinators. UAVs equipped with multispectral imaging and GPS-guided systems enable targeted application, reducing chemical runoff and preserving soil health. This precision spraying approach minimizes environmental contamination, making farming more sustainable and efficient. Beyond agriculture, UAVs contribute to environmental conservation and monitoring efforts. They provide non-invasive surveillance of wildlife, forests, and water bodies, allowing researchers to track deforestation, detect illegal poaching, and monitor climate change impacts. Compared to traditional monitoring methods that require helicopters, fuel-powered vehicles, or human intervention in fragile ecosystems, UAVs offer a low-impact and energy-efficient alternative [25]. They help conservationists collect high-resolution aerial images, assess habitat health, and respond swiftly to environmental threats.

Another major benefit of UAVs is their energy efficiency compared to conventional land and air-based monitoring systems. Traditional agricultural machinery, such as tractors, airplanes, and irrigation systems, consumes vast amounts of fuel, leading to high carbon emissions and environmental degradation. UAVs, primarily powered by electric batteries, offer a cleaner alternative by reducing fuel consumption and lowering greenhouse gas emissions. With further advancements in battery technology and solar-powered UAVs, these devices are becoming even more sustainable, reducing reliance on fossil fuels [26]. UAVs also play a crucial role in soil conservation and land management by preventing excessive soil disturbance. Conventional farming equipment, such as heavy tractors and plows, can compact soil, disrupt natural microbial communities, and accelerate erosion. UAVs eliminate the need for heavy machinery in specific farming operations, ensuring minimal soil disruption while optimizing crop health. Similarly, in forestry and environmental monitoring, UAVs provide aerial assessments without requiring road construction or human encroachment, preserving natural habitats and reducing deforestation risks [25].

Additionally, UAVs are transforming disaster response and environmental hazard management. In the aftermath of natural disasters such as floods, wildfires, and hurricanes, drones provide rapid aerial assessments, helping authorities assess damage, locate survivors, and direct emergency response efforts. Their ability to navigate difficult terrains and access hazardous areas makes them highly effective in mitigating environmental risks and improving disaster resilience. As UAV technology continues to evolve, its role in sustainable agriculture, conservation, and climate research will expand. With enhanced battery life, AI-driven analytics, and improved automation, UAVs will become even more efficient in reducing environmental impact [27]. By integrating UAVs into modern land and resource management strategies, industries can significantly lower their ecological footprint, contributing to a more sustainable and environmentally conscious future.

9.3.1 Minimizing Chemical Use Through Precision Application

The excessive use of chemical fertilizers, pesticides, and herbicides in traditional farming has led to severe environmental consequences, including soil degradation, water pollution, and biodiversity loss. Conventional chemical application methods, such as ground-based boom sprayers and aerial spraying from manned aircraft, often result in uneven distribution, chemical drift, and runoff into nearby ecosystems. This not only increases costs for farmers but also negatively impacts soil health, pollinators, and water bodies. Unmanned Aerial Vehicles (UAVs) provide an efficient and environmentally friendly alternative through precision application, ensuring that chemicals are used only where and when needed, significantly reducing their negative impact on the environment [28].

9.3.1.1 Precision Spraying Technology

UAVs equipped with advanced spraying systems, GPS-guided navigation, and multispectral imaging enable precise application of pesticides and fertilizers. Unlike traditional methods that distribute chemicals uniformly across entire fields, UAVs use real-time data and AI-driven analysis to detect stressed or infected areas and apply treatments only where needed. This targeted approach reduces chemical waste, minimizes soil and water contamination, and enhances crop health. By preventing overapplication and chemical drift, UAVs improve efficiency, protect beneficial organisms like pollinators, and support sustainable agriculture [29]. Their high-precision spraying technology ensures effective pest and disease management while reducing the environmental impact of chemical use.

How UAVs Improve Precision Spraying

Targeted Application—UAVs use AI-driven multispectral imaging to detect early signs of crop stress, diseases, or pest infestations, ensuring pesticides and fertilizers are applied only to affected areas. This minimizes waste, reduces costs, and prevents overexposure to chemicals in healthy regions [30].

Minimized Chemical Drift—Traditional boom sprayers and aerial spraying often result in wind drift, causing unintended chemical dispersion. UAVs, however, fly closer to the crop canopy, delivering controlled, ultra-precise spraying that reduces environmental contamination and increases effectiveness [31].

Reduced Water Pollution Risk—Overuse of agricultural chemicals leads to runoff into nearby water bodies, causing eutrophication, algae blooms, and aquatic ecosystem damage. UAVs prevent excess fertilizers and pesticides from entering rivers, lakes, and groundwater by applying only the necessary amount, safeguarding freshwater resources [32].

Optimized Fertilizer Efficiency—UAVs analyze soil and plant health data in real time, enabling customized fertilizer application based on specific nutrient deficiencies. This ensures maximum nutrient absorption, prevents chemical buildup in soil, and enhances long-term land fertility [33].

Data-Driven Decision-Making—UAVs integrate machine learning algorithms, GPS mapping, and high-resolution imaging to provide farmers with detailed crop health insights. This technology improves pest and disease detection, enhances yield predictions, and enables sustainable, precision-based chemical application with minimal environmental impact. By reducing chemical overuse, protecting soil and water quality, and enhancing agricultural efficiency, UAVs play a crucial role in environmentally sustainable farming practices [34].

9.3.1.2 Preventing Soil Contamination and Water Pollution

Excessive use of chemical fertilizers, pesticides, and herbicides depletes soil fertility and contaminates water sources, posing serious environmental risks. Conventional spraying methods often result in chemical runoff, introducing harmful substances into rivers, lakes, and groundwater. This pollution leads to eutrophication, algal blooms, and aquatic ecosystem degradation, threatening fish, amphibians, and other marine life. Additionally, prolonged chemical exposure disrupts soil microbial communities, reducing nutrient availability and long-term soil health [7]. UAVs offer a sustainable alternative by optimizing chemical use through precision spraying. Equipped with GPS navigation, AI-driven imaging, and real-time data analysis, UAVs apply chemicals only where necessary, significantly reducing overuse and environmental contamination. This minimizes runoff, preserves soil fertility, and protects biodiversity, making UAV technology a vital tool for eco-friendly and sustainable agriculture [35].

Environmental Benefits of UAV-Based Precision Spraying

Preserves Soil Health—Excessive use of pesticides and synthetic fertilizers disrupts the natural balance of soil microbes, reducing nutrient availability and long-term soil fertility. UAV-based precision spraying ensures chemicals are applied only where needed, preventing chemical buildup that can lead to soil degradation, acidification, and loss of organic matter. By maintaining a healthy microbial ecosystem, UAVs support better root growth, improved water retention, and sustainable land use, ensuring that future agricultural productivity is not compromised [36].

Protects Freshwater Resources—Conventional spraying methods result in chemical runoff, where excess pesticides and fertilizers wash into rivers, lakes, and groundwater sources. This leads to eutrophication, causing harmful algal blooms, oxygen depletion, and aquatic life mortality. UAVs minimize runoff by precisely targeting only affected areas, significantly reducing chemical wastage. By protecting freshwater ecosystems, UAVs help maintain clean drinking water supplies, prevent habitat destruction, and support sustainable agriculture without harming nearby water bodies [6].

Supports Biodiversity—Many pollinators and beneficial insects, such as bees, butterflies, and ladybugs, are essential for natural pest control and crop pollination. However, widespread pesticide use has led to declining pollinator populations, threatening ecosystem stability and food security. UAVs reduce unnecessary pesticide exposure by avoiding non-target areas, ensuring that pollinators, birds, and other wildlife are protected from harmful chemical residues [37]. This targeted approach not only preserves biodiversity but also promotes natural ecological balance, allowing beneficial organisms to thrive and contribute to sustainable farming practices. By adopting UAV-based precision spraying, farmers can reduce environmental harm, enhance agricultural efficiency, and support long-term ecosystem sustainability.

These advancements make UAVs a key tool in balancing productivity with environmental conservation, ensuring a greener and more responsible approach to modern agriculture [38].

9.3.1.3 Protecting Pollinators and Ecosystem Balance

Pollinators, including bees, butterflies, and birds, play a vital role in crop pollination, ecosystem stability, and biodiversity conservation. They facilitate fruit and seed production, ensuring high agricultural yields and maintaining natural plant diversity. However, the widespread use of pesticides in conventional farming has led to a significant decline in pollinator populations, endangering global food security and threatening entire ecosystems. UAV-based precision spraying technology offers a sustainable solution by minimizing pesticide exposure to non-target areas [39]. Using AI-driven imaging and GPS navigation, UAVs can identify pest-infested regions and apply pesticides only where necessary, rather than covering entire fields indiscriminately. This targeted approach prevents pollinators from coming into contact with toxic chemicals, reducing their mortality rates and ensuring they continue to perform their essential ecological functions. UAV technology, farmers can achieve effective pest control while protecting beneficial species, leading to a balanced ecosystem, improved crop productivity, and long-term agricultural sustainability [40].

Protecting Pollinators and Ecosystem Balance

Pollinators, including bees, butterflies, and birds, play a vital role in crop pollination, ecosystem stability, and biodiversity conservation ensuring food security by aiding in crop pollination. However, conventional pesticide application methods pose a significant threat to their survival. Traditional spraying techniques, such as tractor-mounted boom sprayers and aerial spraying, often result in widespread pesticide drift, exposing pollinators to toxic chemicals. This has led to declining pollinator populations, which in turn affects crop yields and biodiversity [8]. UAVs (drones) provide a modern, eco-friendly solution by significantly reducing unnecessary pesticide exposure through precision spraying.

Avoiding Unnecessary Spraying

UAVs are equipped with advanced imaging technologies, such as multispectral and hyperspectral cameras, GPS mapping, and AI-driven analysis, allowing them to detect areas of high pollinator activity. By identifying flowering plants, beehives, and other pollinator habitats, UAVs can adjust their spraying patterns to avoid these zones, ensuring that chemicals are applied only where needed [41]. Unlike ground-based sprayers, which coat entire fields, UAVs can target specific pest-affected regions, significantly reducing the risk of pollinators coming into contact with harmful pesticides.

Reducing Toxic Exposure

Many pesticides contain toxic compounds that harm pollinators by affecting their nervous systems, reproduction, and foraging behaviors. When bees and butterflies come into contact with pesticide residues on flowers, leaves, or soil, it can result in reduced colony strength, disorientation, and even mass die-offs [42]. UAVs address this issue by:

- Applying pesticides in precise amounts and locations, preventing unnecessary contamination of non-target areas.
- Using time-sensitive spraying strategies, such as spraying pesticides at night or early in the morning when pollinators are less active, reducing their chances of exposure.
- Implementing eco-friendly alternatives, like biopesticides, which are less harmful to beneficial insects while still effectively controlling pests.

Enhancing Biodiversity Conservation

A healthy ecosystem relies on a diverse range of species, including pollinators and natural pest predators. Overuse of pesticides not only harms pollinators but also disrupts the balance of beneficial insects, such as ladybugs, spiders, and parasitic wasps, which naturally control pest populations. UAV-based spraying helps to:

- Reduce the overall pesticide load in the environment, allowing beneficial species to thrive.
- Support natural pest control mechanisms, reducing the reliance on chemical interventions.
- Promote plant diversity by protecting wild pollinators that contribute to the growth of native vegetation and food crops.

By integrating UAVs into pest management strategies, farmers can achieve effective pest control while preserving pollinator populations, ensuring a balanced and sustainable agricultural system. This approach not only benefits farmers by improving crop yields and reducing pesticide costs but also protects biodiversity, fostering a healthier environment for future generations [43].

9.3.1.4 Increasing Agricultural Productivity with Lower Environmental Impact

UAV-assisted precision agriculture offers significant environmental and economic benefits by optimizing pesticide and fertilizer application. Traditional farming methods often lead to chemical overuse, causing soil degradation, water pollution, and biodiversity loss [44]. UAVs, equipped with GPS navigation and AI-driven imaging, enable targeted spraying, ensuring chemicals are applied only where needed. Research shows that drone-based precision spraying can reduce pesticide use by up to 30% while maintaining the same level of pest control effectiveness. Additionally, fertilizer application becomes more efficient, allowing for higher crop yields with fewer resources. By integrating UAVs into farming, costs decrease, productivity improves, and environmental harm is minimized, making agriculture more sustainable and profitable [45].

Example of UAV-Based Precision Spraying Impact

The integration of UAV-based precision spraying in modern agriculture has demonstrated significant efficiency gains, cost reductions, and environmental benefits. Unlike traditional pesticide application methods, which often involve uniform spraying across entire fields, UAV technology enables targeted application, ensuring chemicals are used only where necessary [46].

Traditional Pesticide Spraying

- Conventional methods, such as tractor-mounted boom sprayers or aerial spraying from manned aircraft, require 100 L of pesticides per hectare to achieve effective pest control.
- A large portion of these chemicals is wasted due to wind drift, runoff, and overapplication, leading to soil degradation, water pollution, and harm to beneficial insects like pollinators.
- Farmers spend more on pesticides, increasing their overall input costs without necessarily improving crop health.

UAV Precision Spraying

- Advanced drone technology, equipped with multispectral imaging, GPS navigation, and AI-driven analysis, detects infested areas and applies pesticides only where needed.
- UAVs fly closer to the crop canopy, reducing pesticide drift and ensuring uniform coverage in targeted zones.
- As a result, UAVs can achieve the same pest control results using just 60 L per hectare, eliminating unnecessary chemical usage.

Outcome: A 40% Reduction in Chemical Use

- **Environmental Benefits**: Less pesticide runoff into rivers, lakes, and groundwater reduces water pollution and preserves aquatic ecosystems.
- **Cost Efficiency**: Farmers save on pesticide expenses, lowering production costs while maintaining yield quality.
- **Soil Health Preservation**: Reduced chemical exposure prevents microbial imbalance and soil degradation, ensuring long-term agricultural sustainability.
- **Biodiversity Protection**: Targeted spraying minimizes pesticide exposure to pollinators, beneficial insects, and non-target organisms, fostering a healthier ecosystem.

UAV-based precision spraying, farmers can achieve optimal pest control while reducing environmental harm, making agriculture more efficient, eco-friendly, and sustainable for future generations. UAVs are transforming modern agriculture by reducing chemical overuse, preventing soil contamination, and protecting water resources [47]. Through precision spraying technology, targeted application, and AI-driven analysis, UAVs minimize pesticide and fertilizer waste, ensuring that only the necessary amount is applied. This approach lowers the risk of chemical runoff, protects pollinators, and enhances soil health, making UAVs a crucial tool for environmentally sustainable farming [48].

9.3.1.5 Energy Efficiency of Drone Operations

Unmanned Aerial Vehicles (UAVs) are revolutionizing industries such as agriculture, environmental monitoring, logistics, and disaster management by offering a highly energy-efficient alternative to traditional machinery. Unlike fuel-powered tractors, helicopters, and airplanes, which require large amounts of fossil fuels, UAVs operate on electric power or hybrid energy systems, reducing carbon emissions, fuel dependency, and operational costs [49]. This shift toward electrified aerial operations is critical for promoting sustainability and energy conservation.

Limitations of Agricultural Tractors and Ground Sprayers: High Fuel Consumption and Environmental Impact

Agricultural tractors and ground sprayers have long been the primary tools for crop spraying, fertilizer application, and field monitoring. However, these machines come with high operational costs, significant energy consumption, and negative environmental effects [50].

9.3.2 High Fuel Consumption

Tractors and ground sprayers rely on diesel fuel, which is both costly and harmful to the environment.

- **Fuel Usage Per Hectare**—Large tractors consume 3–5 L of diesel per hectare when spraying pesticides, herbicides, or fertilizers.
- **Large-Scale Impact**—For farms covering hundreds or thousands of hectares, this results in hundreds of liters of fuel burned daily, significantly increasing operational expenses and carbon emissions.
- **Limited Efficiency**—Traditional spraying methods lack precision, meaning that a larger area is covered with chemicals than necessary, leading to excess fuel use and chemical wastage.

9.3.3 High Maintenance and Repair Costs

Fuel-powered machinery requires constant upkeep, which adds long-term financial burdens on farmers.

- **Frequent Servicing**—Tractors have multiple mechanical components (engines, fuel injectors, hydraulic pumps) that need regular maintenance to function efficiently.
- **Breakdowns and Downtime**—Malfunctions in engine systems or sprayers can lead to delays in farming activities, negatively affecting crop productivity.
- **Spare Parts and Labor Costs**—Repairs often involve expensive replacement parts and skilled labor, which increase the total cost of ownership.

9.3.4 Dependence on Fossil Fuels and Environmental Damage

Traditional farming machinery is heavily reliant on fossil fuels, which has multiple negative environmental consequences:

- **Greenhouse Gas (GHG) Emissions**—Diesel-powered tractors emit carbon dioxide (CO_2), nitrogen oxides (NO_x), and particulate matter, contributing to air pollution and global warming.
- **Soil Compaction and Erosion**—The weight of heavy tractors compresses the soil, reducing water absorption and soil aeration, which negatively impacts crop growth and long-term soil health.
- **Water Contamination**—Excessive pesticide and fertilizer application using ground sprayers increases runoff into nearby water bodies, leading to water pollution and ecosystem disruption.

How UAVs Solve These Issues

Unmanned Aerial Vehicles (UAVs) provide a highly efficient and sustainable alternative to fuel-intensive tractors and ground sprayers [51].

Energy Efficiency—UAVs operate on electric batteries, eliminating fuel costs and reducing carbon emissions.

Lower Maintenance—Drones have fewer mechanical parts, reducing repair costs and downtime.

Precision Spraying—UAVs detect crop stress and apply chemicals only where needed, minimizing chemical overuse and fuel waste.

Zero Soil Compaction—UAVs fly above crops, preventing soil compression and erosion caused by heavy tractors.

Improved Water Conservation—Reduced chemical runoff from UAV spraying helps protect freshwater sources and ecosystems.

By transitioning from fuel-powered farm machinery to UAV-assisted agriculture, farmers can cut costs, enhance productivity, and promote sustainability, making UAVs a game-changing innovation in modern farming [52].

Manned Aircraft (Helicopters and Crop-Dusting Planes)

Manned aircraft, such as helicopters and crop-dusting planes, have traditionally been used for large-scale pesticide and fertilizer application. However, they come with significant drawbacks, including high fuel consumption, expensive operational costs, and environmental harm. These aircraft consume 30–40 L of aviation fuel per hectare, leading to excessive carbon emissions and increased operational expenses. Additionally, their broad spraying techniques often result in chemical drift, causing pesticides and fertilizers to spread beyond the target area, contaminating nearby ecosystems [53]. The reliance on trained pilots and costly maintenance further adds to the financial burden. Moreover, these planes generate noise pollution, disrupting wildlife and pollinators. In contrast, UAVs provide a sustainable alternative, using electric power, precision spraying, and lower operational costs to enhance efficiency while minimizing environmental damage. UAVs, in contrast, operate on electric batteries or hybrid systems, consuming a fraction of the energy required by tractors or airplanes. For example, a battery-powered agricultural drone can cover the same hectare with 90% less energy than a traditional ground-based sprayer [54].

Reduced Carbon Footprint and Greenhouse Gas Emissions

One of the most significant advantages of Unmanned Aerial Vehicles (UAVs) is their minimal environmental impact compared to traditional, fuel-powered agricultural machinery. Conventional tractors, sprayers, and crop-dusting planes rely on diesel and aviation fuel, which release carbon dioxide (CO_2), nitrogen oxides (NO_x), and other harmful pollutants into the atmosphere. These emissions contribute to climate change, air pollution, and soil degradation. In contrast, electric UAVs operate on battery power, producing zero direct emissions, making them a sustainable and eco-friendly alternative [55]. Even hybrid UAVs, which combine electric propulsion with small fuel-based engines, consume significantly less fuel than manned aircraft or ground tractors. This reduction in fuel use helps lower the carbon footprint of agricultural and industrial operations, promoting greener and more efficient farming practices. By adopting UAV technology, farmers and industries can reduce environmental harm while improving productivity and resource efficiency. The reduced reliance on fossil fuels means that UAVs can help farmers, conservationists, and industries lower their overall environmental impact, aligning with global sustainability goals [56].

Smart Energy Management and Extended Flight Time

Modern UAVs are designed with advanced energy-saving technologies, such as:

- **Lightweight Materials**—UAVs are built with carbon fiber and composite materials, reducing power consumption and enhancing battery efficiency.

- **AI-Driven Flight Optimization**—Advanced drones use machine learning algorithms to determine the most energy-efficient flight paths, reducing battery drain and unnecessary energy use.
- **Solar-Powered UAVs**—Some drones are equipped with solar panels, extending flight duration and reducing battery dependency.

Additionally, advancements in battery technology (such as lithium-polymer and hydrogen fuel cells) allow UAVs to operate longer while consuming less power.

Cost Savings and Operational Efficiency

Energy efficiency translates into cost-effectiveness, benefiting farmers, industries, and businesses.

- **Lower Fuel Costs**—Since UAVs rely on electric power rather than expensive fossil fuels, users save significantly on energy costs.
- **Less Maintenance**—Fuel-powered machinery requires frequent servicing, whereas UAVs have fewer mechanical parts, reducing repair costs and downtime.
- **Faster Operations**—UAVs can complete tasks quickly, covering large areas in less time while consuming less energy per operation.

For example, in precision agriculture, a drone can survey 100 ha in just 30 min, whereas a tractor may take hours and consume significant amounts of fuel in the process. The adoption of UAV technology in agriculture, environmental monitoring, and industrial applications reduces energy consumption, lowers operational costs, and minimizes environmental impact [57]. By transitioning from fuel-intensive equipment to energy-efficient UAV systems, industries can achieve greater sustainability, improve efficiency, and contribute to a cleaner, greener future.

9.4 Supporting Sustainable Farming Practices

Unmanned Aerial Vehicles (UAVs) play a crucial role in promoting sustainable farming by improving efficiency, reducing environmental impact, and optimizing resource management. Traditional agricultural methods often lead to soil degradation, excessive chemical use, and high carbon emissions, making it essential to adopt innovative technologies that support long-term sustainability. UAVs enable precision agriculture by using multispectral imaging and AI-driven data analysis to monitor crop health, detect nutrient deficiencies, and identify pest infestations [58]. This targeted approach reduces the need for excessive pesticide and fertilizer application, preventing soil and water contamination. Additionally, UAVs minimize soil disturbance compared to heavy machinery, preserving soil structure and microbial biodiversity, which are essential for healthy crop growth.

Another key advantage of UAVs is their energy efficiency. Unlike fuel-powered tractors and crop-dusting planes, electric UAVs produce zero direct emissions, reducing the carbon footprint of farming operations. Even hybrid UAVs consume significantly less fuel than traditional equipment, making them a more sustainable alternative. By integrating UAVs into precision spraying, irrigation management, and real-time crop monitoring, farmers can increase yields, lower costs, and reduce environmental harm [59]. This eco-friendly approach not only ensures long-term agricultural productivity but also contributes to global efforts in climate resilience and sustainable food production.

9.4.1 UAVs and Organic Farming

Unmanned Aerial Vehicles (UAVs) are transforming organic farming by enhancing efficiency, improving crop health monitoring, and reducing the need for synthetic inputs. Organic farming emphasizes natural soil health, biodiversity conservation, and minimal chemical use, and UAVs support these principles by enabling precision agriculture and sustainable resource management. One of the key advantages of UAVs in organic farming is real-time crop monitoring. Equipped with multispectral and thermal imaging, drones can identify pest infestations, nutrient deficiencies, and soil moisture levels without the need for excessive manual inspections [60]. This helps organic farmers make data-driven decisions to maintain healthy crops without relying on synthetic pesticides or fertilizers.

Additionally, UAVs assist in biological pest control by detecting areas of high pest activity, allowing farmers to introduce natural predators or organic pest management solutions precisely where they are needed. This targeted approach reduces crop losses while avoiding harm to beneficial insects and pollinators. UAVs also support efficient irrigation management by detecting moisture variations in the soil, ensuring that water is used efficiently. This is particularly beneficial in organic farming, where maintaining optimal soil health and avoiding over-irrigation is crucial [61]. By integrating UAV technology, organic farmers can increase productivity, reduce labor costs, and maintain sustainable agricultural practices, ensuring higher yields without compromising environmental integrity.

9.4.2 Enhancing Biodiversity with Drone Technology

Biodiversity is essential for maintaining healthy ecosystems, sustainable agriculture, and environmental balance. However, factors such as deforestation, habitat loss, climate change, and excessive chemical use pose serious threats to global biodiversity. Unmanned Aerial Vehicles (UAVs), commonly known as drones, offer innovative and efficient solutions for monitoring ecosystems, protecting wildlife, and promoting sustainable farming practices [62].

9.4.3 UAVs for Wildlife and Habitat Monitoring

One of the primary ways UAVs support biodiversity is by enabling accurate and non-invasive wildlife monitoring. Traditional methods of tracking animal populations and assessing habitats often require manual surveys, satellite imagery, or manned aircraft, which can be expensive, time-consuming, and disruptive to wildlife. Drones equipped with high-resolution cameras, thermal imaging, and multispectral sensors allow conservationists to survey large areas quickly, track endangered species, and monitor habitat conditions without disturbing natural ecosystems. For example, UAVs are used in forests, wetlands, and coastal regions to monitor biodiversity hotspots. They help detect illegal poaching, deforestation, and habitat destruction, allowing conservationists and authorities to take immediate action [63]. In marine ecosystems, drones assist in tracking whale populations, monitoring coral reef health, and detecting pollution, providing critical data for biodiversity conservation.

9.4.4 Protecting Pollinators and Beneficial Insects

Pollinators such as bees, butterflies, and birds are essential for crop pollination and ecological balance. However, pesticide overuse, habitat loss, and climate change have led to a sharp decline in pollinator populations, threatening food security and ecosystem stability. UAV technology plays a crucial role in protecting pollinators by ensuring targeted and precise pesticide application. Unlike conventional boom sprayers, which apply chemicals uniformly across fields, UAVs use multispectral imaging and AI-driven analysis to detect pest-infested areas [64]. This allows for selective pesticide spraying, reducing unnecessary chemical exposure and protecting pollinator habitats. By minimizing pesticide drift and avoiding excessive spraying, UAVs help maintain a healthier ecosystem with thriving pollinator populations.

9.4.5 Reforestation and Ecosystem Restoration

Deforestation and habitat destruction are major causes of biodiversity loss and climate change. UAVs are now being used in afforestation and reforestation efforts to help restore degraded ecosystems. Traditional tree-planting methods can be slow, labor-intensive, and costly, but UAV-based seed dispersal significantly improves the efficiency of reforestation projects. Drones equipped with seed pods and planting mechanisms can quickly plant thousands of trees over vast areas, even in hard-to-reach terrains. This technology is particularly useful for restoring forests destroyed by wildfires, illegal logging, and desertification [65]. By accelerating reforestation, UAVs help re-establish wildlife habitats, prevent soil erosion, and increase carbon sequestration, contributing to a healthier and more balanced environment.

9.4.6 *Monitoring Climate Change Effects on Biodiversity*

Climate change has led to shifts in species distribution, habitat conditions, and ecosystem dynamics. UAVs provide a valuable tool for monitoring these changes in real-time. By collecting aerial imagery and environmental data, drones help scientists analyze how climate variations impact plant and animal life. For example, UAVs can track changes in vegetation cover, measure glacier retreat, and assess the health of wetlands [66]. These insights are crucial for developing effective conservation strategies and mitigating biodiversity loss caused by global warming.

Drones are revolutionizing biodiversity conservation and sustainable farming by providing efficient, cost-effective, and eco-friendly solutions. Whether it's wildlife monitoring, pollinator protection, reforestation, or climate change analysis, UAV technology plays a crucial role in maintaining a balanced and thriving ecosystem [67]. By integrating UAVs into environmental and agricultural strategies, we can help protect biodiversity, restore habitats, and ensure a sustainable future for generations to come.

9.5 Positive Environmental Impacts of UAVs

Unmanned Aerial Vehicles (UAVs), commonly known as drones, have emerged as valuable tools for environmental protection and sustainability. Their ability to monitor ecosystems, assess environmental changes, and support conservation efforts makes them an essential part of modern ecological management. Unlike traditional aerial monitoring methods that rely on fuel-intensive helicopters or manned aircraft, UAVs offer a low-emission, cost-effective, and efficient alternative with minimal environmental disruption [68].

9.5.1 *Reduced Carbon Footprint*

One of the most significant environmental benefits of UAVs is their energy efficiency and reduced emissions. Conventional aerial surveillance and agricultural spraying methods often involve gasoline-powered aircraft that consume large amounts of fuel and emit carbon dioxide (CO_2), nitrogen oxides (NO_x), and other greenhouse gases (GHGs). UAVs, particularly electric-powered drones, produce zero direct emissions, making them a more sustainable and eco-friendly option for environmental monitoring and resource management [69]. Even hybrid drones, which combine battery and fuel power, consume significantly less energy compared to traditional aircraft.

9.5.2 Wildlife and Ecosystem Conservation

UAVs have revolutionized wildlife conservation and habitat monitoring by enabling researchers to collect high-resolution images, thermal data, and multispectral imaging without disturbing natural ecosystems. Conservationists use drones to track endangered species, monitor illegal deforestation, and combat poaching [70]. Unlike traditional field surveys, which can be time-consuming and invasive, UAVs allow for real-time, non-intrusive monitoring, helping protect species without causing stress or harm to wildlife. For example, drones are widely used to monitor elephant and rhino populations in Africa, track polar bear movements in the Arctic, and assess the health of marine ecosystems by surveying coral reefs. UAVs also assist in mapping deforestation trends and detecting illegal logging, enabling authorities to take timely action against habitat destruction [71].

9.5.3 Climate Change Monitoring and Environmental Assessment

Climate change poses significant challenges to global biodiversity and ecosystem stability. UAVs provide critical data for climate researchers, helping them analyze changes in glacier retreat, sea-level rise, desertification, and extreme weather events. By capturing real-time aerial data, drones enable scientists to study climate patterns with greater accuracy and implement targeted climate mitigation strategies. For instance, UAVs are used to monitor the rate of ice melt in polar regions, track forest fires in real-time, and assess flood damage in disaster-prone areas [72]. Their ability to access hard-to-reach locations makes them indispensable in climate resilience planning and disaster preparedness.

9.5.4 Precision Agriculture for Sustainable Farming

Agriculture is a major contributor to soil degradation, water pollution, and chemical overuse. UAVs support precision farming by optimizing the use of fertilizers, pesticides, and irrigation, reducing the environmental impact of conventional farming. Drones equipped with multispectral cameras and AI-driven analysis help farmers detect crop diseases, soil nutrient deficiencies, and water stress, allowing them to apply resources only where needed. This targeted approach significantly reduces chemical runoff into water bodies, preserves soil health, and minimizes agricultural pollution, contributing to sustainable food production [73]. UAVs have transformed environmental monitoring and conservation by offering a sustainable, efficient, and non-invasive alternative to traditional methods. Their role in reducing carbon emissions, protecting wildlife, monitoring climate change, and promoting sustainable agriculture makes them a key technology for addressing global environmental challenges. By integrating UAV technology into environmental policies and conservation strategies, we can enhance biodiversity protection, mitigate climate change effects, and support long-term ecological sustainability [74].

9.6 Reducing Carbon Footprint with UAVs

The transportation and logistics industries are among the largest contributors to greenhouse gas (GHG) emissions, primarily due to the heavy reliance on fuel-powered vehicles. Unmanned Aerial Vehicles (UAVs), commonly known as drones, are emerging as a promising solution to reduce carbon emissions and enhance sustainable transportation practices [75]. Their ability to operate efficiently with minimal energy consumption makes them a viable alternative for various industries, particularly in last-mile delivery and disaster response.

9.6.1 Sustainable Last-Mile Delivery

One of the most energy-intensive segments of the supply chain is last-mile delivery, where fuel-powered trucks and vans transport goods to consumers. Traffic congestion, idling, and inefficient routing further increase fuel consumption and emissions in urban areas. UAVs offer an eco-friendly alternative by eliminating the need for gasoline-powered vehicles in short-distance deliveries [76]. Major logistics companies like Amazon, UPS, and DHL are actively developing drone delivery systems to reduce their environmental impact. Unlike conventional delivery trucks, which emit significant amounts of CO_2 and NO_x, electric drones operate with zero direct emissions and require significantly less energy per delivery. Studies suggest that UAV-based delivery could cut delivery-related carbon emissions by up to 50%, making them a game-changer in sustainable logistics [77].

9.6.2 Energy-Efficient Emergency Response

UAVs also play a critical role in disaster relief and emergency response, where speed and accessibility are crucial. Traditional rescue and aid missions often rely on fuel-intensive helicopters, trucks, and aircraft, consuming large amounts of diesel and aviation fuel. By contrast, UAVs can deliver medical supplies, food, and emergency aid to inaccessible or disaster-stricken areas while using significantly less energy [78]. For example, UAVs have been deployed to deliver COVID-19 vaccines, blood samples, and essential medicines to remote regions, reducing the need for ground transportation and its associated emissions. In natural disaster scenarios such as earthquakes, floods, and hurricanes, drones provide real-time aerial assessments and aid delivery without excessive fuel consumption, making relief efforts more sustainable and efficient. The adoption of UAV technology in logistics and emergency response has the potential to revolutionize carbon footprint reduction. By replacing fuel-powered vehicles with electric drones, industries can significantly lower emissions, reduce energy consumption, and enhance efficiency [79]. As advancements in battery technology and UAV capabilities continue, drones will play an increasingly important role in creating a cleaner, more sustainable future for transportation and logistics.

9.7 Negative Environmental Impacts of UAVs

While Unmanned Aerial Vehicles (UAVs) offer numerous environmental benefits, they also pose certain ecological challenges that must be addressed for sustainable adoption. The growing use of drones raises concerns about electronic waste (e-waste), energy consumption, and disturbances to wildlife [80].

9.7.1 Electronic Waste (E-Waste) Accumulation

One of the most pressing issues associated with UAVs is the generation of electronic waste. As drone technology rapidly advances, older models quickly become obsolete, leading to increased disposal of UAV components such as batteries, circuit boards, sensors, and plastic casings. Many of these materials contain hazardous substances, including heavy metals and toxic chemicals, which can leach into the environment if not disposed of properly. Additionally, lithium-ion batteries, which power most UAVs, pose fire and pollution risks when discarded irresponsibly [81]. Without proper recycling infrastructure, UAV-related e-waste could contribute to landfill overflow and environmental contamination, undermining the sustainability benefits of drone technology.

9.7.2 Energy Consumption and Carbon Footprint

While electric UAVs produce zero direct emissions, their reliance on battery charging can still have an indirect environmental impact. The electricity used to charge drones often comes from non-renewable energy sources, such as coal or natural gas, which contribute to carbon emissions. As drone usage increases across industries, the cumulative energy demand from large-scale UAV operations could strain power grids and increase dependence on fossil fuels if renewable energy sources are not prioritized [82]. Additionally, drones with higher payload capacities or longer flight times require larger batteries, which consume more energy and resources during manufacturing and operation.

9.7.3 Disturbance to Wildlife and Ecosystems

Another significant concern is the impact of UAVs on wildlife, particularly in protected natural areas. The noise and movement of drones can stress animals, alter natural behaviors, and even cause them to flee their habitats. Bird species, for example, are highly sensitive to unfamiliar aerial objects. Studies have shown that UAVs flying too close to nesting birds can lead to abandoned nests or disrupted mating patterns, affecting overall population health [83]. Similarly, in marine ecosystems, drones used for monitoring may disturb marine mammals, such as seals or whales, who rely on undisturbed environments for breeding and migration. While UAVs present an innovative solution for environmental monitoring, conservation, and agriculture, their negative impacts cannot be ignored. Issues such as e-waste, energy consumption, and wildlife disturbances highlight the need for responsible drone usage and improved sustainability measures. By focusing on eco-friendly battery disposal, renewable energy adoption, and ethical drone operation guidelines, UAV technology can be optimized to minimize harm and maximize environmental benefits [84].

9.8 Strategies for Enhancing UAV Sustainability

To maximize the environmental benefits of UAVs while minimizing their negative impacts, several sustainability strategies should be adopted. One effective approach is the use of eco-friendly materials in drone manufacturing. Many UAVs are constructed using plastic, carbon fiber, and metal alloys, which contribute to e-waste when discarded. By incorporating biodegradable or recyclable materials, such as bioplastics or sustainable composites, the environmental impact of drone disposal can be reduced [85]. Additionally, modular drone designs that allow for easy repairs and upgrades can extend UAV lifespans, decreasing the frequency of replacements and waste generation. Another key strategy involves improving the energy efficiency of UAVs. Most drones rely on lithium-ion batteries, which require significant energy and resources to produce and dispose of. To address this issue, researchers are exploring alternative energy sources, such as solar-powered UAVs, which use photovoltaic panels to harness renewable energy. Hydrogen-fuel-based drones offer another promising solution, as they provide longer flight times with minimal emissions. Hybrid UAVs, which combine electric power with fuel-efficient engines, further enhance operational efficiency by reducing overall energy consumption [86]. Advancements in battery technology, including more efficient and sustainable charging systems, can also contribute to making UAV operations more eco-friendly.

Regulating drone usage in sensitive ecosystems is another important step toward sustainability. UAVs, if not properly managed, can disrupt wildlife by generating noise pollution and disturbing natural habitats. Governments and environmental organizations should establish guidelines that limit drone activity in ecologically sensitive areas, particularly during critical breeding and migration seasons. Implementing safe distance regulations and educating drone operators on responsible flying practices can further minimize disturbances to wildlife. Such regulations help ensure that UAVs continue to support environmental monitoring and conservation without causing unintended harm to ecosystems [87]. Additionally, recycling programs for UAV components can significantly reduce electronic waste. Manufacturers can introduce take-back programs that allow users to return old drones for proper disposal, recycling, or refurbishment. By recovering and repurposing functional components such as batteries, motors, and sensors, the UAV industry can move toward a circular economy, reducing landfill waste and the need for new raw materials. Designing drones with easily separable and recyclable parts can further promote sustainability, ensuring that the valuable resources used in UAV production are not wasted [88]. By integrating these sustainability strategies, UAV technology can continue to provide environmental benefits while minimizing its ecological footprint. The shift toward biodegradable materials, renewable energy sources, responsible drone operation, and effective recycling initiatives ensures that UAVs remain valuable tools in conservation and agriculture without contributing to environmental degradation. These efforts support a long-term vision of using drones as a means to promote sustainable development and environmental stewardship [89].

9.9 Future Prospects of UAVs in Environmental Sustainability

The future of UAVs in environmental sustainability is highly promising, driven by continuous technological advancements that enhance efficiency while minimizing ecological impact. As drone technology evolves, innovations such as artificial intelligence (AI)-powered UAVs are set to revolutionize environmental monitoring and conservation efforts. AI-driven drones can process vast amounts of data in real time, allowing for more precise environmental assessments. By integrating machine learning algorithms, UAVs can analyze changes in vegetation, detect pollution sources, and identify at-risk ecosystems with greater accuracy. Furthermore, AI optimization enables drones to calculate the most energy-efficient flight paths, reducing battery consumption and extending operational duration, thereby contributing to sustainability goals [90]. Another significant development is the integration of UAVs in carbon offset initiatives. With deforestation being a major contributor to climate change, UAVs are increasingly used for reforestation efforts. Specialized drones equipped with seed-dispersal technology can plant trees in deforested or degraded areas at a much faster rate than traditional methods. Additionally, UAVs can monitor forest health, track carbon sequestration rates, and assess the impact of afforestation projects. By providing real-time data on tree growth and ecosystem recovery, drones ensure that carbon capture efforts are effectively managed and contribute to long-term climate mitigation strategies. Moreover, advancements in drone energy sources are expected to further enhance their environmental benefits. Researchers are exploring solar-powered UAVs, which can harness renewable energy to extend flight times and reduce dependence on traditional battery charging methods [91]. Hydrogen-fuel-based drones are also being developed as a sustainable alternative, offering longer operational endurance while generating minimal emissions. The shift toward eco-friendly UAV power sources will further reduce their carbon footprint and promote their use in conservation and environmental management. As regulatory frameworks evolve, UAVs will likely be more widely adopted for large-scale environmental applications. Governments and environmental organizations are working on policies that facilitate responsible drone usage while protecting ecosystems from potential disturbances. Future regulations may promote sustainable drone manufacturing practices, enforce wildlife protection measures, and encourage the recycling of UAV components to minimize e-waste [92]. These efforts will ensure that drones continue to support environmental sustainability without creating unintended ecological harm.

In the coming years, UAVs will play an even greater role in tackling environmental challenges, from monitoring climate change impacts to enhancing biodiversity conservation. Their ability to provide real-time, high-resolution data makes them invaluable tools for informed decision-making in environmental policy and resource management. By integrating AI, renewable energy, and sustainable practices, UAVs will continue to evolve as a critical technology in promoting a greener and more sustainable future. Unmanned Aerial Vehicles (UAVs) offer significant advantages in environmental monitoring, sustainable agriculture, disaster response, and carbon footprint reduction [93]. Their ability to collect real-time data and access remote areas makes them valuable tools for conservation and resource management. However, concerns related to e-waste, energy consumption, and wildlife disturbance must be addressed to ensure their sustainable use. By adopting eco-friendly materials, improving battery technologies, enforcing responsible drone usage policies, and promoting recycling initiatives, UAVs can become more environmentally sustainable [94]. As technology continues to evolve, UAVs are poised to play a crucial role in supporting global sustainability efforts while minimizing their ecological impact.

9.10 Summary

Unmanned Aerial Vehicles (UAVs) play a crucial role in environmental sustainability by optimizing resource use and minimizing ecological footprints. They are particularly useful in agriculture, where precision spraying reduces pesticide and fertilizer overuse, protecting soil and water quality. UAVs also enhance biodiversity conservation by monitoring deforestation, tracking wildlife, and supporting reforestation efforts. Additionally, they improve energy efficiency in transportation and logistics, reducing greenhouse gas emissions. However, UAVs present sustainability challenges such as e-waste accumulation, energy-intensive battery production, and potential disruptions to wildlife. To maximize their positive impact, strategies such as eco-friendly drone materials, recycling programs, and alternative energy sources like solar-powered UAVs should be implemented. With responsible use and continued innovation, UAVs can play a pivotal role in environmental protection and sustainable development. UAVs have emerged as valuable tools in promoting environmental sustainability by optimizing resource management, reducing chemical overuse, minimizing soil disturbance, and supporting climate-resilient farming. Their ability to provide real-time, high-resolution data allows for more efficient decision-making, ultimately benefiting both agricultural productivity and ecosystem conservation.

As technology continues to evolve, UAVs are poised to play a crucial role in supporting global sustainability efforts while minimizing their ecological impact. By integrating drones into sustainable land management strategies, we can enhance food security, protect natural resources, and mitigate the impacts of climate change, ensuring a greener and more resilient future. UAVs are revolutionizing biodiversity conservation and sustainable farming by providing efficient, cost-effective, and eco-friendly solutions. By integrating UAV technology into environmental policies and conservation strategies, we can enhance biodiversity protection, mitigate climate change effects, and support long-term ecological sustainability. In logistics and emergency response, the adoption of electric drones has the potential to revolutionize carbon footprint reduction. As advancements in battery technology and UAV capabilities continue, drones will play an increasingly important role in creating a cleaner, more sustainable future for transportation and logistics.

References

1. Karunathilake E, Le AT, Heo S et al (2023) The path to smart farming: innovations and opportunities in precision agriculture. Agriculture 13:1593
2. Aygun H, Kirmizi M, Turan O (2022) Propeller effects on energy, exergy and sustainability parameters of a small turboprop engine. Energy 249:123759
3. Akdeniz HY, Balli O, Caliskan H (2023) Energy, exergy, thermoecologic, environmental, enviroeconomic and sustainability analyses and assessments of the aircraft engine fueled with biofuel and jet fuel. J Therm Anal Calorim 148:3585–3603
4. Wang Z, Zeng R, Ma P (2024) Determining of gully erosion susceptibility based on UAV and machine learning in Loess Plateau. Geomat Nat Haz Risk 15:2421382
5. Hwang J, Joo K, Moon J (2023) A study on behavioral intentions in the field of eco-friendly drone food delivery services: focusing on demographic characteristics and past experiences. Sustainability 15:6253
6. Zahoor I, Mushtaq A (2023) Water pollution from agricultural activities: a critical global review. Int J Chem Biochem Sci 23:164–176
7. Bisht N, Chauhan PS (2020) Excessive and disproportionate use of chemicals cause soil contamination and nutritional stress. In: Soil contamination. IntechOpen

8. Khalifa SA, Elshafiey EH, Shetaia AA et al (2021) Overview of bee pollination and its economic value for crop production. Insects 12:688
9. Katumo DM, Liang H, Ochola AC et al (2022) Pollinator diversity benefits natural and agricultural ecosystems, environmental health, and human welfare. Plant Divers 44:429–435
10. Khan BA, Nadeem MA, Nawaz H et al (2023) Pesticides: impacts on agriculture productivity, environment, and management strategies. In: Emerging contaminants and plants: interactions, adaptations and remediation technologies. Springer, pp 109–134
11. Mitra B, Chowdhury AR, Dey P et al (2021) Use of agrochemicals in agriculture: alarming issues and solutions. In: Input use efficiency for food and environmental security. pp 85–122
12. Warra AA, Prasad MNV (2020) African perspective of chemical usage in agriculture and horticulture—their impact on human health and environment. In: Agrochemicals detection, treatment and remediation. Elsevier, pp 401–436
13. Pahalvi HN, Rafiya L, Rashid S et al (2021) Chemical fertilizers and their impact on soil health. In: Microbiota and biofertilizers, vol 2: ecofriendly tools for reclamation of degraded soil environs. pp 1–20
14. Mamatha B, Mudigiri C, Ramesh G et al (2024) Enhancing soil health and fertility management for sustainable agriculture: a review. Asian J Soil Sci Plant Nutr 10:182–190
15. Liu K (2024) Parameters optimisation of agricultural machinery for soil compaction mitigation and crop yield enhancement
16. Wen X, Zhen L (2020) Soil erosion control practices in the Chinese Loess Plateau: a systematic review. Environ Dev 34:100493
17. Anderson K (2024) Spatial analysis of soil and plant interactions through UAV-mounted sensor data
18. Pandey S, Kumari N, Mallick L (2024) Review on assessment of land degradation in watershed using geospatial technique based on unmanned aircraft systems. In: Unmanned aircraft systems. pp 263–311
19. Debangshi U (2021) Climate resilient agriculture an approach to reduce the ill-effect of climate change. Int J Recent Adv Multidiscip Top 2:309–315
20. Karri V, Nalluri N (2024) Enhancing resilience to climate change through prospective strategies for climate-resilient agriculture to improve crop yield and food security. Plant Sci Today 11:21–33
21. Sarma HH, Borah SK, Dutta N et al (2024) Innovative approaches for climate-resilient farming: strategies against environmental shifts and climate change. Int J Environ Clim Change 14:217–241
22. Garusinghe GDAU, Perera BAKS, Weerapperuma US (2023) Integrating circular economy principles in modular construction to enhance sustainability. Sustainability 15:11730
23. St-Jean K (2024) Sustainable ground operations of military transport aircraft within the royal Canadian air force
24. Telli K, Kraa O, Himeur Y et al (2023) A comprehensive review of recent research trends on unmanned aerial vehicles (UAVs). Systems 11:400
25. Mohsan SAH, Othman NQH, Li Y et al (2023) Unmanned aerial vehicles (UAVs): practical aspects, applications, open challenges, security issues, and future trends. Intel Serv Robot 16:109–137
26. Librán-Embid F, Klaus F, Tscharntke T, Grass I (2020) Unmanned aerial vehicles for biodiversity-friendly agricultural landscapes—a systematic review. Sci Total Environ 732:139204
27. Bayomi N, Fernandez JE (2023) Eyes in the sky: drones applications in the built environment under climate change challenges. Drones 7:637
28. Ahmed F, Mohanta J, Keshari A, Yadav PS (2022) Recent advances in unmanned aerial vehicles: a review. Arab J Sci Eng 47:7963–7984
29. Vashishth TK, Sharma V, Sharma KK et al (2024) Integration of unmanned aerial vehicles (UAVs) and IoT for crop monitoring and spraying. In: Internet of Things applications and technology. Auerbach Publications, pp 95–117

30. Hanif AS, Han X, Yu S-H (2022) Independent control spraying system for UAV-based precise variable sprayer: a review. Drones 6:383
31. Delavarpour N, Koparan C, Zhang Y et al (2023) A review of the current unmanned aerial vehicle sprayer applications in precision agriculture. J ASABE 66:703–721
32. Chen H, Lan Y, Fritz BK et al (2021) Review of agricultural spraying technologies for plant protection using unmanned aerial vehicle (UAV). Int J Agric Biol Eng 14:38–49
33. Taseer A, Han X (2024) Advancements in variable rate spraying for precise spray requirements in precision agriculture using unmanned aerial spraying systems: a review. Comput Electron Agric 219:108841
34. Rui Z, Zhang Z, Chen T (2024) A review of field plant phenotyping platform, sensing and navigation technology. In: Transforming technologies in smart agriculture: new applications during the past decade. pp 1–46
35. Imran I, Khan AA, Kamal A et al (2024) Utilization and characterization of microbes for heavy metal remediation. Pol J Environ Stud
36. Kamal A, Mian IA, Tariq M (2022) Integration of organic and inorganic nutrients sources for imprint maize yield and soil properties. J Xi'an Shiyou Uni Nat Sci 18:648–662
37. Reddy KN, Nagarjuna T, Tippimath S et al (2025) Safeguarding beneficial insects: strategies and innovations for conservation amidst growing threats
38. Staples B (2023) Please bee-have: a critical look into pollinator decline and its negative impacts on our food security. Braiden Staples
39. Maggi T, Pardo L, Chreil R (2023) Pollinator diversity: a key to ecosystem resilience and food security. Pollinators 6:33–48
40. Van der Sluijs JP, Vaage NS (2016) Pollinators and global food security: the need for holistic global stewardship. Food Ethics 1:75–91
41. Tahir MN, Lan Y, Zhang Y et al (2023) Application of unmanned aerial vehicles in precision agriculture. In: Precision agriculture. Elsevier, pp 55–70
42. Roslim MHM, Juraimi AS, Che'Ya NN et al (2021) Using remote sensing and an unmanned aerial system for weed management in agricultural crops: a review. Agronomy 11:1809
43. Kebe AA, Hameed S, Farooq MS et al (2023) Enhancing crop protection and yield through precision agriculture and integrated pest management: a comprehensive review. Asian J Res Crop Sci 8:443–453
44. Tsouros DC, Bibi S, Sarigiannidis PG (2019) A review on UAV-based applications for precision agriculture. Information 10:349
45. Phang SK, Chiang THA, Happonen A, Chang MML (2023) From satellite to UAV-based remote sensing: a review on precision agriculture. IEEE Access 11:127057–127076
46. Mukhamediev RI, Symagulov A, Kuchin Y et al (2021) Review of some applications of unmanned aerial vehicles technology in the resource-rich country. Appl Sci 11:10171
47. Harle S, Bhagat A, Dash AK (2024) Remote sensing revolution: mapping land productivity and vegetation trends with unmanned aerial vehicles (UAVs). Curr Appl Mater 3:e070224226752
48. Istiak MA, Syeed MM, Hossain MS et al (2023) Adoption of unmanned aerial vehicle (UAV) imagery in agricultural management: a systematic literature review. Eco Inform 78:102305
49. Guebsi R, Mami S, Chokmani K (2024) Drones in precision agriculture: a comprehensive review of applications, technologies, and challenges. Drones 8:686
50. Rahman MFF, Fan S, Zhang Y, Chen L (2021) A comparative study on application of unmanned aerial vehicle systems in agriculture. Agriculture 11:22
51. Mohsan SAH, Khan MA, Noor F et al (2022) Towards the unmanned aerial vehicles (UAVs): a comprehensive review. Drones 6:147
52. Townsend A, Jiya IN, Martinson C et al (2020) A comprehensive review of energy sources for unmanned aerial vehicles, their shortfalls and opportunities for improvements. Heliyon 6
53. Gulak AM (2024) Implementation of drone for spraying herbicide and pesticide
54. Krishna KR (2019) Unmanned aerial vehicle systems in crop production: a compendium. Apple Academic Press
55. Varvemaa J (2024) Carbon footprint for drone-operated power grid smart maintenance

56. Qasem NA, Abdulrahman GA (2024) A recent comprehensive review of fuel cells: history, types, and applications. Int J Energy Res 2024:7271748
57. Richard B, Qi A, Fitt BD (2022) Control of crop diseases through integrated crop management to deliver climate-smart farming systems for low-and high-input crop production. Plant Pathol 71:187–206
58. Islam N, Rashid MM, Pasandideh F et al (2021) A review of applications and communication technologies for internet of things (IoT) and unmanned aerial vehicle (UAV) based sustainable smart farming. Sustainability 13:1821
59. Sarfraz S, Ali F, Hameed A et al (2023) Sustainable agriculture through technological innovations. In: Sustainable agriculture in the era of the OMICs revolution. Springer, pp 223–239
60. Boursianis AD, Papadopoulou MS, Diamantoulakis P et al (2022) Internet of things (IoT) and agricultural unmanned aerial vehicles (UAVs) in smart farming: a comprehensive review. Internet Things 18:100187
61. Lachgar M, Hrimech H, Kartit A (2023) Unmanned aerial vehicle-based applications in smart farming: a systematic review. Int J Adv Comput Sci Appl 14
62. Rai AK, Bana SR, Sachan DS, Singh B (2023) Advancing sustainable agriculture: a comprehensive review for optimizing food production and environmental conservation. Int J Plant Soil Sci 35:417–425
63. Shang M, Xie J (2024) Agricultural sustainable development: soil, water resources, biodiversity, climate change, and technological innovation. Adv Resour Res 4:181–204
64. Bekmurzaeva R, Kalimullin R, Aguzarova F (2024) Application of drones and artificial intelligence to monitor and protect natural ecosystems. EDP Sci 01008
65. Bhambri P, Kautish S (2024) Technological advancements in promoting ecosystem health. In: Digital technologies to implement the UN sustainable development goals. Springer, pp 413–432
66. Pecl GT, Araújo MB, Bell JD et al (2017) Biodiversity redistribution under climate change: impacts on ecosystems and human well-being. Science 355:eaai9214
67. O'Connor B, Bojinski S, Röösli C, Schaepman ME (2020) Monitoring global changes in biodiversity and climate essential as ecological crisis intensifies. Eco Inform 55:101033
68. Elmeseiry N, Alshaer N, Ismail T (2021) A detailed survey and future directions of unmanned aerial vehicles (UAVs) with potential applications. Aerospace 8:363
69. Mohamed N, Al-Jaroodi J, Jawhar I et al (2020) Unmanned aerial vehicles applications in future smart cities. Technol Forecast Soc Chang 153:119293
70. Mangewa LJ, Ndakidemi PA, Munishi LK (2019) Integrating UAV technology in an ecological monitoring system for community wildlife management areas in Tanzania. Sustainability 11:6116
71. Lahoz-Monfort JJ, Magrath MJ (2021) A comprehensive overview of technologies for species and habitat monitoring and conservation. Bioscience 71:1038–1062
72. Kumarasan T, Vinoth M, Durgaram M, Benedict M (2020) Drone-enable wildlife monitoring system: revolutionizing conservation efforst. NeuroQuantology 18:330–339
73. Zaman QU (2023) Precision agriculture technology: a pathway toward sustainable agriculture. In: Precision agriculture. Elsevier, pp 1–17
74. Getahun S, Kefale H, Gelaye Y (2024) Application of precision agriculture technologies for sustainable crop production and environmental sustainability: a systematic review. Sci World J 2024:2126734
75. Sabir RM, Mehmood K, Sarwar A et al (2024) Remote sensing and precision agriculture: a sustainable future. In: Transforming agricultural management for a sustainable future: climate change and machine learning perspectives. Springer, pp 75–103
76. Nath S (2024) A vision of precision agriculture: balance between agricultural sustainability and environmental stewardship. Agron J 116:1126–1143
77. Patro RK, Asare G, Seethapathy K, Sarpong KO (2024) Sustainability in modern logistics. Cari J USA LLC
78. Koskinen O (2022) H_2 ecosystem roadmap for Ostrobothnia, Workpackage 1 report

79. Fink C, Goldman J (2023) AI and sustainability: the opportunity for brands. J AI Robot Workplace Autom 2:202–214
80. Syed F, Gupta SK, Hamood Alsamhi S et al (2021) A survey on recent optimal techniques for securing unmanned aerial vehicles applications. Trans Emerg Telecommun Technol 32:e4133
81. Aydoğan B, Özçelik G, Ünver M (2024) Developing Industry 4.0-based e-waste management: a decision-aided tool equipped with continuous function-valued intuitionistic fuzzy sets. Int J Environ Sci Technol 1–26
82. Wanner D, Hashim HA, Srivastava S, Steinhauer A (2024) UAV avionics safety, certification, accidents, redundancy, integrity, and reliability: a comprehensive review and future trends. Drone Syst Appl 12:1–23
83. Testa S, Troise C, Cincotti S, Camilleri MA (2024) The role of electronic waste management solutions and message framing in influencing consumer behaviours: exploring the crowdfunding context. Bus Strateg Environ 33:917–929
84. Hayat P (2023) Integration of advanced technologies in urban waste management. In: Advancements in urban environmental studies: application of geospatial technology and artificial intelligence in urban studies. Springer, pp 397–418
85. Fang Z, Savkin AV (2024) Strategies for optimized UAV surveillance in various tasks and scenarios: a review. Drones 8:193
86. Majumdar P, Bhattacharya D, Mitra S, Bhushan B (2023) Application of green IoT in agriculture 4.0 and beyond: requirements, challenges and research trends in the era of 5G, LPWANs and Internet of UAV Things. Wirel Pers Commun 131:1767–1816
87. Raj EFI, Manimala K, Rani EFI, Darwin S (2024) Environmental monitoring and conservation using drones. In: Drone applications for Industry 5.0. IGI Global, pp 358–380
88. Singh B (2024) Unmanned aircraft systems (UAS), surveillance, risk management to cybersecurity and legal regulation landscape: unraveling the future analysis, challenges, demand, and benefits in the high sky exploring the strange new world. In: Unmanned aircraft systems. pp 313–354
89. Fascista A (2022) Toward integrated large-scale environmental monitoring using WSN/UAV/Crowdsensing: a review of applications, signal processing, and future perspectives. Sensors 22:1824
90. Javaid S, Saeed N, Qadir Z et al (2023) Communication and control in collaborative UAVs: recent advances and future trends. IEEE Trans Intell Transp Syst 24:5719–5739
91. Al Dhafari LS, Afzal A, Al Abri OK, Khan A (2024) Solar-powered UAVs: a systematic literature review. IEEE, pp 1–8
92. Saravanakumar YN, Sultan MTH, Shahar FS et al (2023) Power sources for unmanned aerial vehicles: a state-of-the art. Appl Sci 13:11932
93. Liller J, Goel R, Aziz A et al (2025) Development of a battery free, solar powered, and energy aware fixed wing unmanned aerial vehicle. Sci Rep 15:6141
94. Mager A, Blass V (2022) From illegal waste dumps to beneficial resources using drone technology and advanced data analysis tools: a feasibility study. Remote Sens 14:3923

Chapter 10
Crop Interaction with UAVS

Abstract Unmanned Aerial Vehicles (UAVs) have become a powerful tool in modern precision agriculture, revolutionizing crop management through advanced imaging, real-time data collection, and automated farming operations. These advancements include vegetation monitoring, water stress detection, disease identification, yield estimation, precision spraying, seeding, pollination assistance, and irrigation management. The integration of UAVs with Artificial Intelligence (AI), the Internet of Things (IoT), and big data analytics is enhancing their role in agriculture, leading to increased efficiency, cost savings, and sustainability. The future of UAVs in farming includes autonomous drone swarms, real-time AI-driven analytics, improved battery technology, and robotic integration. These innovations are expected to further optimize farming operations and enhance global food security. UAVs offer numerous benefits, including precision agriculture, pest detection, irrigation management, and yield estimation. Remote sensing technologies like NDVI, EVI, thermal imaging, and UAV monitoring provide valuable insights for precision agriculture, optimizing irrigation, detecting stress early, preventing disease outbreaks, and enhancing crop productivity. UAV-based spraying offers precision, efficiency, and sustainability by enabling targeted application, reducing chemical use, and improving safety. As technology continues to evolve, UAVs will play an increasingly vital role in addressing global challenges like food security, resource scarcity, and climate change. The integration of UAVs with AI, IoT, big data, and robotics will revolutionize agriculture, making farming more autonomous, data-driven, and efficient.

Keywords Precision agriculture · Thermal imaging · AI-driven analytics · Remote sensing · Yield forecasting · Autonomous drone swarms · Pollination assistance · Automated irrigation

© The Author(s), under exclusive license to Springer Nature Singapore Pte Ltd. 2025

J. Li, *UAV Aerodynamics and Crop Interaction*, Smart Agriculture 13,
https://doi.org/10.1007/978-981-96-8402-1_10

10.1 Introduction

Unmanned Aerial Vehicles (UAVs) are revolutionizing agriculture by providing real-time data on crop health, soil conditions, and environmental factors. These aerial systems enable the collection of this data, leading to more informed decision-making and improved agricultural productivity. UAVs have transformed conventional farming practices by introducing efficient, cost-effective, and highly accurate methods for monitoring crops. Traditional methods rely on ground-based surveys and satellite imagery, which have limitations in terms of accessibility, resolution, and cost. UAVs provide high-resolution imagery and detailed data that can be collected frequently and at a lower cost [1]. One of the primary applications of UAVs in agriculture is remote sensing and monitoring. Equipped with high-resolution RGB, multispectral, and thermal cameras, drones can capture detailed images of crops and assess their health. By analyzing spectral indices such as the Normalized Difference Vegetation Index (NDVI), UAVs help detect plant stress, nutrient deficiencies, and pest infestations before they become visible to the naked eye. This early detection allows farmers to take preventive measures, reducing losses and improving overall crop yield. Precision agriculture is another critical role of UAVs in agriculture. Through aerial mapping, drones create detailed field maps that highlight variations in soil quality, moisture levels, and crop health [2]. This information allows farmers to apply inputs more efficiently, reducing waste and minimizing environmental impact. Variable rate application (VRA) is another significant advantage of using drones in precision agriculture. Drones equipped with sprayers can precisely target specific areas of a field, applying fertilizers or pesticides only where needed, reducing input costs and promoting environmentally sustainable farming practices. Soil and water management is another crucial aspect of UAVs' role in agriculture. By using thermal imaging and infrared sensors, drones can assess soil moisture levels across fields, helping farmers optimize irrigation schedules. They also help identify soil degradation, erosion, and compaction, aiding in land management and soil conservation efforts [3]. Weed and pest detection is another key role of UAVs in agriculture. Equipped with multispectral and hyperspectral sensors, UAVs can differentiate between healthy crops and invasive weeds, allowing for targeted herbicide application and promoting eco-friendly weed management. Yield estimation and harvest planning are essential for planning storage, transportation, and market supply. UAVs provide reliable data on plant height, canopy coverage, and biomass density, which are critical indicators of crop yield. Machine learning algorithms analyze historical data and predict yields with high accuracy, allowing farmers to make informed decisions about harvesting schedules and resource allocation. Post-harvest monitoring and security are also crucial aspects of UAVs' role in agriculture. Thermal cameras can detect temperature variations in storage facilities, preventing spoilage and ensuring proper storage conditions. Additionally, drones assist in farm surveillance and security, protecting crops from theft and damage caused by wildlife or unauthorized intrusions [4]. Despite these challenges, the integration of UAVs in agriculture is transforming the way crops are monitored, managed, and optimized. By providing real-time insights into crop health, soil conditions, and environmental factors, drones enable farmers to make data-driven decisions that enhance productivity and sustainability.

10.2 UAV Applications in Crop Management

Unmanned Aerial Vehicles (UAVs), commonly known as drones, have revolutionized crop management by offering advanced monitoring, precision application, and data-driven decision-making. Equipped with high-resolution cameras, multispectral sensors, and thermal imaging technology, drones provide real-time insights into crop health, growth patterns, and environmental conditions. Farmers can detect early signs of pest infestations, disease outbreaks, nutrient deficiencies, and water stress, enabling timely interventions to minimize yield losses. By capturing high-quality aerial images, drones help identify variability within fields, ensuring better resource allocation and site-specific crop management. One of the most significant applications of UAVs in agriculture is precision spraying, where pesticides, fertilizers, and herbicides are applied with high accuracy, reducing waste and preventing overuse [5]. This targeted approach minimizes chemical runoff, decreases soil and water contamination, and enhances overall sustainability. Unlike traditional ground-based machinery, drones reduce soil compaction and allow access to difficult terrains, making them particularly useful in regions with irregular landscapes. Moreover, UAVs contribute to improved irrigation management by assessing moisture levels and identifying areas requiring more or less water, leading to better water conservation practices. Beyond real-time monitoring, drones also assist in yield prediction and field mapping. By integrating UAV data with artificial intelligence (AI) and machine learning algorithms, farmers can make informed decisions regarding planting, harvesting, and input optimization. This technology enhances operational efficiency, reduces costs, and maximizes agricultural productivity [6]. The use of UAVs in crop management not only improves precision farming but also promotes environmentally friendly practices by reducing resource wastage and chemical dependency. As drone technology continues to evolve, its role in sustainable agriculture will expand, making it an indispensable tool for modern farmers striving for higher yields and reduced environmental impact.

10.2.1 Remote Sensing and Crop Monitoring

Remote sensing and crop monitoring using UAVs have transformed modern agriculture by providing high-resolution images that enable farmers to assess various crop parameters with precision. UAVs equipped with multispectral, hyperspectral, and thermal cameras can detect plant stress, monitor growth stages, and evaluate overall crop health. These sensors facilitate early detection of pest infestations, nutrient deficiencies, and water stress, allowing timely interventions to optimize yield. Multispectral imaging helps in analyzing vegetation indices like the Normalized Difference Vegetation Index (NDVI), which indicates plant vigor and chlorophyll content [7]. Hyperspectral sensors provide more detailed spectral information, enabling the identification of specific crop diseases or nutrient imbalances. Thermal imaging helps in assessing plant transpiration and detecting irrigation issues by identifying temperature variations across the field. By integrating this data with advanced analytics, farmers can make informed decisions on irrigation, fertilization, and pest control strategies, leading to increased efficiency and reduced input costs. UAV-based remote sensing also allows for large-scale field assessments in a fraction of the time required by traditional methods, improving productivity while minimizing environmental impact [8]. These sensors facilitate.

10.2.1.1 Vegetation Indices: NDVI and EVI for Plant Health Assessment

Vegetation indices are essential tools in remote sensing for assessing plant health and monitoring crop conditions. Among them, the Normalized Difference Vegetation Index (NDVI) is one of the most widely used indices. It is calculated using the red and near-infrared (NIR) spectral bands, where healthy vegetation reflects more NIR light while absorbing red light due to chlorophyll. A high NDVI value indicates a dense, healthy plant canopy, while low values suggest stress, disease, or soil exposure. Enhanced Vegetation Index (EVI), on the other hand, improves upon NDVI by incorporating corrections for atmospheric disturbances and soil reflectance, making it more effective in areas with dense vegetation [9]. By analyzing NDVI and EVI over time, farmers and researchers can track crop growth, detect areas with low productivity, and make informed decisions regarding fertilization, irrigation, and pest management. These indices are particularly useful for large-scale agricultural monitoring, allowing for the early detection of plant stress and optimizing yield potential.

Vegetation indices are mathematical combinations of spectral bands that help assess plant health, biomass, and vigor. Two commonly used indices are:

Normalized Difference Vegetation Index (NDVI):

- Formula: $NDVI = (NIR - Red)/(NIR + Red)$
- NDVI uses Near-Infrared (NIR) and Red light wavelengths. Healthy vegetation reflects more NIR and absorbs more Red light. Higher NDVI values indicate healthier vegetation, while lower values suggest stress, disease, or drought.
- Applications:
  - Monitoring crop growth and vigor.
  - Identifying areas with poor vegetation health.
  - Assessing the impact of drought or soil fertility.

Enhanced Vegetation Index (EVI):

- Formula: EVI = G × (NIR − Red)/(NIR + C1 × Red − C2 × Blue + L)
- EVI incorporates additional corrections for atmospheric interference and soil background effects, making it more reliable than NDVI in dense vegetation areas.
- Applications:
  - Monitoring canopy structure and productivity.
  - Detecting subtle variations in plant health.
  - Improving vegetation assessments in tropical or forested regions.

10.2.1.2 Water Stress Detection: Thermal Imaging for Canopy Temperature Monitoring

Water stress is a critical factor affecting crop growth and yield, and detecting it early can help prevent significant losses. Thermal imaging technology is an effective method for identifying water-deficient plants by measuring variations in leaf temperature. Under normal conditions, plants maintain their temperature through transpiration, a process where water evaporates from the leaf surface, cooling the plant. However, when a plant experiences water stress, it closes its stomata to conserve moisture, leading to an increase in leaf temperature [10]. Thermal cameras mounted on drones or satellites can detect these temperature variations, allowing farmers to identify water-stressed areas within a field. This technology is particularly useful for precision irrigation, as it helps optimize water distribution by ensuring that only the required areas receive water, reducing wastage and improving efficiency. Additionally, thermal imaging can be used in drought-prone regions to assess the impact of prolonged dry conditions on crop health [11].

- lack sufficient water for physiological processes, leading to reduced growth and yield.
- Thermal imaging cameras detect infrared radiation emitted by plant leaves, which correlates with temperature.
- **How it works:**
  - Well-hydrated plants maintain cooler leaf temperatures through transpiration.
  - Water-stressed plants close their stomata to conserve water, leading to an increase in leaf temperature.
- **Applications:**
  - Identifying drought-prone areas in fields.
  - Optimizing irrigation schedules by targeting water-deficient zones.
  - Preventing yield loss due to water stress.

10.2.1.3 Disease and Pest Detection: Early Identification of Plant Stress

Early detection of plant diseases and pest infestations is crucial for minimizing yield losses and reducing the need for excessive pesticide use. Remote sensing techniques, such as multispectral and hyperspectral imaging, help identify early signs of plant stress before visible symptoms appear. These imaging techniques analyze the unique spectral signatures of healthy and diseased plants, detecting subtle changes in leaf reflectance caused by infections or pest activity. For example, fungal infections may cause specific chlorophyll degradation patterns, which can be detected using hyperspectral sensors [12]. Similarly, pest infestations often lead to changes in plant pigmentation and structure, which can be identified through aerial imaging. Chlorophyll fluorescence analysis is another method used to assess plant health by measuring the fluorescence emitted by chlorophyll molecules under stress conditions. By integrating these technologies with drone-based monitoring, farmers can conduct large-scale assessments of crop health, allowing for targeted pesticide applications rather than blanket spraying, reducing costs and environmental impact. Early intervention significantly improves crop protection strategies and prevents widespread outbreaks [13].

- Remote sensing helps in the early detection of diseases and pest infestations before visible symptoms appear.
- **Techniques Used:**
  - **Multispectral and Hyperspectral Imaging**: Identifies spectral changes in leaves caused by diseases or pest attacks.
  - **Chlorophyll Fluorescence Analysis**: Detects biochemical stress in plants due to infections.
  - **Drone-Based Monitoring**: UAVs equipped with high-resolution cameras detect areas with unusual color changes, reduced vigor, or defoliation.
- **Applications:**
  - Detecting fungal, bacterial, and viral infections at an early stage.
  - Identifying pest-infested regions for targeted pesticide application.
  - Reducing crop losses by enabling timely intervention.

10.2.1.4 Growth Monitoring: UAVs for Crop Development Tracking

Unmanned Aerial Vehicles (UAVs), commonly known as drones, have revolutionized agricultural monitoring by providing real-time, high-resolution imagery of crop growth. Regular drone surveys allow farmers to track the development stages of crops, from germination to maturity, and identify potential problems such as uneven growth, nutrient deficiencies, or stunted development. Using machine learning algorithms and AI-based analytics, UAVs can automatically detect anomalies, such as lodging (when plants fall over due to wind or weak stems) or signs of nutrient imbalance [14]. This information is crucial for making data-driven decisions on fertilizer application, pest control, and irrigation adjustments. Additionally, growth monitoring helps estimate yield potential, allowing farmers to plan for harvesting and market supply efficiently. By using UAVs for continuous crop surveillance, farmers can respond promptly to emerging issues and ensure optimal crop productivity throughout the growing season [15].

- **Unmanned Aerial Vehicles (UAVs) or Drones** play a crucial role in tracking plant growth stages and identifying abnormalities.
- **How it Works:**
  - UAVs capture high-resolution imagery at regular intervals.
  - AI and machine learning algorithms analyze growth trends.
  - Abnormalities like stunted growth, nutrient deficiencies, or lodging are detected early.
- **Applications:**
  - Estimating crop biomass and yield potential.
  - Monitoring crop uniformity across large fields.
  - Detecting anomalies due to environmental factors like frost or storm damage.

Remote sensing technologies, including NDVI, EVI, thermal imaging, and UAV-based monitoring, have become indispensable in modern precision agriculture. These tools provide farmers with accurate, real-time insights into crop health, water stress, disease outbreaks, and growth patterns, enabling them to take proactive measures to enhance productivity. By integrating these technologies with AI and big data analytics, agricultural decision-making becomes more efficient and sustainable. The ability to detect plant stress early, optimize resource use, and reduce losses is crucial in addressing food security challenges and adapting to climate change. As remote sensing technology continues to evolve, its role in agricultural management will only expand, paving the way for smarter, more sustainable farming practices [16].

10.3 Precision Spraying and Fertilization

Unmanned Aerial Vehicles (UAVs), or drones, have transformed the way pesticides, herbicides, and fertilizers are applied in modern agriculture. Traditional spraying methods, such as ground-based machinery or manual application, often result in excessive chemical usage, uneven distribution, and environmental contamination. In contrast, UAV-based precision spraying ensures that chemicals are applied only where needed, reducing wastage, improving efficiency, and minimizing negative environmental impacts. Equipped with GPS and multispectral sensors, UAVs can identify specific zones within a field that require treatment. These drones use variable rate technology (VRT), which adjusts the amount of pesticide or fertilizer applied based on real-time data. By analyzing crop health using NDVI or thermal imaging,

UAVs can target diseased plants, nutrient-deficient areas, or pest-infested regions, delivering chemicals precisely without affecting healthy crops [17]. This targeted approach leads to cost savings for farmers, as less product is used, and reduces the risk of pesticide resistance among pests. Additionally, UAVs can operate in difficult terrains where traditional machinery cannot reach, such as steep slopes or waterlogged fields. This ability is particularly beneficial for crops like rice, vineyards, and orchards, where precision spraying is essential for maintaining productivity. With advancements in automation and AI integration, UAVs are now capable of autonomous flight, pre-programmed spraying routes, and real-time monitoring, making precision agriculture more accessible and sustainable. By adopting UAV-based spraying and fertilization, farmers can significantly improve crop yields while reducing environmental harm. The precision and efficiency of drones make them a crucial tool in the transition toward sustainable and smart farming practices [18]. Key benefits include:

10.3.1 Targeted Spraying: Precision Application with GPS-Guided Drones

One of the biggest advantages of UAV-based spraying is targeted application, which ensures that chemicals are delivered only to specific areas that require treatment. GPS-guided drones use real-time mapping and georeferencing to create precise spraying routes, minimizing overlap and preventing overuse of pesticides, herbicides, or fertilizers [19]. This technology allows for site-specific application, where only diseased plants, weed-infested zones, or nutrient-deficient areas receive treatment, rather than spraying the entire field indiscriminately. The result is better crop health, reduced chemical drift, and minimal environmental contamination, particularly in sensitive ecosystems near water bodies or protected areas. Additionally, GPS technology helps UAVs navigate challenging terrains, making them ideal for use in vineyards, orchards, and steep farmlands where traditional machinery struggles to operate efficiently. One of the most significant advantages of UAV-based spraying is targeted application, which ensures that chemicals are delivered only to specific areas that require treatment [20]. This precision is achieved through advanced technologies like GPS guidance, real-time mapping, and georeferencing, which enable drones to create highly accurate spraying routes. Here's a deeper dive into how targeted spraying works and its benefits:

How It Works

Real-Time Mapping: UAVs equipped with high-resolution cameras and sensors scan fields to identify areas needing treatment, such as diseased plants, weed-infested zones, or nutrient-deficient patches.

Georeferencing: GPS technology maps these areas with precise coordinates, allowing drones to target specific locations.

Precision Spraying: The drone follows a pre-programmed or real-time-adjusted route, spraying only the identified areas.

Benefits

Minimized Overlap: GPS-guided drones avoid overlapping spray areas, reducing chemical waste and ensuring even coverage.
Reduced Chemical Drift: Precise application minimizes the risk of chemicals drifting to non-target areas, protecting nearby ecosystems and water bodies.
Efficient Use of Resources: Farmers save on chemical costs by applying only what is needed.
Adaptability: UAVs can navigate challenging terrains, such as vineyards, orchards, and steep slopes, where traditional machinery struggles to operate.

Example

In a vineyard, a drone identifies sections of grapevines affected by mildew using multispectral imaging. It then sprays fungicide only on the infected plants, avoiding healthy ones and reducing chemical use by 40% [21].

10.3.2 Reduced Chemical Use: AI-Driven Optimization for Sustainability

UAVs equipped with AI-driven spray systems and advanced nozzles optimize chemical application by adjusting droplet size, spray pressure, and flow rate based on real-time environmental conditions. These drones analyze factors like wind speed, humidity, and crop density to ensure maximum efficiency with minimal chemical use. Variable Rate Technology (VRT) enables drones to deliver precise dosages depending on plant needs, significantly reducing the amount of chemicals wasted. Studies have shown that UAV-based spraying can reduce pesticide and fertilizer usage by up to 30–50% compared to conventional methods, leading to cost savings for farmers and a lower environmental footprint [22]. Additionally, by minimizing excessive chemical applications, UAVs help prevent soil degradation, groundwater contamination, and pesticide resistance in pests and weeds, making them a key tool for sustainable agriculture. UAVs equipped with AI-driven spray systems and advanced nozzles optimize chemical application, ensuring maximum efficiency with minimal environmental impact. Here's how AI and advanced technologies contribute to sustainable spraying [23]:

How It Works

AI-Driven Systems: AI algorithms analyze real-time data, such as wind speed, humidity, and crop density, to adjust spraying parameters like droplet size, spray pressure, and flow rate.

Variable Rate Technology (VRT): This technology allows drones to deliver precise dosages based on the specific needs of different areas within a field.

Benefits

Reduced Chemical Usage: Studies show that UAV-based spraying can reduce pesticide and fertilizer use by 30–50% compared to traditional methods.
Cost Savings: Lower chemical usage translates to significant cost savings for farmers.
Environmental Protection: Minimizing chemical runoff prevents soil degradation, groundwater contamination, and the development of pesticide-resistant pests.
Improved Crop Health: Targeted application ensures that crops receive the right amount of chemicals, promoting healthier growth.

Example

A farmer uses a drone equipped with AI and VRT to spray a wheat field. The drone detects areas with high weed density and applies herbicide only to those spots, reducing overall chemical use by 35% and saving the farmer $500 per season [24].

10.3.3 Safety Improvements: Minimizing Human Exposure to Hazardous Chemicals

Traditional spraying methods often require farmers and laborers to handle hazardous chemicals, exposing them to potential health risks such as respiratory problems, skin irritation, and long-term toxic effects. UAVs eliminate this risk by automating the spraying process, allowing chemicals to be applied remotely without direct human contact. This is especially beneficial for handling highly toxic pesticides or working in fields affected by diseases like fungal outbreaks, where manual spraying could increase the spread of pathogens [25]. Additionally, UAVs can operate in extreme weather conditions, reducing the need for workers to enter dangerous or hard-to-reach areas. By improving safety and reducing labor-intensive spraying tasks, UAVs contribute to better working conditions for farm workers while increasing efficiency and productivity in modern agriculture [26].

Safety Improvements: Minimizing Human Exposure to Hazardous Chemicals

Traditional spraying methods often expose farmers and laborers to hazardous chemicals, posing significant health risks. UAVs address this issue by automating the spraying process, eliminating the need for direct human involvement [27]. Here's how UAVs improve safety:

How It Works

Remote Operation: Farmers can control drones from a safe distance, avoiding exposure to harmful chemicals.
Automated Systems: UAVs follow pre-programmed routes, reducing the need for manual intervention.
Adaptability to Conditions: Drones can operate in extreme weather or hazardous environments, such as fields affected by fungal outbreaks.

Benefits

Reduced Health Risks: Farmers and workers are no longer exposed to toxic chemicals, preventing respiratory problems, skin irritation, and long-term health issues.
Improved Working Conditions: UAVs reduce the need for labor-intensive and dangerous tasks, improving overall working conditions.
Efficiency in Hazardous Areas: Drones can safely spray chemicals in areas that are difficult or dangerous for humans to access, such as steep slopes or dense vegetation.

Example

In a rice field affected by a fungal outbreak, a drone sprays fungicide without requiring workers to enter the contaminated area. This prevents the spread of the fungus and protects workers from exposure to harmful chemicals [28].

Additional Details on UAV Spraying

Spray Calibration

Importance: Proper calibration ensures that drones apply the correct amount of chemicals, avoiding over- or under-application.
Process: Calibration involves adjusting parameters like nozzle size, spray pressure, and flight speed based on field conditions and crop requirements.
Tools: Farmers use software and sensors to calibrate drones for optimal performance.

Legal Regulations

Global Variations: Regulations for UAV spraying vary by country. For example, the U.S. Federal Aviation Administration (FAA) requires certification for commercial drone operations, while the European Union has strict guidelines on chemical use [29].
Compliance: Farmers must ensure that their drones meet local regulations, including safety standards, chemical usage limits, and flight restrictions.

Case Studies

Japan: UAVs are widely used for rice farming, where they spray pesticides with precision, reducing chemical use by 50% and labor costs by 70%.

Brazil: Drones are used in sugarcane fields to apply herbicides, improving efficiency and reducing environmental impact.
India: Smallholder farmers use affordable drones to spray crops, increasing yields and reducing chemical costs.

UAV-based spraying represents a transformative advancement in agriculture, offering precision, efficiency, and sustainability. By enabling targeted application, reducing chemical use, and improving safety, drones are helping farmers optimize crop management while minimizing environmental impact. As technology continues to evolve, UAVs will play an increasingly vital role in addressing global challenges like food security, resource scarcity, and climate change. Whether through AI-driven optimization, advanced spray calibration, or compliance with legal regulations, UAVs are paving the way for a smarter, safer, and more sustainable future in agriculture [30].

10.4 Soil and Field Analysis Using UAVs

Soil health and field conditions is essential for optimizing crop production and ensuring sustainable farming practices. Remote sensing technologies and UAVs (drones) have transformed soil and field analysis by providing real-time, high-resolution data on various soil properties such as composition, moisture levels, nutrient availability, and topography. By using multispectral, hyperspectral, and thermal imaging sensors, drones can quickly map and assess soil variability, helping farmers make informed decisions about soil management and crop planning [31]. One of the critical applications of UAV-based soil analysis is soil moisture detection, which ensures efficient irrigation management. Thermal imaging and infrared sensors detect temperature variations in the soil, indicating dry or overly saturated areas, allowing for precise water distribution. Soil nutrient mapping is another crucial aspect where drones equipped with spectral sensors analyze reflectance patterns to determine nutrient deficiencies, enabling site-specific fertilization. UAVs also assist in topographic mapping, creating detailed elevation models that help in designing drainage systems and erosion control strategies [32].

Example:

A farmer notices uneven crop growth in different sections of a 50-acre cornfield. Using a UAV equipped with multispectral sensors, the farmer maps soil variations and discovers that certain areas lack nitrogen. With this data, the farmer applies fertilizers only where needed, reducing costs and improving yield without overusing chemicals [33].

Formula:

A commonly used formula in soil analysis is Soil Moisture Index (SMI):

$$SMI = (NIR - SWIR)/(NIR + SWIR)$$

where:

- NIR = Near-Infrared Reflectance
- SWIR = Short-Wave Infrared Reflectance.

This index helps determine soil moisture levels, assisting in irrigation planning.

Importance:

1. **Enhanced Soil Fertility**: UAVs help detect nutrient deficiencies, improving fertilization efficiency.
2. **Water Conservation**: By identifying moisture levels, drones assist in targeted irrigation, reducing water waste.
3. **Erosion Control**: Topographic mapping aids in designing effective drainage and soil conservation measures.
4. **Cost and Resource Efficiency**: Optimized input application minimizes excess fertilizer, pesticide, and water usage.
5. **Sustainable Farming**: UAV-driven soil analysis reduces environmental impact while increasing productivity.

10.4.1 Soil Moisture Mapping: Optimizing Irrigation Efficiency

Soil moisture plays a critical role in plant growth, and improper water distribution can lead to drought stress, reduced yields, or waterlogging, negatively impacting crop health. Traditional irrigation methods often result in overwatering or underwatering, leading to inefficient water use and increased costs. UAVs equipped with thermal sensors and multispectral cameras provide an advanced solution by accurately mapping soil moisture variations across a field in real-time [34]. Thermal imaging detects temperature differences in the soil, indicating moisture levels—drier areas appear warmer, while wetter regions appear cooler. Additionally, Near-Infrared (NIR) and Shortwave Infrared (SWIR) sensors analyze how much water the soil retains by assessing reflectance patterns. Normalized Difference Water Index (NDWI) is commonly used to measure soil moisture levels and guide irrigation decisions [35].

Example:

A farmer managing a 100-acre wheat field experiences irregular crop growth. Using a UAV equipped with thermal and NIR sensors, the farmer detects moisture-deficient zones and areas with excessive water retention. With this data, the farmer adjusts irrigation schedules, preventing drought stress and waterlogging, leading to improved crop uniformity and increased yield [36].

Formula:

One of the key formulas for soil moisture estimation is **NDWI (Normalized Difference Water Index):**

$$\mathrm{NDWI} = (\mathrm{NIR} - \mathrm{SWIR})/(\mathrm{NIR} + \mathrm{SWIR})$$

where:

- **NIR** = Near-Infrared Reflectance
- **SWIR** = Shortwave Infrared Reflectance.

A higher NDWI value indicates wetter soil, while a lower value suggests dry conditions.

Importance:

1. **Precision Irrigation**: Ensures that only the necessary areas receive water, reducing overuse and improving efficiency.
2. **Water Conservation**: Helps prevent excessive irrigation, conserving valuable water resources.
3. **Cost Savings**: Reduces energy and water costs associated with irrigation systems.
4. **Prevention of Soil Degradation**: Minimizes soil erosion, nutrient leaching, and waterlogging issues.
5. **Sustainable Agriculture**: Enhances resource efficiency while maintaining high crop productivity.

By integrating UAV-based soil moisture mapping into irrigation management, farmers can achieve higher yields, lower costs, and improved sustainability, making agriculture more resilient to climate variability [37].

10.4.2 Soil Composition and Nutrient Mapping: Smart Fertilization

Healthy crop growth relies on the availability of essential nutrients such as nitrogen (N), phosphorus (P), and potassium (K), along with organic matter and micronutrients. Traditional soil testing methods involve physical sampling and laboratory analysis, which can be time-consuming, labor-intensive, and expensive. However, UAV-based remote sensing and hyperspectral imaging provide a rapid, non-invasive alternative for assessing soil composition and nutrient levels with high precision [38]. Hyperspectral and multispectral sensors on UAVs analyze spectral reflectance patterns to detect variations in soil organic matter, NPK levels, and pH balance. By mapping these variations, farmers can determine areas that require additional fertilization and those with sufficient nutrients. Variable Rate Technology (VRT) integrates this data to enable UAV-guided precision fertilization, ensuring that fertilizers are applied only where needed, reducing excess application, lowering costs, and minimizing environmental pollution [39].

Example:

A farmer managing a 200-acre cornfield observes uneven growth patterns. Instead of conducting manual soil tests across the entire field, the farmer deploys a UAV with hyperspectral imaging to map soil nutrient levels. The data reveals nitrogen deficiency in specific areas, allowing the farmer to apply fertilizers only to deficient zones using VRT-equipped drones. This approach optimizes fertilizer use, enhances crop uniformity, and reduces costs [40].

Importance:

1. **Precision Fertilization**: Ensures optimal application of nutrients, improving plant health and yield.
2. **Cost Efficiency**: Reduces fertilizer wastage, lowering input costs.
3. **Environmental Sustainability**: Prevents nutrient runoff into water bodies, reducing soil and water pollution.
4. **Time-Saving**: Eliminates the need for extensive manual soil sampling.
5. **Improved Soil Health**: Promotes balanced nutrient distribution, preventing over-fertilization and soil degradation.

By integrating UAV-based soil composition and nutrient mapping, farmers can adopt smart fertilization strategies, leading to higher yields, lower costs, and environmentally friendly farming practices [41].

10.4.3 Field Topography and Drainage Analysis

Field elevation and drainage patterns are critical factors affecting soil health, water management, and crop productivity. Poor drainage can lead to waterlogging, root diseases, reduced oxygen availability, and nutrient leaching, all of which negatively impact plant growth. UAVs equipped with LiDAR (Light Detection and Ranging) sensors and photogrammetry provide high-resolution 3D topographic maps, enabling farmers to analyze elevation changes, slope gradients, and water flow patterns with precision [42]. By detecting low-lying areas prone to water accumulation, these maps help farmers design efficient drainage systems to prevent excess moisture retention. Proper drainage solutions, such as subsurface drainage tiles, contour plowing, or raised beds, can be implemented to improve water flow and enhance soil aeration. Moreover, UAV-generated topographic data is valuable for erosion control, as it helps identify steep slopes or bare soil patches that are susceptible to runoff. By implementing soil conservation practices like terracing, grass waterways, or contour farming, farmers can prevent soil degradation and sustain long-term productivity [43].

Example:

A farmer managing a 150-acre soybean field notices persistent waterlogging in certain areas after heavy rains. By deploying a UAV with LiDAR sensors, the farmer generates a 3D elevation map that highlights drainage inefficiencies. Based on this data, a subsurface drainage system is installed in the low-lying zones, effectively reducing water accumulation and improving crop yields [44].

Importance:

1. **Improved Drainage Management**: Prevents waterlogging and enhances root aeration.
2. **Erosion Control**: Helps design soil conservation structures like terraces and contour farming.
3. **Precision Land Leveling**: Ensures uniform crop growth and irrigation efficiency.
4. **Increased Crop Yield**: Reduces plant stress caused by excessive or insufficient water.
5. **Sustainable Soil Management**: Minimizes nutrient loss and maintains long-term soil productivity.

By integrating UAV-based field topography and drainage analysis, farmers can optimize water distribution, prevent soil erosion, and enhance overall farm productivity, contributing to sustainable and efficient agriculture [45].

10.4.4 Detecting Soil Degradation and Compaction

Soil compaction is a major issue in agriculture, as it restricts root penetration, reduces water infiltration, limits nutrient uptake, and decreases soil aeration, ultimately leading to poor crop performance and lower yields. Compacted soils also increase runoff and erosion risks, further degrading soil quality. Traditional soil testing methods for compaction, such as penetrometers and bulk density measurements, can be time-consuming and labor-intensive [46]. Remote sensing technologies, including radar-based imaging, multispectral sensors, and thermal cameras, provide a faster and more efficient method for detecting compacted areas. UAVs equipped with these sensors assess soil structure by analyzing surface texture, temperature variations, and vegetative stress patterns. Compacted soil tends to retain heat longer due to reduced porosity and moisture retention capacity, making thermal imaging a key tool in identifying compacted zones. Additionally, UAVs can detect early signs of plant stress, such as low NDVI (Normalized Difference Vegetation Index) values, which may indicate restricted root growth due to compaction [47].

Example:

A farmer notices stunted crop growth in certain areas of a 100-acre wheat field. Using a UAV with radar-based imaging and thermal sensors, the farmer detects high soil density zones where water infiltration is poor. Based on this data, the farmer performs deep tillage and aeration in the affected areas, improving soil structure and enhancing crop performance [48].

Importance:

1. **Improved Root Growth**: Reduces mechanical resistance, allowing roots to penetrate deeper.
2. **Enhanced Water Infiltration**: Prevents surface runoff and improves soil moisture retention.
3. **Better Nutrient Uptake**: Loosening the soil structure helps plants absorb essential nutrients more efficiently.
4. **Prevention of Yield Loss**: Early detection and mitigation of compaction ensure consistent crop productivity.
5. **Sustainable Soil Management**: Encourages conservation practices like **crop rotation, deep tillage, and cover cropping**.

By utilizing UAV-based soil compaction detection, farmers can implement targeted soil improvement strategies, ensuring fertile, well-aerated, and productive soils for long-term agricultural sustainability [49].

10.4.5 Precision Tillage and Land Preparation

UAV-based soil analysis helps farmers determine the best tillage practices for their specific field conditions. By analyzing soil structure, moisture content, and organic matter distribution, UAVs assist in deciding whether no-till, reduced tillage, or conventional tillage methods would be most effective. This prevents unnecessary soil disturbance, reduces fuel costs, and enhances carbon sequestration, promoting sustainable land management practices [34]. Soil and field analysis using UAVs and remote sensing technologies provides farmers with critical insights into soil health, moisture levels, nutrient distribution, and field topography. By leveraging these advanced tools, farmers can enhance soil fertility, optimize irrigation and fertilization, improve drainage, and prevent soil degradation. These precision agriculture techniques lead to higher crop yields, reduced environmental impact, and more efficient resource utilization, paving the way for sustainable farming in the future [50].

By capturing high-resolution multispectral and thermal images, UAVs can analyze soil compaction levels, water retention capacity, and fertility variations. This data helps farmers decide whether to use:

- **No-Till Farming**—Ideal for fields with **high organic matter**, reducing soil disturbance and enhancing **carbon sequestration**.
- **Reduced Tillage**—Suitable for **moderately compacted soils**, minimizing erosion and fuel consumption.
- **Conventional Tillage**—Used in **highly compacted soils** to break up hardpan layers and improve root penetration.

By avoiding unnecessary tillage, farmers can reduce fuel and labor costs, preserve soil structure, and enhance microbial activity, leading to improved long-term soil health and sustainability [51].

Example:

A farmer managing a 250-acre soybean field wants to determine the best tillage strategy. By using UAV-based soil compaction and moisture mapping, the farmer identifies areas with minimal compaction, suitable for no-till farming, and zones with hardpan layers, requiring deep tillage. This precision-based approach reduces fuel usage, prevents erosion, and enhances soil productivity [52].

Importance:

1. **Optimized Tillage Practices**: Prevents over-tillage, maintaining soil health and fertility.
2. **Fuel and Cost Efficiency**: Reduces diesel consumption and labor costs associated with excessive plowing.
3. **Carbon Sequestration**: Encourages no-till and reduced tillage practices, enhancing soil carbon storage and reducing CO_2 emissions.
4. **Prevention of Soil Erosion**: Minimizes topsoil loss, preserving soil nutrients and moisture levels.
5. **Sustainable Land Management**: Ensures long-term productivity while reducing the environmental impact of tillage operations.

10.5 Yield Estimation and Forecasting

Yield estimation and forecasting are crucial in modern agriculture, helping farmers, agronomists, and policymakers predict crop production, optimize resource allocation, and make informed decisions about market supply. Traditional methods rely on manual sampling and historical data, which can be time-consuming and prone to errors. However, with the advent of UAVs, remote sensing, and AI-powered analytics, farmers can now estimate crop yields with greater accuracy, efficiency, and speed. These technologies provide real-time insights into crop health, biomass, growth stages, and environmental conditions, allowing for proactive decision-making and improved farm management [53]. UAVs equipped with multispectral, hyperspectral, and thermal sensors capture high-resolution images of crops at different growth stages, allowing AI-powered software to assess canopy cover, plant height, chlorophyll content, and biomass accumulation, key indicators of crop productivity. Advanced machine learning algorithms process this data and generate yield prediction models, providing farmers with early insights into expected harvest outcomes. This level of precision allows for better planning of storage, logistics, and market distribution, reducing post-harvest losses [54].

UAV-based yield estimation can detect stress factors such as drought, nutrient deficiencies, pest infestations, and diseases before they significantly impact production.

AI-driven analytics can identify subtle variations in plant health through vegetation indices like NDVI (Normalized Difference Vegetation Index) and EVI (Enhanced Vegetation Index), helping farmers take corrective actions to prevent yield losses. This proactive approach leads to higher efficiency in input application (fertilizers, pesticides, and irrigation), ultimately improving overall crop output [55]. Yield mapping and field variability analysis allow farmers to identify high-performing and low-performing zones within a field, enabling them to implement precision farming techniques and make data-driven decisions about crop rotation, land use, and soil conservation strategies. Yield forecasting helps farmers determine the optimal time for harvesting, ensuring that crops are harvested at their peak maturity for maximum quality and market value. Accurate yield estimation is valuable not only for farmers but also for insurance companies and financial institutions that provide crop insurance and agricultural loans. UAV-based yield data helps insurers assess potential risks and damages caused by extreme weather events, pests, or diseases, enabling them to offer fair and data-driven compensation to farmers [56].

Benefits include:

Benefits of UAV-Based Soil and Field Analysis

Yield prediction is one of the most valuable applications of UAV technology in agriculture. By leveraging advanced sensors, imaging technologies, and data analytics, UAVs provide farmers with accurate and timely insights into crop yields. These predictions offer numerous benefits, including improved decision-making, cost savings, and optimized harvesting processes [57]. Here's a detailed look at the key benefits:

10.5.1 Predictive Analysis

UVs play a crucial role in predicting crop growth, soil health, and yield potential, allowing farmers to make informed decisions about harvesting, resource allocation, and market supply. With advanced sensors such as multispectral, hyperspectral, and thermal cameras, UAVs collect high-resolution images that provide valuable insights into vegetation indices (NDVI, EVI), chlorophyll content, and plant stress levels. These indices help farmers assess crop health, detect early signs of diseases, and predict yield variations across different field zones [58]. By analyzing historical UAV data, farmers can track seasonal trends and soil fertility fluctuations, improving their ability to forecast expected harvest quality and quantity. This predictive capability is particularly useful in climate-sensitive regions, where fluctuating temperatures and irregular rainfall patterns can impact crop development and harvesting schedules. Farmers can integrate UAV data with machine learning algorithms to develop yield prediction models, ensuring optimal market planning and supply chain management [59].

For example, a rice farmer in South Asia uses UAVs to monitor moisture levels and crop maturity across large fields. By identifying mature sections early, the farmer can schedule harvesting operations efficiently, preventing grain spoilage and yield losses due to unpredictable weather conditions. In commercial farming, agribusinesses use UAVs for large-scale predictive analysis, allowing them to plan export shipments and distribution networks based on real-time crop growth assessments. Overall, UAV-based predictive analysis enhances farm efficiency, reduces uncertainty, and ensures better decision-making [60]. Farmers can adapt to changing conditions proactively, improving economic returns and food security while reducing the risk of overproduction or supply shortages in agricultural markets. UAVs equipped with advanced sensors and AI-powered analytics enable predictive analysis, helping farmers make data-driven decisions about harvest timing, market supply, and resource allocation.

How It Works

UAVs collect data on crop health, growth stages, and environmental conditions using multispectral, hyperspectral, and thermal imaging.
AI algorithms analyze this data to predict yields and identify trends.

Benefits

Harvest Timing: Farmers can determine the optimal time to harvest, ensuring maximum crop quality and yield.
Market Planning: Accurate yield predictions help farmers plan for market demand, reducing the risk of overproduction or shortages.
Resource Allocation: Farmers can allocate resources like labor, storage, and transportation more effectively.

Example

A wheat farmer uses a drone to monitor crop growth and predict yields. Based on the data, the farmer schedules the harvest for the optimal time, ensuring high-quality grain and maximizing profits [61].

10.5.2 Cost Reduction

The integration of UAVs in soil and field analysis significantly reduces operational costs by minimizing manual labor, fuel consumption, and inefficient input applications. Traditional field surveys and soil testing require extensive manpower and expensive laboratory analysis, making them costly and time-consuming, especially for large-scale farms. UAVs, on the other hand, provide instantaneous and automated field assessments, cutting down the time and expenses associated with physical inspections and sample collection. One of the biggest cost-saving aspects of UAVs is precision agriculture, where farmers apply water, fertilizers, and pesticides only where needed [62]. Equipped with high-resolution cameras and AI-driven analytics, UAVs map out soil nutrient deficiencies and moisture variability, enabling site-specific treatments rather than uniform applications across the entire field. This prevents overuse of fertilizers and chemicals, reducing both input costs and environmental contamination. A corn farmer in the U.S. Midwest switched from traditional manual crop scouting to UAV-based remote sensing and saved over 30% on fertilizer and pesticide costs while increasing yields by 15%. Additionally, UAVs eliminate the need for excessive tractor use, which reduces diesel fuel consumption and maintenance costs [63]. For small-scale farmers in developing countries, drones provide a low-cost alternative to expensive satellite imaging or ground-based soil analysis. Moreover, UAVs contribute to long-term economic sustainability by preventing crop losses due to undetected pests, diseases, or soil degradation. Farmers can monitor crop health frequently without hiring additional labor, ensuring that issues are identified and resolved early. The combination of cost-effective field monitoring, optimized input usage, and reduced fuel dependency makes UAVs an essential tool for profitable and efficient farming operations. UAVs reduce the need for manual surveys and labor-intensive data collection, leading to significant cost savings for farmers [64].

How It Works

UAVs cover large areas quickly, collecting data in a fraction of the time required for manual surveys.
Automated data processing eliminates the need for extensive manual analysis.

Benefits

Labor Savings: Reduces the need for field workers to conduct surveys, lowering labor costs.
Time Efficiency: Farmers can gather and analyze data in real-time, speeding up decision-making.
Resource Optimization: Precise yield predictions help farmers avoid over-investment in resources like fertilizers and pesticides.

Example

A corn farmer uses a drone to survey 500 acres of land in just a few hours, saving weeks of manual labor and reducing survey costs by 60%.

10.5.3 Optimized Harvesting

UAVs enhance harvesting efficiency by identifying the most productive field zones, ensuring that high-yield areas are prioritized during harvest. By using biomass estimation, chlorophyll analysis, and plant maturity mapping, UAVs help farmers determine the right time to harvest crops for maximum quality and yield. Traditional harvesting schedules are often based on calendar estimates, which may not accurately reflect crop readiness across different parts of a field. UAVs provide precise data on crop maturity, enabling site-specific harvesting plans that minimize losses due to premature or delayed cutting. For example, in grape vineyards, UAVs analyze color changes in grape clusters, indicating when they have reached optimal sugar content for winemaking [65]. In grain farming, UAVs detect moisture content variations in wheat and barley fields, ensuring mechanized harvesting occurs at the ideal dryness level to prevent spoilage or excessive grain cracking. UAV technology also plays a vital role in harvest logistics and workforce planning. By identifying high-density crop zones, farmers can allocate resources efficiently, reducing labor costs and minimizing the need for unnecessary field passes with heavy machinery. This results in lower fuel consumption and reduced soil compaction, which helps maintain long-term soil health. A sugarcane farm in Brazil adopted UAV-based harvest optimization techniques and increased sugar yield by 20% while reducing harvesting costs by 25% [66]. The UAVs mapped crop uniformity and lodging patterns, allowing for strategic harvesting sequences that minimized wastage. Similarly, UAV-based yield estimation helps agricultural cooperatives and food processing industries plan transportation, storage, and supply chain logistics efficiently, preventing post-harvest bottlenecks. By optimizing harvesting operations, UAVs contribute to higher productivity, reduced waste, and improved product quality. These advancements ensure higher profits for farmers, improved market supply stability, and a more sustainable food production system [67]. The adoption of UAV-based precision agriculture brings numerous benefits, including:

More accurate decision-making through predictive analysis

Lower operational costs, reduced input waste, and improved resource efficiency.

Optimized harvesting strategies for maximum yield and better crop quality By integrating UAV technology into modern farming, agricultural operations become smarter, more efficient, and environmentally sustainable. Farmers can maximize productivity while reducing costs, ensuring long-term profitability and food security in a rapidly evolving agricultural landscape [68]. UAVs help farmers identify the best areas for harvesting, ensuring that resources are used efficiently and yields are maximized.

How It Works

UAVs create detailed maps of crop health and maturity, highlighting areas ready for harvest.

Farmers can prioritize harvesting in high-yield areas, reducing waste and improving efficiency.

Benefits

Increased Efficiency: Farmers can focus on high-yield areas, reducing the time and effort required for harvesting.

Reduced Waste: Precise harvesting minimizes crop damage and loss.

Improved Profitability: Maximizing yields from the most productive areas increases overall profitability.

Example

A vineyard owner uses a drone to identify grape clusters with the highest sugar content. The harvest team focuses on these areas, producing higher-quality wine and increasing profits by 20% [69].

Additional Benefits of Yield Prediction Using UAVs

10.5.4 Early Problem Detection

UAVs can detect issues like pest infestations, nutrient deficiencies, and water stress early, allowing farmers to take corrective actions before yields are affected.

Example: A drone detects a fungal infection in a soybean field, enabling the farmer to apply fungicide before the infection spreads.

10.5.5 Sustainability

By optimizing resource use and reducing waste, UAV-based yield prediction promotes sustainable farming practices.

Example: A farmer uses yield predictions to apply fertilizers only where needed, reducing chemical runoff and environmental impact.

10.5.6 Improved Crop Management

Yield predictions provide insights into crop performance, helping farmers refine their management practices for future seasons.

Example: A farmer analyzes yield data to identify the most productive crop varieties and planting patterns.

10.6 Using Multispectral and Thermal Imaging

The integration of multispectral and thermal imaging in agriculture has transformed modern farming by providing detailed, data-driven insights into crop health, soil conditions, and water management. These imaging technologies, often mounted on

UAVs (drones) or satellites, allow farmers to detect plant stress, moisture deficiencies, and temperature variations that are invisible to the naked eye. By utilizing these advanced tools, farmers can optimize irrigation, fertilization, and pest control, leading to higher yields, reduced costs, and improved sustainability [70].

10.6.1 Multispectral Imaging: Assessing Plant Health and Growth

Multispectral imaging captures light reflections across multiple wavelength bands, including visible (RGB), near-infrared (NIR), and shortwave infrared (SWIR). This technology enables precise monitoring of plant physiology and health by detecting variations in chlorophyll content, leaf structure, and water absorption. One of the most important applications of multispectral imaging is the calculation of vegetation indices, such as the Normalized Difference Vegetation Index (NDVI), which measures plant greenness and overall vitality [71]. Other indices like the Enhanced Vegetation Index (EVI) improve upon NDVI by minimizing atmospheric interference, while the Canopy Chlorophyll Content Index (CCCI) helps assess nitrogen levels in crops. By using multispectral imaging, farmers can detect nutrient deficiencies, pest infestations, and disease outbreaks before visible symptoms appear. This early detection allows for targeted interventions, reducing the need for excessive pesticide or fertilizer use. Additionally, multispectral imaging plays a crucial role in monitoring crop growth stages, enabling farmers to adjust planting schedules, irrigation strategies, and harvest timing for maximum productivity [72].

Example of Multispectral Imaging in Agriculture

Case Study: Detecting Crop Health in a Wheat Field Using NDVI

In recent years, precision agriculture has emerged as a powerful approach to optimizing crop production by integrating advanced technologies into traditional farming practices. One such innovation is the use of the Normalized Difference Vegetation Index (NDVI) to assess crop health and detect stress at various growth stages. This case study highlights the application of NDVI in a 15-hectare wheat field located in Punjab, Pakistan, during the Rabi growing season.

The primary objective was to monitor the health status of the wheat crop and detect spatial variability in plant vigor using NDVI data collected via an Unmanned Aerial Vehicle (UAV). The UAV was equipped with a multispectral camera (RedEdge-MX) and flown over the field at three critical stages of crop development: tillering, heading, and grain filling. Flights were scheduled during midday under clear skies to ensure uniform lighting and optimal reflectance conditions. The captured images were stitched and processed using Pix4D software, with radiometric calibration applied to ensure accuracy. NDVI values were computed using the standard formula: (NIR − Red)/(NIR + Red), where NIR and Red refer to near-infrared and red light reflectance, respectively. The resulting NDVI maps revealed clear spatial patterns of crop vigor, with NDVI values ranging from 0.3 to 0.85. Areas with values below 0.4 were identified as zones of poor crop health, likely affected by water stress, pest infestation, or nutrient deficiency. In contrast, zones with NDVI values above 0.7 indicated healthy, vigorous growth. These findings were validated through on-ground surveys, where plant height, chlorophyll content (measured with a SPAD meter), and biomass samples were collected from different NDVI zones. A strong positive correlation ($R^2 = 0.82$) was observed between NDVI values and SPAD readings, confirming the reliability of NDVI in detecting physiological stress and plant vigor.

The analysis supported targeted interventions in specific zones. For instance, poor-performing areas were recommended for nitrogen supplementation and improved irrigation practices, especially where waterlogging was evident. Pest control measures were prioritized in sections where NDVI had dropped between the heading and grain-filling stages, suggesting potential biotic stress. This case study underscores the effectiveness of NDVI as a decision-support tool in precision wheat farming. By providing a rapid, non-destructive, and spatially explicit assessment of crop health, NDVI allows farmers and agronomists to detect problems early, reduce unnecessary input costs, and implement zone-based management strategies that enhance yield and sustainability.

Case Study: Monitoring Rice Crop Health Using NDVI-Based UAV Imaging

Rice, being one of the most important staple crops in Asia, requires careful monitoring throughout its growth cycle to ensure optimum yield and quality. In this case study, a precision agriculture approach was employed to monitor crop health in a 20-hectare rice field located in Jiangxi Province, China. The objective was to assess spatial variability in rice vigor and identify areas of biotic and abiotic stress using the Normalized Difference Vegetation Index (NDVI), derived from UAV-based multispectral imaging. A UAV equipped with a five-band multispectral camera was deployed at three critical growth stages: tillering, panicle initiation, and grain filling. Each flight was conducted under cloud-free skies around solar noon to minimize atmospheric variability and shadows. The acquired imagery was processed using Pix4Dmapper, and orthomosaics were generated to calculate NDVI values, using the standard equation: $(NIR - Red)/(NIR + Red)$.

NDVI maps revealed noticeable variability across the field. During the tillering stage, NDVI ranged between 0.45 and 0.80, with several patches falling below 0.55, indicating poor vigor likely due to uneven seedling establishment and early nitrogen deficiency. As the season progressed to panicle initiation, NDVI increased across most areas, but localized declines were observed in the lower eastern quadrant of the field. Ground truthing in these areas revealed signs of potassium deficiency, water stress due to uneven field leveling, and minor pest activity. During the grain filling stage, a few zones maintained high NDVI values (above 0.75), while others showed a sharp decline to below 0.5, which correlated with spikelet sterility and panicle discoloration observed during field inspections.

These spatial insights allowed the farm management team to implement site-specific interventions. Nitrogen top-dressing was adjusted based on early-season NDVI values, and deficient areas were re-evaluated for micronutrient applications. Drainage channels were realigned in waterlogged areas, and insecticidal treatments were applied where pest-induced NDVI drop was noted. Yield data collected at harvest confirmed the effectiveness of this targeted approach—zones with stable high NDVI throughout the season recorded significantly higher grain yield (by 0.8–1.2 tons/ha) compared to stressed areas.

This case study demonstrates the practical value of UAV-based NDVI imaging in managing rice cultivation. It not only facilitated real-time crop monitoring but also enabled a more sustainable, input-efficient, and data-driven farming approach. By integrating UAV technology with agronomic decision-making, rice farmers can address within-field variability, minimize losses, and improve overall production efficiency under varying environmental and soil conditions.

Real-World Benefits of Multispectral Imaging:

Early Stress and Disease Detection: **Farmers** can detect plant stress before visible symptoms appear, allowing early intervention.
Optimized Fertilizer and Water Use: Identifies areas needing more nutrients or irrigation, reducing waste and costs.
Improved Yield Estimation: Provides accurate crop growth analysis for better harvest planning and market supply predictions.

10.6.2 Thermal Imaging: Water Stress Detection and Irrigation Management

Thermal imaging detects temperature variations in crops and soil, providing valuable information about water stress, irrigation efficiency, and heat stress in plants. By measuring canopy temperature, thermal cameras can identify areas where crops are experiencing drought stress or overwatering, allowing for precise adjustments to irrigation systems. In precision agriculture, thermal imaging is used to create soil moisture maps, which help farmers determine optimal irrigation schedules and prevent water wastage [73]. In large-scale farming, center-pivot irrigation systems can be adjusted based on real-time thermal imaging data, ensuring that each section of the field receives the right amount of water. This approach not only conserves valuable water resources but also prevents root diseases and soil degradation caused by overwatering. Another crucial application of thermal imaging is the detection of heat stress in crops. High canopy temperatures indicate that plants are struggling to maintain transpiration and nutrient uptake, which can severely impact yield [74]. By identifying these stress zones early, farmers can take preventative measures such as mulching, adjusting irrigation frequency, or implementing shade nets to protect vulnerable crops.

Formula for Crop Water Stress Index (CWSI)

The Crop Water Stress Index (CWSI) is a widely used index to quantify plant water stress based on canopy temperature measurements. The standard formula for CWSI is:

$$CWSI = \frac{T_c - T_{wet}}{T_{dry} - T_{wet}}$$

where:

T_c Canopy temperature (measured by the thermal sensor).
T_{wet} Temperature of a well-watered reference crop.
T_d Temperature of a completely dry reference crop.

High CWSI (close to 1): Severe water stress, requiring immediate irrigation.
Low CWSI (close to 0): Well-hydrated plants with sufficient soil moisture.

Example of Thermal Imaging in Agriculture

Case Study: Detecting Water Stress in a Vineyard

A vineyard in California uses UAV-mounted thermal cameras to assess water stress in grapevines.

Step 1: Data Collection

- A thermal drone flies over the vineyard, capturing canopy temperature variations.
- Areas with higher temperatures indicate drought-stressed vines.

Step 2: CWSI Calculation

- Using the CWSI formula, the farmer identifies sections of the vineyard experiencing high water stress.
- Stressed vines have CWSI values above 0.7, requiring immediate irrigation.

Step 3: Optimizing Irrigation

- Instead of watering the entire field, the farmer applies targeted irrigation to stressed areas.
- This conserves water, reduces costs, and improves grape quality.

Outcome: The vineyard reduces water usage by 30% while maintaining high grape yield and quality.

Benefits of Thermal Imaging in Irrigation Management

Early Water Stress Detection: Identifies drought-prone areas before visible signs appear.
Precision Irrigation: Ensures water is applied only where needed, reducing waste.

Cost Savings: Lowers water and electricity costs, making irrigation more efficient.
Improved Crop Quality: Prevents wilting and yield loss, ensuring healthy plant growth.

10.6.3 Integrating Multispectral and Thermal Imaging for Precision Agriculture

When combined, multispectral and thermal imaging provide a comprehensive understanding of plant health, soil conditions, and environmental factors. While multispectral imaging detects chlorophyll levels, nutrient deficiencies, and disease symptoms, thermal imaging highlights water stress, soil moisture variations, and temperature-related crop stress [75]. The integration of both technologies enables:

Precision Fertilization: By analyzing vegetation indices and soil health, farmers can apply fertilizers only where needed, reducing waste and environmental pollution.
Targeted Irrigation: Thermal imaging ensures that water is distributed efficiently, preventing waterlogging and drought stress.
Yield Prediction and Harvest Optimization: By monitoring growth patterns and crop conditions, farmers can plan harvesting schedules for maximum yield and quality. For example, a vineyard in California uses multispectral imaging to monitor grape maturity and thermal imaging to detect water stress, ensuring that vines receive adequate hydration without over-irrigation [9]. Similarly, large-scale wheat farms in Australia integrate these imaging techniques to optimize nitrogen application and improve drought resilience, leading to higher efficiency and reduced input costs.

Example: Smart Farm Management with Integrated Imaging

Case Study: Monitoring Corn Fields in Nebraska

A corn farmer in Nebraska uses a drone equipped with both multispectral and thermal cameras to monitor crop health and water stress.

Step 1: Multispectral Imaging for Crop Health

- The drone captures NDVI data, revealing areas with low chlorophyll content (possible nutrient deficiency or disease).
- The farmer identifies yellow zones on the NDVI map, indicating stressed crops.

Step 2: Thermal Imaging for Water Stress Detection

- The thermal sensor measures canopy temperature to assess water stress levels.
- High CWSI values (above 0.7) indicate severely water-stressed areas.

Step 3: Precision Agriculture Decision-Making

The farmer **cross-references NDVI and CWSI data** to determine the cause of stress.

If NDVI is low and CWSI is high: Plants lack water → Adjust irrigation.
If NDVI is low but CWSI is normal: Likely nutrient deficiency → Apply targeted fertilization.
If both NDVI and CWSI are normal: Healthy crops → No intervention needed.

Outcome: The farm reduces water use by 25%, optimizes fertilizer application, and increases yield by 15%.

Benefits of Integrating Multispectral and Thermal Imaging

More Accurate Crop Stress Detection: Identifies whether stress is caused by water deficiency, disease, or nutrient imbalance.

Optimized Irrigation and Fertilization: Reduces water and fertilizer waste, lowering production costs.

Early Pest and Disease Identification: Allows farmers to act before stress becomes visible, improving crop protection.

Higher Yields with Less Resource Use: Supports sustainable and cost-effective farm management.

Integrating multispectral and thermal imaging offers a powerful precision agriculture tool. This combination enhances farm efficiency, conserves resources, and improves overall crop productivity. The adoption of multispectral and thermal imaging in agriculture has led to significant improvements in crop monitoring, resource management, and sustainability. By leveraging these advanced technologies, farmers can make data-driven decisions that enhance productivity, conserve resources, and reduce environmental impact. As UAVs and remote sensing technologies continue to evolve, the future of precision agriculture will be driven by real-time imaging, artificial intelligence, and automation, paving the way for smarter, more efficient, and climate-resilient farming [76].

10.7 Seeding, Planting, and Pollination Assistance with UAVs

Unmanned Aerial Vehicles (UAVs) are transforming modern agriculture by automating essential farming processes such as seeding, planting, and pollination. These aerial systems offer an efficient way to disperse seeds across large or hard-to-reach areas, significantly reducing the time and labor required for manual planting. This technology is especially useful in steep terrains, wetlands, and degraded lands where conventional planting methods are difficult. By ensuring uniform seed distribution, UAVs contribute to higher germination rates and improved crop yields. One of the major advantages of UAV-based seeding is cost reduction [77]. Traditional planting requires significant manpower, but drones equipped with seed dispensers can automate the process, minimizing labor expenses while maximizing efficiency. In erosion-prone farmlands, UAVs are used to plant cover crops that help stabilize soil, reduce runoff, and improve soil fertility. Additionally, drones are being deployed for reforestation efforts, accelerating tree planting in deforested areas. Beyond planting, UAVs are also being explored as a pollination solution, particularly in response to the declining bee population. Pollination is crucial for fruit and seed production, and with fewer natural pollinators, UAVs equipped with pollen dispensers can artificially pollinate crops. These drones can target specific flowering crops, ensuring timely and efficient pollen transfer, especially in orchards and greenhouse farms [78]. A successful example of UAV-assisted pollination comes from Japan, where researchers have developed drones capable of pollinating fruit trees. These drones mimic bees by carrying pollen and distributing it among flowers, ensuring stable crop production. As technology advances, UAVs will play an increasingly important role in sustainable agriculture, improving efficiency while conserving natural resources.

10.7.1 UAVs for Seeding and Planting

UAV-based seeding technology is revolutionizing modern agriculture by allowing farmers to plant crops in challenging terrains where traditional methods are impractical. Drones equipped with advanced seed dispensers can cover vast landscapes quickly and uniformly, ensuring precise seed distribution. This technology is particularly beneficial for planting cover crops, which help prevent soil erosion, enhance soil fertility, and improve moisture retention. Additionally, UAVs play a crucial role in reforestation and afforestation efforts, enabling large-scale tree planting in degraded and deforested areas [79].

One of the most significant advantages of UAV-based seeding is its ability to reduce labor costs. Traditional planting methods require extensive manpower and time, but UAVs automate the process, significantly cutting down on expenses while improving efficiency. Furthermore, UAVs offer precision planting, ensuring seeds are placed at optimal distances, which enhances germination rates, crop uniformity, and overall yield potential [80]. This level of accuracy helps maximize resource use while minimizing seed waste.

A compelling example of UAV-based seeding comes from Brazil, where agricultural drones have been deployed to disperse cover crops in erosion-prone areas. These crops help stabilize the soil, reduce runoff, and improve organic content, leading to healthier farmland and increased agricultural productivity. By integrating UAV technology, farmers can enhance sustainability, boost efficiency, and improve long-term soil health, making precision seeding a game-changer in modern agriculture [81].

10.7.2 UAVs for Pollination Assistance

Pollination is an essential process for crop reproduction and fruit formation, yet the decline in bee populations due to climate change, pesticide exposure, habitat loss, and disease has become a significant challenge for global agriculture. Without adequate pollination, many fruit, vegetable, and seed crops experience reduced yields, threatening food security. To counteract this issue, researchers are developing UAV-based pollination systems that can artificially transfer pollen between flowers, ensuring successful fertilization and stable crop production. Equipped with pollen dispensers, UAVs can carry and distribute pollen with exceptional precision, targeting specific crops at their peak flowering stage. This technology is particularly beneficial for greenhouses, orchards, and farms where natural pollinators are scarce [82]. Unlike traditional hand-pollination methods, UAVs provide a controlled and efficient alternative, allowing farmers to maintain consistent yields even in areas experiencing pollinator decline. Additionally, UAV-assisted pollination reduces labor costs and time, making it a scalable and practical solution for commercial agriculture.

One of the most promising examples of UAV-assisted pollination comes from Japan, where researchers have developed drones equipped with tiny pollen-coated brushes or automated dispensers. These UAVs hover over flowers, carefully transferring pollen in a way that mimics the natural role of bees. Early trials have demonstrated high pollination success rates, suggesting that UAVs could become a viable, large-scale solution for precision pollination in the near future. As technology advances, UAV pollination could play a crucial role in mitigating the effects of declining bee populations while enhancing crop productivity and sustainability [83].

UAVs are revolutionizing agricultural seeding, planting, and pollination by making these processes more efficient, precise, and less labor-intensive. In seeding and planting, they enable fast and uniform seed distribution, reducing costs and improving productivity. In pollination, they offer a potential solution to bee population decline, ensuring fruit and seed production stability. As technology continues to advance, UAVs will play an even greater role in sustainable and precision agriculture, helping farmers increase yields while conserving resources [84].

10.8 Benefits of UAV-Crop Interaction

The integration of Unmanned Aerial Vehicles (UAVs) in agriculture offers transformative benefits, enhancing efficiency, cost-effectiveness, and sustainability. By providing real-time data and precision farming capabilities, UAVs help farmers make informed decisions that improve overall crop health, optimize resource use, and reduce environmental impact [85].

10.8.1 Increased Efficiency and Productivity

One of the most significant advantages of UAVs in agriculture is their ability to cover large areas quickly while collecting high-resolution aerial data. With advanced imaging and remote sensing technologies, UAVs provide real-time insights into crop conditions, identifying issues such as pest infestations, nutrient deficiencies, or water stress. This allows for timely interventions, preventing potential yield losses and ensuring optimal crop performance [78]. The ability to rapidly assess vast fields also reduces the need for manual field scouting, saving both time and labor.

For example, large-scale farms in the U.S. and China use UAVs for precision spraying and crop monitoring, reducing labor costs while increasing overall efficiency. Farmers can assess vast fields in minutes rather than days, ensuring that resources are directed exactly where needed [86].

10.8.2 Reduced Resource Use and Costs

UAVs play a critical role in precision agriculture, helping farmers minimize the overuse of water, fertilizers, and pesticides. By analyzing crop health using multispectral and thermal imaging, UAVs enable targeted application of inputs, ensuring that resources are used only where needed. This reduces input costs while maximizing yield efficiency. Additionally, UAVs eliminate the need for expensive and time-consuming manual inspections, cutting operational costs significantly [87].

- **Water Management**: UAVs can assess soil moisture levels using thermal imaging, allowing farmers to adjust irrigation schedules based on actual crop needs. This reduces water waste and ensures that only the necessary amount is applied.
- **Fertilizer Application**: UAVs analyze crop nutrient levels through multispectral imaging, enabling variable rate application (VRA) of fertilizers. This approach enhances nutrient absorption while preventing overuse, which can lead to soil degradation and runoff contamination.
- **Pest and Weed Control**: Traditional pesticide spraying methods often result in overuse and chemical drift, affecting non-target areas. UAVs, however, apply chemicals with pinpoint accuracy, reducing the amount of pesticide required and limiting environmental impact.

By reducing input costs while maintaining or increasing yields, UAVs improve profitability for farmers. For example, in India, UAV-assisted spraying has cut pesticide use by up to 30%, significantly lowering expenses while improving crop health [88].

10.8.3 Enhanced Sustainability

Sustainability is a growing concern in modern agriculture, and UAVs contribute to environmentally friendly farming by optimizing resource use and reducing chemical runoff. By enabling precise spraying of fertilizers and pesticides, UAVs help prevent soil and water contamination, protecting ecosystems. Additionally, UAVs support sustainable irrigation management by detecting water-stressed areas, ensuring that water is used efficiently [89]. Through these practices, UAVs promote long-term soil health, biodiversity conservation, and reduced carbon footprints, making agriculture more resilient and sustainable.

Minimizing Chemical Runoff: Precision spraying ensures that pesticides and fertilizers are applied only where needed, reducing the risk of soil and water contamination. This protects nearby ecosystems, including rivers, lakes, and groundwater sources [90].

Reducing Carbon Footprint: Traditional farming methods rely on heavy machinery such as tractors and aircraft for spraying, which consume large amounts of fuel. UAVs, being electric-powered, offer a low-emission alternative, contributing to climate-friendly farming. Protecting Biodiversity: Excessive pesticide use harms beneficial insects like bees and butterflies, which play a crucial role in pollination. By reducing unnecessary chemical application, UAVs help preserve pollinator populations and maintain biodiversity in agricultural landscapes [91]. As UAV technology continues to evolve, its role in precision agriculture will become even more significant, leading to higher yields, lower costs, and improved environmental stewardship.

10.9 Future Prospects

The future of UAVs in agriculture looks promising, with advancements in Artificial Intelligence (AI), the Internet of Things (IoT), and big data analytics set to revolutionize farming practices. One significant development is the introduction of autonomous UAV swarms, where multiple drones operate simultaneously to perform various tasks, such as monitoring crop health, spraying pesticides, and planting seeds with minimal human intervention. Additionally, AI-driven real-time data processing will enhance decision-making by instantly analyzing field conditions and providing actionable insights, allowing farmers to respond to issues such as pest infestations or irrigation needs without delay [92].

Another key innovation is the improvement in battery technology, enabling UAVs to achieve longer flight durations and cover larger agricultural areas in a single mission. This advancement is particularly beneficial for large-scale farms, where frequent drone deployment is necessary for precision agriculture. Moreover, the integration of UAVs with robotic systems is expected to transform automated farming, with drones working in coordination with ground-based robots for tasks such as harvesting, soil analysis, and autonomous weeding. These advancements will lead to more efficient, cost-effective, and sustainable farming methods, ensuring increased agricultural productivity while reducing environmental impact [93]. Future developments may include:

10.9.1 Autonomous UAV Swarms

One of the most promising future developments is the deployment of autonomous UAV swarms—groups of AI-coordinated drones that can work together simultaneously. Unlike individual UAVs, which operate separately, swarms of drones can divide tasks efficiently, such as:

- Simultaneous crop health monitoring, spraying, and mapping, allowing for faster and more comprehensive field coverage.
- Automated decision-making, where drones share information and adjust operations in real-time based on environmental conditions.
- Faster response to crop threats like pest infestations, nutrient deficiencies, and water stress, reducing potential yield losses.

For example, researchers are developing AI-driven swarm technology where UAVs communicate and collaborate to spray crops more efficiently while minimizing chemical waste. These systems could eliminate the need for manual drone operation, further increasing productivity [94].

10.9.2 Real-Time Data Processing

As UAVs become more sophisticated, AI-driven analytics will play a key role in real-time data processing and decision-making. Currently, UAVs collect vast amounts of multispectral, thermal, and LiDAR data, but processing this information can take hours or even days [95]. Future advancements will enable:

- Onboard AI systems that analyze crop health instantly, providing real-time insights for farmers.
- Predictive analytics, allowing UAVs to forecast yield, detect diseases before symptoms appear, and recommend corrective actions.
- Integration with smart farm management platforms, where UAV data can be combined with IoT sensors to create automated irrigation, fertilization, and pest control plans.

For instance, AI-powered UAVs could autonomously detect an outbreak of fungal disease in a specific crop section and immediately deploy targeted spraying, preventing its spread without human intervention [96].

10.9.3 Improved Battery Technology

One of the biggest current limitations of UAVs is limited flight time due to battery constraints. Most agricultural drones can only fly for 20–40 min per charge, which restricts their ability to cover large farmland areas [97]. However, future advancements in battery technology and alternative power sources will enable:

- Longer flight durations, allowing UAVs to cover vast agricultural fields without frequent recharging.
- Solar-powered drones, which can extend operational time by harvesting solar energy while in flight.
- Wireless charging stations in fields, enabling UAVs to recharge automatically without needing human intervention.

For example, hydrogen fuel cells are being explored as a lightweight and energy-dense alternative to lithium batteries, potentially allowing UAVs to fly for several hours without recharging [98].

10.9.4 Integration with Robotics

The future of UAVs in agriculture is not just limited to aerial applications. Integration with ground-based robotic systems will create fully automated farms, where drones and robots work together for precision farming [99]. Potential developments include:

- UAVs identifying crop diseases or nutrient deficiencies and sending signals to robotic tractors for immediate soil treatment.
- Drones monitoring livestock while robotic feeders deliver food based on animal needs.
- AI-driven robotic arms working alongside UAVs for automated fruit picking and harvesting.

For example, in Japan, researchers are developing UAV-assisted robotic harvesting systems that detect ripened fruits from the air and signal ground robots to pick them, reducing labor costs and improving efficiency [100].

The integration of UAVs with AI, IoT, big data, and robotics is expected to revolutionize agriculture by making farming more autonomous, data-driven, and efficient. This will increase food production, reduce resource waste, and enhance environmental sustainability. UAVs have transformed agricultural practices by improving crop monitoring, spraying, soil analysis, and yield prediction [101]. Despite challenges, continuous technological advancements are making UAVs more efficient and accessible. The future of UAVs in agriculture is promising, with innovations expected to further optimize farming operations and enhance global food security.

10.10 Summary

UAVs have revolutionized agricultural practices by improving crop monitoring, spraying, soil analysis, and yield prediction. Despite challenges, continuous technological advancements are making UAVs more efficient and accessible. The future of UAVs in agriculture is promising, with innovations expected to further optimize farming operations and enhance global food security. UAVs offer numerous benefits, including precision agriculture, pest detection, irrigation management, and yield estimation. They provide real-time insights into crop health, soil conditions, and environmental factors, enabling farmers to make data-driven decisions that enhance productivity and sustainability. As drone technology continues to evolve, its role in sustainable and smart farming will only expand, driving the future of agricultural innovation. Remote sensing technologies, such as NDVI, EVI, thermal imaging, and UAV monitoring, provide valuable insights for precision agriculture. These tools help optimize irrigation, detect stress early, prevent disease outbreaks, and enhance overall crop productivity. By integrating these technologies with AI and big data analytics, agricultural decision-making becomes more efficient and sustainable. UAV-based spraying represents a transformative advancement in agriculture, offering precision, efficiency, and sustainability. By enabling targeted application, reducing chemical use, and improving safety, drones help farmers optimize crop management while minimizing environmental impact.

As technology continues to evolve, UAVs will play an increasingly vital role in addressing global challenges like food security, resource scarcity, and climate change. Through AI-driven optimization, advanced spray calibration, or compliance with legal regulations, UAVs are paving the way for a smarter, safer, and more sustainable future in agriculture. The integration of UAVs, remote sensing, and AI-powered analytics in yield estimation and forecasting is revolutionizing agriculture by providing real-time, data-driven insights into crop performance. These technologies enable higher accuracy in predicting yields, early stress detection, precision input management, and better market planning, leading to increased profitability and sustainability in farming. As AI and UAV technologies continue to advance, their role in global food security and smart agriculture will become even more significant. UAV-based yield estimation and forecasting provide predictive insights, cost savings, and efficiency improvements, making agriculture more profitable and sustainable. With AI-driven analytics and precision data collection, farmers can reduce input costs, maximize yields, and improve market planning, ensuring higher productivity and lower environmental impact. Yield prediction using UAVs offers numerous benefits for farmers, including predictive analysis, cost reduction, and optimized harvesting.

Multispectral and thermal imaging are game-changers in precision agriculture, allowing for early stress detection, optimized irrigation, improved resource management, and enhanced crop protection. When combined with UAVs and AI-driven analytics, these technologies enable smarter, data-driven farming, leading to higher yields, reduced input costs, and sustainable agricultural practices. The adoption of multispectral and thermal imaging in agriculture has led to significant improvements in crop monitoring, resource management, and sustainability. By leveraging these advanced technologies, farmers can make data-driven decisions that enhance productivity, conserve resources, and reduce environmental impact. As UAVs and remote sensing technologies continue to evolve, the future of precision agriculture will be driven by real-time imaging, artificial intelligence, and automation, paving the way for smarter, more efficient, and climate-resilient farming. The integration of UAVs with AI, IoT, big data, and robotics will revolutionize agriculture, making farming more autonomous, data-driven, and efficient. These advancements will increase food production, reduce resource waste, and enhance environmental sustainability, ensuring farmers can meet growing global food demand efficiently.

References

1. Maesano M, Khoury S, Nakhle F et al (2020) UAV-based LiDAR for high-throughput determination of plant height and above-ground biomass of the bioenergy grass Arundo donax. Remote Sensing 12:3464
2. Wei H-E, Grafton M, Bretherton M et al (2022) Evaluation of the use of UAV-derived vegetation indices and environmental variables for grapevine water status monitoring based on machine learning algorithms and SHAP analysis. Remote Sensing 14:5918
3. Ling J, Zeng Z, Shi Q et al (2023) Estimating winter wheat LAI using hyperspectral UAV data and an iterative hybrid method. IEEE J Sel Top Appl Earth Observations Remote Sensing 16:8782–8794
4. Li J, Zhang W, Ren J et al (2024) A multi-area task path-planning algorithm for agricultural drones based on improved double deep Q-learning net. Agriculture 14:1294
5. Rahman MFF, Fan S, Zhang Y, Chen L (2021) A comparative study on application of unmanned aerial vehicle systems in agriculture. Agriculture 11:22
6. Roslim MHM, Juraimi AS, Che'Ya NN et al (2021) Using remote sensing and an unmanned aerial system for weed management in agricultural crops: a review. Agronomy 11:1809
7. Omia E, Bae H, Park E et al (2023) Remote sensing in field crop monitoring: a comprehensive review of sensor systems, data analyses and recent advances. Remote Sensing 15:354
8. Khanal S, Kc K, Fulton JP et al (2020) Remote sensing in agriculture—accomplishments, limitations, and opportunities. Remote sensing 12:3783
9. Awais M, Li W, Cheema M et al (2022) UAV-based remote sensing in plant stress imagine using high-resolution thermal sensor for digital agriculture practices: a meta-review. Int J Environ Sci Technol 1–18
10. Bhandari M (2016) Use of infrared thermal imaging for estimating canopy temperature in wheat and maize
11. Pradawet C, Khongdee N, Pansak W et al (2023) Thermal imaging for assessment of maize water stress and yield prediction under drought conditions. J Agron Crop Sci 209:56–70
12. Abdullah HM, Mohana NT, Khan BM et al (2023) Present and future scopes and challenges of plant pest and disease (P&D) monitoring: remote sensing, image processing, and artificial intelligence perspectives. Remote Sens Appl Soc Environ 32:100996
13. Elangovan MM, Nigam R, Sankarganesh E, Srivastava G (2024) Plant diseases: diagnosis, management, and control. Academic Guru Publishing House
14. Neupane K, Baysal-Gurel F (2021) Automatic identification and monitoring of plant diseases using unmanned aerial vehicles: a review. Remote Sensing 13:3841
15. Manfreda S, Eyal BD (2023) Unmanned aerial systems for monitoring soil, vegetation, and riverine environments
16. Ronchetti G (2020) UAV surveys for crop monitoring and management in precision agriculture

17. Taseer A, Han X (2024) Advancements in variable rate spraying for precise spray requirements in precision agriculture using unmanned aerial spraying systems: a review. Comput Electron Agric 219:108841
18. Wang L, Huang X, Li W et al (2022) Progress in agricultural unmanned aerial vehicles (UAVs) applied in China and prospects for Poland. Agriculture 12:397
19. Daniel Lawrence I, Pavitra ARR, Karu R, Saravanan M (2025) An advanced application of UAV—drone technologies in precision agriculture for seed dropping, fertilizers and pesticides spraying and field monitoring. In: Computer vision in smart agriculture and crop management. pp 139–157
20. Das A, Kadawla K, Nath H et al (2024) Drone-based intelligent spraying of pesticides: current challenges and its future prospects. In: Applications of computer vision and drone technology in agriculture 4.0. Springer, pp 199–223
21. Guebsi R, Mami S, Chokmani K (2024) Drones in precision agriculture: a comprehensive review of applications, technologies, and challenges. Drones 8:686
22. Nahiyoon SA, Ren Z, Wei P et al (2024) Recent development trends in plant protection UAVs: a journey from conventional practices to cutting-edge technologies—a comprehensive review. Drones 8:457
23. Toscano F, Fiorentino C, Santana LS, Magalhães RR (2025) Recent developments and future prospects in the integration of machine learning in mechanized systems for autonomous spraying: a brief review
24. Mittu B, Chaturvedi A, Singh M (2025) Use of pesticide management using AI. In: Artificial intelligence in the food industry. CRC Press, pp 260–276
25. Nag P, Gite L, Nag P, Gite L (2020) Pesticides and chemical toxicity—challenges in farming. In: Human-centered agriculture: ergonomics and human factors applied. pp 239–271
26. Dhananjayan V, Jayakumar S, Ravichandran B (2019) Conventional methods of pesticide application in agricultural field and fate of the pesticides in the environment and human health. In: Controlled release of pesticides for sustainable agriculture. Springer, pp 1–39
27. Elete TY, Odujobi O, Nwulu EO, Onyeke FO (2024) Safety-first innovations: advancing HSE standards in coating and painting operations. Int J Eng Res Dev 20:290–298
28. Fenton K, Simske S, Luu J (2023) Mitigation of chemical reporting liabilities through systematic modernization of chemical hazard and safety data management systems. ACS Omega 8:4928–4936
29. Savas T (2024) Comparative evaluation of current regulations regarding agricultural unmanned aerial vehicles used for spraying purposes in Türkiye. Int J Sustain Aviat 10:24–42
30. Rauhala A, Tuomela A, Leviäkangas P (2023) An overview of unmanned aircraft systems (UAS) governance and regulatory frameworks in the European Union (EU). In: Unmanned aerial systems in agriculture. pp 269–285
31. Getahun S, Kefale H, Gelaye Y (2024) Application of precision agriculture technologies for sustainable crop production and environmental sustainability: a systematic review. Sci World J 2024:2126734
32. Mamatha B, Mudigiri C, Ramesh G et al (2024) Enhancing soil health and fertility management for sustainable agriculture: a review. Asian J Soil Sci Plant Nutr 10:182–190
33. Raza I, Zubair M, Zaib M et al (2023) Precision nutrient application techniques to improve soil fertility and crop yield: a review with future prospect. Int Res J Educ Tecnol
34. Näsi R, Mikkola H, Honkavaara E et al (2023) Can basic soil quality indicators and topography explain the spatial variability in agricultural fields observed from drone orthomosaics? Agronomy 13:669
35. Seyar MH, Ahamed T (2024) Optimization of soil-based irrigation scheduling through the integration of machine learning, remote sensing, and soil moisture sensor technology. In: IoT and AI in agriculture: smart automation systems for increasing agricultural productivity to achieve SDGs and society 5.0. Springer, pp 275–299
36. Liang Z, Liu X, Xiong J, Xiao J (2020) Water allocation and integrative management of precision irrigation: a systematic review. Water 12:3135

37. Vishwakarma DK, Kumar R, Abed SA et al (2023) Modeling of soil moisture movement and wetting behavior under point-source trickle irrigation. Sci Rep 13:14981
38. Kamal A, Mian I, Akbar W et al (2023) Effects of soil depth and altitude on soil texture and soil quality index. Appl Ecol Environ Res 21
39. Ali M, Khan F, Khan I et al (2018) Soil and water conservation practices in district Swabi, KP, Pakistan. Adv Crop Sci Tech 6:2
40. Kamal A, Mian IA, Tariq M (2023) Sustainable maize farming: organic soil amendments boost maize production and soil quality. Plant Bull 2:41–58
41. Khan S, Shah Z, Mian IA et al (2020) Soil fertility, N_2 fixation and yield of chickpea as influenced by long-term biochar application under mung–chickpea cropping system. Sustainability 12:9008
42. Vlotman W, Smedema L, Rycroft D (2020) Modern land drainage: planning, design and management of agricultural drainage systems. CRC Press
43. Hagage M, Abdulaziz AM, Elbeih SF, Hewaidy AGA (2024) Monitoring soil salinization and waterlogging in the northeastern Nile Delta linked to shallow saline groundwater and irrigation water quality. Sci Rep 14:27838
44. Qamar R, Shafaat S, Javeed HMR (2023) Management of crops in water-logged soil. In: Disaster risk reduction in agriculture. Springer, pp 233–275
45. Kumar N, Kumar A, Jeena N et al (2020) Factors influencing soil ecosystem and agricultural productivity at higher altitudes. In: Microbiological advancements for higher altitude agro-ecosystems and sustainability. pp 55–70
46. Bedadi B, Beyene S, Erkossa T, Fekadu E (2023) Soil management. In: The soils of Ethiopia. Springer, pp 193–234
47. Ali W, Ali M, Kamal A et al (2019) Maize yield response under various phosphorus sources and their ratios. Eur J Exp Biol 9:5
48. Selim MM (2020) Introduction to the integrated nutrient management strategies and their contribution to yield and soil properties. Int J Agron 2020
49. Imran I, Khan AA, Kamal A et al (2024) Utilization and characterization of microbes for heavy metal remediation. Pol J Environ Stud
50. Rebelo C, Nascimento J (2021) Measurement of soil tillage using UAV high-resolution 3D data. Remote Sensing 13:4336
51. Kannan P, Abishek J, Pazhanivelan S, Firnass MMRA (2025) Farming from above: the role of UAVs in modern agricultural practices. In: Digital farming and smart agriculture for sustainable future. Chapman and Hall/CRC, pp 235–254
52. Gokool S, Mahomed M, Kunz R et al (2023) Crop monitoring in smallholder farms using unmanned aerial vehicles to facilitate precision agriculture practices: a scoping review and bibliometric analysis. Sustainability 15:3557
53. Vahidi M, Shafian S, Frame WH (2025) Precision soil moisture monitoring through drone-based hyperspectral imaging and PCA-driven machine learning. Sensors 25:782
54. Chitsiko RJ, Mutanga O, Dube T, Kutywayo D (2022) Review of current models and approaches used for maize crop yield forecasting in sub-Saharan Africa and their potential use in early warning systems. Phys Chem Earth Parts A/B/C 127:103199
55. Bharadiya JP, Tzenios NT, Reddy M (2023) Forecasting of crop yield using remote sensing data, agrarian factors and machine learning approaches. J Eng Res Rep 24:29–44
56. Paudel D, Boogaard H, de Wit A et al (2021) Machine learning for large-scale crop yield forecasting. Agric Syst 187:103016
57. Rashid M, Bari BS, Yusup Y et al (2021) A comprehensive review of crop yield prediction using machine learning approaches with special emphasis on palm oil yield prediction. IEEE Access 9:63406–63439
58. Elbasi E, Zaki C, Topcu AE et al (2023) Crop prediction model using machine learning algorithms. Appl Sci 13:9288
59. Cen H, Wan L (2023) Crop yield estimation and prediction. In: Encyclopedia of digital agricultural technologies. Springer, pp 214–226

60. Malashin I, Tynchenko V, Gantimurov A et al (2024) Predicting sustainable crop yields: deep learning and explainable AI tools. Sustainability 16:9437
61. Mohyuddin G, Khan MA, Haseeb A et al (2024) Evaluation of machine learning approaches for precision farming in smart agriculture system—a comprehensive review. IEEE Access
62. Mishra S, Mishra D, Santra GH (2016) Applications of machine learning techniques in agricultural crop production: a review paper. Indian J Sci Technol 9:1–14
63. Barnhart IH (2024) Use of artificial intelligence to locate and treat weeds in Midwestern United States corn (Zea mays) and soybean (Glycine max) cropping systems. Kansas State University
64. Gaither EL (2023) An assessment of UAV technology and Shattercane population dynamics to improve pest management
65. Ming R, Jiang R, Luo H et al (2023) Comparative analysis of different uav swarm control methods on unmanned farms. Agronomy 13:2499
66. Yu F, Bai J, Jin Z et al (2022) Research on precise fertilization method of rice tillering stage based on UAV hyperspectral remote sensing prescription map. Agronomy 12:2893
67. Yang C, Yang G, Wang H et al (2025) Identifying key traits for screening high-yield Soybean varieties by combining UAV-based and field phenotyping. Remote Sensing 17:690
68. Yang G, Li Y, Yuan S et al (2024) Enhancing direct-seeded rice yield prediction using UAV-derived features acquired during the reproductive phase. Precision Agric 25:1014–1037
69. Zhu H, Lu X, Zhang K et al (2023) Optimum basic seedling density and yield and quality characteristics of unmanned aerial seeding rice. Agronomy 13:1980
70. Trybała P, Morelli L, Remondino F et al (2024) Under-canopy drone 3D surveys for wild fruit hotspot mapping. Drones 8:577
71. Adinarayana S, Raju MG, Srirangam DP et al (2024) Enhancing resource management in precision farming through AI-based irrigation optimization. In: How machine learning is innovating today's world: a concise technical guide. pp 221–251
72. Padhiary M, Sethi LN, Kumar A (2024) Enhancing hill farming efficiency using unmanned agricultural vehicles: a comprehensive review. Trans Indian Natl Acad Eng 9:253–268
73. Quebrajo L, Pérez-Ruiz M, Pérez-Urrestarazu L et al (2018) Linking thermal imaging and soil remote sensing to enhance irrigation management of sugar beet. Biosys Eng 165:77–87
74. Pineda M, Barón M, Pérez-Bueno M-L (2020) Thermal imaging for plant stress detection and phenotyping. Remote Sensing 13:68
75. Gerhards M, Schlerf M, Mallick K, Udelhoven T (2019) Challenges and future perspectives of multi-/Hyperspectral thermal infrared remote sensing for crop water-stress detection: a review. Remote Sensing 11:1240
76. Xing Y, Wang X (2024) Precision agriculture and water conservation strategies for sustainable crop production in arid regions. Plants 13:3184
77. Boursianis AD, Papadopoulou MS, Diamantoulakis P et al (2022) Internet of things (IoT) and agricultural unmanned aerial vehicles (UAVs) in smart farming: a comprehensive review. Internet Things 18:100187
78. Velusamy P, Rajendran S, Mahendran RK et al (2021) Unmanned aerial vehicles (UAV) in precision agriculture: applications and challenges. Energies 15:217
79. Librán-Embid F, Klaus F, Tscharntke T, Grass I (2020) Unmanned aerial vehicles for biodiversity-friendly agricultural landscapes—a systematic review. Sci Total Environ 732:139204
80. Jang G, Kim J, Yu J-K et al (2020) Cost-effective unmanned aerial vehicle (UAV) platform for field plant breeding application. Remote Sensing 12:998
81. Mohan M, Richardson G, Gopan G et al (2021) UAV-supported forest regeneration: current trends, challenges and implications. Remote Sensing 13:2596
82. Manzoor SH, Kabir MH, Zhang Z (2023) UAV-based apple flowers pollination system. In: Towards unmanned apple orchard production cycle: recent new technologies. Springer, pp 211–236
83. Wu S, Liu J, Lei X et al (2022) Research progress on efficient pollination technology of crops. Agronomy 12:2872

84. Wu P, Lei X, Zeng J et al (2024) Research progress in mechanized and intelligentized pollination technologies for fruit and vegetable crops. Int J Agric Biol Eng 17:11–21
85. Alyafei MA, Al Dakheel A, Almoosa M, Ahmed ZF (2022) Innovative and effective spray method for artificial pollination of date palm using drone. HortScience 57:1298–1305
86. Olson D, Anderson J (2021) Review on unmanned aerial vehicles, remote sensors, imagery processing, and their applications in agriculture. Agron J 113:971–992
87. Mohsan SAH, Khan MA, Noor F et al (2022) Towards the unmanned aerial vehicles (UAVs): a comprehensive review. Drones 6:147
88. Xia X, Fattah SMM, Babar MA (2023) A survey on UAV-enabled edge computing: resource management perspective. ACM Comput Surv 56:1–36
89. Xiao Z, Chen Y, Jiang H et al (2022) Resource management in UAV-assisted MEC: state-of-the-art and open challenges. Wireless Netw 28:3305–3322
90. Sahni RK, Kumar SP, Thorat D et al (2024) Drone spraying system for efficient agrochemical application in precision agriculture. In: Applications of computer vision and drone technology in agriculture 4.0. Springer, pp 225–244
91. Allmendinger A, Spaeth M, Saile M et al (2022) Precision chemical weed management strategies: a review and a design of a new CNN-based modular spot sprayer. Agronomy 12:1620
92. Bhat SA, Huang N-F (2021) Big data and ai revolution in precision agriculture: survey and challenges. IEEE Access 9:110209–110222
93. Majumdar P, Mitra S, Bhattacharya D, Bhushan B (2024) Enhancing sustainable 5G powered agriculture 4.0: summary of low power connectivity, internet of UAV things, AI solutions and research trends. In: Multimedia tools and applications. pp 1–45
94. El Alaoui M, Amraoui KE, Masmoudi L et al (2024) Unleashing the potential of IoT, artificial intelligence, and UAVs in contemporary agriculture: a comprehensive review. J Terrramech 115:100986
95. Sekhon SS, Kumar V, Patel A, Parmar BS (2025) Technological advances in smart and sustainable agriculture: the role of Internet of Things, artificial intelligence, big data analysis, machine learning and deep learning. In: Food and Industry 50: transforming the food system for a sustainable future. pp 61–71
96. Thilakarathne NN, Bakar MSA, Abas PE, Yassin H (2025) Internet of things enabled smart agriculture: current status, latest advancements, challenges and countermeasures. Heliyon 11
97. Mahobiya C, Iyer S, Verma M et al (2024) Navigating the future: unmanned aerial systems in IoT paradigms. In: Unmanned aircraft systems. pp 355–385
98. Khadatkar A, Mehta C, Sawant C (2022) Application of robotics in changing the future of agriculture
99. Lytridis C, Kaburlasos VG, Pachidis T et al (2021) An overview of cooperative robotics in agriculture. Agronomy 11:1818
100. Oliveira LF, Moreira AP, Silva MF (2021) Advances in agriculture robotics: a state-of-the-art review and challenges ahead. Robotics 10:52
101. Vc SS, Hareendran A, Albaaji GF (2024) Precision farming for sustainability: an agricultural intelligence model. Comput Electron Agric 226:109386

Chapter 11
UAV and Irrigation and Water Resource Management

Abstract Irrigation and water resource management are crucial for sustainable agriculture, ensuring efficient water use and crop productivity. With increasing water scarcity and climate variability, advanced technologies like UAVs, AI-driven analytics, and smart irrigation systems have revolutionized water management practices. These innovations enhance irrigation efficiency, reduce water wastage, and promote conservation efforts. By integrating data-driven approaches, farmers can optimize water distribution, minimize resource depletion, and improve drought resilience. UAVs provide real-time data on soil moisture, crop health, and water distribution, helping identify drought-prone areas and optimize irrigation efficiency. They also assist in irrigation system planning, analyzing terrain, optimizing water flow, and identifying inefficiencies in irrigation networks. By integrating UAV technology with smart irrigation systems, agriculture can achieve greater efficiency, reduce environmental impact, and ensure water security for the future. The future of UAV-based irrigation lies in AI, IoT, and climate adaptation strategies. AI-driven smart irrigation systems analyze real-time data to optimize water distribution, preventing over- or under-irrigation. IoT-based irrigation networks integrate sensors and automated controllers, allowing remote monitoring and control of irrigation systems. UAVs provide aerial imagery to assess soil moisture, detect plant stress, and optimize water application. However, overcoming cost, regulatory, and technical barriers is crucial for widespread adoption.

Keywords Irrigation · Water resource management · UAV technology · Smart irrigation · Precision agriculture · Water conservation · Groundwater monitoring · AI in agriculture · Climate resilience · Sustainable farming

11.1 Introduction

Water is a vital resource for agriculture, industry, and human survival, making its effective management crucial for sustainable development. Irrigation and Water Resource Management focuses on the efficient use, conservation, and distribution of

© The Author(s), under exclusive license to Springer Nature Singapore Pte Ltd. 2025
J. Li, *UAV Aerodynamics and Crop Interaction*, Smart Agriculture 13,
https://doi.org/10.1007/978-981-96-8402-1_11

water to meet the growing demands of agriculture, domestic use, and industry while ensuring environmental sustainability. Irrigation refers to the artificial application of water to crops to supplement natural rainfall, ensuring consistent agricultural productivity. It is particularly important in arid and semi-arid regions where rainfall is insufficient [1]. Different irrigation techniques are used to optimize water use, including surface irrigation, drip irrigation, sprinkler irrigation, and subsurface irrigation. Each method has its advantages and is selected based on soil type, crop requirements, and water availability. Water Resource Management involves the sustainable planning and utilization of water to balance supply and demand while considering environmental conservation. It includes key practices such as watershed management, which focuses on protecting water sources; water conservation, which minimizes wastage and promotes efficient use; flood control and drainage, which help manage excess water and prevent disasters; and groundwater management, which ensures the sustainable use and recharge of underground water sources [2]. Climate change, population growth, and industrial expansion have intensified the need for effective water management strategies to prevent water scarcity and degradation. Proper irrigation and water resource management are essential for maintaining agricultural productivity, ensuring food security, and supporting economic stability. Sustainable water use not only benefits present generations but also preserves resources for the future. Therefore, implementing advanced irrigation techniques, adopting water conservation strategies, and improving policy frameworks are necessary steps toward a balanced and efficient water management system that meets both human and environmental needs [3].

11.2 UAVs for Irrigation Efficiency

Unmanned Aerial Vehicles (UAVs), commonly known as drones, are revolutionizing modern agriculture by enhancing irrigation efficiency. These advanced aerial systems provide real-time data and precision monitoring, allowing farmers to optimize water use, reduce wastage, and improve crop health. One of the key applications of UAVs in irrigation is remote sensing, where drones equipped with multispectral and thermal cameras detect variations in soil moisture and plant water stress. By analyzing these images, farmers can identify dry areas that require more water and adjust irrigation schedules accordingly, leading to precision irrigation [4]. This targeted approach minimizes overwatering and prevents water runoff, conserving water resources and reducing costs. Additionally, UAVs facilitate automated irrigation system monitoring by inspecting pipelines, sprinklers, and irrigation channels. They can quickly identify leaks, blockages, or malfunctions, preventing water loss and ensuring efficient distribution. Some advanced UAV systems are also integrated with artificial intelligence (AI) to analyze historical weather and soil data, helping farmers make data-driven irrigation decisions. Furthermore, drones contribute to sustainable water management by supporting smart irrigation technologies such as Variable Rate Irrigation (VRI) [5]. By providing accurate field data, UAVs help in customizing irrigation applications based on specific field zones, ensuring optimal water distribution and maximizing crop yield. The use of UAVs in irrigation not only improves efficiency but also promotes environmental sustainability by reducing water wastage and minimizing the risk of soil degradation caused by excessive watering. As drone technology continues to evolve, its integration with precision agriculture is expected to play a crucial role in addressing global water challenges and enhancing food security.

11.2.1 Identifying Drought-Prone Areas

Unmanned Aerial Vehicles (UAVs) play a crucial role in identifying drought-prone areas by providing high-resolution imagery and real-time data. Equipped with multispectral and thermal sensors, drones can assess soil moisture levels, detect vegetation stress, and map areas experiencing water deficiency. By analyzing these aerial images, farmers and water resource managers can predict drought conditions, allocate water efficiently, and implement early mitigation strategies to prevent crop loss and land degradation [6].

How UAVs Identify Drought-Prone Areas

UAVs (drones) are equipped with advanced imaging technologies that allow farmers and researchers to detect early signs of drought stress in agricultural fields. Two key technologies—multispectral imaging and thermal imaging—are widely used for this purpose.

11.2.1.1 Multispectral Imaging for Drought Detection

Multispectral cameras capture images in multiple bands of light, including visible (RGB), near-infrared (NIR), and shortwave infrared (SWIR), which the human eye cannot see. These images provide detailed insights into plant and soil conditions, helping to identify drought-prone areas [7].

How Multispectral Imaging Works:

Detects variations in vegetation health:

- Healthy plants reflect more near-infrared (NIR) light and absorb more red light.
- Stressed or drought-affected plants reflect less NIR, indicating reduced chlorophyll content and poor health.

Generates Vegetation Index Maps:

- **Normalized Difference Vegetation Index (NDVI)**: Measures plant vigor and helps detect drought stress before visible symptoms appear.
- **Normalized Difference Water Index (NDWI)**: Identifies areas with low water content in soil and vegetation.

Early Warning System:

- Multispectral imaging helps farmers identify which parts of the field are drying out, allowing for targeted irrigation and water conservation strategies.

11.2.1.2 Thermal Imaging for Soil and Plant Moisture Levels

Thermal imaging cameras detect temperature variations in the field, helping identify water stress in plants and soil. Since water absorbs heat more slowly than dry soil, drought-affected areas appear warmer in thermal images.

How Thermal Imaging Works:

Detects Water Deficiency in Soil:

- Moist soil stays cooler due to water evaporation.
- Dry soil heats up faster, indicating areas that need irrigation.

Identifies Heat Stress in Plants:

- When plants lack water, they **close their stomata** to reduce moisture loss, leading to increased leaf temperatures.
- UAV thermal cameras detect **hotspots** in fields where crops are under heat stress, signaling early drought conditions.

Improves Water Management Efficiency:

- Thermal imaging helps farmers adjust irrigation schedules and water distribution, reducing overuse and waste.

Benefits of UAV-Based Drought Monitoring:

Early detection of drought stress before visible damage occurs is crucial for maintaining crop health and productivity. By identifying water deficiencies early, farmers can take preventive measures to minimize potential losses. Precision irrigation systems play a key role by delivering water only to areas that require it, reducing waste and optimizing water use. This targeted approach not only conserves valuable resources but also lowers operational costs. Effective moisture management through precision irrigation enhances soil health and promotes better plant growth. As a result, farmers can achieve higher crop yields while improving sustainability and ensuring efficient use of water resources [7]. By combining multispectral and thermal imaging, UAVs provide a powerful tool for identifying drought-prone areas and optimizing irrigation strategies.

Mapping Soil Moisture Variability Using UAVs

Soil moisture is a critical factor in crop growth, and uneven moisture distribution can lead to poor yields and water waste. UAVs equipped with advanced sensors help map soil moisture variability to ensure optimal irrigation [8].

11.2.1.3 UAVs Capture High-Resolution Images to Create Soil Moisture Maps

Unmanned Aerial Vehicles (UAVs) equipped with advanced imaging and sensor technologies play a crucial role in soil moisture mapping and agricultural monitoring. These high-tech systems provide detailed, real-time data on soil conditions, enabling better water management, improved crop yields, and early detection of moisture-related issues [9].

High-Resolution Aerial Imaging:

UAVs perform high-resolution aerial imaging by flying over agricultural fields and capturing detailed images using multispectral, hyperspectral, and thermal sensors. These advanced sensors collect comprehensive data on soil composition, water content, and vegetation health, allowing for a thorough analysis of field conditions. Multispectral and hyperspectral imaging captures data across multiple wavelengths of light, revealing subtle differences in plant health and moisture levels that are invisible to the human eye. Thermal sensors detect surface temperature variations, helping identify areas experiencing heat stress or low soil moisture [10]. This high-precision imaging allows for early identification of water stress, facilitating timely interventions to prevent crop losses.

Sensor Technologies Used for Soil Moisture Mapping:

UAVs are equipped with a variety of advanced sensors that provide critical data for soil moisture assessment:

1. Multispectral Cameras: These cameras detect changes in plant reflectance, which indicate plant stress due to insufficient moisture. They can identify areas where crops are struggling, even before visual signs of stress appear.
2. Thermal Cameras: By measuring surface temperature, thermal sensors identify hot, dry areas where soil moisture is low. Warmer regions in the imagery indicate insufficient water availability, which can guide targeted irrigation efforts.
3. LiDAR and Near-Infrared (NIR) Sensors: Light Detection and Ranging (LiDAR) and NIR sensors provide detailed information on soil texture, density, and water retention capacity. These sensors help create three-dimensional models of the terrain, highlighting areas prone to drought or water retention.

Each of these sensor technologies provides unique insights, and when combined, they offer a comprehensive view of soil moisture distribution across the field.

Georeferenced Data for Mapping:

The data collected by UAVs is processed and analyzed using Geographic Information Systems (GIS) and AI-based models to produce detailed soil moisture distribution maps. These maps are georeferenced, meaning each data point is tied to a specific location on the field [11].

These maps provide valuable insights, including:

- Water-Deficient Zones: Areas where soil moisture levels are critically low, requiring immediate irrigation.
- Well-Irrigated Areas: Regions with optimal moisture levels, ensuring efficient water use.
- Potential Waterlogging Regions: Zones where excess water accumulates, posing a risk of root damage and reduced crop productivity.

The integration of UAV data with GIS and AI allows for the creation of dynamic, real-time maps that can be updated regularly. This information helps farmers and agricultural managers optimize irrigation schedules, reduce water waste, and maintain soil health. UAV-based high-resolution imaging and advanced sensor technologies are transforming modern agricultural practices. By providing accurate soil moisture data and detailed georeferenced maps, these systems enable precise water management, early detection of problem areas, and more sustainable farming [12]. The use of UAVs in soil moisture mapping not only increases crop productivity but also supports long-term environmental conservation and efficient resource utilization.

11.2.1.4 Soil Moisture Maps Help Farmers Optimize Irrigation

UAVs for Efficient Water Management in Agriculture

Unmanned Aerial Vehicles (UAVs) play a critical role in improving agricultural water management by providing detailed, real-time insights into soil moisture levels. Through advanced imaging and sensor technologies, UAVs help farmers optimize irrigation practices, reduce water waste, and maintain crop health [13].

Identifying Water-Stressed Areas:

One of the key applications of UAVs is identifying water-stressed areas. UAVs equipped with multispectral, hyperspectral, and thermal sensors capture high-resolution images that reveal variations in soil moisture and vegetation health. This data is processed into moisture maps, which clearly highlight zones experiencing drought stress. These moisture maps allow farmers to identify areas where crops are not receiving adequate water, enabling them to take corrective irrigation measures. By addressing water stress promptly, farmers can prevent crop damage, maintain plant health, and improve overall yield [14]. Early detection of drought stress also helps reduce the long-term effects of water scarcity on agricultural productivity.

Precision Irrigation Application:

UAV data enhances precision irrigation, allowing farmers to apply water only where it is needed rather than using a uniform irrigation approach. This targeted method optimizes water use, improves efficiency, and minimizes waste. With precise moisture information, farmers can focus irrigation efforts on dry zones while avoiding unnecessary watering in areas with adequate moisture. Furthermore, UAV data can be integrated with smart irrigation systems, enabling automated water flow adjustments. These systems use real-time UAV data to control irrigation schedules and water delivery. For example, if UAVs detect moisture deficits in certain areas, the smart system increases water flow to those zones while reducing it in areas already sufficiently irrigated [15]. This automated, data-driven approach saves water, lowers energy costs, and enhances crop yields.

Preventing Over-Irrigation and Waterlogging:

In addition to detecting drought-stressed areas, UAVs also help in preventing over-irrigation and waterlogging. Excessive water application can cause problems such as root rot, fungal infections, and soil erosion, all of which threaten crop health and productivity. UAVs equipped with thermal and near-infrared (NIR) sensors identify regions receiving excess moisture by detecting cooler, wetter zones and areas where water is pooling. By providing accurate moisture distribution data, UAVs help farmers adjust irrigation practices to prevent overwatering. This ensures uniform crop health and yield consistency, reducing the risk of localized crop failures due to water imbalance [16]. Proper water management not only improves plant health but also protects the soil's structure and prevents nutrient loss through erosion.

UAV technology is transforming modern agricultural water management by providing detailed moisture maps and enabling precision irrigation. By identifying both water-stressed and over-irrigated areas, UAVs allow farmers to make data-driven decisions that optimize water use, prevent crop damage, and increase efficiency. Integration with smart irrigation systems further enhances these benefits through automated adjustments, ensuring sustainable water management and improved crop productivity [17]. Through accurate monitoring and timely intervention, UAVs support both higher agricultural yields and environmental conservation.

Benefits of UAV-Based Soil Moisture Mapping:

Precision irrigation offers numerous benefits, making it a vital practice for modern agriculture. One of its key advantages is saving water resources by applying water precisely where and when it is needed. This targeted approach prevents overuse and minimizes water waste, ensuring efficient utilization of limited resources. By reducing excess water application, precision irrigation helps conserve groundwater and surface water, contributing to long-term water sustainability. Another significant benefit is that it increases crop yields by providing crops with the right amount of moisture at the right time. Proper hydration supports optimal plant growth, improves nutrient uptake, and reduces stress caused by drought conditions [18]. This results in healthier crops and higher yields, enhancing farm productivity and profitability. Precision irrigation also reduces costs by lowering expenses related to water, energy, and labor. By using water more efficiently, farmers decrease their water consumption and reduce the energy required to pump and distribute it. Automated systems further cut labor costs by minimizing the need for manual irrigation. Moreover, this method supports sustainable farming by conserving natural resources and preventing environmental degradation. Reduced water runoff and soil erosion protect ecosystems while maintaining soil fertility [8]. Overall, precision irrigation promotes sustainable agricultural practices, ensuring long-term productivity and environmental protection.

Predictive Analytics and Machine Learning in Drought Prediction

Advancements in Artificial Intelligence (AI) and Machine Learning (ML) have transformed how UAVs analyze environmental data for predicting drought conditions. By utilizing long-term environmental monitoring, these technologies help in proactive water resource management and drought mitigation strategies [19].

11.2.1.5 UAVs Collect Long-Term Environmental Data for AI-Based Drought Prediction

UAVs and AI-Based Drought Prediction

Unmanned Aerial Vehicles (UAVs) have become essential tools for collecting long-term environmental data to enhance AI-based drought prediction. Their ability to gather precise, real-time information over large agricultural areas helps detect early signs of drought, enabling proactive measures to mitigate its impacts. By combining advanced data collection technologies with artificial intelligence (AI) and machine learning, UAVs significantly improve the accuracy and efficiency of drought monitoring and prediction [20].

Continuous Data Collection

One of the primary benefits of UAVs is their ability to perform continuous and comprehensive environmental monitoring. UAVs are equipped with specialized sensors, including multispectral, hyperspectral, and thermal sensors, which collect a wide range of data. This includes soil moisture levels, vegetation health, temperature fluctuations, and atmospheric conditions. By capturing high-resolution images and real-time environmental readings, UAVs identify subtle changes in crop health and soil moisture that may indicate early drought stress. Frequent monitoring allows for the detection of water stress trends before visible damage occurs. For instance, thermal sensors can measure the surface temperature of crops, revealing areas where plants are experiencing heat stress due to insufficient moisture [21]. Multispectral and hyperspectral imaging provides insights into plant health by analyzing reflected light across different wavelengths, identifying early signs of drought-related stress that are invisible to the human eye.

Integration with Historical Climate Data:

To improve the accuracy of drought predictions, UAV-collected data is integrated with historical climate information. This includes satellite imagery, weather forecasts, and past drought records, providing a more comprehensive understanding of environmental patterns. Combining real-time UAV data with these historical datasets enables scientists and agricultural experts to analyze long-term trends and identify areas at higher risk of drought. By using geospatial mapping, UAV data is correlated with historical climate records to develop predictive models. This integration helps identify vulnerable regions, forecast potential drought severity, and implement water management strategies. Moreover, combining different data sources increases the accuracy of predictions by accounting for local climate variability and seasonal changes [22]. Such insights are crucial for farmers, policymakers, and water resource managers who need to make timely decisions to prevent crop losses and manage water resources effectively.

AI and Machine Learning Models Process the Data

The large volumes of data collected by UAVs are processed and analyzed by AI and machine learning models. These models identify patterns and relationships in key environmental variables such as temperature, humidity, precipitation, and evapotranspiration rates. By continuously learning from new data, machine learning algorithms improve over time, making more accurate drought predictions as they process larger datasets. AI-based models can also detect subtle correlations between environmental factors that might not be immediately obvious through traditional analysis. For example, by analyzing past drought events, machine learning models can recognize specific conditions that precede drought onset. This predictive capability allows for the early detection of drought risks, giving farmers the opportunity to adjust irrigation schedules, apply soil moisture retention techniques, and plan for water conservation [23]. The integration of UAVs and AI in drought prediction is transforming modern agriculture. UAVs provide detailed, real-time environmental data, while AI and machine learning models analyze complex patterns to deliver accurate drought forecasts. This combination not only enhances early detection but also supports sustainable water management practices. By using these advanced technologies, farmers and policymakers can make informed decisions to mitigate the effects of drought, protect crop yields, and ensure long-term agricultural sustainability.

11.2.1.6 Using AI Predictions for Proactive Water Management

Early Warning Systems for Farmers:

- AI-driven UAV models forecast potential drought conditions weeks or months in advance.
- Farmers receive alerts and irrigation recommendations, allowing them to take preventive actions.

Optimized Water Allocation and Conservation:

- Machine learning models help governments and water management agencies distribute water resources efficiently.
- Water conservation plans are designed based on UAV-predicted drought severity levels.

Precision Farming Adjustments:

- AI-powered UAVs recommend crop rotation strategies, soil amendments, and alternative irrigation techniques based on drought forecasts.

- This helps in reducing crop losses and maximizing yield under water-scarce conditions.

Benefits of AI-Driven Predictive Analytics in Agriculture

Early Drought Detection: Enables farmers to take preventive actions before water shortages worsen.

Efficient Water Resource Management: Helps governments and agencies allocate water based on predicted needs.

Cost Savings: Reduces losses due to crop failure, inefficient irrigation, and water wastage.

Climate Resilience: Supports sustainable farming practices by helping farmers adapt to changing weather patterns.

11.2.2 Drone-Assisted Irrigation System Planning

Drones enhance irrigation planning by collecting precise topographical and hydrological data, allowing for better system design. By surveying fields and generating detailed 3D maps, UAVs help in determining optimal locations for irrigation channels, pipelines, and sprinkler systems. This data-driven approach minimizes water wastage, ensures uniform coverage, and reduces operational costs. Additionally, drone technology supports site-specific irrigation planning, enabling farmers to customize irrigation layouts based on soil type, slope, and crop water requirements. Drones are instrumental in designing efficient irrigation systems by surveying land, analyzing topography, and identifying the best water distribution methods.

How UAVs Assist in Irrigation Planning:

Aerial Land Surveys:

1. UAVs conduct fast and precise topographic mapping to analyze field slopes, drainage patterns, and elevation changes.
2. This data helps determine the best locations for irrigation channels, sprinklers, and pipelines.

Water Flow and Drainage Analysis:

1. UAV-generated Digital Elevation Models (DEMs) simulate water flow patterns.
2. By detecting natural water retention zones and runoff pathways, farmers can design irrigation layouts that minimize water loss.

Automated Irrigation Mapping:

1. Drones generate real-time vegetation index maps to identify areas needing more or less water.

2. This information is integrated into precision irrigation systems for automated water distribution.

Benefits:

11.2.2.1 Helps Create Water-Efficient Irrigation Networks

UAVs (Unmanned Aerial Vehicles) play a crucial role in designing and maintaining water-efficient irrigation networks by providing real-time aerial data on soil moisture levels, crop health, and water distribution patterns. Traditional irrigation systems often result in water wastage due to uneven distribution, runoff, or leakage. Drones equipped with thermal, multispectral, and hyperspectral cameras can detect variations in soil moisture, allowing farmers to precisely target areas that need irrigation while avoiding unnecessary water application. This data-driven approach helps optimize irrigation networks, ensuring that water is used efficiently and sustainably [24]. UAVs can also assist in planning new irrigation infrastructure, such as drip irrigation layouts, sprinkler placements, and canal systems, by mapping terrain and evaluating water flow dynamics.

11.2.2.2 Reduces Labor Costs and Manual Field Assessments

Traditional irrigation monitoring requires extensive labor, with farmers or irrigation specialists manually inspecting fields to assess soil moisture and detect issues such as leaks, clogged pipes, or over-irrigation. This process is time-consuming, costly, and prone to human error. UAVs eliminate the need for frequent manual field visits by capturing high-resolution imagery and thermal data from above. Drones can cover large agricultural areas in a fraction of the time, allowing farmers to receive instant insights into irrigation performance [25]. Automated UAV flights can be programmed to regularly survey fields, providing continuous monitoring without the need for constant human intervention. This reduces operational costs, improves accuracy, and allows farmers to focus on strategic decision-making rather than routine inspections.

11.2.2.3 Enhances Crop Uniformity by Ensuring Even Water Distribution

One of the biggest challenges in irrigation management is achieving uniform water distribution across a field. Uneven irrigation can result in patchy crop growth, where some areas receive excessive water, leading to waterlogging and disease, while others remain under-irrigated, causing drought stress and yield reduction. UAVs help detect these inconsistencies in real time by using thermal and multispectral imaging to highlight variations in soil moisture and plant health. By analyzing this data, farmers can adjust irrigation schedules and water flow to ensure that each section of the field receives the appropriate amount of water [26]. This not only improves crop uniformity and yield quality but also minimizes water waste, reduces nutrient leaching, and enhances overall farm productivity.

UAV technology is revolutionizing irrigation management by making water application more precise, cost-effective, and sustainable. By helping create efficient irrigation networks, reducing labor costs, and ensuring uniform water distribution, drones contribute to improved water conservation, increased crop yields, and lower environmental impact. As technology continues to advance, integrating UAVs with AI-driven irrigation systems and IoT sensors will further enhance precision agriculture, ensuring long-term sustainability in water resource management [27].

11.2.3 Optimizing Water Distribution with UAV Data

UAVs improve irrigation efficiency by providing real-time insights into water distribution across agricultural fields. By capturing thermal and hyperspectral images, drones can detect uneven irrigation patterns, identify overwatered or under-irrigated zones, and recommend necessary adjustments. This enables precision irrigation, where water is applied only where needed, reducing water loss due to evaporation, runoff, and infiltration. By integrating UAV data with Geographic Information Systems (GIS), farmers can develop irrigation maps to enhance overall water management and crop productivity [28]. By using data collected from UAVs, farmers can adjust irrigation schedules, minimize water wastage, and improve overall water-use efficiency.

How UAV Data Optimizes Irrigation:

Precision Watering Strategies:

1. UAVs capture soil and plant health data, guiding farmers in applying the right amount of water to each area.
2. This approach prevents over-irrigation, which can lead to soil erosion and nutrient leaching.

Real-Time Water Stress Detection:

1. Thermal cameras detect heat stress in crops, signaling water shortages.
2. Farmers receive instant alerts to adjust irrigation settings before damage occurs.

Adjusting Irrigation Schedules Dynamically:

1. UAVs monitor weather conditions, soil moisture, and crop growth stages to recommend irrigation adjustments.
2. By syncing drone data with automated irrigation controllers, farmers can respond to changing water needs efficiently.

Benefits:

11.2.3.1 Prevents Over- and Under-Irrigation, Improving Crop Yields

Over-irrigation can lead to waterlogging, root diseases, and nutrient leaching, while under-irrigation can cause drought stress, stunted growth, and lower yields. UAVs equipped with thermal and multispectral sensors help monitor soil moisture levels in real-time, identifying areas that receive too much or too little water. By analyzing this data, farmers can adjust irrigation schedules and water distribution precisely, ensuring that crops receive the optimal amount of water [29]. This leads to healthier plants, improved nutrient uptake, and higher yields, making farming more productive and efficient.

11.2.3.2 Reduces Water Costs and Promotes Sustainable Farming

Water is one of the most expensive and limited resources in agriculture. Inefficient irrigation systems often lead to excessive water use, increasing operational costs and depleting water supplies. UAV technology enables precision irrigation, ensuring that only the necessary amount of water is applied, reducing waste. This not only lowers water bills but also supports sustainable farming practices by conserving water resources, reducing runoff, and minimizing the environmental impact of irrigation [30]. By preventing water overuse, UAVs help maintain groundwater levels and protect ecosystems.

11.2.3.3 Increases Farm Productivity by Automating Irrigation Adjustments

Traditional irrigation management requires manual field inspections and adjustments, which are time-consuming and labor-intensive. UAVs automate irrigation monitoring by collecting real-time data on soil moisture, plant health, and weather conditions. When integrated with AI-driven irrigation systems and IoT sensors, drones can automatically trigger irrigation adjustments, ensuring that water is applied efficiently without constant human intervention. This saves time, reduces labor costs, and allows farmers to focus on other critical aspects of farm management, ultimately boosting overall farm productivity [31].

By preventing over- and under-irrigation, reducing water costs, and automating irrigation adjustments, UAVs play a key role in making agriculture more efficient, cost-effective, and environmentally sustainable. These advancements help farmers increase yields, conserve resources, and improve long-term farm profitability.

11.2.4 Integrating UAVs with Smart Irrigation Systems

The integration of UAVs with smart irrigation systems enables automated and data-driven water management. Drones can feed real-time data into Internet of Things (IoT)-based irrigation networks, allowing smart controllers to adjust water flow based on crop needs and environmental conditions. Additionally, UAVs can be combined with Variable Rate Irrigation (VRI) technology to optimize water application at different field zones. This synergy between UAV technology and smart irrigation systems enhances resource efficiency, promotes sustainability, and ensures long-term agricultural resilience against water scarcity. The future of irrigation lies in combining UAV technology with IoT (Internet of Things)-based smart irrigation systems for fully automated, data-driven water management [32].

How UAVs Work with Smart Irrigation Systems:

Wireless Connectivity and Data Sharing:

1. UAVs collect real-time soil moisture and plant stress data, transmitting it to IoT-based irrigation controllers.
2. Smart irrigation systems automatically adjust water distribution based on drone feedback.

AI-Driven Decision Making:

1. Machine learning models process UAV data to predict future water needs based on weather patterns and crop growth.
2. This AI-driven approach ensures optimal water delivery at the right time.

Case Study: UAV-Enabled Drip Irrigation Systems:

1. In some farms, UAVs are used to analyze soil conditions and control drip irrigation systems remotely.
2. By automating drip irrigation, farmers save up to 50% of water compared to traditional methods.

Benefits:

11.2.4.1 Enhances Irrigation Precision and Reduces Water Waste

Traditional irrigation methods often lead to inefficient water use, resulting in over-watering, runoff, or under-irrigation that negatively impacts crop growth. UAVs equipped with thermal and multispectral sensors provide high-resolution aerial imagery to assess soil moisture levels, detect plant stress, and map water distribution patterns. This data allows farmers to apply water exactly where it is needed, preventing waste and ensuring that every crop receives the optimal amount of moisture [33]. By eliminating excess water usage, UAVs contribute to water conservation, lower irrigation costs, and improved crop health, making farming more sustainable.

11.2.4.2 Increases Efficiency by Integrating UAVs with Smart Technology

When combined with AI-driven irrigation systems, IoT sensors, and automated irrigation controllers, UAVs enhance efficiency by providing real-time data and automated decision-making. These integrated systems can analyze weather forecasts, soil conditions, and plant water needs to adjust irrigation schedules dynamically. This automation reduces manual labor, minimizes human errors, and ensures that irrigation is optimized based on real-time environmental conditions [34]. By leveraging data analytics and automation, farmers can maximize crop productivity while minimizing resource waste.

11.2.4.3 Provides Farmers with Real-Time Insights for Better Decision-Making

Access to accurate, real-time information is essential for effective farm management. UAVs continuously monitor fields, collecting valuable data on water usage, crop health, and irrigation performance. This information is instantly processed and delivered to farmers via mobile apps or cloud-based platforms, allowing them to make informed decisions about irrigation adjustments, pest control, and resource allocation. With these insights, farmers can identify problems early, take preventive actions, and ensure optimal crop growth and yield improvement [35].

UAVs are revolutionizing irrigation by enhancing precision, integrating with smart technology, and providing real-time insights. These advancements help farmers conserve water, increase efficiency, and make data-driven decisions, leading to higher productivity, lower costs, and more sustainable agricultural practices.

11.3 Monitoring Water Use in Agriculture

Efficient water use is essential for sustainable agriculture, and modern technologies, including UAVs, play a vital role in monitoring and optimizing irrigation. Monitoring water use in agriculture involves assessing soil moisture, tracking irrigation efficiency, and detecting water stress in crops to prevent overuse and ensure sustainable water management. Drones equipped with multispectral and thermal sensors provide real-time data on soil moisture variations, allowing farmers to identify areas that need more or less water. This helps in precision irrigation, reducing unnecessary water consumption and preventing waterlogging or drought conditions. UAVs also assist in detecting leaks, blockages, or inefficiencies in irrigation systems, ensuring proper water distribution Another key aspect of monitoring water use is analyzing crop water stress levels using UAVs [36]. By capturing thermal images, drones can identify plants experiencing heat stress due to insufficient water. This enables farmers to adjust irrigation schedules and optimize water distribution to maintain crop health and maximize yields. Furthermore, UAVs contribute to data-driven decision-making by integrating with Geographic Information Systems (GIS) and smart irrigation networks. This integration enhances predictive analytics, helping farmers adapt to changing environmental conditions and improve water conservation strategies. By leveraging UAV technology, agriculture can achieve greater efficiency, sustainability, and resilience in water management [37].

11.3.1 Evaluating Water Consumption in Real Time

Real-time monitoring of water consumption is essential for efficient irrigation management. UAVs equipped with advanced sensors collect real-time data on water usage across agricultural fields. By analyzing thermal and multispectral imagery, farmers can track irrigation patterns and identify areas receiving excess or insufficient water. This helps in adjusting irrigation schedules dynamically, ensuring optimal water use and improving overall crop health while conserving resources [38].

How UAVs Monitor Water Consumption:

- **Thermal Imaging**: Drones equipped with thermal sensors detect plant heat stress, a key indicator of insufficient water supply. Stressed plants appear warmer in thermal images due to reduced transpiration, helping farmers pinpoint areas that need immediate irrigation.
- **Multispectral and Hyperspectral Imaging**: These imaging techniques analyze different wavelengths of light reflected by crops. Healthy, well-watered plants reflect more near-infrared (NIR) light, while water-stressed plants reflect less, allowing farmers to identify drought-affected zones early.
- **IoT and AI Integration**: Drones can communicate with smart irrigation systems, adjusting water distribution automatically based on real-time crop hydration levels.

Benefits:

Prevents Over-Irrigation and Under-Irrigation, Leading to Healthier Crops

Over-irrigation can cause root rot, nutrient leaching, and waterlogging, while under-irrigation leads to drought stress, stunted growth, and yield loss. UAVs equipped with thermal, multispectral, and hyperspectral sensors monitor soil moisture levels and crop health in real time. By identifying areas receiving too much or too little water, UAVs enable precision irrigation, ensuring that each crop gets the right amount of moisture [39]. This results in stronger, healthier plants, improved nutrient absorption, and higher agricultural yields.

Reduces Water Costs by Ensuring Irrigation is Applied Only Where Needed

Water is one of the most valuable and expensive resources in agriculture. Inefficient irrigation practices lead to excessive water consumption and high operational costs. UAVs help reduce water waste by mapping soil moisture levels and directing water only to areas that need it. This optimizes irrigation efficiency, leading to lower water bills, reduced energy costs for pumping, and overall savings for farmers [40].

Supports Sustainable Water Use, Especially in Drought-Prone Regions

In regions facing water scarcity and unpredictable rainfall, sustainable irrigation is critical for maintaining crop productivity. UAV technology helps monitor drought-prone areas, assess water distribution efficiency, and prevent unnecessary water loss. By promoting targeted irrigation and water conservation strategies, UAVs contribute to long-term sustainability, protecting water resources while ensuring food security. By preventing over- and under-irrigation, reducing water costs, and promoting sustainable water use, UAVs are transforming agricultural irrigation [41]. These benefits help increase productivity, conserve resources, and support eco-friendly farming practices, making them essential tools for modern precision agriculture.

11.3.2 Detecting Leaks and Inefficiencies in Irrigation Networks

UAVs play a crucial role in detecting leaks, blockages, and inefficiencies in irrigation systems. Using thermal imaging and high-resolution cameras, drones can identify leaks in pipelines, malfunctioning sprinklers, and uneven water distribution. Early detection of such issues prevents water wastage, reduces maintenance costs, and ensures efficient water delivery to crops [42]. Drone-based inspections also minimize the need for manual field surveys, saving time and labor.

How UAVs Detect Irrigation Leaks and Inefficiencies:

- **Thermal Imaging for Leak Detection**: When irrigation pipes leak, water seeps into the surrounding soil, cooling the affected area. Thermal cameras on UAVs detect these temperature differences, identifying leaks, broken pipes, or malfunctioning drip irrigation lines.
- **Multispectral Analysis for Uneven Water Distribution**: If some areas in a field receive more water than others, crops grow unevenly. UAVs capture high-resolution multispectral images, allowing farmers to identify irregular irrigation patterns and adjust sprinklers or drippers accordingly.
- **Aerial Mapping for System Analysis**: UAVs provide a complete top-down view of irrigation networks, making it easier to spot waterlogged zones, dry patches, or clogged irrigation emitters that require maintenance.

Benefits:

Advanced leak detection systems offer significant benefits for irrigation management. They help reduce water loss by identifying hidden leaks, conserving valuable resources. By improving the performance of irrigation systems, these technologies also extend the lifespan of equipment, ensuring more efficient operation. Additionally, early detection of problems minimizes repair costs and lowers water bills, providing long-term financial savings [43]. Implementing such systems enhances overall efficiency while promoting sustainable water use.

11.3.3 Remote Sensing for Soil Moisture Analysis

Soil moisture is a critical factor in determining irrigation needs, and UAVs with remote sensing capabilities provide accurate soil moisture analysis. By capturing multispectral and infrared images, drones assess soil water content and identify dry or over-saturated areas. This information allows farmers to fine-tune irrigation schedules, apply water precisely where needed, and prevent issues like water logging or drought stress.

How UAVs Analyze Soil Moisture:

- **Near-Infrared (NIR) and Shortwave Infrared (SWIR) Sensors**: These sensors detect moisture variations in soil by analyzing how much light is absorbed or reflected. Moist soil absorbs more infrared light, while dry soil reflects more, allowing UAVs to create accurate moisture distribution maps.
- **Thermal Imaging for Soil Dryness Detection**: Warmer soil indicates low moisture content, while cooler soil suggests adequate moisture levels. UAVs capture these temperature differences, allowing farmers to adjust irrigation schedules based on soil moisture conditions.
- **LiDAR (Light Detection and Ranging) for Water Retention Analysis**: By mapping elevation and topography, LiDAR helps farmers understand how water moves and settles in their fields, identifying areas prone to waterlogging or drought stress.

Benefits:

Real-time soil moisture monitoring plays a crucial role in optimizing irrigation scheduling. By providing accurate, up-to-date data, it allows farmers to apply water only when necessary, reducing waste and conserving resources. This precision prevents overwatering, which can lead to nutrient leaching and soil degradation, ensuring that essential nutrients remain available to crops. Consistent soil moisture levels promote uniform crop growth, improving yield quality and productivity. Additionally, optimizing irrigation schedules enhances water-use efficiency, lowers operational costs, and supports sustainable agricultural practices [44]. Implementing real-time monitoring systems helps maintain a balance between crop needs and resource conservation, benefiting both the environment and farm profitability.

11.3.4 Reducing Water Waste Through Precision Agriculture

Precision agriculture, powered by UAVs, helps minimize water waste by delivering data-driven insights for efficient irrigation. By mapping soil moisture levels, detecting crop water stress, and integrating data with smart irrigation systems, drones enable targeted water application. This approach reduces runoff, prevents over-irrigation, and maximizes water-use efficiency. Implementing UAV-based precision irrigation ensures sustainable water management while enhancing agricultural productivity and environmental conservation.

How UAVs Improve Water Efficiency in Precision Agriculture:

- **Targeted Irrigation with UAV Data**: UAVs collect real-time crop health and soil moisture data, guiding Variable-Rate Irrigation (VRI) systems to apply different water amounts to different field zones based on need.
- **AI-Based Water Use Predictions**: Machine learning models analyze historical data, weather forecasts, and soil conditions to optimize irrigation schedules.
- **Automated Sprinkler and Drip System Control**: UAV data is integrated with **smart** irrigation controllers, which adjust water distribution automatically, preventing waste.

Benefits:

UAVs play a crucial role in optimizing water use in agriculture by ensuring that irrigation is applied only where necessary, thereby lowering overall water consumption while still supporting optimal crop growth. By using real-time aerial imaging and soil moisture analysis, drones help farmers precisely determine which areas require water, preventing over-irrigation and minimizing waste. This targeted approach not only conserves water but also reduces the energy costs associated with pumping and distributing water, as less water needs to be transported across fields. Additionally, UAV-assisted irrigation promotes sustainable and environmentally friendly farming practices by reducing water runoff, preventing soil erosion, and limiting the overuse of natural water sources [45]. By integrating UAV technology into precision irrigation systems, farmers can achieve higher crop yields while maintaining long-term water sustainability and reducing their overall environmental footprint.

UAVs are transforming agricultural water management by providing real-time insights into water consumption, detecting inefficiencies, analyzing soil moisture, and enabling precision irrigation. By leveraging these technologies, farmers can enhance crop yields, reduce operational expenses, and implement sustainable irrigation strategies [46]. The integration of UAVs with AI, IoT, and remote sensing ensures that every drop of water is used efficiently, making modern agriculture more productive, cost-effective, and environmentally responsible.

11.4 Sustainable Water Management Practices

Sustainable water management is crucial for conserving resources, maintaining soil health, and ensuring long-term agricultural productivity. Advanced technologies like UAVs play a significant role in improving water use efficiency and promoting conservation strategies. One key approach is implementing smart irrigation systems, where UAVs and IoT sensors provide real-time data to optimize water distribution. Automated irrigation schedules and Variable Rate Irrigation (VRI) help reduce water waste by adjusting water application based on soil and crop needs [47]. Additionally, UAVs support water conservation techniques by assessing soil moisture retention and monitoring the effectiveness of practices like mulching, cover cropping, and rainwater harvesting. These techniques reduce evaporation and enhance soil water-holding capacity.

Furthermore, UAVs assist in groundwater recharge and management by identifying recharge zones and assessing the impact of agricultural water use on aquifers. This enables farmers to implement managed aquifer recharge (MAR) strategies, such as percolation ponds and recharge wells, to sustain groundwater levels. Drones also play a crucial role in monitoring and reducing agricultural runoff by detecting erosion-prone areas and analyzing drainage patterns. This information helps design buffer strips, contour farming systems, and sustainable drainage structures to prevent soil loss and water pollution [48]. By integrating UAV technology with sustainable water management practices, farmers can enhance efficiency, reduce environmental impact, and ensure long-term water security.

11.4.1 Conservation Strategies for Water-Scarce Regions

In regions facing water scarcity, effective conservation strategies are essential to maximize water availability for agriculture, drinking, and industrial use. Traditional methods such as drip irrigation, which delivers water directly to plant roots, mulching, which reduces evaporation, and rainwater harvesting, which captures runoff for reuse, help optimize water use. UAVs enhance these efforts by providing high-resolution aerial data to identify drought-prone areas, monitor soil moisture levels, and evaluate the effectiveness of conservation techniques [49]. By analyzing UAV-generated maps, farmers can implement precision irrigation, ensuring that water is applied efficiently to areas that need it most, reducing waste and improving crop yields.

Importance of Water Conservation in Agriculture

Regions suffering from low rainfall and declining groundwater levels, such as California's Central Valley (USA), Rajasthan (India), and parts of the Middle East and Africa, rely heavily on conservation strategies to sustain agriculture. Farmers in these regions use drip irrigation, rainwater harvesting, and soil moisture management techniques, but these methods can be further optimized with UAV technology [50].

How UAVs Aid in Water Conservation

Precision Irrigation with UAV Data

1. UAVs with multispectral and thermal sensors map crop hydration levels and identify areas that require more or less water.
2. For example, in Australia's Murray-Darling Basin, drones have been used to detect over-irrigated areas, helping farmers adjust water distribution accordingly.

Drought-Resistant Crop Management

1. UAVs monitor crop stress in real time, allowing farmers to determine which varieties can survive with minimal water.
2. In India's Maharashtra region, farmers use UAV data to select millets and pulses, which thrive in dry conditions compared to water-intensive crops like rice.

Soil Moisture Retention Techniques

1. UAVs create detailed soil moisture maps, identifying areas needing mulching, cover crops, or soil amendments.
2. In South Africa, vineyards use UAVs to assess soil moisture retention and apply organic compost more efficiently.

Water Recycling and Reuse

1. UAVs help monitor drainage and runoff collection systems, ensuring that recycled water is effectively stored and distributed for irrigation.
2. In Israel, UAV-assisted water recycling projects have reduced agricultural water waste by 30%.

Key Benefits:

UAV technology significantly enhances water conservation in drought-prone areas by providing real-time data on soil moisture levels and crop water needs. By identifying water-stressed regions, drones enable targeted irrigation, ensuring that limited water resources are used efficiently and effectively. This precision irrigation approach reduces over-reliance on freshwater resources, helping farmers optimize their water use and minimize waste. By preventing excessive water withdrawals from lakes, rivers, and groundwater reserves, UAVs contribute to long-term water sustainability and protect vital ecosystems. Additionally, when integrated with AI-driven analytics, UAVs support climate-resilient agriculture by analyzing weather patterns, predicting drought conditions, and automating irrigation decisions [51]. This proactive approach allows farmers to adapt to changing climatic conditions, improve crop resilience, and maintain consistent yields, even in challenging environments. Through these advancements, UAVs play a vital role in ensuring sustainable and resource-efficient farming practices.

11.4.2 Role of UAVs in Groundwater Monitoring

Groundwater is a crucial water source for agriculture, but over-extraction leads to depletion and long-term sustainability challenges. UAVs equipped with thermal and multispectral sensors help monitor groundwater by detecting soil moisture variations and identifying natural recharge zones. Drones can survey aquifer-dependent ecosystems, assessing the impact of groundwater withdrawal on vegetation, rivers, and lakes. By integrating UAV data with remote sensing and GIS mapping, researchers and water resource managers can track groundwater trends, prevent over-extraction, and develop sustainable recharge strategies, such as creating artificial recharge wells, percolation ponds, and check dams to replenish underground water reserves [52].

Why Groundwater Monitoring is Crucial

Groundwater is a primary source of irrigation in many agricultural regions, but over-extraction, pollution, and declining recharge rates threaten its availability. Traditional methods like borehole measurements and satellite imagery lack real-time precision, making it difficult to assess groundwater depletion accurately [53]. UAVs provide high-resolution data, detect seepage areas, and support recharge planning.

How UAVs Monitor Groundwater

Remote Sensing of Vegetation Health

1. Groundwater depletion often results in stressed vegetation. UAVs equipped with NDVI (Normalized Difference Vegetation Index) cameras identify stress patterns before they become visible to the human eye.
2. Example: In California, drones detected vegetation stress in almond orchards, leading to adjustments in irrigation schedules and reduced water consumption.

Thermal and Multispectral Imaging

1. UAVs detect temperature variations in soil and water bodies, helping locate underground water seepage and evaporation zones.
2. Example: In Saudi Arabia, UAV thermal imaging helped farmers identify declining groundwater levels in date palm plantations, enabling better water allocation.

Groundwater Recharge Zone Identification

1. UAVs help locate areas where rainwater can effectively percolate into underground aquifers.
2. Example: In Mexico's Yucatán Peninsula, UAVs were used to identify karst recharge zones, preventing over-extraction in vulnerable areas.

Key Benefits:

UAV technology plays a crucial role in improving groundwater conservation strategies by providing real-time data on soil moisture levels, vegetation health, and water distribution patterns. By monitoring groundwater recharge areas and detecting over-extraction zones, drones help farmers and policymakers develop more efficient water management plans. UAVs also enable data-driven irrigation planning by integrating aerial imaging, AI-powered analytics, and IoT sensors to optimize water use based on crop needs and environmental conditions [54]. This precision-based approach ensures that irrigation is applied efficiently, minimizing waste and preventing excessive water withdrawals. Additionally, UAV-assisted monitoring reduces over-extraction and prevents water table depletion by identifying areas at risk of groundwater decline. By promoting sustainable water use, UAVs help maintain groundwater reserves for long-term agricultural productivity and environmental stability [55].

11.4.3 Rainwater Harvesting and UAV-Based Assessment

Rainwater harvesting is a key strategy for enhancing water availability, particularly in arid and semi-arid regions. UAVs play a crucial role in improving rainwater harvesting efficiency by conducting terrain analysis, identifying suitable sites for rainwater collection structures, and monitoring their performance. Drones equipped with LiDAR (Light Detection and Ranging) and infrared sensors can map topography, slopes, and water flow paths, helping engineers design check dams, storage ponds, and rooftop harvesting systems in optimal locations [56]. Additionally, UAVs can regularly inspect these structures, detecting erosion, silt accumulation, or leaks, ensuring that harvesting systems operate effectively and contribute to long-term water conservation.

The Need for Rainwater Harvesting

Rainwater harvesting is an effective method for storing water for irrigation during dry periods. However, poor planning and infrastructure gaps often reduce its effectiveness. UAVs help assess storage potential, optimize catchment areas, and monitor system efficiency [57].

How UAVs Optimize Rainwater Harvesting

Mapping Catchment Areas

1. UAVs survey landscapes to identify ideal locations for rainwater collection.
2. Example: In Rajasthan, India, UAVs identified low-lying areas with high runoff potential, leading to the construction of efficient check dams and ponds.

Analyzing Rooftop Water Collection

1. UAVs evaluate roof slope and material to determine rainwater harvesting efficiency.
2. Example: In urban Kenya, drones mapped schools and community centers for potential rooftop harvesting projects.

Monitoring Water Storage Capacity

1. UAVs monitor reservoirs, ponds, and water tanks, ensuring they are properly maintained and free of leaks.
2. Example: In Brazil, UAVs helped monitor Amazonian water storage systems, reducing water loss by 15%.

Key Benefits:

UAV technology plays a vital role in maximizing rainwater harvesting potential by identifying optimal collection areas and assessing surface runoff patterns. Using high-resolution aerial imaging and topographic mapping, drones help farmers and engineers determine the best locations for rainwater collection, ensuring efficient capture and utilization of natural precipitation. Additionally, UAV data helps design efficient storage systems by evaluating terrain elevation, soil infiltration rates, and existing water infrastructure. This enables the construction of reservoirs, ponds, and rainwater storage tanks that effectively retain and distribute water for irrigation and other agricultural uses [40]. By enhancing rainwater harvesting and storage, UAVs reduce dependency on external water sources, such as groundwater and municipal supplies, leading to greater water security, cost savings, and long-term sustainability in agriculture.

11.4.4 Policy and Regulations for Efficient Water Use

Effective water management requires strong policies and regulations to ensure sustainable use and prevent over-exploitation. UAVs support data-driven policy-making by monitoring illegal water extraction, evaluating irrigation efficiency, and detecting unsustainable land-use changes. For example, drones can identify unauthorized boreholes, excessive groundwater pumping, and inefficient irrigation methods, enabling regulatory authorities to enforce conservation laws [58]. Additionally, UAV data can help governments and environmental agencies design water-efficient policies, promote incentives for sustainable irrigation technologies, and support precision agriculture initiatives that optimize water use. By integrating UAV technology into water management frameworks, policymakers can improve compliance, encourage responsible water use, and ensure long-term sustainability.

The Role of Policy in Sustainable Water Management

Governments and regulatory bodies are implementing laws and incentives to promote efficient water use. UAVs help enforce policies, ensure compliance, and guide decision-making [59].

How UAVs Support Water Regulations

Monitoring Compliance with Water Use Laws

1. UAVs track water extraction and irrigation practices to ensure farmers comply with regulations.
2. Example: In Spain, UAVs were used to detect illegal water withdrawals from the Guadalquivir River.

Supporting Government Incentives

1. UAV data helps determine eligibility for subsidies related to precision irrigation and conservation projects.
2. Example: In the EU, UAV-assisted assessments are used to distribute CAP (Common Agricultural Policy) funds to sustainable farms.

Disaster Preparedness and Response

1. UAVs aid in monitoring drought severity and planning emergency water distribution.
2. Example: In Africa's Sahel region, UAVs helped map desertification zones, guiding long-term water policies.

Key Benefits:

UAV technology plays a crucial role in ensuring efficient enforcement of water regulations by providing real-time aerial surveillance and monitoring of agricultural water use. By detecting illegal water extraction, over-irrigation, and non-compliance with conservation policies, drones help regulatory agencies enforce water management laws more effectively. Additionally, UAV data supports government planning for drought resilience by identifying water-stressed regions, tracking groundwater levels, and assessing the effectiveness of conservation strategies. This data-driven approach enables policymakers to develop targeted drought mitigation plans and allocate resources efficiently [60]. Furthermore, UAV-assisted monitoring helps farmers access financial incentives for sustainable water use by providing accurate documentation of their conservation efforts. Governments and environmental agencies can use UAV data to verify compliance with water-saving practices, making it easier for farmers to qualify for subsidies, tax benefits, and funding for sustainable irrigation projects. Through these benefits, UAVs contribute to better water governance, improved agricultural resilience, and long-term sustainability [61].

Sustainable water management is essential for addressing global water challenges, ensuring food security, and protecting ecosystems. UAV technology has emerged as a powerful tool in optimizing water use, monitoring conservation efforts, and supporting evidence-based policy decisions. By adopting UAV-based water conservation strategies, groundwater monitoring systems, rainwater harvesting assessments, and regulatory frameworks, agriculture and water resource sectors can significantly enhance efficiency and sustainability [62]. The integration of advanced technology with smart water management practices will play a crucial role in mitigating water scarcity and securing water resources for future generations.

11.5 Future Trends in Irrigation Technology

The future of irrigation is being shaped by cutting-edge technologies such as artificial intelligence (AI), the Internet of Things (IoT), automation, and UAV-based monitoring, all aimed at improving water efficiency and sustainability in agriculture. AI-driven smart irrigation systems are revolutionizing water management by analyzing real-time data from UAVs, soil moisture sensors, and climate models to optimize irrigation schedules. These systems use machine learning algorithms to predict plant water needs, adjust water distribution dynamically, and prevent both over- and under-irrigation, ensuring optimal crop growth while conserving water. In addition, IoT-based irrigation networks are becoming increasingly common, integrating wireless soil moisture sensors, weather stations, and automated irrigation controllers [63]. These interconnected devices allow farmers to monitor and control irrigation remotely via mobile applications, reducing manual labor and ensuring efficient water usage. UAVs further enhance precision irrigation by providing high-resolution aerial imagery to assess soil moisture variability, detect plant stress, and identify inefficient irrigation zones. Another emerging trend is Variable Rate Irrigation (VRI), which allows irrigation systems to apply different amounts of water to different areas of a field based on soil type, crop variety, and moisture levels, significantly reducing water wastage. As climate change and water scarcity intensify, these technological advancements will play a crucial role in transforming traditional irrigation methods into smart, data-driven, and highly efficient systems that promote sustainable agriculture and long-term water conservation [64].

11.5.1 AI and Machine Learning in UAV-Based Irrigation

Artificial intelligence (AI) and machine learning are transforming UAV-based irrigation by enabling data-driven decision-making and automation. UAVs equipped with

AI-powered image recognition and data analytics can assess crop health, detect water stress, and predict irrigation needs with high precision. Machine learning algorithms analyze past and real-time data from drones, soil moisture sensors, and weather models to optimize irrigation schedules, reducing water wastage and improving crop yields [65]. This advanced technology ensures that water is applied only where and when needed, enhancing efficiency and sustainability.

The Role of AI and ML in UAV Irrigation

AI and ML improve UAV-based irrigation by analyzing vast datasets, identifying patterns, and optimizing water distribution. Drones equipped with AI-powered sensors can assess soil moisture levels, detect water stress in crops, and automate irrigation adjustments [66].

How AI and ML Improve Irrigation Efficiency

Predictive Analytics for Water Demand

1. AI algorithms process weather forecasts, soil conditions, and plant growth data to predict future irrigation needs.
2. Example: In California's almond farms, AI-based UAVs analyze historical climate patterns to recommend irrigation schedules, reducing water use by 25% while maintaining high yields.

Automated Detection of Crop Water Stress

1. UAVs with machine learning models process thermal and multispectral images to detect early signs of water deficiency.
2. Example: In India's sugarcane plantations, ML-powered UAVs identified areas suffering from drought stress, allowing farmers to provide targeted irrigation, saving thousands of liters of water per hectare.

Precision Irrigation Control Using AI

1. AI-powered UAVs adjust water distribution dynamically based on real-time field conditions.
2. Example: In Australia's wheat fields, drones equipped with AI and soil moisture sensors controlled irrigation levels, reducing water usage by 30% without impacting crop production.

Key Benefits:

UAV technology enhances irrigation efficiency through AI-driven analysis by collecting real-time data on soil moisture, crop health, and weather conditions. Using machine learning algorithms, UAVs process this data to optimize irrigation schedules, ensuring that water is distributed precisely where and when it is needed. This reduces waste and maximizes crop productivity [67]. Additionally, UAV systems reduce human intervention by automating decision-making, integrating with smart irrigation networks and IoT sensors to adjust water flow based on environmental conditions. This automation not only saves labor costs but also minimizes human error, making irrigation management more efficient. Furthermore, UAVs improve drought resilience by predicting water shortages, analyzing historical climate patterns, and detecting early signs of soil dryness. By providing farmers and policymakers with early warnings, drones enable proactive water conservation measures, ensuring sustainable resource use and long-term agricultural stability in drought-prone regions [68].

11.5.2 Integration of IoT with UAV Irrigation Systems

The integration of the Internet of Things (IoT) with UAV-based irrigation systems is revolutionizing precision agriculture. IoT devices, such as soil moisture sensors, automated irrigation controllers, and climate monitoring stations, continuously collect and transmit data to UAVs and cloud-based platforms. Drones use this information to generate high-resolution maps, monitor irrigation effectiveness, and automate water distribution. Farmers can access real-time data via mobile applications, enabling remote management and reducing labor costs [69]. This seamless integration improves irrigation efficiency, conserves water, and enhances decision-making in agricultural water management.

How IoT Enhances UAV-Based Irrigation

The Internet of Things (IoT) connects UAVs with ground sensors, weather stations, and automated irrigation systems, creating an intelligent real-time water management network. This data-sharing ecosystem ensures that water is used efficiently, reducing losses due to over-irrigation or evaporation [70].

Applications of IoT in UAV Irrigation

Real-Time Soil Moisture Monitoring

1. IoT-enabled sensors continuously monitor soil moisture levels and send data to UAVs.
2. Example: In Israel's Negev desert, IoT-integrated UAVs helped manage drip irrigation, reducing water use by 40% in arid regions.

Automated UAV-Irrigation System Coordination

1. UAVs and IoT-controlled irrigation systems work together for precision irrigation.
2. Example: In China's rice fields, IoT-linked UAVs regulated water levels based on weather forecasts, preventing excess water usage.

Remote Irrigation Management with Mobile Applications

1. Farmers can monitor and control irrigation from their smartphones.
2. Example: In the Netherlands, IoT-connected UAVs sent alerts on irrigation needs, allowing farmers to adjust water use remotely.

Key Benefits of UAV Technology in Irrigation

UAV technology enables remote water management, allowing farmers to monitor and control irrigation systems without the need for constant manual labor. By using real-time aerial imaging and AI-powered analytics, drones help identify moisture levels, assess crop health, and automate water distribution, significantly reducing physical effort and operational costs [71]. Additionally, UAVs minimize water wastage by providing precise irrigation—ensuring that water is applied only where and when it is needed. This targeted approach prevents overwatering, reduces runoff, and conserves valuable water resources. Furthermore, UAV-driven irrigation systems enhance farm efficiency through automation by integrating with IoT sensors, AI models, and smart irrigation controllers. This automation optimizes water distribution, improves crop yield, and allows farmers to focus on other critical aspects of farm management [72]. By leveraging UAV technology, agriculture becomes more sustainable, cost-effective, and resource-efficient.

11.5.3 Climate Change Adaptation in Water Resource Management

Climate change is intensifying water scarcity and increasing the unpredictability of rainfall, making efficient water management more critical than ever. UAV technology plays a key role in climate adaptation by providing real-time monitoring of drought conditions, soil moisture levels, and water availability. Drones assist in mapping climate-resilient irrigation strategies, identifying areas at risk of water stress, and optimizing water storage and distribution [73]. By integrating UAV data with predictive climate models, farmers and policymakers can develop proactive strategies to mitigate the impacts of climate change on agriculture and ensure long-term water security.

Why Climate Change Adaptation is Crucial

Climate change is causing unpredictable rainfall, prolonged droughts, and extreme weather events, making traditional irrigation methods inefficient. UAVs provide a solution for adaptive water management by analyzing climate patterns and optimizing irrigation strategies [74].

How UAVs Support Climate Adaptation

Early Detection of Drought Conditions

1. UAVs collect historical and real-time soil moisture data to detect drought risks.
2. Example: In Ethiopia's maize farms, UAVs identified early signs of drought stress, allowing farmers to shift to drought-resistant crops.

Floodwater Management for Irrigation

1. UAVs assess water accumulation and flood risks, helping redirect excess water for irrigation.
2. Example: In Bangladesh's rice-growing regions, UAV-collected flood data helped store and redistribute water to dry areas.

Climate-Adaptive Precision Irrigation

1. UAVs adjust irrigation patterns based on changing climate conditions.
2. Example: In Texas cotton farms, UAVs helped transition from traditional flood irrigation to precision irrigation, reducing water waste.

Key Benefits of UAV Technology in Irrigation

UAV technology improves climate resilience by adapting irrigation to changing weather conditions. By analyzing real-time weather data, soil moisture levels, and crop health, UAVs help farmers adjust their irrigation strategies in response to droughts, rainfall variations, and temperature fluctuations. This proactive approach ensures that crops receive the optimal amount of water, even under extreme climatic conditions [75]. Additionally, UAVs reduce water wastage by optimizing water use through precision irrigation. By identifying areas that need water and preventing over-irrigation, drones help conserve water resources, lower costs, and improve sustainability. Furthermore, UAV-driven irrigation enhances food security by ensuring sustainable farming practices. By promoting efficient water management, reducing environmental impact, and increasing crop yields, UAV technology supports long-term agricultural productivity, helping farmers maintain stable food production while preserving natural resources [76].

11.5.4 Challenges and Opportunities in UAV Adoption for Irrigation

Despite the benefits of UAVs in irrigation, several challenges hinder widespread adoption. High costs of drones, limited technical expertise, regulatory restrictions, and data processing complexities are key barriers. However, advancements in UAV affordability, AI-driven automation, and government support for smart agriculture present significant opportunities. Training programs, subsidies, and improved drone regulations can encourage farmers to integrate UAVs into irrigation practices [77]. As UAV technology continues to evolve, addressing these challenges will unlock its full potential in optimizing water use, reducing environmental impact, and enhancing agricultural productivity.

Challenges in UAV Adoption

High Initial Costs

1. UAVs with AI, IoT, and sensors require significant investment, which can be a barrier for small farmers.
2. Solution: Government subsidies and UAV leasing programs can lower costs.

Regulatory Restrictions

1. Many countries have strict drone operation laws, limiting agricultural UAV use.
2. Solution: Policymakers should develop clear regulations for UAV-based irrigation.

Technical Expertise and Training

1. Operating UAVs requires specialized knowledge, which may be lacking among farmers.
2. Solution: Educational programs and training workshops can help farmers adopt UAV technology.

Data Integration and Connectivity Issues

1. Remote farms often suffer from poor internet connectivity, affecting real-time UAV operation.
2. Solution: Offline UAV processing and low-bandwidth IoT solutions can improve reliability.

Opportunities for UAV-Based Irrigation

Decreasing UAV Costs

1. Advances in UAV technology are making drones cheaper and more accessible.

Increased Government and Private Sector Investment

1. Governments and agricultural firms are funding precision irrigation projects using UAVs.

Advanced AI, IoT, and Blockchain Integration

1. Future UAV irrigation systems will be fully automated, requiring minimal human intervention.

Key Benefits of UAV Technology in Irrigation

UAV technology creates a data-driven, efficient irrigation system by collecting and analyzing real-time data on soil moisture, weather conditions, and crop health. This allows farmers to make informed decisions and apply water precisely where and when it is needed, reducing waste and improving overall efficiency. Additionally,

UAV-assisted irrigation ensures long-term sustainability and profitability by optimizing water use, lowering operational costs, and increasing crop yields [78]. By integrating AI-driven analytics and automation, farmers can maintain high productivity while minimizing resource depletion. Furthermore, UAVs promote water conservation and global food security by preventing over-irrigation, reducing water runoff, and maximizing agricultural output. As water scarcity becomes a growing global concern, drone technology plays a vital role in ensuring sustainable farming practices, supporting climate resilience, and contributing to stable food production for future generations [79].

UAV-based irrigation, powered by AI, IoT, and climate adaptation strategies, is transforming modern agriculture. By analyzing real-time data, predicting irrigation needs, and reducing water wastage, UAVs help farmers optimize water use while ensuring crop productivity. However, high costs, regulatory challenges, and technical barriers remain hurdles to widespread adoption [80]. With ongoing technological advancements, increased investments, and supportive policies, UAVs will play a crucial role in sustainable water resource management, ensuring efficient and climate-resilient farming in the future.

11.6 Summary

UAVs (Unmanned Aerial Vehicles) are revolutionizing irrigation efficiency by providing real-time data on soil moisture, crop health, and water distribution. They help identify drought-prone areas by mapping soil conditions and detecting water stress in crops. UAVs assist in irrigation system planning by analyzing terrain, optimizing water flow, and identifying inefficiencies in irrigation networks. By using thermal and multispectral imaging, drones can ensure optimal water distribution, reducing waste and improving crop yields. Additionally, UAVs can be integrated with smart irrigation systems, automating water application based on real-time data. This technology enhances precision irrigation, conserves water, and supports sustainable agriculture.

Sustainable water management is essential for conserving resources and ensuring long-term agricultural productivity. UAVs contribute by enhancing water conservation strategies in water-scarce regions, helping monitor soil moisture, assess drought conditions, and optimize irrigation efficiency. Drones play a crucial role in groundwater monitoring, detecting moisture variations and identifying recharge zones to prevent over-extraction. UAV-based assessments also improve rainwater harvesting by identifying suitable collection sites and monitoring the effectiveness of water storage systems. UAV data supports policy and regulatory enforcement, ensuring efficient water use and compliance with conservation guidelines. By integrating UAV technology with smart water management strategies, agriculture can achieve greater efficiency, reduce environmental impact, and ensure water security for the future. By preventing over- and under-irrigation, reducing water costs, and promoting sustainable water use, UAVs are transforming agricultural irrigation. These benefits help increase productivity, conserve resources, and support eco-friendly farming practices, making them essential tools for modern precision agriculture.

The future of irrigation technology is advancing with AI, IoT, UAVs, and automation to enhance water efficiency and sustainability in agriculture. AI-driven smart irrigation systems analyze real-time data to optimize water distribution, preventing over- or under-irrigation. IoT-based irrigation networks integrate sensors and automated controllers, allowing remote monitoring and control of irrigation systems. UAVs provide aerial imagery to assess soil moisture, detect plant stress, and optimize water application. Variable Rate Irrigation (VRI) further improves efficiency by adjusting water distribution based on soil and crop needs. The future of UAV-based irrigation lies in AI, IoT, and climate adaptation strategies. With real-time monitoring, predictive analytics, and automated decision-making, UAVs are revolutionizing water resource management. However, overcoming cost, regulatory, and technical barriers is crucial for widespread adoption. As governments and private sectors invest in UAV technologies, agriculture is moving towards a smarter, more sustainable future where water use is optimized for maximum efficiency and productivity.

References

1. Mndela Y, Ndou N, Nyamugama A (2023) Irrigation scheduling for small-scale crops based on crop water content patterns derived from UAV multispectral imagery. Sustainability 15:12034
2. Wabela K, Hammani A, Abdelilah T et al (2022) Optimization of irrigation scheduling for improved irrigation water management in Bilate Watershed, Rift Valley, Ethiopia. Water 14:3960
3. Ahansal Y, Bouziani M, Yaagoubi R et al (2022) Towards smart irrigation: a literature review on the use of geospatial technologies and machine learning in the management of water resources in arboriculture. Agronomy 12:297
4. Mishra V, Avtar R, Prathiba A et al (2023) Uncrewed aerial systems in water resource management and monitoring: a review of sensors, applications, software, and issues. Adv Civ Eng 2023:3544724
5. Peng J, Nieto H, Andersen MN et al (2023) Accurate estimates of land surface energy fluxes and irrigation requirements from UAV-based thermal and multispectral sensors. ISPRS J Photogramm Remote Sens 198:238–254
6. Nesterenko G (2024) Evaluation of the effectiveness of the use of an irrigation system using UAVS. J Agric Environ
7. Massaoudi A, Berguiga A, Harchay A et al (2023) Spectral and energy efficiency trade-off in UAV-based olive irrigation systems. Appl Sci 13:10739
8. Khose SB, Mailapalli DR (2024) Spatial mapping of soil moisture content using very-high resolution UAV-based multispectral image analytics. Smart Agric Technol 8:100467
9. Garcia-Vasquez AC, Mokari E, Samani Z, Fernald A (2022) Using UAV-thermal imaging to calculate crop water use and irrigation efficiency in a flood-irrigated pecan orchard. Agric Water Manag 272:107824
10. Ortas I (2023) A solution to soil, water and environmental crises. In: A solution to the crises of soil, water, and climate in plant production, vol 18
11. Gago J, Douthe C, Coopman RE et al (2015) UAVs challenge to assess water stress for sustainable agriculture. Agric Water Manag 153:9–19
12. Safavi S (2024) Leveraging GIS for environmental planning. Glob Chall Environ Clim Change 22:308

13. Chatrabhuj MK, Mishra U, Omar PJ (2024) Integration of remote sensing data and GIS technologies in river management system. Discover Geosci 2:67
14. Rana R, Bhambri P (2025) The role of IoT in shaping the future of geospatial AI. In: Recent trends in geospatial AI. IGI Global Scientific Publishing, pp 177–216
15. Quamar MM, Al-Ramadan B, Khan K et al (2023) Advancements and applications of drone-integrated geographic information system technology—a review. Remote Sensing 15:5039
16. Garré S, Janssens P, Reynaert S et al (2020) Variable rate irrigation and nitrogen fertilization in potato; engage the spatial variation (potential)
17. Miller GR, Rajkai K, Ács F et al (2018) Water management for sustainable agriculture. Burleigh Dodds Science Publishing
18. Lendzioch T, Langhammer J, Vlček L, Minařík R (2021) Mapping the groundwater level and soil moisture of a montane peat bog using UAV monitoring and machine learning. Remote Sensing 13:907
19. Zhu S, Cui N, Zhou J et al (2023) Digital mapping of root-zone soil moisture using UAV-based multispectral data in a kiwifruit orchard of northwest China. Remote Sensing 15:646
20. Guan Y, Grote K (2023) Assessing the potential of UAV-based multispectral and thermal data to estimate soil water content using geophysical methods. Remote Sensing 16:61
21. Cheng Q, Su Q, Binley A et al (2023) Estimation of surface soil moisture by a multi-elevation UAV-based ground penetrating radar. Water Resour Res 59:e2022WR032621
22. Ge X, Ding J, Jin X et al (2021) Estimating agricultural soil moisture content through UAV-based hyperspectral images in the arid region. Remote Sensing 13:1562
23. Alqurashi FA, Alsolami F, Abdel-Khalek S et al (2022) Machine learning techniques in internet of UAVs for smart cities applications. J Intell Fuzzy Syst 42:3203–3226
24. Joshi R, Pandey K (2024) IoT-enabled UAV: a comprehensive review of technological change in Indian farming. In: Unmanned aircraft systems, pp 93–135
25. Makam S, Komatineni BK, Meena SS, Meena U (2024) Unmanned aerial vehicles (UAVs): an adoptable technology for precise and smart farming. Discov Internet Things 4:12
26. Ahmed Z, Gui D, Murtaza G et al (2023) An overview of smart irrigation management for improving water productivity under climate change in drylands. Agronomy 13:2113
27. Toscano F, Fiorentino C, Capece N et al (2024) Unmanned aerial vehicle for precision agriculture: a review. IEEE Access
28. Rahmah MM, Ahmed AA, Belrzaeg M (2025) Drone-based precision agriculture enhancing crop monitoring and management. Open Eur J Appl Sci (OEJAS) 36–50
29. Guebsi R, Mami S, Chokmani K (2024) Drones in precision agriculture: a comprehensive review of applications, technologies, and challenges. Drones 8:686
30. Koech R, Langat P (2018) Improving irrigation water use efficiency: a review of advances, challenges and opportunities in the Australian context. Water 10:1771
31. Faridi H, Ghoreishi B, Faridi H (2023) Intelligent irrigation and automation. In: Handbook of irrigation hydrology and management. CRC Press, pp 295–320
32. Kloppers B (2022) Developing a framework for a configuration of an irrigation system for small-scale farmers
33. Edan Y, Adamides G, Oberti R (2023) Agriculture automation. In: Springer handbook of automation. pp 1055–1078
34. Tian H, Wang T, Liu Y et al (2020) Computer vision technology in agricultural automation—a review. Inf Process Agric 7:1–19
35. Vashishth TK, Sharma V, Sharma KK et al (2024) Integration of unmanned aerial vehicles (UAVs) and IoT for crop monitoring and spraying. In: Internet of Things applications and technology. Auerbach Publications, pp 95–117
36. Tiwari AK, Mishra H, Nishad DC, Pandey A (2023) Sustainable water management in agriculture: irrigation techniques and water conservation. Dr Ajay B Jadhao 53
37. Saad A, Gamatié A (2020) Water management in agriculture: a survey on current challenges and technological solutions. IEEE Access 8:38082–38097
38. Ray S, Majumder S (2024) Water management in agriculture: innovations for efficient irrigation. Mod Agron 169–185

39. Khan N, Ray RL, Sargani GR et al (2021) Current progress and future prospects of agriculture technology: gateway to sustainable agriculture. Sustainability 13:4883
40. Bwambale E, Abagale FK, Anornu GK (2022) Smart irrigation monitoring and control strategies for improving water use efficiency in precision agriculture: a review. Agric Water Manag 260:107324
41. Wattanapanich C, Imjai T, Kefyalew F et al (2024) Integration of Internet of Things (IoT) and machine learning for management of ground water banks in drought-prone areas: a case study from IMJAI Organic Garden, Thailand. Eng Sci 31:1248
42. Cody RA, Narasimhan S (2020) A field implementation of linear prediction for leak-monitoring in water distribution networks. Adv Eng Inform 45:101103
43. Mashhadi N, Shahrour I, Attoue N et al (2021) Use of machine learning for leak detection and localization in water distribution systems. Smart Cities 4:1293–1315
44. Alves Coelho J, Glória A, Sebastião P (2020) Precise water leak detection using machine learning and real-time sensor data. IoT 1:474–493
45. Violino S, Figorilli S, Ferrigno M et al (2023) A data-driven bibliometric review on precision irrigation. Smart Agric Technol 5:100320
46. Adewusi AO, Asuzu OF, Olorunsogo T et al (2024) AI in precision agriculture: a review of technologies for sustainable farming practices. World J Adv Res Rev 21:2276–2285
47. Eissa M (2024) Precision agriculture using artificial intelligence and robotics. J Res Agric Food Sci 1:35–35
48. Mgendi G (2024) Unlocking the potential of precision agriculture for sustainable farming. Discov Agric 2:87
49. Nath S (2024) A vision of precision agriculture: balance between agricultural sustainability and environmental stewardship. Agron J 116:1126–1143
50. Padhiary M, Hoque A, Prasad G et al (2025) Precision agriculture and AI-driven resource optimization for sustainable land and resource management. In: Smart water technology for sustainable management in modern cities. IGI Global Scientific Publishing, pp 197–232
51. Vellingiri A, Kokila R, Nisha P et al (2025) Harnessing GPS, sensors, and drones to minimize environmental impact: precision agriculture. In: Designing sustainable internet of things solutions for smart industries. IGI Global, pp 77–108
52. Sigopi M, Shoko C, Dube T (2024) Advancements in remote sensing technologies for accurate monitoring and management of surface water resources in Africa: an overview, limitations, and future directions. Geocarto Int 39:2347935
53. Shu P, Aslam RW, Naz I et al (2025) Deep learning-based super-resolution of remote sensing images for enhanced groundwater quality assessment and environmental monitoring in urban areas. IEEE J Sel Top Appl Earth Observations Remote Sensing
54. Sheik AG, Kumar A, Sharanya AG et al (2024) Machine learning-based monitoring and design of managed aquifer rechargers for sustainable groundwater management: scope and challenges. Environ Sci Pollut Res 1–34
55. Gantayat RR, Elumalai V, Li P (2025) Future research imperatives in hydrogeology. In: Sustainable groundwater and environment: challenges and solutions. Springer, pp 365–385
56. Aarif KOM, Alam A, Hotak Y (2025) Smart Sensor Technologies Shaping the Future of Precision Agriculture: Recent Advances and Future Outlooks. Journal of Sensors 2025:2460098
57. Paul K, Chatterjee SS, Pai P et al (2022) Viable smart sensors and their application in data driven agriculture. Comput Electron Agric 198:107096
58. Sabir RM, Sarwar A, Shoaib M et al (2024) Managing water resources for sustainable agricultural production. In: Transforming agricultural management for a sustainable future: climate change and machine learning perspectives. Springer, pp 47–74
59. Shaikh M, Birajdar F (2024) Advancements in remote sensing and GIS for sustainable groundwater monitoring: applications, challenges, and future directions. Int J Res Eng Sci Manage 7:16–24
60. Eid MH, Shebl A, Eissa M et al (2024) Comprehensive approach integrating remote sensing, machine learning, and physicochemical parameters to detect hydrodynamic conditions and groundwater quality deterioration in non-rechargeable aquifer systems. Heliyon 10

61. Bwambale E, Naangmenyele Z, Iradukunda P et al (2022) Towards precision irrigation management: a review of GIS, remote sensing and emerging technologies. Cogent Eng 9:2100573
62. Romman ZA, Al-Smadi B, Weshah A (2024) A statistical analysis of the Effectiveness of groundwater related policies in mitigating over-extraction in arid regions: challenges and impacts. Groundw Sustain Dev 26:101203
63. Kannan P, Abishek J, Pazhanivelan S, Firnass MMRA (2025) Farming from above: the role of UAVs in modern agricultural practices. In: Digital farming and smart agriculture for sustainable future. Chapman and Hall/CRC, pp 235–254
64. Mehedi IM, Hanif MS, Bilal M et al (2024) Remote sensing and decision support system applications in precision agriculture: challenges and possibilities. IEEE Access
65. Mowla MN, Mowla N, Shah AS et al (2023) Internet of Things and wireless sensor networks for smart agriculture applications: a survey. IEEE Access 11:145813–145852
66. Louta M, Banti K, Karampelia I (2024) Emerging technologies for sustainable agriculture: the power of humans and the way ahead. IEEE Access
67. Balyan S, Jangir H, Tripathi SN et al (2024) Seeding a sustainable future: navigating the digital horizon of smart agriculture. Sustainability 16:475
68. Taha MF, Mao H, Zhang Z et al (2025) Emerging technologies for precision crop management towards agriculture 5.0: a comprehensive overview. Agriculture 15:582
69. Nahiyoon SA, Ren Z, Wei P et al (2024) Recent development trends in plant protection UAVs: a journey from conventional practices to cutting-edge technologies—a comprehensive review. Drones 8:457
70. Fuentes-Peñailillo F, Gutter K, Vega R, Silva GC (2024) Transformative technologies in digital agriculture: leveraging Internet of Things, remote sensing, and artificial intelligence for smart crop management. J Sens Actuator Netw 13:39
71. Jatav HS, Minkina T, Singh SK et al (2024) Environmental nexus for resource management. CRC Press
72. Sharma K, Shivandu SK (2024) Integrating artificial intelligence and internet of things (IoT) for enhanced crop monitoring and management in precision agriculture. Sens Int 100292
73. Reddy PP (2024) Smart farming technologies to attain food and nutrition security. CRC Press
74. Singh VP, Singh R, Paul PK et al (2024) Hydrological processes modelling and data analysis. Water Sci Technol Libr 127
75. Hamad TA (2024) Irrigation planning using geographic information system and remote sensing of the Baranti plain in Northern Iraq
76. Gupta V, Garg A, Agrawal S (2024) Remote monitoring and control systems in agriculture and farming. In: Recent trends in artificial intelligence towards a smart world: applications in industries and sectors. Springer, pp 279–294
77. Yacoob A, Gokool S, Clulow A et al (2024) Leveraging unmanned aerial vehicle technologies to facilitate precision water management in smallholder farms: a scoping review and bibliometric analysis. Drones 8:476
78. Khan S, Mazhar T, Shahzad T et al (2024) Future of sustainable farming: exploring opportunities and overcoming barriers in drone-IoT integration. Discov Sustain 5:1–22
79. Mohsan SAH, Othman NQH, Li Y et al (2023) Unmanned aerial vehicles (UAVs): practical aspects, applications, open challenges, security issues, and future trends. Intel Serv Robot 16:109–137
80. Emimi M, Khaleel M, Alkrash A (2023) The current opportunities and challenges in drone technology. Int J Electr Eng Sustain 74–89

Chapter 12
Environmental and Economic Impact of UAV Technology in Agriculture

Abstract This chapter explores the environmental and economic impact of UAV technology in agriculture, focusing on its role in improving efficiency, sustainability, and cost-effectiveness. UAVs offer advanced capabilities such as precision monitoring, aerial imaging, and targeted input application, which reduce resource consumption and environmental degradation. Economically, UAVs lower operational costs, enhance yield forecasting, and provide data-driven decision-making to optimize agricultural practices. Government subsidies and incentives are examined to reduce the financial burden of UAV adoption, fostering technological innovation and sustainable practices. UAV technology plays a crucial role in modern precision farming by optimizing resource use, reducing input costs, and minimizing environmental impacts. Advanced mapping, sensor integration, and automated delivery systems enable precise application of agricultural inputs, minimizing pollution and conserving resources. Waste reduction, composting organic waste, sustainable transportation, and eco-friendly consumption contribute to reducing environmental impacts. UAV technology supports sustainable agriculture by improving precision, reducing resource consumption, and minimizing environmental impacts. Advanced sensors monitor crop health in real time, identifying pests, diseases, and nutrient deficiencies, allowing for targeted interventions and sustainable practices like Integrated Pest Management. UAVs also reduce soil degradation and erosion by avoiding physical contact with the ground and enabling precise application of cover crop seeds and soil amendments. UAV technology offers significant ROI by improving efficiency, reducing costs, and increasing agricultural productivity. With advancements in technology and increasing affordability, UAV adoption is becoming an economically viable and sustainable investment for modern farmers.

Keywords UAV technology · Agriculture · Environmental impact · Economic benefits · Precision agriculture · Sustainability · Government subsidies · Data-driven decision-making · Resource optimization · Technological innovation

© The Author(s), under exclusive license to Springer Nature Singapore Pte Ltd. 2025
J. Li, *UAV Aerodynamics and Crop Interaction*, Smart Agriculture 13,
https://doi.org/10.1007/978-981-96-8402-1_12

12.1 Introduction

Unmanned Aerial Vehicles (UAVs) have become a transformative technology in agriculture, offering significant environmental and economic benefits. As the global agricultural sector faces increasing demands for food production while grappling with environmental sustainability challenges, UAV technology has become an indispensable tool for precision agriculture [1]. These advanced aerial systems enable farmers to monitor crops, assess field conditions, and optimize resource use, leading to improved yields, reduced environmental impact, and cost savings. One of the most profound environmental impacts of UAV technology in agriculture is the reduction in the use of chemical inputs such as pesticides and fertilizers. Traditional agricultural practices often rely on blanket applications of these chemicals, leading to overuse and environmental degradation. UAVs equipped with advanced imaging systems allow for precise monitoring of crop health, enabling targeted application of fertilizers and pesticides, reducing excess use and minimizing runoff into surrounding ecosystems. This helps mitigate soil and water pollution, safeguard biodiversity, and enhance the overall health of agricultural landscapes [2]. UAVs also contribute to water conservation efforts by providing accurate data on soil moisture levels and crop water requirements, enabling farmers to identify areas experiencing drought stress and implement targeted irrigation strategies. This supports sustainable agriculture and aligns with global efforts to combat climate change and preserve natural resources. From an economic perspective, UAV technology offers substantial cost savings and increased operational efficiency for farmers [3]. Traditional methods of field monitoring and crop assessment are often labor-intensive and time-consuming, but UAVs can cover large areas swiftly, providing comprehensive and accurate data. UAV technology facilitates early detection of pest infestations and crop diseases, enabling timely intervention and preventing significant yield losses. Despite these advantages, the widespread adoption of UAV technology in agriculture presents challenges, such as regulatory frameworks varying across regions and initial investment in UAV equipment and technical expertise [4]. However, as UAV technology continues to evolve and become more accessible, its environmental and economic benefits will likely play a vital role in shaping the future of sustainable agriculture.

12.2 Reducing Environmental Footprints

Reducing environmental footprints involves minimizing the negative impact of human activities on the environment by conserving resources, reducing pollution, and promoting sustainable practices. One effective way to achieve this is through energy conservation, such as using energy-efficient appliances, switching to renewable energy sources like solar or wind, and reducing unnecessary power consumption. Waste reduction also plays a crucial role, which can be achieved by following the principles of reduce, reuse, and recycle, minimizing single-use plastics, and composting organic waste. Sustainable transportation methods, like using public transit, carpooling, biking, or choosing electric vehicles, help lower greenhouse gas emissions [5]. Water conservation is another vital aspect, which includes using water-efficient fixtures, repairing leaks, and adopting practices like rainwater harvesting. Additionally, sustainable food choices, such as supporting local and organic products, reducing food waste, and opting for plant-based diets, can significantly reduce environmental harm. By adopting these practices, individuals and communities can reduce their ecological impact and contribute to a healthier and more sustainable planet.

12.2.1 Energy Conservation

Energy conservation is an essential strategy for reducing environmental footprints and preserving natural resources. It involves using energy more efficiently and adopting practices that lower overall consumption. One effective method is utilizing energy-efficient appliances, such as LED bulbs and ENERGY STAR-rated devices, which consume less electricity while maintaining performance. Transitioning to renewable energy sources like solar and wind also reduces reliance on fossil fuels, cutting greenhouse gas emissions. Furthermore, simple daily habits, such as unplugging devices when not in use, turning off lights, and optimizing heating and cooling systems, play a significant role in conserving energy [6]. By adopting these practices, individuals can reduce energy waste, lower utility costs, and contribute to a more sustainable future.

12.2.2 Water Conservation

Water conservation is crucial for reducing environmental impact and ensuring sustainable use of water resources. With global water scarcity increasing, adopting effective conservation methods is essential for future generations. Fixing leaks and using water-efficient fixtures can significantly reduce daily water consumption. Rainwater harvesting for non-potable uses like irrigation and cleaning reduces demand on municipal water systems and conserves treated drinking water. Advanced irrigation systems, like drip irrigation and sprinkler systems, can minimize water loss by delivering water directly to plant roots, making irrigation more efficient and sustainable [7]. Landscaping practices, such as xeriscaping, using drought-resistant plants and minimizing lawn areas, can also reduce outdoor water use. Mindful water use in daily activities, such as turning off taps while brushing teeth, running dishwashers with full loads, and watering gardens during cooler times, further reduces water consumption. Water conservation is a shared responsibility that requires conscious efforts at individual, community, and global levels. By adopting water-saving technologies and sustainable practices, we can protect this valuable resource, reduce environmental harm, and ensure water availability for future generations [8].

12.2.3 Waste Reduction

Waste reduction is crucial for minimizing pollution, conserving natural resources, and reducing the strain on landfills. One effective method is avoiding single-use plastics and excessive packaging, which helps decrease the amount of waste generated. Choosing reusable items, such as shopping bags, water bottles, and containers, extends product lifespans and reduces the need for disposable alternatives. Recycling materials like paper, plastic, metal, and glass is another key strategy, as it diverts waste from landfills, conserves raw materials, and reduces energy consumption during production [9]. Composting organic waste, such as food scraps and yard trimmings, transforms biodegradable materials into nutrient-rich soil, supporting sustainable agriculture while reducing landfill contributions. By adopting these practices, individuals and communities can significantly lower their environmental impact and promote a more sustainable future. Waste reduction minimizes pollution and conserves resources by avoiding single-use plastics, reusing items, and recycling materials. Composting organic waste reduces landfill contributions and promotes soil health. These actions collectively reduce environmental harm and support sustainability [10].

12.2.4 Sustainable Transportation

Sustainable transportation plays a key role in reducing greenhouse gas emissions and minimizing the environmental impact of human mobility. Personal vehicles are a major source of carbon emissions, but choosing alternative transportation methods can significantly lower this footprint. Using public transportation, such as buses and trains, reduces the number of individual vehicles on the road, leading to decreased fuel consumption and air pollution. Carpooling is another effective strategy, as sharing rides with others reduces the overall number of vehicles in use. Biking and walking are the most eco-friendly options, as they produce no emissions and promote better personal health. Driving fuel-efficient or electric vehicles also contributes to sustainability by reducing dependence on fossil fuels and lowering greenhouse gas emissions. Electric vehicles (EVs), in particular, produce zero tailpipe emissions and become even more sustainable when powered by renewable energy sources [11]. Additionally, minimizing air travel helps reduce carbon emissions, as airplanes produce significant greenhouse gases. When flying is unavoidable, participating in carbon offset programs can help balance the environmental impact by investing in projects that reduce or capture emissions, such as reforestation and renewable energy initiatives. Sustainable transportation practices, such as public transit, carpooling, biking, and driving fuel-efficient or electric vehicles, can significantly reduce environmental impact. These practices conserve natural resources and contribute to a cleaner, healthier planet. Small daily changes, combined with policy and technological advancements, are crucial for long-term environmental sustainability [12]. Minimizing air travel and participating in carbon offset programs further contribute to a cleaner future.

12.2.5 Sustainable Agriculture and Food Choices

Sustainable agriculture and food choices are crucial for reducing environmental footprints and preserving natural resources. Supporting local and organic farms reduces the need for long-distance food transportation and minimizes the use of synthetic pesticides and fertilizers. Locally-sourced foods require less energy for storage and distribution, while organic farming practices promote soil health and biodiversity. Reducing meat consumption and adopting plant-based diets significantly decrease the carbon footprint associated with food production, as livestock farming contributes to greenhouse gas emissions, deforestation, and water usage. Minimizing food waste is another crucial aspect of sustainable food practices [13]. Better meal planning, proper food storage, and consuming leftovers can prevent waste and contribute to methane emissions. Composting organic waste converts food scraps into nutrient-rich soil, reducing landfill contributions and enhancing soil quality for future agricultural use. Informed food choices and sustainable agricultural practices can reduce environmental impact, promote biodiversity, and contribute to a more resilient food system. Small changes in daily habits and community initiatives can have a significant positive effect on the planet's health [14].

12.2.6 Eco-friendly Consumption

Eco-friendly consumption is a conscious approach to purchasing that promotes sustainability and reduces environmental harm. It involves choosing sustainable products with eco-labels and fair-trade certifications, which support ethical manufacturing processes and environmentally responsible practices. Supporting companies that prioritize reducing their environmental impact encourages corporate sustainability and encourages businesses to adopt greener practices. Durable goods, such as reusable bags, bottles, and household items, reduce waste and demand for raw materials. Repairing and repurposing items extends their lifecycle and minimizes landfill waste. Consumers can also make informed choices about their purchases, choosing items made from recycled or sustainably sourced materials and avoiding excessive packaging [15]. Mindful consumption, such as buying only what is necessary, conserves resources and reduces environmental strain caused by mass production and waste. By adopting eco-friendly consumption habits, individuals contribute to a more sustainable future by reducing pollution, conserving natural resources, and supporting businesses committed to environmental responsibility.

12.2.7 Protecting Ecosystems

Protecting and restoring natural ecosystems is crucial for environmental sustainability and biodiversity preservation. Ecosystems offer essential services like clean air, water purification, and climate regulation, which are vital for human well-being and the planet's health. Participating in reforestation and habitat restoration efforts helps repair damaged ecosystems, restore biodiversity, and combat climate change. Planting native trees and restoring degraded lands supports wildlife return and improves natural habitat resilience [16]. Consumers can also protect ecosystems by avoiding products that contribute to deforestation, such as palm oil, soy, or timber. Choosing sustainably certified products, such as those with Rainforest Alliance or Forest Stewardship Council labels, reduces pressure on forests and supports responsible sourcing. Preserving green spaces, such as parks, wetlands, and wildlife reserves, supports biodiversity by providing shelter and food for various species. Reducing pollution and limiting harmful chemicals also helps maintain healthy ecosystems. Individual and collective actions can ensure the survival of diverse species and the long-term health of the environment. Sustainable practices, such as supporting conservation initiatives and making responsible consumer choices, are essential for preserving the planet's natural resources for future generations [17].

12.2.8 Advocacy and Education

Advocacy and education are crucial for driving systemic change and promoting environmental sustainability. Individuals can influence decision-makers to adopt regulations that protect natural resources and combat climate change. Supporting initiatives like renewable energy investments, pollution control measures, and wildlife conservation policies helps create a legal framework for long-term environmental protection. Engaging in public forums, signing petitions, and contacting policymakers are effective ways to advocate for environmental causes. Education plays a crucial role in fostering awareness and encouraging sustainable behaviors. Learning about environmental issues like climate change, biodiversity loss, and pollution allows individuals to make informed decisions and inspire others [18]. Sharing knowledge through workshops, social media, and community outreach promotes collective action and empowers more people to adopt eco-friendly habits. Supporting organizations dedicated to environmental conservation amplifies these efforts. Volunteering, donating, or participating in awareness campaigns strengthens their ability to drive positive change. Through advocacy and education, individuals and communities can influence public policies, promote sustainable lifestyles, and contribute to a global movement for environmental stewardship, ensuring a healthier, more sustainable future for generations to come [19].

12.3 Precision Application of Inputs Using UAVs in Agriculture

Unmanned Aerial Vehicles (UAVs), or drones, have revolutionized agricultural practices by enabling precise application of essential inputs like fertilizers, pesticides, and herbicides. Traditional methods often lead to excessive use, uneven distribution, and increased costs. UAV technology addresses these challenges by providing a targeted and efficient approach to input delivery. UAVs can map fields with exceptional accuracy, identifying areas that require specific treatment. Multispectral and thermal sensors on UAVs can detect crop health issues, soil moisture levels, and pest infestations, allowing them to deliver inputs precisely to affected zones, reducing the risk of over-application and chemical runoff. This targeted approach conserves resources and enhances crop yield by ensuring plants receive the right amount of nutrients and protection [20]. UAV-based precision application also reduces labor and operational costs, as they cover large areas quickly with minimal human intervention. This improves efficiency, saves time, and reduces physical risks associated with manual input application. UAV technology supports sustainable agriculture by reducing the environmental footprint, mitigating risks of soil degradation and water pollution [21]. This sustainable approach aligns with modern agricultural goals of increasing productivity while preserving natural resources.

12.4 Minimizing Soil and Water Pollution

The use of Unmanned Aerial Vehicles (UAVs) in agricultural input applications is a significant step towards reducing soil and water pollution. Traditional broadcasting methods for pesticides, herbicides, and fertilizers are often inefficient, leading to over-application and uneven distribution, which can contaminate aquatic ecosystems and soil. UAV technology offers precise and targeted application of agrochemicals, reducing wastage and environmental impact. Advanced sensors and GPS-guided systems enable UAVs to accurately map fields, identify zones requiring treatment, and selectively apply inputs. This site-specific approach not only minimizes pollution but also enhances soil health by preventing harmful residue accumulation [22]. UAVs also play a vital role in crop health monitoring through remote sensing and multispectral imaging, enabling early detection of pests, diseases, and nutrient deficiencies, enabling timely intervention with minimal chemical use. They can also access areas difficult for traditional machinery, ensuring uniform application and preventing erosion. The adoption of UAVs in precision agriculture not only improves environmental sustainability but also offers economic benefits, reducing input usage and preserving natural resources for future agricultural productivity [23].

12.5 Promoting Sustainable Agricultural Practices

Unmanned Aerial Vehicle (UAV) technology is revolutionizing agriculture by promoting sustainable practices that balance productivity with environmental stewardship. UAVs collect high-resolution data through advanced sensors, enabling farmers to monitor crop health, assess soil conditions, and detect pest infestations with accuracy. This data allows for precise application of inputs like water, fertilizers, and pesticides, reducing overall consumption and chemical runoff. Precision application not only improves crop yields but also protects soil and water resources from overuse and contamination [24]. UAVs also support sustainable land management by facilitating regular field monitoring without heavy machinery, allowing for early detection of issues like erosion, nutrient deficiencies, or disease outbreaks. This proactive approach reduces crop loss and enhances farm efficiency. UAVs also contribute to reducing the carbon footprint of agricultural operations by replacing conventional methods with fuel-intensive machinery, reducing energy consumption and greenhouse gas emissions. Precision irrigation techniques conserve water by targeting specific areas and reducing waste. UAV technology plays a crucial role in advancing sustainable agriculture by improving precision, conserving resources, and minimizing environmental impacts [25].

12.5.1 Precision Input Application

UAVs are revolutionizing precision agriculture by enabling the precise and targeted delivery of agricultural inputs like fertilizers, pesticides, and herbicides. Traditional methods often result in overuse of chemicals, leading to increased costs, environmental degradation, and soil contamination. UAVs equipped with advanced sensors and GPS technology can deliver inputs precisely where they are needed, optimizing efficiency and minimizing waste. This approach reduces input wastage, agrochemical runoff, and water pollution by applying chemicals only to affected areas. It also preserves biodiversity and ecological balance by limiting harmful substances' spread to non-target areas. UAV-based input application also benefits soil health by preventing toxic residue accumulation, ensuring long-term agricultural productivity [26]. This method also allows for site-specific management, allowing farmers to address variations in soil conditions and crop needs with greater accuracy. UAVs enhance sustainable farming practices by providing a more efficient, precise, and environmentally responsible approach to input application, promoting crop yields, resource efficiency, and the long-term health of agricultural lands.

12.5.2 Monitoring and Managing Crop Health

Advanced UAVs are revolutionizing crop health monitoring by providing real-time, high-resolution data. These drones can quickly identify areas affected by pests, diseases, and nutrient deficiencies, allowing for targeted interventions and reduced crop losses. This approach minimizes chemical runoff, protects ecosystems, and reduces the environmental impact of agricultural practices. UAV-based crop monitoring also supports Integrated Pest Management (IPM), a sustainable agricultural strategy that combines biological, cultural, physical, and chemical tools to manage pests in an eco-friendly way. By providing detailed data on pest activity and crop health, UAVs help farmers make informed decisions about when and where to use biological controls and apply pesticides only when necessary. Furthermore, UAV technology enhances farm operations efficiency by enabling frequent and consistent monitoring over large areas. This proactive approach improves crop resilience and minimizes disease spread [27]. UAVs are a vital tool in sustainable agriculture, enabling precise monitoring and management of crop health. Their real-time data supports smarter decision-making, reduces chemical use, and promotes environmentally responsible farming practices.

12.5.3 Reducing Soil Degradation and Erosion

UAV technology significantly reduces soil degradation and erosion by minimizing physical disturbance and enabling targeted soil management practices. Traditional agricultural machinery, such as tractors and heavy sprayers, can compact soil, weakening its ability to absorb water and air, and accelerating erosion. UAVs operate without direct contact with the ground, preserving soil integrity and reducing mechanical stress. One effective method is the targeted application of cover crop seeds, which protect bare soil by providing a vegetative cover. UAVs can efficiently sow seeds in vulnerable areas, preventing soil loss and improving soil health [28]. UAVs can also deliver soil amendments, such as organic fertilizers and biochar, directly to areas with low soil fertility, enhancing soil structure and boosting organic matter content. This method reduces overuse of inputs and prevents excess chemicals from degrading the soil. High-resolution aerial data from UAVs helps farmers monitor soil conditions and identify erosion risks, enabling early intervention. UAVs support sustainable soil management by preventing compaction, reducing erosion, and promoting soil regeneration. Their precision and non-intrusive operation help maintain healthy soils, essential for long-term agricultural productivity and environmental sustainability [29].

12.5.4 Water Conservation

UAV technology is revolutionizing water conservation in agriculture by enabling precise monitoring and targeted irrigation. Traditional methods often lead to over-irrigation and under-irrigation, but UAVs equipped with thermal and infrared sensors can detect variations in soil moisture levels, providing farmers with real-time data. This allows for water application only where it is needed, reducing water wastage. UAVs also help farmers implement precision irrigation techniques, optimizing water distribution and ensuring crops receive adequate hydration without excess use. This approach conserves freshwater resources and reduces the energy required for irrigation systems. Reducing over-irrigation is crucial for preventing environmental issues like soil salinization and nutrient leaching. UAV-based irrigation monitoring helps mitigate these problems by ensuring water is delivered only where and when needed, preserving soil quality and local ecosystems [30]. UAVs also support sustainable practices by providing detailed maps of water distribution patterns, enabling farmers to adjust irrigation schedules and adopt water-saving techniques like drip irrigation or variable-rate irrigation (VRI). This proactive water management reduces operating costs and enhances the long-term sustainability of agricultural production. Overall, UAV technology is transforming water conservation in agriculture by providing precise moisture monitoring and efficient irrigation.

12.5.5 Biodiversity Preservation

UAV technology is crucial in preserving biodiversity by reducing the need for extensive chemical applications in agriculture. Traditional methods often harm non-target organisms, such as beneficial insects, pollinators, and soil microorganisms. UAVs allow for targeted delivery of these inputs, minimizing damage to ecosystems and promoting biodiversity conservation. By applying chemicals only where needed, UAVs minimize the exposure of beneficial species, such as bees, butterflies, and natural predators, which play crucial roles in pollination and pest control. This precision reduces ecological disruption and maintains the delicate balance of agricultural ecosystems. UAVs also contribute to integrated pest management strategies, prioritizing biological and mechanical controls over chemical treatments [31]. By providing real-time data on pest activity and plant health, UAVs enable farmers to deploy targeted interventions, reducing reliance on broad-spectrum pesticides. Additionally, UAVs assist in habitat monitoring and conservation efforts by surveying areas adjacent to farmland, identifying vulnerable ecosystems, and planting cover crops and wildflower strips. This technology promotes ecological balance while allowing agricultural operations to coexist with natural ecosystems.

12.5.6 Reducing Carbon Footprint

UAV technology is a significant tool in reducing the carbon footprint of agricultural operations by minimizing the use of heavy, fuel-intensive machinery. Traditional farming methods, which rely on tractors and other large equipment, consume significant amounts of fossil fuels, contributing to greenhouse gas emissions. UAVs offer a more energy-efficient alternative for performing these tasks with greater precision and minimal environmental impact. By enabling targeted input application and aerial monitoring, UAVs reduce the number of field passes required, resulting in lower emissions. For example, drones can accurately deliver fertilizers and pesticides to affected areas, conserving inputs and reducing energy requirements for production, transportation, and application [32]. UAVs also support precision agriculture practices, optimizing resource use and minimizing environmental impact. For example, drones equipped with multispectral and thermal sensors can identify specific areas needing water or nutrients, reducing wasteful over-application and improving farm efficiency. Additionally, UAVs contribute to climate change mitigation by reducing soil compaction caused by heavy machinery, preserving soil structure, and promoting healthier soils as carbon sinks. In summary, UAV technology enhances sustainable agriculture by lowering energy consumption, reducing the need for fossil fuel-driven machinery, and promoting efficient input use [33].

12.5.7 Enhancing Data-Driven Decision-Making

UAV technology is revolutionizing agriculture by providing precise, high-resolution data on crop growth, soil conditions, and environmental factors. Equipped with advanced sensors like multispectral, thermal, and LiDAR, UAVs collect detailed information across large agricultural areas quickly and efficiently, enabling farmers to monitor field conditions in real time, identify potential issues, and make informed decisions to improve productivity and sustainability. UAV-based data collection supports precision agriculture practices, allowing farmers to target specific areas for optimal resource use, reducing waste and environmental impact. UAV insights also facilitate the implementation of sustainable agricultural practices, such as crop rotation, conservation tillage, and organic amendments. By providing accurate and timely information, UAVs help farmers customize their practices to align with each field's unique needs, enhancing long-term soil fertility and ecological balance [34]. Furthermore, UAVs improve the efficiency of field monitoring over time by allowing for frequent and consistent data collection. This longitudinal data helps farmers track changes in crop performance, assess the effectiveness of management practices, and adjust strategies as needed. This proactive approach minimizes crop loss, reduces operational costs, and supports sustainable production systems. UAVs empower farmers to make evidence-based decisions by delivering accurate, real-time data, enhancing resource efficiency, supporting sustainable practices, and promoting long-term agricultural productivity while reducing environmental impacts [35].

12.5.8 Promoting Economic and Social Sustainability

UAV technology is a crucial tool in promoting economic and social sustainability in agriculture. It reduces input costs by enabling precise application of fertilizers, pesticides, and herbicides, minimizing waste and lowering overall expenditure. UAVs also improve crop management by providing real-time data on crop health, soil conditions, and environmental factors, allowing farmers to address problems like nutrient deficiencies and pest infestations promptly. This early detection and intervention improve crop quality and output, leading to greater profitability. UAV services provide an affordable entry point for smallholder farmers, enabling them to adopt advanced agricultural practices without significant financial burdens. They also facilitate better planning and risk management by providing insights into changing weather patterns and crop conditions, especially in regions vulnerable to climate change [36]. From a social sustainability perspective, UAV technology contributes to food security by increasing agricultural efficiency and reducing environmental degradation. Sustainable farming practices, such as reduced chemical use and improved soil management, protect the environment and ensure long-term food production. Adoption of UAVs creates new employment opportunities in rural areas, including drone operation, data analysis, and technical support [37]. Overall, UAV technology enhances the economic viability of agriculture while fostering social sustainability through increased productivity, cost reduction, and improved rural livelihoods.

12.6 Promoting Economic and Social Sustainability

UAV technology is a crucial tool in promoting economic and social sustainability in agriculture. It reduces input costs by enabling precise application of fertilizers, pesticides, and herbicides, minimizing waste and lowering overall expenditure. UAVs also improve crop management by providing real-time data on crop health, soil conditions, and environmental factors, allowing farmers to address problems like nutrient deficiencies and pest infestations promptly. This early detection and intervention improve crop quality and output, leading to greater profitability. UAV services provide an affordable entry point for smallholder farmers, enabling them to adopt advanced agricultural practices without significant financial burdens [38]. They also facilitate better planning and risk management by providing insights into changing weather patterns and crop conditions, especially in regions vulnerable to climate change. From a social sustainability perspective, UAV technology contributes to food security by increasing agricultural efficiency and reducing environmental degradation. Sustainable farming practices, such as reduced chemical use and improved soil management, protect the environment and ensure long-term food production. Adoption of UAVs creates new employment opportunities in rural areas, including drone operation, data analysis, and technical support. Overall, UAV technology enhances the economic viability of agriculture while fostering social sustainability through increased productivity, cost reduction, and improved rural livelihoods [39].

12.7 Supporting Ecosystem Balance with Technology

UAVs are a vital tool in supporting ecosystem balance by enabling precision agriculture and reducing the environmental impact of conventional farming. Their advanced monitoring capabilities and real-time data collection enable farmers to adopt sustainable practices while maintaining productivity and preserving natural ecosystems. UAVs can monitor and protect vulnerable areas, such as soil erosion zones, by identifying preventative measures like cover cropping and contour farming. They also aid in biodiversity conservation by tracking plant and animal activity across agricultural lands, protecting wildlife habitats and ensuring farming operations do not disrupt natural ecosystems [40]. UAVs can identify areas with high pollinator activity, preserving these zones and supporting essential species like bees and butterflies. UAVs also provide precise surveillance of pest populations, allowing for targeted pesticide application, reducing chemical use, protecting beneficial insects, and limiting environmental contamination. They also assist in water conservation by monitoring soil moisture and guiding efficient irrigation practices, reducing water waste and preventing problems like soil salinization [41].

Importance of UAVs in Supporting Ecosystem Balance

1. **Soil Conservation**: UAVs identify erosion-prone areas and guide sustainable land management.
2. **Biodiversity Preservation**: UAVs monitor wildlife and pollinator activity, enabling farmers to protect natural habitats.
3. **Reduced Chemical Use**: Through targeted input application, UAVs minimize pesticide use and reduce environmental pollution.
4. **Water Management**: UAVs monitor moisture levels, optimizing irrigation and conserving water resources.
5. **Sustainable Practices**: Data-driven insights from UAVs support eco-friendly farming practices and long-term agricultural sustainability.

UAVs play a vital role in balancing agricultural productivity with environmental stewardship. By facilitating precision agriculture and promoting sustainable practices, UAV technology helps protect ecosystems, conserve resources, and ensure the long-term health of agricultural landscapes [8].

12.7.1 Identifying and Mitigating Erosion Risks

UAVs (Unmanned Aerial Vehicles) are transforming erosion risk management by providing accurate and real-time data on soil conditions. Equipped with high-resolution cameras and multispectral sensors, UAVs can capture detailed images of agricultural landscapes, allowing for the identification of areas vulnerable to soil erosion. This advanced technology enables farmers to monitor erosion patterns, assess soil degradation, and implement effective mitigation strategies. One key advantage of UAVs is their ability to map erosion-prone areas with precision. By analyzing elevation changes, vegetation cover, and moisture levels, UAVs identify regions susceptible to water and wind erosion. For example, UAV surveys can detect exposed slopes, compacted soils, or poorly vegetated areas that are more likely to experience erosion [42]. This data allows farmers to take proactive measures to prevent soil loss and maintain long-term soil health. Farmers can use UAV-generated insights to implement targeted erosion control measures. These include:

1. **Planting Cover Crops**: UAVs help identify bare or fragile areas where cover crops can be sown to protect the soil. Cover crops prevent erosion by stabilizing the soil, reducing surface runoff, and improving organic matter content.
2. **Constructing Buffer Strips**: UAV data can pinpoint erosion hotspots near **water bodies**, allowing farmers to install vegetative buffer strips. These strips filter sediment and prevent pollutants from reaching streams and rivers, thereby protecting water quality.
3. **Practicing No-Till Farming**: UAVs detect areas with high soil disturbance where **no-till techniques** can be adopted. This practice minimizes soil disruption, preserves soil structure, and reduces erosion caused by wind and water.

Moreover, UAVs can monitor the effectiveness of these mitigation measures over time. By conducting regular aerial surveys, farmers can track changes in erosion patterns and adjust management practices accordingly. This continuous monitoring helps maintain land productivity while reducing the environmental impact of soil erosion [43].

Importance of UAVs in Mitigating Erosion Risks

1. **Accurate Risk Identification**: UAVs provide precise mapping of erosion-prone areas, allowing for targeted intervention.
2. **Improved Soil Conservation**: UAV data supports the implementation of erosion control measures, preserving soil health and fertility.
3. **Water Quality Protection**: UAV-guided buffer strips and reduced runoff help protect nearby water bodies from sediment pollution.
4. **Sustainable Land Management**: By promoting conservation practices like no-till and cover cropping, UAVs support long-term agricultural sustainability.
5. **Ongoing Monitoring**: UAVs offer continuous surveillance of erosion trends, enabling farmers to evaluate and refine their erosion prevention strategies.

UAV technology enhances erosion risk management by providing detailed data to identify and mitigate soil erosion. This targeted approach reduces soil loss, protects water quality, and ensures the sustainability of agricultural landscapes [44].

12.7.2 Monitoring Biodiversity and Wildlife Movement

UAVs (Unmanned Aerial Vehicles) play a significant role in monitoring biodiversity and tracking wildlife movement across agricultural landscapes. Equipped with high-resolution cameras, thermal imaging, and multispectral sensors, UAVs capture detailed images that provide valuable insights into the presence, behavior, and distribution of plant and animal species. This technology allows for comprehensive monitoring while minimizing human disturbance to sensitive ecosystems [45]. One of the key applications of UAVs is tracking wildlife movement. By conducting regular aerial surveys, UAVs can observe the migration patterns of animals, identify breeding areas, and monitor how agricultural practices affect local wildlife. For example, UAVs have been used to track the movement of pollinators like bees, which are essential for crop fertilization, and larger species such as deer or birds, which can indicate ecosystem health. UAVs also identify habitat loss caused by agricultural expansion, deforestation, or land-use changes. By comparing aerial imagery over time, farmers can detect shrinking natural areas and assess how their activities impact biodiversity [46]. This data helps inform conservation strategies and ensures that critical habitats are protected. With these insights, farmers can implement measures to preserve and restore biodiversity, such as:

1. **Creating Wildlife Corridors**: UAVs help identify areas where wildlife corridors can be established. These corridors allow animals to move safely between habitats, reducing the impact of fragmented landscapes.
2. **Preserving Natural Habitats**: UAV mapping supports the preservation of wetlands, forests, and other natural areas that are essential for biodiversity.
3. **Adopting Eco-Friendly Land Practices**: UAV data allows farmers to adjust crop rotation patterns, reduce pesticide usage in sensitive zones, and implement buffer strips that protect native species.

Moreover, UAVs provide real-time monitoring, allowing for quick responses to environmental changes. For instance, if a habitat is at risk due to increased agricultural activity, UAVs can detect these changes early and guide conservation efforts before irreversible damage occurs [47].

Importance of UAVs in Monitoring Biodiversity and Wildlife Movement

1. **Enhanced Observation**: UAVs provide comprehensive, real-time data on plant and animal species across large areas.
2. **Wildlife Protection**: Monitoring movement patterns helps protect species from habitat destruction and supports wildlife conservation.

3. **Informed Decision-Making**: Accurate biodiversity data allows farmers to adopt sustainable land-use practices that minimize ecological harm.
4. **Habitat Preservation**: UAVs help identify and monitor critical habitats, ensuring their protection from agricultural expansion.
5. **Adaptive Management**: Continuous UAV surveillance allows for responsive and adaptive conservation strategies based on changing environmental conditions.

UAVs are powerful tools for monitoring biodiversity and wildlife movement, enabling farmers to balance agricultural productivity with environmental stewardship. This technology supports biodiversity conservation by identifying habitat loss, guiding the creation of wildlife corridors, and promoting sustainable land-use practices [48].

12.7.3 Supporting Integrated Pest Management (IPM)

UAVs are a vital tool in Integrated Pest Management (IPM) by providing real-time surveillance of pest populations and crop health. They use advanced technologies like thermal imaging, multispectral analysis, and high-resolution cameras to detect pest infestations, disease outbreaks, and crop stress with precision. This early detection allows farmers to implement targeted interventions, reducing the need for widespread chemical treatments. UAVs can identify pest hotspots, capturing detailed images that reveal areas affected by pests or diseases. This information helps farmers apply localized treatments, reducing chemical use and operational costs. UAVs also support biological pest control by tracking pest and predator populations, allowing farmers to integrate natural enemies into pest management strategies [49]. They facilitate timely interventions, preventing pest outbreaks from spreading across fields. For example, if a UAV detects an infestation of aphids, farmers can immediately apply spot treatments with biological controls or eco-friendly pesticides, minimizing crop damage and limiting environmental contamination. Regular aerial surveys provide data on pest population dynamics and treatment success, allowing farmers to refine their management practices [50].

Importance of UAVs in Supporting IPM

1. **Early Pest Detection**: UAVs identify pest hotspots and disease outbreaks before visible signs appear, enabling timely action.
2. **Targeted Interventions**: UAV data allows for precise, localized pest control, reducing the need for broad-spectrum pesticides.
3. **Environmental Protection**: Reduced pesticide use minimizes soil, water, and air contamination while preserving biodiversity.
4. **Promoting Beneficial Insects**: UAVs support the use of biological pest control by protecting pollinators and natural predators.
5. **Improved Monitoring**: Continuous UAV surveillance tracks pest activity and the effectiveness of IPM strategies, enabling adaptive management.

UAVs significantly enhance Integrated Pest Management by providing real-time, accurate pest surveillance and enabling targeted, eco-friendly interventions. This technology reduces chemical dependency, protects beneficial organisms, and supports sustainable, environmentally responsible agriculture [51].

12.7.4 Optimizing Resource Use

Unmanned Aerial Vehicles (UAVs) are revolutionizing agriculture by providing detailed data on crop health, soil conditions, and environmental variables. This information enables farmers to optimize resource use, reduce waste, and improve efficiency. UAVs support precision agriculture by applying water, fertilizers, and other inputs precisely where and when needed, minimizing overuse and environmental harm. UAVs equipped with multispectral and thermal sensors detect variations in plant health and soil moisture, allowing for targeted fertilizer application. In irrigation management, UAVs identify low moisture levels and adjust irrigation schedules to conserve water resources [52]. They also monitor environmental factors like temperature, humidity, and wind patterns, guiding precise pesticide spraying and minimizing chemical exposure to non-target areas. By reducing excess input use, UAVs lower production costs, allowing farmers to maintain or improve crop yields while reducing pollution. Optimized input use also reduces the ecological footprint of farming operations, protecting natural ecosystems. Overall, UAVs are revolutionizing agriculture by providing valuable insights into crop health, soil conditions, and environmental factors [53].

Importance of UAVs in Optimizing Resource Use

1. **Precision Input Delivery**: UAVs ensure accurate application of water, fertilizers, and pesticides, reducing resource waste.
2. **Cost Efficiency**: Optimized input use lowers operational costs by minimizing excess chemical and water consumption.
3. **Environmental Protection**: Precision methods prevent chemical runoff, protecting soil health and aquatic ecosystems.
4. **Water Conservation**: UAVs identify moisture-deficient zones, allowing for targeted and efficient irrigation.
5. **Sustainable Practices**: Improved resource management supports long-term agricultural sustainability and ecosystem balance.

UAV technology enhances resource efficiency by providing real-time, accurate data for input management. This reduces waste, lowers costs, and protects the environment, making agricultural practices more sustainable and productive [27].

12.7.5 Enhancing Conservation Efforts

UAV technology is crucial in supporting conservation agriculture by providing real-time data for sustainable land management. It allows for better decision-making and compliance with environmental standards. UAVs monitor reforestation projects, assess the health of conservation buffers, and evaluate the effectiveness of eco-friendly agricultural practices. This data helps farmers and policymakers maintain critical areas and improve water quality. UAVs also evaluate the effectiveness of sustainable farming practices, such as no-till farming, cover cropping, and crop rotation. By comparing soil quality, crop health, and vegetation cover over time, UAVs provide insight into how these practices impact long-term land productivity and ecosystem health [54]. UAVs also enforce environmental regulations by providing objective evidence of compliance with sustainable land-use policies. This aerial surveillance improves transparency and helps regulatory bodies enforce environmental laws. Overall, UAVs play a vital role in supporting conservation agriculture and ensuring sustainable land use.

Importance of UAVs in Enhancing Conservation Efforts

1. **Accurate Monitoring**: UAVs track reforestation progress and conservation buffer health with precise data collection.
2. **Sustainable Land Management**: UAVs assess the impact of eco-friendly practices, supporting informed land-use decisions.
3. **Environmental Compliance**: UAVs provide evidence to verify adherence to environmental policies and regulations.
4. **Erosion and Runoff Control**: UAV data helps evaluate and maintain conservation buffers, reducing soil and water pollution.
5. **Efficient Resource Allocation**: UAVs identify areas needing intervention, ensuring targeted efforts for maximum conservation impact.

UAV technology enhances conservation efforts by providing comprehensive data to monitor reforestation, evaluate conservation practices, and enforce environmental regulations. This supports sustainable land management and promotes long-term ecological health. UAVs enhance precision agriculture, reduce environmental impacts, and promote eco-friendly practices. They identify erosion-prone areas, monitor biodiversity, and track wildlife movements [4]. They also support Integrated Pest Management (IPM) by providing accurate surveillance and reducing chemical pesticide use. This technology promotes agricultural sustainability while preserving ecological health.

12.8 Economic Feasibility of UAV Farming

The adoption of UAV (Unmanned Aerial Vehicle) technology in agriculture offers both economic benefits and cost challenges. UAVs enhance productivity, optimize resource use, and reduce input costs, making them an attractive option for modern farmers. Evaluating the economic feasibility of UAV farming involves analyzing the initial investment, operational costs, potential savings, and long-term returns [45].

12.8.1 Initial Investment Costs

The cost of acquiring UAVs varies based on their capabilities and features. Basic agricultural drones for monitoring and imaging range from $1000 to $5000, while advanced models with multispectral sensors and precision spraying capabilities can cost $10,000 to $50,000 [51]. Additional expenses include:

- **Software and Data Analytics**: UAVs require specialized software for data analysis, with subscription costs ranging from $500 to $5000 annually.
- **Training and Licensing**: Operators need training for safe and legal UAV operation. Depending on local regulations, certification costs can range from $200 to $1000.

12.8.2 Operational and Maintenance Costs

Regular maintenance is essential to ensure UAV efficiency and longevity. Key operational costs include:

- **Battery Replacement**: UAV batteries typically last 100–300 charging cycles and may cost $100 to $500 each.
- **Repairs and Upgrades**: Mechanical repairs and technology upgrades can add $500 to $2000 annually.
- **Labor Costs**: While UAVs reduce manual labor, skilled personnel are needed for flight planning, data analysis, and maintenance.

12.8.3 Cost Savings and Efficiency Gains

UAVs contribute to significant cost reductions through precision agriculture, including:

- **Reduced Input Costs**: UAVs apply pesticides, fertilizers, and water with greater accuracy, reducing input usage by 20–40%.

- **Improved Yields**: Better monitoring and early detection of crop issues increase yields by 10–25%, boosting revenue.
- **Labor Savings**: UAVs reduce the need for manual field inspections, saving 30–50% on labor costs.

12.8.4 Increased Productivity and ROI

UAVs improve productivity by providing faster, more accurate data collection. The return on investment (ROI) depends on farm size, crop type, and technology usage. On large-scale farms, UAVs can achieve an ROI within 2–4 years through increased yields and cost savings [55].

Example Calculation:

For a 1000-acre farm:

- **Annual Input Savings**: $15,000–$25,000 (through reduced chemical use)
- **Increased Yield Revenue**: $20,000–$40,000
- **Total Annual Savings/Profit**: $35,000–$65,000

12.8.5 Scalability and Adaptability

UAV technology is a versatile and scalable solution for agricultural operations, enhancing productivity and sustainability. It can perform tasks such as crop monitoring, livestock management, and environmental conservation. For small farms, lightweight and affordable UAV models offer cost-effective solutions for precision agriculture, improving resource management, increased yields, and reduced operational costs. In large-scale farms, advanced UAVs with longer flight ranges and higher payload capacities provide efficient coverage, allowing for large-area mapping, precision agricultural input delivery, and extensive data collection [56]. UAVs are adaptable to various agricultural activities, such as crop monitoring using multispectral and thermal sensors, livestock management tracking animal movement, and environmental conservation monitoring soil erosion and biodiversity. These versatile UAVs can be customized based on different climates and farming systems, supporting precision agriculture, climate-smart farming, and sustainable land management across diverse agricultural landscapes [57]. Overall, UAVs offer a cost-effective and efficient solution for a wide range of agricultural operations.

Importance of Scalability and Adaptability in UAV Technology

1. **Accessibility for All Farm Sizes**: Small farms benefit from affordable models, while large farms use advanced drones for large-scale monitoring.
2. **Diverse Applications**: UAVs adapt to crop monitoring, livestock management, and environmental conservation.

3. **Cost Efficiency**: UAV scalability reduces operational costs through precision input delivery and efficient resource use.
4. **Flexibility in Different Environments**: UAVs operate effectively across varied climates and farming systems.
5. **Enhanced Productivity**: Adaptable UAVs support diverse agricultural tasks, improving decision-making and yield optimization.

UAV technology is both scalable and adaptable, providing tailored solutions for farms of all sizes and supporting a broad range of agricultural and environmental applications [58]. This versatility enhances efficiency, sustainability, and agricultural productivity.

12.8.6 Economic Challenges and Barriers

UAV technology offers significant benefits to agriculture, but its adoption faces several economic challenges, particularly for small-scale farmers and those operating in regions with strict regulations and limited resources. The high upfront cost of advanced agricultural drones, equipped with multispectral sensors, GPS technology, and data analytics software, is a major barrier. Small-scale farmers may struggle to invest in such technology due to limited financial resources [59]. Regulatory compliance is another financial challenge, as farmers must comply with aviation regulations, licensing fees, and operational guidelines, which can be prohibitive for smaller farms. Technological maintenance is an ongoing expense, with UAVs requiring regular software updates, sensor calibration, and hardware repairs to maintain accuracy and efficiency. Advanced drones with complex components are prone to wear and tear, adding to the financial burden on farmers. Data management and storage present hidden costs, as large volumes of data collected by UAVs require secure storage systems and specialized software for analysis. Farmers may need to invest in cloud services or hire external experts to interpret the information, increasing operational expenses [60].

Key Economic Challenges and Barriers

1. **High Upfront Costs**: Advanced UAV systems are expensive, limiting accessibility for small-scale farmers.
2. **Regulatory Compliance**: Meeting aviation and agricultural regulations requires permits and incurs legal fees.
3. **Maintenance Expenses**: Regular software updates, hardware repairs, and sensor calibration increase long-term costs.
4. **Training and Expertise**: Farmers may need to invest in technical training or hire experts to analyze UAV data.
5. **Data Management Costs**: Storing and processing large volumes of UAV-collected data requires additional resources.

UAV technology enhances agricultural efficiency and sustainability, but economic barriers like high initial investments, regulatory compliance costs, and ongoing maintenance can hinder its widespread adoption. To make UAVs more accessible, subsidies, policy support, and technological advancements are needed. UAV farming is economically feasible due to reduced input costs, increased yields, and labor efficiency [61]. Long-term cost savings and improved productivity typically yield a positive ROI within 2–4 years. UAVs are suitable for both small and large agricultural operations.

12.9 Economic Advantages for Farmers

UAV technology offers farmers significant economic benefits by enhancing productivity, reducing operational costs, and improving resource efficiency. It provides accurate, real-time data and enables precision agriculture, supporting smarter decision-making and cost-effective farming practices. UAVs enable targeted application of fertilizers, pesticides, and herbicides, minimizing waste and environmental impact. They can also monitor crop health, identify nutrient deficiencies, and detect pest infestations early, preventing crop loss and ensuring higher yields. UAV technology also lowers labor costs by covering large areas quickly and providing faster and more accurate data collection [4]. Advanced sensors measure soil moisture levels, allowing farmers to apply water only where it is needed, conserving water resources and reducing energy consumption for irrigation systems. UAV-based data analytics help farmers plan crop rotations, assess soil health, and optimize planting schedules, leading to more efficient land management and better resource allocation. This leads to long-term profitability and increased market competitiveness. By adopting modern precision agriculture techniques, farmers can produce higher-quality crops, meet consumer demand for sustainable practices, and qualify for certification programs that reward environmentally friendly operations, opening new market opportunities and increasing profit margins [62].

Key Economic Advantages of UAVs for Farmers

1. **Reduced Input Costs**: Precision application lowers the use of fertilizers, pesticides, and water.
2. **Increased Yields**: Early detection of crop issues improves productivity and prevents losses.
3. **Lower Labor Costs**: UAVs automate tasks, reducing the need for manual fieldwork.
4. **Efficient Resource Use**: Optimized irrigation and input application cut costs and conserve resources.
5. **Improved Decision-Making**: Data-driven insights enhance planning and long-term profitability.
6. **Market Competitiveness**: Sustainable practices and higher-quality crops open new market opportunities.

UAV technology provides economic advantages by reducing operational expenses, increasing crop yields, and enhancing efficiency. These benefits contribute to sustainable profitability and improved financial outcomes for farmers across various agricultural sectors [43].

12.9.1 Cost Savings on Labor

UAV technology offers significant labor cost savings by automating labor-intensive agricultural tasks, such as crop monitoring, spraying, and data collection. Traditional farming relies heavily on manual labor, which can be time-consuming and expensive. UAVs streamline these processes, improving efficiency and reducing the need for a large workforce, especially in large-scale farming operations. Automated crop monitoring, performed by UAVs equipped with advanced imaging sensors, provides real-time data on crop health, soil moisture levels, and pest infestations, eliminating the need for manual field inspections [63]. Precision spraying, performed by UAVs, reduces labor costs by performing targeted aerial spraying, covering large fields efficiently and minimizing human exposure to hazardous chemicals. UAVs also facilitate data collection and analysis, collecting high-resolution images and environmental data in a fraction of the time, providing faster, more accurate insights for decision-making. By reducing seasonal labor reliance, UAVs help mitigate labor shortages, especially during critical planting and harvesting periods, particularly in areas with scarce or expensive agricultural labor.

Key Labor Cost Savings from UAV Technology:

1. **Automated Crop Monitoring**: Reduces the need for manual field inspections.
2. **Precision Spraying**: Lowers labor requirements for chemical application.
3. **Data Collection**: Automates environmental and crop performance monitoring.
4. **Reduced Seasonal Labor**: Minimizes dependence on temporary workers.
5. **Increased Efficiency**: Completes tasks faster and with greater accuracy.

UAV technology significantly reduces labor costs by automating critical agricultural tasks. This not only improves operational efficiency but also lowers expenses, making farming more cost-effective and sustainable in the long run [64].

12.9.2 Lower Input Costs with UAV Precision

UAV technology significantly reduces input costs in agriculture by enabling precise and targeted application of essential resources like fertilizers, pesticides, and water. Traditional methods often result in overuse and wastage due to uneven distribution and manual inaccuracies. UAVs, equipped with advanced sensors and GPS technology, deliver inputs exactly where and when needed, minimizing excess use and lowering operational expenses. Precision spraying reduces the volume of fertilizers and pesticides used, reducing purchasing costs and environmental contamination. UAVs also help in irrigation management by monitoring soil moisture levels across fields, allowing farmers to apply water only to areas in need, preventing over-irrigation and reducing water consumption. This lower water use reduces energy costs for pumping systems and helps mitigate issues like soil salinization and nutrient leaching. UAVs also contribute to better crop yields, increased profit margins for farmers, and lower transportation and storage costs [1].

Key Benefits of UAV Precision in Reducing Input Costs:

1. **Minimized Input Wastage**: Targeted application reduces the overuse of fertilizers, pesticides, and water.
2. **Lower Chemical Expenses**: Precise delivery decreases the volume of inputs required.
3. **Optimized Water Use**: UAV data guides efficient irrigation, saving water and energy.
4. **Improved Crop Health**: Reduced chemical stress enhances yields and product quality.
5. **Cost Efficiency**: Less input use results in lower purchase, storage, and transport expenses.

UAV precision lowers input costs by reducing waste, improving efficiency, and maintaining crop productivity. This targeted approach enhances financial sustainability while promoting environmentally responsible agricultural practices [65].

ROI Analysis of Drone Usage:

Investing in Unmanned Aerial Vehicle (UAV) technology offers a significant return on investment (ROI) by improving yield quality, lowering operational expenses, and enhancing overall efficiency. The initial investment can be substantial, but the long-term savings and increased productivity typically outweigh the upfront costs. UAVs provide precise data on crop health, soil moisture, and pest activity, enabling farmers to make informed decisions and detect issues early. This real-time monitoring reduces crop loss and ensures higher-quality harvests. UAVs also reduce operational costs by automating processes like crop scouting, pesticide spraying, and irrigation monitoring, reducing labor expenses, input wastage, and fuel consumption [66]. This results in faster and more accurate farm management, optimizing resource allocation, preventing over-application of fertilizers and pesticides, and minimizing environmental impact. Despite the initial cost of purchasing and maintaining UAV systems, many farmers experience rapid payback within a few growing seasons due to the combined effects of input savings, higher yields, and labor cost reductions.

Key Factors Driving ROI from UAV Usage:

1. **Increased Yield Quality**: Early detection of issues improves harvest quality and quantity.

2. **Reduced Labor Costs**: Automation lowers the need for manual labor and field inspections.
3. **Input Optimization**: Targeted application minimizes chemical and water use.
4. **Time Efficiency**: Faster data collection accelerates decision-making and response times.
5. **Environmental Benefits**: Precision practices reduce pollution and resource depletion.

UAV technology offers significant ROI by improving efficiency, reducing costs, and increasing agricultural productivity. With advancements in technology and increasing affordability, UAV adoption is becoming an economically viable and sustainable investment for modern farmers [67].

12.10 Cost Analysis of Drone Adoption

The adoption of UAVs involves initial costs like drones, software, and training, but long-term gains like reduced labor, lower input costs, and enhanced productivity offset these costs. Over time, increased operational efficiency and improved crop management justify the initial investment. Evaluating ROI for small and large farms requires strategic planning to balance initial expenses with expected savings, leasing options, and government incentives [68]. Large farms benefit from economies of scale, where UAV technology enhances productivity and reduces per-acre costs.

12.11 Business Models for UAV-Based Agriculture

The adoption of UAV (Unmanned Aerial Vehicle) technology in agriculture has led to the development of various business models designed to meet the diverse needs of farmers and agricultural enterprises. These models provide cost-effective, data-driven solutions that improve efficiency, lower operational costs, and support sustainable farming. The service-based model allows farmers to access UAV technology through specialized companies that offer services like aerial surveys and precision spraying without the need for equipment ownership. In the equipment sales and leasing model, farmers can either purchase or lease UAVs, providing flexibility and affordability, particularly for larger agricultural operations [69]. The data analytics subscription model offers real-time insights by using UAV-collected data for crop monitoring, yield forecasting, and pest detection, helping farmers make informed decisions. A hybrid model combines these approaches, offering customized solutions that integrate services, leasing, and data analytics. Lastly, the cooperative or shared ownership model allows farmer groups to share UAV resources, making the technology accessible and affordable for smallholder farmers. Each model enhances precision agriculture, reduces costs, and supports sustainable resource management, making UAV technology an essential tool for modern agricultural practices [70].

12.11.1 Service-Based Model

The service-based model in UAV agriculture provides drone-as-a-service (DaaS) services to farmers, allowing them to access advanced UAV capabilities without the burden of ownership. This model is particularly cost-effective for small and medium-sized farms, as it eliminates upfront investment and maintenance costs. DaaS providers offer a range of agricultural applications, such as aerial surveys, crop health monitoring, pesticide application, and data analysis. They often use state-of-the-art UAVs equipped with high-resolution sensors and advanced imaging technologies, allowing farmers to access the latest innovations without upgrading their equipment. An example of this model is a drone service company providing precision spraying and field mapping on a per-acre basis. This allows farmers to optimize input usage, reduce waste, and improve crop yields without investing in UAV ownership. The service-based model democratizes access to UAV technology, enabling farms of various sizes to implement data-driven decision-making and sustainable agricultural practices. By offering scalable, affordable, and customized services, this model helps farmers enhance efficiency, reduce costs, and improve productivity without the financial strain of direct UAV ownership [71].

12.11.2 Equipment Sales and Leasing Model

The equipment sales and leasing model in UAV agriculture enables manufacturers and distributors to sell or lease drones and associated software to farmers, offering flexible payment options that lower the financial barriers to adoption. Farmers can choose to purchase UAVs for permanent use or lease them for specific periods, depending on their operational needs and budget. This model is particularly beneficial for small and medium-sized farms, as it reduces upfront capital expenditure while providing access to advanced technology. For large farms and agribusinesses, it facilitates long-term integration of UAVs into daily operations, improving efficiency and productivity [72]. Leasing options allow farmers to use UAV technology during critical periods, such as the growing season, without making a large initial investment. Maintenance services and software updates are often included in leasing agreements, ensuring farmers always have access to the latest UAV innovations. This model offers a scalable and affordable way for farmers to adopt UAV technology, balancing cost management with technological advancement, allowing farms of all sizes to benefit from precision agriculture while minimizing financial risk and ensuring operational flexibility [73].

12.11.3 Data Analytics Subscription Model

The data analytics subscription model in UAV agriculture uses advanced analytics and UAV technology to provide farmers with data-driven insights. Farmers subscribe to a cloud-based platform that processes data collected by UAVs into actionable intelligence, including crop performance analysis, yield prediction, pest detection, and soil health monitoring. This data enables farmers to make informed decisions to optimize resource use, increase yields, and reduce operational costs. The model offers continuous access to the latest data without requiring farmers to perform complex analyses themselves [74]. Subscribers receive regular reports and real-time monitoring updates, enabling proactive crop management. The scalable nature of this model allows it to serve farms of various sizes, offering customized insights to address specific agricultural needs. An example of this model is a UAV subscription service that provides monthly field scans and customized reports, enabling farmers to optimize irrigation, pest management, and planting strategies without large upfront investments in analytical software or specialized staff.

12.11.4 Hybrid Model

The hybrid business model in UAV agriculture combines service provision, equipment leasing, and data analytics to provide comprehensive solutions. This model is flexible and scalable, making it suitable for agribusinesses, large farms, and cooperatives with diverse needs. Farmers can customize solutions based on farm size, crop type, and operational requirements. They can lease UAVs for specific tasks, utilize on-demand services like precision spraying, and subscribe to data analytics platforms for real-time insights. This integrated approach streamlines agricultural processes, improves resource efficiency, and optimizes crop management. The hybrid model reduces the financial burden by offering bundled services at a lower cost than purchasing each component separately [75]. It also allows for technological adaptability, enabling farmers to scale their UAV usage as their operations grow. A key advantage of the hybrid model is its holistic solution, offering UAV leasing during peak growing seasons, precision spraying for targeted pest control, and advanced analytics packages to monitor crop health and predict yields.

12.11.5 Cooperative or Shared Ownership Model

The cooperative or shared ownership model in UAV agriculture allows farmers to collectively invest in advanced agricultural technology, sharing costs and benefits. This model is particularly beneficial for smallholder and marginal farmers who may not have the financial capacity to purchase UAVs independently. By pooling resources, farmers can access advanced agricultural technology while minimizing individual expenses. Under this model, UAVs are shared among cooperative members for tasks like aerial monitoring, precision input application, and crop health analysis [76]. This promotes cost efficiency by distributing purchase, maintenance, and operational expenses across multiple users. It also encourages collaboration and knowledge-sharing among farmers, leading to better resource management and sustainable agricultural practices. The cooperative model provides access to technology that would otherwise be financially out of reach for individual farmers, allowing members to scale operations according to seasonal needs without incurring heavy upfront costs. Shared training and technical support within the cooperative ensure all members benefit from the UAV's capabilities and optimize its usage [77].

Key Benefits of UAV-Based Business Models:

- **Cost Accessibility**: Flexible leasing and service models lower financial barriers.
- **Efficiency and Precision**: Data analytics optimize input use and enhance productivity.
- **Scalability**: Models adapt to farm sizes, from smallholders to large commercial farms.
- **Innovation**: Access to cutting-edge UAV technology through leasing and service packages.

UAV-based agriculture business models—ranging from service-based and equipment leasing to data subscriptions—offer flexible, scalable solutions. These models empower farmers to improve efficiency, reduce costs, and adopt sustainable practices while minimizing capital investment [78].

12.11.6 Drone Leasing and Service Providers

Drone leasing and third-party service providers offer a cost-effective solution for farmers who cannot afford to purchase UAVs outright. These services provide access to advanced drone technology without a significant upfront investment, making them suitable for small and medium-sized farms. Lease options include maintenance, software updates, and technical support, ensuring farmers always operate with cutting-edge technology. Third-party service providers offer specialized expertise, enabling accurate data collection and analysis without requiring farmers to develop technical skills [79]. Drones are used for crop monitoring, precision spraying, soil analysis, and yield forecasting, optimizing input usage, improving crop health, and increasing productivity. Leasing reduces capital expenditure and allows farmers to allocate resources more efficiently. By leveraging drone leasing and third-party services, farmers can adopt advanced agricultural practices, enhance operational efficiency, and remain competitive in a technology-driven industry [80].

12.11.7 Cost-Effective Access to Advanced Technology

Drone leasing is a cost-effective way for farmers to access advanced UAV technology, such as high-resolution imaging and precision spraying, without the significant upfront cost. This model is particularly beneficial for small- and medium-sized farms, where the high upfront cost of drones can be a barrier. Leasing fees range from $500 to $3000 per season, depending on the drone's features and service scope. It includes maintenance and software updates, ensuring farmers have access to the latest technology without extra expenses. This model also mitigates the risk of technology obsolescence, allowing farmers to stay updated with rapidly evolving UAV advancements [81]. It also offers flexibility, allowing farmers to scale drone usage based on seasonal needs and experiment with drone applications before making a long-term investment.

12.11.8 Flexible Usage and Scalability

Drone leasing services offer farmers the flexibility to adjust their use of UAVs based on seasonal agricultural demands. Instead of investing in unused drones during off-seasons, they can lease equipment during critical periods like planting, crop monitoring, and harvesting. This not only reduces financial burden but also minimizes maintenance responsibilities. Drone leasing allows farmers to access advanced drone technology when it is most needed without incurring year-round ownership costs. Service providers typically offer upgraded UAV models with the latest imaging and precision capabilities, ensuring farmers benefit from cutting-edge innovations. This scalability is particularly useful for operations with fluctuating workloads, allowing them to increase or decrease drone usage as needed. Farmers can experiment with different drone models and technologies through leasing arrangements, enhancing productivity and resource management without significant capital investment [82]. Drone leasing supports efficient farm management, reduces equipment downtime, and enables farmers to scale operations effectively to meet seasonal challenges.

12.11.9 Reduced Financial Risk

Drone leasing is a cost-effective solution for farmers, allowing them to access the latest advancements without long-term commitment. This approach reduces financial risks associated with technology obsolescence and maintenance, as drones evolve rapidly. Leasing companies handle routine maintenance, software updates, and repairs, ensuring drones operate at peak performance throughout their usage period. The pay-as-you-go model allows farmers to test UAV solutions in real-world conditions before investing in permanent equipment, allowing them to assess whether drone technology meets specific agricultural needs without the financial burden of ownership. This approach is particularly beneficial for small- and medium-sized farms, as it reduces the need for substantial upfront capital and provides flexibility to upgrade to newer models as technology improves [83]. The pay-as-you-go model also allows farmers to evaluate drones' effectiveness in improving crop yields and resource management.

12.11.10 Professional Expertise and Data Analysis

Third-party service providers offer farmers more than just UAV hardware—they bring specialized expertise in data collection, analysis, and advanced agricultural insights. This combination of technology and knowledge allows farmers to leverage UAV capabilities without requiring extensive in-house technical skills [84]. These services can significantly enhance agricultural productivity through the following applications:

- **Aerial Mapping and Imaging**: High-resolution images provide detailed insights into crop health, soil moisture levels, and pest infestations. This data helps farmers identify problem areas early and take corrective actions.
- **Precision Spraying**: Drones equipped with advanced spraying systems allow for the targeted application of pesticides, herbicides, and fertilizers, reducing waste and minimizing environmental impact.
- **Yield Forecasting**: UAVs collect data that can be analyzed to predict harvest outputs, enabling better field management and more accurate production planning.

Outsourcing tasks to professional service providers allows farmers to access advanced technology and expert analysis, enhancing decision-making and operational efficiency. This model eliminates the need for farmers to develop specialized UAV knowledge, while still benefiting from data-driven agricultural practices [85]. Their expertise in aerial mapping, precision spraying, and yield forecasting supports informed decision-making, improving efficiency and yields.

12.11.11 Types of Drone Leasing Models

Farmers have access to various drone leasing and service models tailored to their specific agricultural needs. These flexible options allow them to adopt UAV technology without the financial commitment of ownership [20]. The primary leasing models include:

- **Short-Term Leasing**: This model is ideal for seasonal monitoring and specialized tasks like pest detection or crop health assessments. Farmers can lease drones during critical periods, reducing costs when UAVs are not needed.

- **Long-Term Leasing**: Designed for year-round agricultural operations, this model provides continuous drone access with comprehensive service packages, including maintenance and software updates. It's beneficial for large farms requiring regular drone use across multiple growing cycles.
- **Pay-Per-Service**: In this model, farmers pay only for specific UAV operations—such as aerial spraying, crop mapping, or soil analysis—without handling the equipment themselves. This option offers maximum flexibility and is ideal for farms that need occasional drone services without long-term commitments.

Farmers can choose from three drone leasing models: Short-Term Leasing for seasonal tasks, Long-Term Leasing for year-round access, and Pay-Per-Service for individual UAV operations. Each model offers flexibility, cost-efficiency, and access to advanced technology without full ownership burden, allowing them to optimize their agricultural practices while minimizing costs and technological risks [86].

12.11.12 Economic and Operational Benefits

Drone leasing and third-party services offer significant economic and operational advantages for modern agricultural practices. By reducing financial barriers and improving efficiency, these models make advanced UAV technology accessible to a wider range of farmers [87]. Key benefits include:

- **Lower Upfront Costs**: Farmers can access cutting-edge UAV technology without a large capital investment, allowing them to allocate funds to other critical areas of their operations.
- **Operational Efficiency**: Drones provide faster field coverage and precise data collection, enabling better decision-making in areas like crop monitoring, pest control, and resource management. This leads to improved productivity and reduced labor costs.
- **Technology Access**: Leasing and third-party services offer regular access to the latest UAV models and advanced data analytics tools. Farmers benefit from new technological innovations without the risk of owning outdated equipment.
- **Reduced Downtime**: Service providers handle maintenance, software updates, and repairs, ensuring that drones operate efficiently and minimizing disruptions to farm operations.

Drone leasing and third-party services offer farmers an affordable and flexible solution to adopt drone technology, enhancing agricultural productivity, optimizing resource use, and reducing financial risk. These services provide lower upfront costs, enhanced operational efficiency, access to the latest technology, and reduced downtime through professional maintenance. By using leasing options, farmers can optimize their practices, improve productivity, and adapt to technological advancements, thereby reducing financial and technological risks [82]. This allows them to manage crops effectively and sustainably, ultimately improving their overall agricultural practices.

12.11.13 Government Subsidies and Incentives

Governments worldwide are promoting the adoption of Unmanned Aerial Vehicles (UAV) technology in agriculture to reduce financial burdens, promote technological innovation, and encourage sustainable practices. UAVs play a crucial role in modernizing agriculture by improving efficiency, reducing resource consumption, and enhancing crop yields through precision techniques. Direct subsidies, which cover 20–50% of total expenses, offset the initial costs of purchasing or leasing UAV equipment [88]. Grants provide non-repayable funds for UAV-based research, development, and precision agriculture applications, encouraging innovation. Low-interest loans and tax credits and deductions also help farmers finance UAV purchases under favorable repayment terms, reducing immediate financial pressure. Governments also invest in training and capacity-building programs to ensure farmers can effectively operate UAVs and interpret data collected. These programs, often conducted in collaboration with research institutions and agricultural agencies, provide farmers with the skills to maximize UAV benefits while adhering to safety and regulatory standards. Government subsidies and incentives facilitate the transition to data-driven, precision agriculture, enhancing farm productivity and promoting environmental sustainability by enabling more efficient use of water, fertilizers, and pesticides [89]. As technology evolves, policymakers are likely to expand these initiatives to further accelerate UAV adoption and drive sustainable agricultural practices globally.

12.11.14 Types of Government Support Programs

Governments are providing various financial support programs to encourage the adoption of Unmanned Aerial Vehicles (UAVs) in agriculture. These programs aim to reduce the financial burden on farmers while promoting technological innovation and sustainable agricultural practices. The types of support include direct subsidies, grants, low-interest loans, tax incentives, and training programs. Direct subsidies cover 20–50% of the total expense of purchasing or leasing UAVs and related technologies, making advanced UAV systems more affordable. Grants offer non-repayable funds for research, development, and UAV implementation, often supporting pilot projects that demonstrate the benefits of UAVs. Low-interest loans provide below-market interest rates to help farmers finance UAV purchases, particularly for small- and medium-sized farms. Tax credits and deductions further reduce costs by allowing farmers to claim tax incentives for UAV investments [90]. Training and capacity-building programs ensure that farmers and agricultural professionals are equipped to use UAVs effectively. These programs provide hands-on training in

UAV operation, data analysis, and regulatory compliance. Governments may partner with universities, research institutions, and agricultural agencies to offer subsidized or free training. These initiatives help farmers maximize the benefits of UAV technology while ensuring safe and responsible use.

12.11.15 Eligibility Criteria for Subsidies

UAV-related incentives are often awarded to farmers who use UAV technology effectively and in line with sustainability and regulatory standards. These criteria are crucial for successful applications and compliance. Farm size and type are common eligibility criteria, as many subsidy programs target small- to medium-scale farms that may struggle to afford UAV technology without financial assistance. Programs also focus on specific agricultural sectors, such as crop farming, horticulture, or livestock monitoring, where UAVs can have the greatest impact. Environmental commitment is another key requirement, as programs often prioritize farmers who use UAVs to implement sustainable practices, such as precision input application, soil conservation, and water management [91]. Applicants may need to submit plans outlining how UAVs will reduce environmental impact, aligning with government goals for sustainable and climate-smart agriculture. Furthermore, farmers must adhere to UAV operation standards, including flight safety protocols, data privacy laws, and environmental guidelines. This may involve obtaining drone pilot certification, following airspace regulations, and maintaining accurate records of UAV use. Programs may require periodic audits to ensure continued compliance. Failure to meet these standards can result in penalties or loss of funding [92].

12.11.16 Examples of UAV Subsidy Programs

- **United States**: Programs under the Farm Service Agency (FSA) and Environmental Quality Incentives Program (EQIP) provide financial support for precision agriculture technologies, including UAVs.
- **European Union**: The Common Agricultural Policy (CAP) offers funding for digital agriculture projects, including drone implementation, to enhance sustainability.
- **China**: Government initiatives promote UAV adoption for pest control and precision spraying, especially for smallholder farmers.
- **India**: The Sub-Mission on Agricultural Mechanization (SMAM) scheme provides subsidies for modern technology adoption, including UAVs.

12.11.17 Economic Impact of Subsidies on UAV Adoption

Government subsidies are instrumental in reducing the initial costs associated with the adoption of Unmanned Aerial Vehicles (UAVs) in agriculture. These subsidies help offset the high initial investment required for purchasing or leasing drones, making advanced UAV technology more affordable and accessible. By providing financial assistance, governments enable farmers to adopt precision agriculture techniques that improve productivity and sustainability. The subsidies can save farmers between 20 and 50% on UAV equipment and related expenses, making it easier for small- and medium-sized farms to access cutting-edge drone technology. This lower initial expense also improves the return on investment (ROI) for farmers, as UAVs enhance operational efficiency through faster data collection, precision spraying, and accurate monitoring, leading to higher yields and lower input costs [93]. Furthermore, subsidies encourage innovation by promoting the adoption of data-driven agricultural practices, fostering technological advancement, decision-making capabilities, and the shift towards sustainable agricultural practices.

12.11.18 Research Available Programs

Farmers seeking UAV-related subsidies should start by researching national, regional, and local funding opportunities. Governments worldwide recognize the value of UAV technology in improving agricultural productivity and sustainability, and many have established financial assistance programs to encourage its adoption. These programs offer various forms of support, including grants, low-interest loans, tax credits, and pilot funding. Farmers can find information on these subsidies through government websites, agricultural ministries, and local extension offices. Key programs, such as the USDA Environmental Quality Incentives Program (EQIP) in the United States or the Common Agricultural Policy (CAP) in the European Union, provide financial assistance for implementing precision agriculture technologies, including drones. To secure funding, farmers must identify the most suitable program based on their specific needs and eligibility criteria [94]. Programs may target particular farming activities such as precision spraying, aerial mapping, or crop health monitoring. Some initiatives focus on promoting sustainable practices, meaning applicants must demonstrate how UAVs will reduce chemical usage, conserve water, or improve efficiency. Farmers should also monitor deadlines and application windows, as many programs operate on a fixed annual schedule. By thoroughly researching and understanding these programs, farmers can strategically access funding to adopt UAV technology and improve their operational efficiency while reducing financial risks. A successful subsidy application requires comprehensive documentation that clearly outlines the farmer's needs and how UAV technology will be used. The first step is to compile basic farm information, including the farm's size, location, crop types, and current agricultural practices. Farmers must also prepare a UAV usage plan that details how the drone will enhance agricultural operations, including specific applications such as crop monitoring, pest detection, precision spraying, and yield forecasting, as well as expected benefits like improved efficiency, reduced resource use, and environmental sustainability [95]. Environmental compliance documentation is another critical component, as many subsidy programs prioritize sustainable agriculture, so applicants must demonstrate how drone usage aligns with environmental goals. Financial documentation is also required, including detailed cost estimates for drone acquisition or leasing, maintenance, and training. After receiving UAV-related subsidies, farmers must comply with program regulations and maintain detailed records of drone operations. Most subsidy programs require farmers to demonstrate that UAVs are used according to the approved usage plan and deliver the intended agricultural benefits. This involves documenting drone flight logs, tracking operational outcomes, and submitting regular reports to verify ongoing compliance [96]. Non-compliance can result in penalties, including repayment of funds or disqualification from future programs. Farmers should stay informed about program requirements, maintain accurate records, and submit timely reports to ensure continued funding and support future applications for advanced agricultural technology subsidies.

12.12 Summary

UAV technology plays a crucial role in modern precision farming by optimizing resource use, reducing input costs, and minimizing environmental impact. It supports sustainable agriculture by improving precision, reducing resource consumption, and minimizing environmental impacts. UAVs enable precise application of agricultural inputs through advanced mapping, sensor integration, and automated delivery systems. This technology optimizes resource use, reduces input costs, minimizes environmental impact, and enhances overall agricultural efficiency. Waste reduction minimizes pollution and conserves resources by avoiding single-use plastics, reusing items, and recycling materials. Composting organic waste reduces landfill contributions and promotes soil health. Sustainable transportation reduces greenhouse gas emissions by using public transit, carpooling, biking, and driving fuel-efficient or electric vehicles. Minimizing air travel and participating in carbon offset programs further reduce environmental impacts. Sustainable agriculture and food choices reduce environmental impacts by supporting local and organic farms, reducing meat consumption, and adopting plant-based diets. Minimizing food waste through better planning and composting further lessens the environmental burden and promotes a healthier planet. Eco-friendly consumption involves choosing sustainable products, supporting environmentally responsible companies, and prioritizing durable goods over disposable items. These practices reduce waste, conserve resources, and promote corporate sustainability, contributing to a healthier and more sustainable planet.

Protecting ecosystems involves participating in reforestation, avoiding products linked to deforestation, and preserving green spaces. Advocacy and education drive systemic change by promoting environmental policies, fostering public awareness, and supporting conservation organizations. UAV technology supports sustainable agriculture by improving precision, reducing resource consumption, and minimizing environmental impacts. Through advanced data collection and targeted application, drones enhance productivity while protecting natural resources and ecosystems. Advanced sensors monitor crop health in real time, identifying pests, diseases, and nutrient deficiencies, allowing for targeted interventions, reducing chemical use and supporting sustainable practices like Integrated Pest Management (IPM). This precision improves crop yields while minimizing environmental impact. UAVs reduce soil degradation and erosion by avoiding physical contact with the ground, preserving soil structure, and enabling the precise application of cover crop seeds and soil amendments. They aid water conservation by using thermal and infrared sensors to detect soil moisture levels, enabling targeted irrigation. UAVs preserve biodiversity by enabling targeted chemical applications, protecting beneficial insects and soil microorganisms, and facilitating sustainable farming practices while maintaining natural ecosystems. UAV technology reduces the carbon footprint in agriculture by decreasing the use of fuel-intensive machinery, optimizing input application, and supporting precision agriculture. It collects high-resolution data on crop growth, soil conditions, and environmental factors, supporting evidence-based decision-making.

UAV technology promotes economic and social sustainability by reducing input costs, increasing yields, and improving profitability. It provides affordable precision agriculture solutions for smallholder farmers, enhances food security, and supports sustainable rural livelihoods while reducing environmental degradation. UAV technology enhances erosion risk management by providing detailed data to identify and mitigate soil erosion, reducing soil loss, protecting water quality, and ensuring the sustainability of agricultural landscapes. In summary, UAV technology offers significant ROI by improving efficiency, reducing costs, and increasing agricultural productivity. With advancements in technology and increasing affordability, UAV adoption is becoming an economically viable and sustainable investment for modern farmers. The hybrid business model combines UAV leasing, service provision, and data analytics to deliver comprehensive, scalable solutions. It offers flexibility, cost savings, and advanced insights, making it ideal for agribusinesses and cooperatives seeking customized UAV solutions. The cooperative or shared ownership model enables farmers to collectively invest in UAV technology, making advanced agricultural tools more affordable. It distributes costs, promotes collaboration, and ensures access to precision agriculture for smallholder farmers, enhancing efficiency and sustainability. Drone leasing and service providers offer cost-effective access to UAV technology, allowing farmers to test solutions and scale operations without substantial capital investment. This model supports flexibility, reduces financial risk, and encourages the adoption of precision agriculture through on-demand access to advanced drone capabilities. By using leasing options, farmers can optimize their agricultural practices, improve productivity, and adapt to changing technological advancements.

Drone leasing provides farmers with affordable access to advanced UAV technology, reducing upfront costs and offering flexibility. With fees ranging from \$500 to \$3000 per season, it includes maintenance and upgrades, allowing small- and medium-sized farms to adopt modern agricultural practices while mitigating financial risk and technology obsolescence. Drone leasing also reduces financial risks by avoiding costs related to technology obsolescence and maintenance. The pay-as-you-go model allows farmers to test UAV technology before purchasing, providing a flexible and cost-effective solution for modern agricultural needs. Third-party UAV service providers offer professional expertise in aerial mapping, precision spraying, and yield forecasting, supporting informed decision-making, improving efficiency and yields without requiring farmers to invest in specialized in-house knowledge or technology. Farmers can choose from three drone leasing models: Short-Term Leasing for seasonal tasks, Long-Term Leasing for year-round access, and Pay-Per-Service for individual UAV operations. Government subsidies and incentives, including grants, low-interest loans, tax benefits, and R&D support, lower the cost of adopting UAV technology in agriculture. Programs in countries like the United States, EU nations, China, and India help farmers improve productivity, reduce resource use, and promote sustainability. Eligibility for UAV-related incentives typically depends on farm size and type, environmental commitment, and compliance with regulations. Government support programs play a key role in encouraging UAV adoption by reducing financial barriers and promoting sustainable and technological innovation. These initiatives not only increase farm productivity but also drive technological innovation and sustainable agricultural practices.

References

1. Quan X, Guo Q, Ma J, Doluschitz R (2023) The economic effects of unmanned aerial vehicles in pesticide application: evidence from Chinese grain farmers. Precision Agric 24:1965–1981
2. Yaqot M, Menezes BC, Al-Ansari T (2021) Unmanned aerial vehicles in precision agriculture towards circular economy: a process system engineering (PSE) assessment. In: Computer aided chemical engineering. Elsevier, pp 1559–1565
3. Yablokova A, Kovalev D, Kovalev I et al (2024) Environmental safety problems of swarm use of UAVs in precision agriculture. EDP Sciences, p 04018
4. Mukhamediev RI, Symagulov A, Kuchin Y et al (2021) Review of some applications of unmanned aerial vehicles technology in the resource-rich country. Appl Sci 11:10171
5. Yaqot M, Menezes BC (2021) Unmanned aerial vehicle (UAV) in precision agriculture: business information technology towards farming as a service. IEEE, pp 1–7
6. Nath S (2024) A vision of precision agriculture: balance between agricultural sustainability and environmental stewardship. Agron J 116:1126–1143
7. Shang M, Xie J (2024) Agricultural sustainable development: soil, water resources, biodiversity, climate change, and technological innovation. Adv Resour Res 4:181–204
8. Dayıoğlu MA, Turker U (2021) Digital transformation for sustainable future-agriculture 4.0: a review. J Agric Sci 27:373–399
9. Nahiyoon SA, Ren Z, Wei P et al (2024) Recent development trends in plant protection UAVs: a journey from conventional practices to cutting-edge technologies—a comprehensive review. Drones 8:457
10. Sabir RM, Mehmood K, Sarwar A et al (2024) Remote sensing and precision agriculture: a sustainable future. In: Transforming agricultural management for a sustainable future: climate change and machine learning perspectives. Springer, pp 75–103

11. Arya A, Bachheti A, Bachheti RK et al (2024) Role of artificial intelligence in minimizing carbon footprint: a systematic review of recent insights. In: Biorefinery and industry 40: empowering sustainability. pp 365–386
12. Jagadesh M, Dash M, Singh SK et al (2024) Carbon footprints in agriculture: challenges, mitigation strategies, and future prospects. In: Carbon footprint assessments: case studies and best practices. Springer, pp 17–43
13. Yuan X, Li S, Chen J et al (2024) Impacts of global climate change on agricultural production: a comprehensive review. Agronomy 14:1360
14. Dwivedi YK, Hughes L, Kar AK et al (2022) Climate change and COP26: are digital technologies and information management part of the problem or the solution? An editorial reflection and call to action. Int J Inf Manage 63:102456
15. Pricopoaia O, Lupaşc A, Mihai IO (2024) Implications of innovative strategies for sustainable entrepreneurship—solutions to combat climate change. Sustainability 16:9742
16. Dey P, Mahapatra B, Mitra B et al (2024) Potential nexus approach for sustainable soil productivity resource management. In: Environmental nexus for resource management. CRC Press, pp 243–273
17. Yingjun Z, Jahan S, Qamruzzaman M (2024) Technological innovation, trade openness, natural resources, and environmental sustainability in Egypt and Turkey: evidence from load capacity factor and inverted load capacity factor with Fourier functions. Sustainability 2071–1050:16
18. Güçlü Z, Senel BN (2024) The role of logistics in achieving SDG 15: protecting life on land. In: Effective logistics for sustainable development goals. IGI Global, pp 212–233
19. Reddy VV, Kumar P, Rao A et al (2023) Waste to wealth generation: innovative methodologies in resource utilization and minimization in circular economy. EDP Sciences, p 01035
20. Xiaobin W, Wu F, Alharthi M et al (2024) Natural resources, trade and fintech in the era of digitalization: a study of economies involved in Belt and Road Initiative. Resour Policy 93:105082
21. Balasundram SK, Shamshiri RR, Sridhara S, Rizan N (2023) The role of digital agriculture in mitigating climate change and ensuring food security: an overview. Sustainability 15:5325
22. Mohsan SAH, Othman NQH, Li Y et al (2023) Unmanned aerial vehicles (UAVs): practical aspects, applications, open challenges, security issues, and future trends. Intel Serv Robot 16:109–137
23. Emon MMH, Khan T, Rahman MA et al (2025) GreenTech revolution: navigating challenges and seizing opportunities. In: AI and green technology applications in society. IGI Global Scientific Publishing, pp 63–90
24. Emimi M, Khaleel M, Alkrash A (2023) The current opportunities and challenges in drone technology. Int J Electr Eng Sustain 74–89
25. Mohamed N, Al-Jaroodi J, Jawhar I et al (2020) Unmanned aerial vehicles applications in future smart cities. Technol Forecast Soc Chang 153:119293
26. Raza I, Zubair M, Zaib M et al (2023) Precision nutrient application techniques to improve soil fertility and crop yield: a review with future prospect. Int Res J Educ Tecnol
27. Padhiary M, Kumar R, Sethi LN (2024) Navigating the future of agriculture: a comprehensive review of automatic all-terrain vehicles in precision farming. J Inst Eng (India) Ser A 105:767–782
28. Karunathilake E, Le AT, Heo S et al (2023) The path to smart farming: innovations and opportunities in precision agriculture. Agriculture 13:1593
29. Munir A, Salah MA, Ali M et al (2024) Advancing agriculture: harnessing smart nanoparticles for precision fertilization. BioNanoScience 14:3846–3863
30. Guebsi R, Mami S, Chokmani K (2024) Drones in precision agriculture: a comprehensive review of applications, technologies, and challenges. Drones 8:686
31. Getahun S, Kefale H, Gelaye Y (2024) Application of precision agriculture technologies for sustainable crop production and environmental sustainability: a systematic review. Sci World J 2024:2126734
32. Taseer A, Han X (2024) Advancements in variable rate spraying for precise spray requirements in precision agriculture using unmanned aerial spraying systems: a review. Comput Electron Agric 219:108841

33. Sharma S (2023) Precision agriculture: reviewing the advancements technologies and applications in precision agriculture for improved crop productivity and resource management. Rev Food Agric 4:45–49
34. Emami Y, Zhou H, Almeida L, Li K (2025) Diffusion models for smarter UAVs: decision-making and modeling. arXiv preprint arXiv:250105819
35. Sun X, Hu Y, Qin Y, Zhang Y (2024) Risk assessment of unmanned aerial vehicle accidents based on data-driven Bayesian networks. Reliab Eng Syst Saf 248:110185
36. Wang X, Liu H, Gao Q (2023) Data-driven decision making and near-optimal path planning for multiagent system in games. IEEE J Miniat Air Space Syst 4:320–328
37. Li K, Ni W, Emami Y, Dressler F (2022) Data-driven flight control of internet-of-drones for sensor data aggregation using multi-agent deep reinforcement learning. IEEE Wirel Commun 29:18–23
38. Dehghan M, Khosravian E (2024) Advancing situation awareness systems: evaluating decision-making methods with UAV applications. Manage Strat Eng Sci 6:122–133
39. Khan N, Ray RL, Sargani GR et al (2021) Current progress and future prospects of agriculture technology: gateway to sustainable agriculture. Sustainability 13:4883
40. Raheem D, Dayoub M, Birech R, Nakiyemba A (2021) The contribution of cereal grains to food security and sustainability in Africa: potential application of UAV in Ghana, Nigeria, Uganda, and Namibia. Urban Sci 5:8
41. Rai AK, Bana SR, Sachan DS, Singh B (2023) Advancing sustainable agriculture: a comprehensive review for optimizing food production and environmental conservation. Int J Plant Soil Sci 35:417–425
42. Sarfraz S, Ali F, Hameed A et al (2023) Sustainable agriculture through technological innovations. In: Sustainable agriculture in the era of the OMICs revolution. Springer, pp 223–239
43. Islam N, Rashid MM, Pasandideh F et al (2021) A review of applications and communication technologies for internet of things (IoT) and unmanned aerial vehicle (UAV) based sustainable smart farming. Sustainability 13:1821
44. Chamuah A, Singh R (2020) Securing sustainability in Indian agriculture through civilian UAV: a responsible innovation perspective. SN Appl Sci 2:1–10
45. Gryech I, Vinogradov E, Saboor A et al (2024) A systematic literature review on the role of UAV-enabled communications in advancing the UN's sustainable development goals. Front Commun Net 5:1286073
46. Fascista A (2022) Toward integrated large-scale environmental monitoring using WSN/UAV/Crowdsensing: a review of applications, signal processing, and future perspectives. Sensors 22:1824
47. Millner N, Newport B, Sandbrook C, Simlai T (2024) Between monitoring and surveillance: geographies of emerging drone technologies in contemporary conservation. Progr Environ Geogr 3:17–39
48. Chisom ON, Biu PW, Umoh AA et al (2024) Reviewing the role of AI in environmental monitoring and conservation: a data-driven revolution for our planet. World J Adv Res Rev 21:161–171
49. Kandrot S, Hayes S, Holloway P (2022) Applications of uncrewed aerial vehicles (UAV) technology to support integrated coastal zone management and the UN sustainable development goals at the coast. Estuaries Coasts 45:1230–1249
50. Librán-Embid F, Klaus F, Tscharntke T, Grass I (2020) Unmanned aerial vehicles for biodiversity-friendly agricultural landscapes—a systematic review. Sci Total Environ 732:139204
51. Bahuguna R, Pandey S, Singh R et al (2023) Enabling technologies for wildlife conservation. IEEE, pp 217–220
52. Abubakar AI, Ahmad I, Omeke KG et al (2023) A survey on energy optimization techniques in UAV-based cellular networks: from conventional to machine learning approaches. Drones 7:214

53. Sediles Martinez AJ (2022) Unmanned aerial vehicles (UAVs) as a non-invasive optimization tool for the exploration and management of raw materials
54. Suanpang P, Jamjuntr P (2024) Optimizing autonomous UAV navigation with d* algorithm for sustainable development. Sustainability 16:7867
55. Lu W, Lou J, Ababio BK et al (2024) Digital technologies for construction sustainability: status quo, challenges, and future prospects. npj Mater Sustain 2:10
56. Adinarayana S, Raju MG, Srirangam DP et al (2024) Enhancing resource management in precision farming through AI-based irrigation optimization. In: How machine learning is innovating today's world: a concise technical guide. pp 221–251
57. Akintuyi OB (2024) Adaptive AI in precision agriculture: a review: investigating the use of self-learning algorithms in optimizing farm operations based on real-time data. Res J Multidisc Stud 7:016–030
58. Şimşek Ö (2025) Harvesting sustainability: innovations and practices in modern agriculture. Green Technol Sustain 100192
59. Roslim MHM, Juraimi AS, Che'Ya NN et al (2021) Using remote sensing and an unmanned aerial system for weed management in agricultural crops: a review. Agronomy 11:1809
60. Alahmari N, Mehmood R, Alzahrani A et al (2023) Autonomous and sustainable service economies: data-driven optimization of design and operations through discovery of multi-perspective parameters. Sustainability 15:16003
61. Krishna KR (2019) Unmanned aerial vehicle systems in crop production: a compendium. Apple Academic Press
62. Boursianis AD, Papadopoulou MS, Diamantoulakis P et al (2022) Internet of things (IoT) and agricultural unmanned aerial vehicles (UAVs) in smart farming: a comprehensive review. Internet Things 18:100187
63. Nhamo L, Magidi J, Nyamugama A et al (2020) Prospects of improving agricultural and water productivity through unmanned aerial vehicles. Agriculture 10:256
64. Papadopoulos G, Arduini S, Uyar H et al (2024) Economic and environmental benefits of digital agricultural technologies in crop production: a review. Smart Agric Technol 100441
65. Sinha J (2020) Aerial robot for smart farming and enhancing farmers' net benefit. Indian J Agri Sci 90:258–267
66. Suvittawat A (2024) Investigating farmers' perceptions of drone technology in Thailand: exploring expectations, product quality, perceived value, and adoption in agriculture. Agriculture 14:2183
67. Abdulai A-R, Pulido-Castanon J, Duncan ER et al (2024) Will agricultural digitalization deliver relative advantages in quality of work, productivity, profitability, return on investments, and reliability? Perceptions of Canadian producers. Cogent Food Agric 10:2422529
68. Mishra H, Mishra D (2024) Economic evaluation of UAV-based soil sampling approaches. In: Applications of computer vision and drone technology in agriculture 4.0. Springer, pp 271–291
69. Ndlovu HS, Odindi J, Sibanda M, Mutanga O (2024) A systematic review on the application of UAV-based thermal remote sensing for assessing and monitoring crop water status in crop farming systems. Int J Remote Sens 45:4923–4960
70. Cui B, Liang L, Ji B et al (2024) Exploring the YOLO-FT deep learning algorithm for UAV-based smart agriculture detection in communication networks. IEEE Trans Netw Serv Manage
71. Ecke S, Dempewolf J, Frey J et al (2022) UAV-based forest health monitoring: a systematic review. Remote Sensing 14:3205
72. Toscano F, Fiorentino C, Capece N et al (2024) Unmanned aerial vehicle for precision agriculture: a review. IEEE Access
73. Tahir MN, Lan Y, Zhang Y et al (2023) Application of unmanned aerial vehicles in precision agriculture. In: Precision agriculture. Elsevier, pp 55–70
74. Phang SK, Chiang THA, Happonen A, Chang MML (2023) From satellite to UAV-based remote sensing: a review on precision agriculture. IEEE Access 11:127057–127076
75. Wang L, Huang X, Li W et al (2022) Progress in agricultural unmanned aerial vehicles (UAVs) applied in China and prospects for Poland. Agriculture 12:397

76. Istiak MA, Syeed MM, Hossain MS et al (2023) Adoption of unmanned aerial vehicle (UAV) imagery in agricultural management: a systematic literature review. Eco Inform 78:102305
77. Merz M, Pedro D, Skliros V et al (2022) Autonomous UAS-based agriculture applications: general overview and relevant European case studies. Drones 6:128
78. Velusamy P, Rajendran S, Mahendran RK et al (2021) Unmanned Aerial Vehicles (UAV) in precision agriculture: applications and challenges. Energies 15:217
79. Puppala H, Peddinti PR, Tamvada JP et al (2023) Barriers to the adoption of new technologies in rural areas: the case of unmanned aerial vehicles for precision agriculture in India. Technol Soc 74:102335
80. Rahman MFF, Fan S, Zhang Y, Chen L (2021) A comparative study on application of unmanned aerial vehicle systems in agriculture. Agriculture 11:22
81. Singh P, Singh P (2023) Drones in Indian agriculture: trends, challenges, and policy implications
82. Tezei JF, de Freitas CR, Neto AL et al (2024) Strategic analysis for a truck-drone hybrid model in last mile delivery. Int Contemp Manage Rev 5:31–53
83. Gross A, Varangis P, Biallas M (2023) Financing needs and solutions for an agribusiness supply chain. Smallholders 143
84. Boman D, Geraghty G (2024) Emerging business models in integrated terrestrial and non-terrestrial networks a study of connectivity provisioning in rural and remote areas and autonomous flying taxi services
85. Purtell CT, Manuj I, Pohlen TL et al (2025) Innovators and transformers: envisioning a revolution in middle mile logistics with extended range cargo drones. Int J Phys Distr Logistics Manage
86. Filiopoulou E, Bardaki C, Nikolaidou M, Michalakelis C (2025) Drone-as-a-service for last-mile delivery: evidence of economic viability. Econ Transp 41:100398
87. Widhalm P, Ritzinger U, Prüggler N et al (2025) Community drones: a concept study on shared drone services. Drones (2504–446X) 9
88. Reynoso Vanderhorst HD, Heesom D, Suresh S et al (2023) Barriers and cost model of implementing unmanned aerial system (UAS) services in a decentralised system: case of the Dominican Republic. Constr Innov 23:833–857
89. Dang TN, Manzoor A, Tun YK et al (2023) A contract-theory-based incentive mechanism for UAV-enabled VR-based services in 5G and beyond. IEEE Internet Things J 10:16465–16479
90. Chamuah A, Singh R (2022) Responsible governance of civilian unmanned aerial vehicle (UAV) innovations for Indian crop insurance applications. J Responsible Technol 9:100025
91. Pathak H, Kumar G, Mohapatra S et al (2020) Use of drones in agriculture: potentials, problems and policy needs. ICAR-Nat Inst Abiotic Stress Manage 300:4–15
92. Tan E, Pattyn V, Flores CC, Crompvoets J (2021) A capacity assessment framework for the fit-for-purpose land administration systems: the use of unmanned aerial vehicle (UAV) in Rwanda and Kenya. Land Use Policy 102:105244
93. Onososen AO, Musonda I, Onatayo D et al (2023) Impediments to construction site digitalisation using unmanned aerial vehicles (UAVs). Drones 7:45
94. Jahani H, Khosravi Y, Kargar B et al (2024) Exploring the role of drones and UAVs in logistics and supply chain management: a novel text-based literature review. Int J Prod Res 1–25
95. Trappey AJ, Lin G-B, Chen H-K, Chen M-C (2023) A comprehensive analysis of global patent landscape for recent R&D in agricultural drone technologies. World Patent Inf 74:102216
96. Teshaboyev O, Rasulov S (2025) Popularizing the use of unmanned aerial vehicles in agriculture in Uzbekistan. Educ Res Univ Sci 4:45–52

Chapter 13
UAV Technology in Agriculture: Legal, Ethical, and Future Perspectives

Abstract UAV technology is revolutionizing agriculture by enhancing precision, efficiency, and sustainability. However, challenges such as airspace regulations, data protection laws, and operational compliance are essential for responsible drone usage. Ethical concerns include data privacy, labor displacement, environmental impact, and technology access. The future of UAV technology in agriculture is characterized by autonomous drone systems, AI integration, and renewable energy solutions. These advancements promise improved efficiency, scalability, and environmental conservation. Regulatory compliance is crucial for agricultural UAV use, including airspace laws, altitude limits, and restricted areas. Ethical implications include environmental responsibility, worker protection, fair access, and transparent data practices. Balancing automation with job preservation and fostering inclusive access to emerging agricultural technologies is crucial. Staying updated with emerging UAV technologies is essential for maintaining competitive and sustainable practices. Swarm drone technology enhances agricultural productivity by enabling faster, more precise, and cost-effective operations. Renewable energy-powered drones represent a promising step toward sustainable agricultural innovation, reducing carbon footprints, lowering operational costs, and providing extended surveillance capabilities. Next-generation farming leverages drones, robotics, and IoT to create a connected, automated, and sustainable agricultural future. Overcoming operational challenges requires technological innovation, resilient infrastructure, and supportive policies. Harmonized policies, clear operational guidelines, and continuous dialogue between authorities and the agricultural sector can unlock the full potential of UAV technology while safeguarding public interests and airspace integrity.

Keywords UAV technology · Agricultural drones · Legal regulations · Ethical considerations · Airspace laws · Sustainability · Precision agriculture · AI in UAVs · Autonomous drones

© The Author(s), under exclusive license to Springer Nature Singapore Pte Ltd. 2025

J. Li, *UAV Aerodynamics and Crop Interaction*, Smart Agriculture 13,
https://doi.org/10.1007/978-981-96-8402-1_13

13.1 Introduction

Unmanned Aerial Vehicles (UAVs), commonly known as drones, are transforming modern agricultural practices by enhancing efficiency, precision, and sustainability. These advanced technologies play a crucial role in precision agriculture, enabling farmers to monitor crops, assess soil health, and optimize resource use. UAVs are equipped with high-resolution cameras, multispectral sensors, and advanced data analytics capabilities, allowing for real-time decision-making and improved yield outcomes. The adoption of UAV technology in agriculture raises significant legal and ethical considerations. Regulatory frameworks govern drone usage to ensure safety, privacy, and compliance with airspace regulations. Ethical concerns involve the responsible use of data, potential impacts on rural employment, and the equitable access to technology [1]. As UAV applications expand, it is vital to address these challenges while fostering innovation and sustainable agricultural practices. Future perspectives on UAV technology in agriculture highlight ongoing advancements in artificial intelligence, automation, and data integration. These innovations promise enhanced predictive capabilities, improved crop management, and increased operational efficiency. As the agricultural sector navigates these technological transformations, a balanced approach addressing legal, ethical, and technical dimensions will be essential to harness the full potential of UAVs for sustainable and equitable food production systems [2].

13.2 Legal and Ethical Considerations in UAV Technology for Agriculture

The use of UAV (drone) technology in agriculture is governed by legal frameworks and ethical responsibilities to ensure safe and responsible operations. Legally, drone operators must comply with regulations set by national aviation authorities, such as the FAA in the U.S. or PCAA in Pakistan, which typically require proper licensing and registration for commercial use. Flight operations are often restricted to visual line of sight (VLOS) unless specific authorization is granted for beyond visual line of sight (BVLOS). Additionally, operators must respect no-fly zones near sensitive areas like airports and military installations [3].

Data protection is another crucial legal concern, requiring compliance with privacy laws when collecting aerial images of private property and obtaining informed consent when necessary. Environmental regulations also apply, particularly in the use of drones for pesticide application, where adherence to safety guidelines is essential to prevent ecological harm. Furthermore, drone operators are encouraged to have appropriate insurance to cover potential liabilities in case of accidents or equipment damage. Ethically, the use of UAVs in agriculture involves respecting privacy, minimizing environmental impact, and ensuring transparency in data collection and usage

[4]. Responsible drone use includes considering the well-being of local communities, safeguarding sensitive information, and reducing any unintended harm to ecosystems.

Legal Considerations in UAV Technology for Agriculture

UAV operators must comply with various aviation laws, including the FAA in the USA, CAA in the UK, and PCAA in Pakistan. These regulations cover drone registration, licensing, altitude restrictions, and no-fly zones near sensitive locations. Pilot certification is often mandatory for commercial operations. Privacy laws, such as the GDPR in the EU, protect personal information captured during drone operations. Airspace and property rights must be respected, with some regions allowing drone activity without proper authorization. Operators are legally responsible for any damage or harm caused by their drones, and many countries mandate third-party liability insurance to cover potential accidents [5]. Environmental regulations are crucial when using UAVs for agricultural tasks, such as pesticide spraying, to avoid contaminating ecosystems and ensure sustainable practices.

Ethical Considerations in UAV Technology for Agriculture

The ethical considerations in using UAV technology for agriculture include respecting privacy and surveillance, ensuring transparency and accountability, minimizing environmental impact, ensuring equitable access to technology, and protecting sensitive agricultural data from unauthorized access or cyberattacks. Operators should obtain informed consent before capturing images or data from private property, balancing data collection with individual privacy rights. Clear and transparent policies regarding data management are necessary, outlining who owns and controls the data, how it is stored, and the duration of retention. Responsible drone use promotes sustainable agricultural practices. Equity and access to technology are also crucial, as drone technology may disproportionately benefit large-scale farmers while smallholders are excluded due to high costs [5]. Data security is a critical ethical responsibility, requiring encryption and secure storage practices to safeguard private and commercially valuable information.

Best Practices for Legal and Ethical UAV Use in Agriculture

To ensure the responsible use of UAV technology in agriculture, operators must follow best practices that align with legal and ethical standards. First, it is essential to follow local regulations by staying informed about aviation laws and agricultural policies established by national and international regulatory bodies. Regularly reviewing updates helps ensure legal compliance and avoids penalties. Respecting privacy is another key practice—operators should avoid collecting personal data without informed consent and take measures to protect private property rights. Transparency is crucial; clearly communicating how data is collected, used, and stored builds trust with stakeholders, including farmers and local communities [6]. Additionally, operators should minimize harm by using drones in ways that protect the environment and community welfare, such as reducing disturbances to wildlife and adhering to environmental regulations. Lastly, ensuring equity means promoting fair access to UAV technology by supporting initiatives that make drone services affordable and accessible to smallholder farmers, fostering more inclusive agricultural benefits.

13.2.1 Regulations for UAV Operations in Agriculture

The use of UAVs (Unmanned Aerial Vehicles) in agriculture is subject to stringent regulatory frameworks to ensure safe and responsible usage. These regulations vary by region but generally require drone operators to register their devices, obtain licenses, and adhere to specific flight guidelines. Compliance with aviation authorities such as the Federal Aviation Administration (FAA) in the United States or the Civil Aviation Authority (CAA) in other regions is essential [7].

13.2.1.1 Regulatory Framework for UAV Operations in Agriculture

Agricultural UAV operations are governed by regulations that ensure safety, privacy, and environmental protection. Drone registration and licensing are mandatory in most countries, requiring operators to register their drones with national aviation authorities. For drones above a specific weight threshold—such as 250 g in the USA—a Remote Pilot License (RPL) is often required. Operational flight restrictions include altitude limits, usually 400 feet (120 m) above ground level, and adherence to no-fly zones around sensitive areas like airports and military installations. Most regions also enforce the Visual Line of Sight (VLOS) rule, requiring drones to remain within the operator's unaided visual range unless Beyond Visual Line of Sight (BVLOS) permission is granted [8].

Safety and insurance requirements play a crucial role in UAV operations. Many countries require third-party liability insurance to cover potential damages, while preflight checks and risk assessments are essential for safe operation. Specific regulations apply to agricultural drones, especially those used for pesticide spraying, which may require additional certification and must comply with environmental protection laws. Data privacy is another critical area—operators collecting imagery or sensor data on private property must follow data protection laws and, in some cases, obtain informed consent from landowners [9]. Agricultural UAV regulations focus on registration, operational limits, safety measures, and privacy compliance. Following these guidelines ensures the safe, legal, and ethical use of drones in agricultural practices.

13.2.1.2 Region-Specific UAV Regulations in Agriculture

United States (FAA—Federal Aviation Administration)

Agricultural UAV operations in the U.S. are governed by FAA regulations. Part 107 Certification is mandatory for commercial drone use, including agricultural applications. Drones exceeding 55 lbs (25 kg) require special airworthiness certification for legal operation. Night operations are permitted if drones are equipped with anti-collision lighting and operators receive appropriate training [10].

European Union (EASA—European Union Aviation Safety Agency)

The EASA classifies drones under three categories: Open, Specific, and Certified, with most agricultural UAVs falling under the Specific category, requiring Operational Authorization. Additionally, any drone capturing farm data must comply with the General Data Protection Regulation (GDPR) to protect privacy [11].

United Kingdom (CAA—Civil Aviation Authority)

In the UK, agricultural drone operations require an Operational Authorization (OA) for drones over 25 kg or for specialized tasks. Specific flight zone regulations apply when operating near urban areas or sensitive locations to ensure public safety and privacy [12].

Pakistan (PCAA—Pakistan Civil Aviation Authority)

The PCAA mandates drone registration for all commercial UAVs. Operational limitations prohibit flying near sensitive areas without prior authorization, ensuring both security and compliance with national airspace regulations [13].

Australia (CASA—Civil Aviation Safety Authority)

Agricultural drone operators must obtain a Remote Pilot License (RePL). However, micro-drones weighing under 2 kg and used for specific farm tasks may qualify for exemptions, allowing more flexible regulations [14]. UAV regulations in agriculture vary by region but generally require licensing, operational authorizations, and compliance with data protection and airspace restrictions. Understanding these region-specific rules is crucial for legal and safe drone use.

13.2.1.3 Best Practices for Regulatory Compliance in Agricultural UAV Use

To maintain regulatory compliance in agricultural UAV operations, it is essential to follow best practices that ensure legal adherence and safe drone usage. Staying updated on local and international UAV regulations is crucial, as policies frequently change. Regularly reviewing and implementing these updates helps avoid legal violations. Documentation is another key practice—operators should maintain accurate records of UAV licenses, flight logs, and maintenance checks to demonstrate compliance during inspections. Training is vital for ensuring safe and responsible UAV operations. Operators should complete approved training programs to understand technical and legal aspects thoroughly [15]. Lastly, safety first should be a guiding principle—following established operational procedures and prioritizing public safety minimizes the risk of accidents and legal liability.

13.2.2 Airspace Laws and Flight Permissions

Drones operating in agricultural settings must follow airspace laws to prevent interference with manned aircraft and other UAVs. Flight permissions often include altitude limits, no-fly zones near airports, and designated agricultural zones where drone operations are permissible. Farmers and drone operators must stay informed about changes in airspace regulations to maintain legal compliance. Agricultural UAV operations must comply with airspace laws to prevent conflicts with manned aircraft and other drones. These regulations typically enforce altitude limits—often 400 feet (120 m) above ground level—and restrict flights in no-fly zones near sensitive areas such as airports, military bases, and urban zones [16]. Some regions designate specific agricultural zones where drone use is permitted, providing flexibility for farming activities. To maintain legal compliance, farmers and UAV operators must stay updated on airspace regulations and secure appropriate flight permissions when required. Adherence to airspace laws, including respecting altitude limits and restricted areas, is essential for the safe and legal operation of agricultural drones [17]. Regularly monitoring regulatory changes ensures compliance and reduces the risk of airspace violations.

13.2.3 Airspace Classification and UAV Operations

Airspace is categorized into controlled and uncontrolled zones, each with specific regulations for UAV operations. Controlled airspace includes areas managed by air traffic control (ATC), such as airports and urban areas, where UAV operators must obtain special permission to fly. This ensures drones do not interfere with manned aircraft. Uncontrolled airspace does not require ATC clearance, allowing more flexibility for drone use. However, operators must still follow general rules, including maintaining Visual Line of Sight (VLOS) and adhering to altitude limits—typically 400 feet (120 m) above ground level [18].

UAV operators must understand airspace classifications to ensure legal compliance. Controlled zones require authorization, while uncontrolled zones provide greater operational freedom under standard regulations.

Airspace is typically divided into controlled and uncontrolled zones, with specific rules for each:

Airspace class	Description	Drone operation rules
Controlled airspace	Surrounds airports and areas with heavy air traffic	Requires special authorization for UAV flights
Uncontrolled airspace	Lower-level airspace away from major aviation routes	Generally open for UAVs but with altitude limits
Restricted areas	Includes military zones and sensitive locations	Prohibited without specific government approval
Agricultural zones	Designated for farming operations and crop monitoring	UAV use is permitted but must follow guidelines
No-fly zones	Areas like airports, government buildings, and national parks	UAVs are banned or require special clearance

13.2.4 Key Airspace Laws for UAVs in Agriculture

Altitude Restrictions

1. In most countries, UAVs are limited to flying 400 feet (120 m) above ground level to prevent interference with manned aircraft.
2. Flying above this limit requires special flight authorization from the relevant aviation authority.

Flight Boundaries

1. UAVs are typically prohibited from flying within 5 km (3 miles) of airports, though exact distances may vary by country.
2. Operators must also avoid restricted areas like national parks, military bases, and critical infrastructure without prior approval.

Beyond Visual Line of Sight (BVLOS)

1. Standard regulations require drones to remain within the operator's unassisted line of sight to ensure safe operation.
2. For BVLOS operations—such as monitoring large agricultural areas—special permission is required, often with additional safety protocols in place.

Agricultural UAV operators must comply with altitude limits, respect restricted zones, and obtain special authorization for BVLOS flights to ensure safe and legal drone operations [19].

13.2.5 Obtaining Flight Permissions for Agricultural UAVs

Register the Drone

1. Drone registration is mandatory with national aviation authorities such as the Federal Aviation Administration (FAA) in the USA, the European Union Aviation Safety Agency (EASA), and the Pakistan Civil Aviation Authority (PCAA). This process ensures legal accountability and compliance with airspace regulations.

Apply for Operational Authorization

1. Operational authorization is required for specialized UAV activities, including flights in controlled airspace, Beyond Visual Line of Sight (BVLOS) operations, and agricultural-specific uses.
2. Examples of required authorizations include:
    1. FAA Part 107 Waiver (USA)—for advanced operations exceeding standard drone rules.
    2. Specific Category Authorization (EU)—for UAV operations that fall outside the open-category regulations.

Request Temporary Flight Permissions

1. For short-term UAV use near sensitive areas (e.g., around airports or military zones), operators can apply for temporary airspace approval from the appropriate aviation authority. This allows limited and controlled drone access for specific tasks.

Agricultural UAV operators must register drones, obtain operational authorization for advanced tasks, and apply for temporary flight permissions when operating in restricted areas. Following these steps ensures legal compliance and safe UAV operations [20].

13.2.6 Country-Specific Airspace Regulations

United States (FAA)

1. **Maximum Altitude**: UAVs are restricted to 400 feet (120 m) above ground level.
2. **LAANC System**: The Low Altitude Authorization and Notification Capability (LAANC) system allows real-time airspace authorization for agricultural drone operations in controlled airspace.

European Union (EASA)

1. **UAS Operator ID**: Commercial UAV operators must obtain a UAS Operator ID for legal drone use.

2. **BVLOS Operations**: Beyond Visual Line of Sight (BVLOS) flights require a Specific Operations Risk Assessment (SORA) to assess and mitigate risks.

Pakistan (PCAA)

1. **Pre-Flight Approval**: Operators must obtain pre-flight approval for UAV operations near airports and urban zones to avoid airspace conflicts.
2. **Pesticide Spraying**: Drones used for pesticide spraying need special permission to ensure compliance with safety and environmental regulations.

Airspace regulations for agricultural UAVs vary by country. The USA, EU, and Pakistan enforce altitude limits, require authorization systems, and mandate special permissions for advanced operations, ensuring safe and regulated drone use [21].

13.2.7 Best Practices for Airspace Compliance

Check Airspace Maps: Use digital tools like Aloft, DroneDeploy, or official CAA apps to access real-time airspace information. These tools help identify no-fly zones, controlled airspace, and other restricted areas.
Maintain Flight Logs: Keep detailed records of flight operations, including dates, locations, and permissions. These logs provide legal documentation and support regulatory compliance in case of audits or inspections.
Stay Informed: Regularly monitor updates from your country's aviation authority (e.g., FAA, EASA, PCAA) to remain aware of policy changes, new regulations, and emerging guidelines affecting UAV operations.

Maintaining compliance with airspace laws involves checking airspace maps, documenting flights, and staying updated on regulatory changes. These practices ensure safe, legal, and efficient agricultural UAV operations [22].

13.2.8 Safety Protocols for Drone Usage

Safety is a fundamental concern in UAV operations, particularly in agriculture, where drones play a crucial role in tasks such as crop monitoring, soil analysis, and pesticide application. To minimize risks associated with drone operations—such as equipment malfunctions, data loss, and flight accidents—operators must adhere to standardized safety protocols. These protocols include comprehensive pre-flight checks, routine maintenance, and emergency procedures, all of which ensure safe, efficient, and legally compliant drone usage. Pre-flight checks are essential to verify the drone's readiness before takeoff. Operators must inspect the drone's battery levels, propellers, GPS calibration, and communication systems to prevent mid-flight failures [23]. Physical damage should be assessed, and the surrounding environment must be reviewed to ensure there are no obstacles or restricted zones along the flight path. Additionally, operators should confirm that the weather conditions are suitable for drone operation and that the UAV's software is updated to meet current safety standards.

Maintenance routines are equally important to ensure long-term drone functionality. Regular inspections of motors, sensors, cameras, and control systems help identify potential faults before they cause serious issues. Updating firmware and maintaining a detailed maintenance log ensures compliance with regulatory requirements and provides a record for troubleshooting and audits. Regular maintenance enhances drone performance and extends its operational life while reducing the likelihood of unexpected malfunctions. In addition to pre-flight checks and maintenance, emergency procedures play a crucial role in handling unforeseen situations. Operators must establish a clear emergency response plan for situations such as signal loss, mechanical failure, or uncontrolled flight. Practicing manual overrides and ensuring familiarity with emergency landing protocols can prevent property damage and protect public safety [24]. It is also essential to have a protocol in place for shutting down the drone if it poses a risk to people or the environment. Comprehensive safety protocols—including pre-flight inspections, routine maintenance, and emergency preparedness—is vital for ensuring the safe, efficient, and compliant use of drones in agriculture. These practices not only reduce operational risks but also help maintain public trust and ensure adherence to legal and regulatory standards. By prioritizing safety, drone operators can optimize agricultural productivity while safeguarding both human and environmental well-being [25].

13.2.9 Ethical Implications of Drone Farming

The increasing use of drones in agriculture raises significant ethical concerns related to environmental impact, labor displacement, and equitable access to technology. While drones provide numerous benefits, such as improved efficiency, precision monitoring, and resource optimization, their adoption must be carefully managed to minimize negative social and environmental consequences. Ethical drone use emphasizes sustainability, transparency, and fair distribution of technological advancements. One major ethical concern is the environmental impact of drone farming. Although drones reduce the overuse of pesticides and fertilizers through targeted application, their production, operation, and battery disposal contribute to electronic waste and carbon emissions. Ethical UAV deployment should focus on minimizing ecological harm by adhering to environmental regulations and adopting eco-friendly practices, such as using recyclable materials and optimizing flight efficiency to reduce energy consumption [26]. Another ethical issue involves labor displacement. As drones automate tasks such as crop monitoring and pesticide spraying, there is a risk of job loss for agricultural workers who traditionally perform these roles. This technological shift disproportionately affects rural communities where farming is a primary livelihood. To address this, it is essential to invest in worker retraining programs and provide opportunities for skill development in drone operation and data analysis, ensuring that agricultural innovation does not come at the expense of human labor. Equitable access to drone technology is also a critical ethical consideration. Large-scale farms with greater financial resources can more easily adopt UAV systems, widening the gap between industrial and smallholder farmers. This digital divide can lead to unequal productivity and reduced competitiveness for smaller agricultural enterprises [27]. Promoting affordable drone solutions, offering subsidies, and encouraging open-access drone programs can help democratize UAV technology and ensure its benefits are shared equitably across all farming sectors.

Transparency is fundamental to ethical drone use, especially regarding data collection. Agricultural drones gather sensitive information on soil health, crop yields, and land usage, raising privacy concerns. Operators must ensure informed consent, secure data storage, and clear communication about how the data is used. Ethical practices require compliance with data protection regulations and fostering trust with local communities. The ethical implications of drone farming extend beyond technological advancements to include environmental responsibility, worker protection, fair access, and transparent data practices [28]. By addressing these concerns, drone operators can foster sustainable, inclusive, and ethical agricultural development.

13.2.10 Environmental Impact of Drone Farming

Drones play a dual role in agriculture, offering both environmental benefits and challenges. On the positive side, precision spraying reduces the overuse of pesticides, leading to less chemical runoff and minimizing contamination of nearby ecosystems. Additionally, drones decrease soil compaction compared to heavy machinery, preserving soil health and improving crop yields. They also contribute to water conservation by enabling targeted irrigation, which reduces water waste. However, the negative impact of drone farming cannot be overlooked. The manufacturing process of drones, including the use of rare materials, and their energy consumption during operation contribute to carbon emissions and generate electronic waste [29]. Batteries used in drones pose an additional concern due to their limited lifespan and disposal challenges.

From an ethical standpoint, it is crucial to adopt eco-friendly practices to minimize these negative effects. This includes using energy-efficient drones, promoting recyclable materials, and adhering to sustainable manufacturing methods. Drone operators should also comply with environmental regulations and adopt strategies to reduce ecological disturbance, such as minimizing flights near wildlife areas. While drones in agriculture reduce chemical overuse and soil disturbance, their production and operation raise environmental concerns [30]. Ethical drone use requires prioritizing energy efficiency, sustainable practices, and minimizing environmental harm.

13.2.11 Labor Displacement in Drone Farming

The integration of drones in agriculture brings both efficiency gains and employment challenges, particularly in rural areas. Automation of tasks such as crop monitoring, pesticide spraying, and field mapping significantly reduces the need for manual labor, allowing farmers to increase productivity while lowering operational costs. This technological efficiency helps address labor shortages during peak seasons and improves the precision and speed of agricultural practices. However, these advancements pose a risk of job displacement, especially for traditional farm workers who depend on agricultural employment for their livelihoods. The reduced demand for manual tasks may lead to unemployment and economic insecurity in rural communities where alternative job opportunities are limited [31]. This shift disproportionately affects low-skilled workers, who may face challenges adapting to technologically advanced agricultural methods.

From an ethical perspective, it is essential to balance technology adoption with job preservation. This can be achieved by supporting worker retraining programs to equip displaced workers with new skills related to drone operation, data analysis, and precision agriculture. Collaboration between government agencies, private sectors, and educational institutions can foster inclusive innovation, ensuring that technological progress benefits both farmers and rural labor forces. While drones increase efficiency in agriculture, they also pose a risk of labor displacement [32]. Ethical drone use involves balancing automation with job preservation by promoting worker retraining and fostering inclusive access to emerging agricultural technologies.

13.2.12 Data Privacy and Surveillance in Drone Farming

Drones in agriculture collect large volumes of data, including crop health, soil conditions, and field mapping. While this data enhances precision farming and improves productivity, it also raises serious privacy concerns. One primary risk is unauthorized surveillance, where drones may inadvertently capture information from private farms or neighboring properties without permission. This encroachment on personal and commercial privacy can lead to legal disputes and ethical violations, especially when sensitive information is collected without explicit consent. Another major concern involves data ownership. Farmers who collaborate with third-party UAV services may lose control over their agricultural data. Companies may store, analyze, or sell this data without providing clear information on how it is used [33]. This loss of ownership can reduce farmer autonomy, create power imbalances, and raise questions about who ultimately benefits from data-driven insights.

Ethically, drone operators must adopt transparent data practices. This includes obtaining informed consent before collecting data and clearly explaining how the data will be used, stored, and shared. Farmers should retain ownership rights and have access to their own data. Implementing secure storage solutions and following data protection regulations—such as the General Data Protection Regulation (GDPR) in the European Union—can safeguard sensitive information and maintain trust between drone operators and agricultural communities. While drones provide valuable agricultural data, they also pose privacy risks and raise concerns about data ownership. Ethical drone use requires informed consent, transparent data handling, and ensuring farmers retain control over their information [34].

13.2.13 Equity and Technology Access

Equity and Technology Access in Drone Farming

The high costs of UAV (Unmanned Aerial Vehicle) technology present a significant barrier for small and resource-poor farmers, raising concerns about equity in agriculture. Wealthier farms, with the resources to invest in advanced drone systems, gain greater productivity and operational efficiency, widening the digital divide between large-scale and small-scale producers. This technological advantage allows larger farms to increase yields, reduce input costs, and make data-driven decisions, while smaller farms struggle to keep pace, exacerbating economic inequality in rural areas. Affordability issues are a key factor driving this disparity. Sophisticated drone systems, especially those equipped for precision agriculture tasks like crop monitoring, soil analysis, and pesticide spraying, require significant financial investment. For many smallholder farmers, these costs are prohibitive, preventing them from accessing the productivity gains and sustainability benefits that drones offer [35]. Without targeted intervention, the adoption of UAV technology may favor large farms, leaving smaller agricultural operations at a competitive disadvantage.

From an ethical standpoint, it is crucial to promote equitable access to drone technology. This can be achieved through government subsidies, low-interest financing, and public–private partnerships to reduce costs for marginalized farmers. Community drone-sharing programs and cooperative models allow smallholders to pool resources and access drone services without the burden of full ownership. Additionally, providing training programs on drone operation and data analysis empowers small-scale farmers to adopt and benefit from UAV technology effectively. The high costs of drone technology deepen the digital divide, limiting access for small farmers and favoring wealthier producers [36]. Ethical drone use involves reducing barriers through subsidies, community sharing, and training to ensure fair access and equitable agricultural advancement.

13.2.14 Transparency and Accountability

Transparency and accountability are fundamental to ensuring the ethical use of drones in agriculture. As UAV (Unmanned Aerial Vehicle) technology becomes increasingly integrated into agricultural practices, concerns about misuse, privacy violations, and environmental harm arise. Without proper oversight and regulation, drones may negatively impact local communities, ecosystems, and food systems. Establishing clear guidelines and promoting open communication are essential to mitigate these risks and foster trust between technology users and the public. Responsibility is a crucial aspect of drone accountability. Legal frameworks should clearly define who is liable for drone-related incidents, including accidents, data breaches, and environmental damage. This involves identifying operator responsibilities—such as adhering to flight regulations, maintaining equipment, and protecting collected data. It also extends to manufacturers, who must ensure their drones meet safety standards and privacy safeguards [37]. In cases of misuse or malfunctions, having clear legal recourse allows affected parties to seek compensation and ensures operators are held accountable.

Transparency in drone farming means openly sharing how drones operate and the impact they have on agricultural and local ecosystems. This includes providing public access to information about where drones are deployed, what data is collected, and how that data is used. Stakeholder engagement is also critical—farmers, policymakers, and local communities should be informed of drone activities and have opportunities to voice concerns. Public disclosure of UAV operations helps build trust, enabling communities to monitor drone use and ensure responsible practices are followed. A critical area for transparency involves data handling. Drones collect large amounts of agricultural data, such as crop health, soil composition, and field conditions. It is essential to establish clear policies on data ownership, ensuring that farmers retain control over their information. Third-party UAV operators must be transparent about how data is stored, who can access it, and whether it will be shared or sold [38]. Implementing secure data practices—such as encryption and compliance with data protection laws—further protects sensitive agricultural information.

Ethical considerations in drone farming demand comprehensive regulations and public awareness campaigns. Regulatory bodies should enforce consistent standards for drone registration, pilot certification, and flight permissions. Special attention must be given to drones used for pesticide spraying or other sensitive operations to prevent environmental harm. Encouraging community participation in policy discussions ensures that regulations reflect local needs and ethical concerns. To strengthen accountability, drone operators should maintain detailed flight logs and data records, making them available for regulatory review. Independent audits and impact assessments can further verify compliance with safety, privacy, and environmental standards [39]. By fostering a culture of transparency, the agricultural sector can demonstrate a commitment to responsible drone use and mitigate potential ethical challenges. Ensuring transparency and accountability in agricultural drone use requires clear guidelines for responsibility, public disclosure of operations, and robust regulatory frameworks. By establishing legal accountability and promoting open communication, stakeholders can protect public interests, minimize misuse, and encourage ethical and sustainable drone practices.

13.2.15 Ethical Drone Farming Principles

Ethical drone farming requires a balanced approach that prioritizes sustainability, equity, privacy, transparency, and accountability. Sustainability involves minimizing environmental harm by adopting eco-friendly practices, such as using energy-efficient drones, responsibly managing battery disposal, and complying with environmental regulations. Equity emphasizes ensuring fair access to drone technology for all farmers, including smallholders, through affordable solutions, government subsidies, and training programs. Privacy is another critical concern, requiring operators to protect farmers' data by securing information, obtaining informed consent, and adhering to data protection laws. Transparency involves being open about drone operations and their societal impacts, providing clear communication regarding data collection, storage, and usage, and fostering trust through community engagement [40]. Lastly, accountability ensures that drone operators are held legally and ethically responsible for their actions, with clear guidelines on their duties and obligations. Ethical drone farming requires a commitment to sustainability, equity, privacy, transparency, and accountability to ensure responsible and fair use of UAV technology while minimizing social and environmental risks.

13.2.16 Sustainability

Ethical drone farming prioritizes environmental responsibility by reducing ecological impacts:

- **Eco-Friendly Practices**: Use energy-efficient drones and adopt precision agriculture to reduce resource waste.
- **Responsible Disposal**: Properly manage battery waste and electronic components to reduce environmental pollution.
- **Compliance**: Follow environmental regulations and minimize harm to ecosystems during drone operations.

13.2.17 Equity

Ensuring fair access to drone technology is essential for social justice in agriculture:

- **Affordable Solutions**: Support low-cost UAVs, government subsidies, and financial aid for small-scale farmers.
- **Inclusive Access**: Promote training programs to ensure all farmers, including smallholders, can adopt and benefit from UAV technology.
- **Bridging the Digital Divide**: Address disparities by creating shared drone cooperatives for under-resourced communities.

13.2.18 Privacy

Protecting data security and respecting personal privacy is a core ethical principle:

- **Data Protection**: Implement encryption and secure storage for sensitive agricultural data.
- **Informed Consent**: Obtain explicit permission before collecting or sharing farm-related information.
- **Legal Compliance**: Adhere to national and international data protection laws (e.g., GDPR).

13.2.19 Transparency

Open communication about drone practices fosters public trust and community engagement:

- **Clear Communication**: Provide detailed information on how data is collected, stored, and used.
- **Public Engagement**: Involve local communities in decision-making and educate stakeholders on drone applications.
- **Operational Clarity**: Ensure drone use is documented and disclosed to relevant authorities and affected parties.

13.2.20 Accountability

Ethical drone use requires holding operators responsible for their actions and outcomes:

- **Legal Responsibility**: Establish clear guidelines for drone operators regarding their duties.
- **Liability Framework**: Ensure compensation mechanisms are in place for any damages caused by UAV operations.
- **Ethical Oversight**: Implement audits and compliance checks to monitor drone activity and enforce ethical standards.

13.2.21 Privacy Concerns in UAV Data Collection

Agricultural drones gather extensive data through advanced technologies such as imaging, thermal sensing, and geographic mapping. While this data enhances crop monitoring, precision agriculture, and resource management, it also poses significant privacy risks. Unauthorized data collection can infringe on the privacy of neighboring properties and sensitive areas. Ensuring ethical drone use requires compliance with data protection laws, such as the General Data Protection Regulation (GDPR) in the European Union, which mandates the secure collection, storage, and use of personal and environmental data. Informed consent is essential when capturing data in shared or privately-owned spaces. Farmers and drone operators must communicate how collected information will be used and provide options to opt out if desired [41]. Transparent data practices not only protect privacy but also build trust between agricultural stakeholders and the public. Additionally, anonymizing data and using encryption techniques can further safeguard sensitive information from unauthorized access or misuse.

Ethical considerations in UAV data collection emphasize the importance of balancing technological advancement with individual privacy rights. This includes clear policies on data ownership, secure handling, and ensuring that no surveillance occurs without proper authorization. Protecting privacy in UAV data collection involves following legal frameworks, obtaining informed consent, and ensuring secure data handling to maintain ethical standards and public trust [42].

13.2.22 Balancing Technology with Traditional Practices

The adoption of UAV technology should complement, not replace, traditional agricultural practices. Integrating drones with conventional farming requires a balanced approach that respects cultural values while leveraging modern innovations. Training programs and community engagement can bridge the gap between technological advancements and established farming knowledge [43].

13.2.23 Types of Data Collected by Agricultural UAVs

Agricultural drones gather various forms of sensitive data, including:

Data type	Purpose	Privacy concern
Aerial imagery	Monitor crop health, field conditions	May capture adjacent private properties
Thermal imaging	Detect irrigation patterns and soil moisture	Reveals land use patterns beyond intended areas
Geospatial data (GPS)	Map farm boundaries and optimize logistics	Risk of location tracking and property surveillance
Multispectral data	Assess plant health and nutrient levels	Unintentional capture of neighboring lands

13.2.24 Key Privacy Concerns in UAV Data Collection

Ethical Concerns in Agricultural UAV Use

Agricultural UAVs present several ethical challenges related to privacy, data ownership, informed consent, and data security. Unauthorized surveillance is a major concern, as drones may inadvertently capture private property or personal activities of individuals living near agricultural zones, leading to potential violations of personal privacy and property rights. Data ownership and control also raise critical questions about who owns and can access the collected data—whether it is the farmers, drone service providers, or third parties [44]. Without clear agreements, farmers risk losing autonomy if their data is shared or sold without their consent. Informed consent is another crucial issue, particularly when drones collect data on shared agricultural land or near public areas. Ethical breaches can occur if neighbors or adjacent landowners are not explicitly informed and their permission is not obtained. Additionally, data security risks are significant, as sensitive agricultural data is vulnerable to cyberattacks and unauthorized access. If not properly secured, data leaks could expose confidential information such as farm yields, supply levels, or land values. Addressing these ethical concerns is essential to ensure responsible and fair use of UAV technology in agriculture [45].

13.2.25 Legal Frameworks for UAV Data Protection

Agricultural drone operations must comply with national and international **data protection laws**:

Region	Applicable law	Key requirement
United States	FAA Part 107 & Privacy Act	Consent and geofencing in private areas
European Union	General Data Protection Regulation (GDPR)	Informed consent and data minimization
United Kingdom	Data Protection Act (DPA 2018)	Secure storage and lawful data collection
Pakistan	Personal Data Protection Bill (PDPB)	Limited access and user rights to data
Australia	Privacy Act 1988	Individual consent and data confidentiality

13.2.26 Best Practices for Ethical UAV Data Collection

To address ethical concerns in agricultural UAV operations, it is essential to follow key practices that protect privacy, data ownership, and security. Obtaining informed consent is crucial—operators should seek written consent before capturing data from shared or adjacent properties and notify neighboring landowners about planned drone activities. Securing data storage and access involves using encryption to protect all UAV-collected data and restricting access to authorized personnel only, ensuring sensitive information remains confidential. Implementing data minimization helps reduce privacy risks by collecting only the data necessary for agricultural purposes and using geo-fencing technology to prevent drones from crossing into private zones. Transparency and accountability are vital; operators should clearly disclose how data is collected, stored, and shared while establishing a data use policy that allows farmers to retain ownership of their data [46]. By following these guidelines, agricultural UAV operations can balance technological innovation with ethical responsibility.

Agricultural UAVs improve crop management by providing precise data for better decision-making; however, they also raise ethical concerns regarding privacy, data ownership, and security. Ethical data collection involves several key principles. Informed consent is crucial, requiring UAV operators to inform and obtain permission from affected parties before collecting data. Data security must be maintained by using secure methods to collect, store, and manage sensitive information, ensuring protection against unauthorized access. Legal compliance is essential, as operators must adhere to both national and international data protection laws to safeguard individuals' rights. Lastly, transparency is necessary, meaning operators should be open and clear about how the collected data will be used and shared [47]. By following these guidelines, agricultural UAV operations can balance technological advancements with ethical responsibilities.

13.3 Challenges and Future of UAVs in Agriculture

The integration of UAVs (Unmanned Aerial Vehicles) in agriculture has transformed modern farming by providing accurate data collection and real-time monitoring. Drones equipped with advanced sensors can capture aerial imagery, thermal readings, and geographic information, allowing farmers to make informed decisions regarding crop health, irrigation, and land management. This technology improves efficiency, reduces resource wastage, and enhances yield predictions. However, the widespread use of UAVs raises critical privacy concerns. One major issue involves the unauthorized surveillance of private properties. As drones capture vast areas during agricultural operations, they may inadvertently record images of neighboring lands or personal activities. This intrusion can violate property rights and lead to ethical and legal disputes. Additionally, data ownership remains a contentious issue [48]. When third-party UAV service providers collect and analyze data, farmers may lose control over their sensitive agricultural information. This raises concerns about how the data is used, stored, or potentially shared with external entities without the farmers' consent.

Informed consent is another crucial aspect of privacy in UAV data collection. When drones operate near shared or private spaces, it is essential to obtain permission from affected parties. Without transparent communication, drone operations can breach ethical standards and erode trust between farmers and local communities. Moreover, data security is paramount to prevent unauthorized access or cyberattacks on confidential information. As UAV adoption grows, adhering to national and international data protection laws becomes vital. Frameworks like the GDPR (EU) and FAA (USA) outline guidelines for lawful data collection, transparency, and user rights [49]. Ensuring ethical UAV practices requires a balance between technological advancement and privacy protection through clear legal frameworks, data minimization, and responsible data handling.

13.3.1 Emerging Technologies and Trends

Advancements in UAV (Unmanned Aerial Vehicle) technology are reshaping agricultural practices by increasing efficiency, precision, and sustainability. Emerging innovations such as multispectral imaging, AI-driven analysis, and real-time data transmission are revolutionizing how farmers monitor and manage their crops. These cutting-edge technologies enable more accurate decision-making, reduce resource waste, and enhance overall productivity in agricultural operations. Multispectral and Hyperspectral Imaging are among the most significant innovations in UAV technology. These imaging systems allow drones to capture data beyond the visible spectrum, providing detailed insights into plant health, soil moisture, and nutrient levels [50]. Farmers can use this data to detect crop stress, identify diseases, and optimize irrigation schedules, leading to improved yields and reduced chemical inputs.

Artificial Intelligence (AI) and Machine Learning (ML) are also playing a crucial role in modern UAV applications. AI-powered drones can analyze large datasets, recognize patterns, and deliver actionable insights. For instance, drones equipped with AI algorithms can autonomously identify weeds, assess crop maturity, and even predict harvest times. This automation reduces labor costs and improves the accuracy of agricultural monitoring. Real-Time Data Transmission is another emerging trend transforming agricultural UAVs. With the integration of 5G technology and advanced communication systems, drones can relay real-time information directly to farm management systems. This allows farmers to make immediate decisions regarding irrigation, pesticide application, or disease control. Fast and accurate data transfer is particularly valuable for large-scale farming operations where timely responses are critical [51]. Other promising innovations include drone swarming technology, where multiple drones work together to cover extensive fields, and blockchain-based data storage, which ensures secure and transparent management of agricultural data.

13.3.2 Swarm Drones for Large-Scale Operations

Swarm drone technology is transforming large-scale agricultural operations by enabling the coordinated deployment of multiple drones to perform tasks simultaneously. This innovative approach enhances efficiency in critical agricultural processes such as crop monitoring, pesticide application, and data collection, making it a valuable tool for modern precision farming. Unlike single-drone operations, swarm technology allows for faster data acquisition and broader coverage, which is particularly advantageous for extensive farmlands where timely interventions are crucial. In crop monitoring, swarm drones equipped with advanced imaging sensors can survey large areas in a fraction of the time required by individual drones. These drones work collaboratively to capture multispectral, thermal, and geospatial data, providing farmers with comprehensive insights into crop health, soil conditions, and pest infestations [52]. The simultaneous operation of multiple drones enables real-time assessment of agricultural fields, allowing for quicker identification and resolution of potential issues.

Swarm drones also revolutionize pesticide and fertilizer application by improving both speed and precision. Coordinated drone fleets can cover large fields uniformly, ensuring even distribution while minimizing chemical usage. This targeted application reduces environmental impact and lowers input costs. Furthermore, automated swarm operations decrease the reliance on manual labor, addressing labor shortages and reducing overall operational expenses. Technological advancements such as AI-driven coordination and 5G connectivity play a crucial role in optimizing swarm drone performance. These systems enable drones to communicate with one another in real-time, adjusting their flight paths and tasks to avoid collisions and maximize efficiency [53]. Additionally, swarm drones can operate in difficult terrains where conventional machinery may struggle, offering greater accessibility for farmers managing remote or irregularly shaped fields. Swarm drone technology enhances agricultural productivity by enabling faster, more precise, and cost-effective operations. As this technology continues to evolve, it is expected to become a cornerstone of large-scale agricultural management, driving sustainable and efficient farming practices.

13.3.3 Renewable Energy-Powered Drones

Renewable energy-powered drones are emerging as a sustainable innovation in agricultural technology. These drones, powered by solar energy, hybrid systems, or other renewable sources, offer an environmentally friendly alternative to conventional battery or fuel-based UAVs. By reducing reliance on fossil fuels, renewable energy-powered drones align with global efforts to promote sustainable agriculture and climate-conscious practices. Solar-powered drones utilize photovoltaic cells to capture solar energy, allowing for extended flight times without frequent recharging.

This capability is particularly beneficial for long-duration agricultural surveillance, such as monitoring vast farmlands, tracking crop health, and managing irrigation. Solar UAVs can operate for hours or even days under favorable sunlight conditions, providing continuous data collection and reducing downtime associated with battery recharging [54]. Additionally, hybrid-powered drones combine solar energy with battery backup or biofuels, ensuring operational flexibility in areas with inconsistent sunlight.

In agricultural applications, renewable energy-powered drones enhance precision farming by enabling frequent and cost-effective monitoring. Farmers can deploy these drones for tasks like aerial mapping, pest detection, and crop health analysis without the environmental impact associated with traditional fuel-powered systems. Moreover, their ability to cover large areas with minimal energy consumption supports resource efficiency and reduces greenhouse gas emissions. Another advantage is the reduced operational cost over time. While the initial investment in renewable energy-powered drones may be higher, their longer flight endurance and lower energy expenses contribute to significant long-term savings. These drones also support agricultural practices in remote regions, where access to conventional power sources is limited. Renewable energy-powered drones represent a promising step toward sustainable agricultural innovation [55]. By reducing carbon footprints, lowering operational costs, and providing extended surveillance capabilities, they play a crucial role in advancing eco-friendly and efficient farming practices. As renewable technology continues to evolve, these drones will likely become essential tools for modern agriculture.

13.3.4 Vision for Next-Generation Farming

Next-generation farming envisions a future where advanced technologies like drones, robotics, and the Internet of Things (IoT) work together to create fully automated agricultural systems. This interconnected ecosystem is designed to enhance efficiency, precision, and sustainability while reducing the need for human intervention. By integrating cutting-edge innovations, next-generation farming aims to address global challenges such as food security, resource conservation, and climate change. In this vision, drones play a pivotal role by providing real-time data through advanced imaging and sensor technologies. These UAVs monitor crop health, detect pests, assess soil conditions, and support precision spraying, ensuring targeted interventions and reducing input waste. Drones will work alongside autonomous tractors and robotic harvesters, enabling continuous farm operations with minimal human oversight [56]. For example, drones could identify areas needing attention while ground-based robots carry out tasks like weeding, fertilizing, or harvesting.

The IoT connects all these devices, allowing real-time communication and data sharing across the entire farm. Smart sensors placed throughout the fields collect information on temperature, humidity, soil moisture, and plant growth. This data is analyzed using artificial intelligence (AI) to optimize irrigation schedules, fertilizer application, and crop rotation plans. Farmers can monitor and control these processes remotely via smartphone apps or cloud-based platforms, improving decision-making and resource allocation. A core benefit of this vision is sustainability—automated systems reduce water usage, chemical runoff, and energy consumption by applying inputs precisely where and when they are needed. Furthermore, labor shortages in rural areas could be mitigated by autonomous machinery capable of performing repetitive tasks year-round. Next-generation farming leverages drones, robotics, and IoT to create a connected, automated, and sustainable agricultural future [57]. This vision not only increases productivity and resource efficiency but also paves the way for climate-resilient and data-driven farming practices.

13.3.5 Fully Automated Drone Systems

The future of UAV technology in agriculture is moving toward fully autonomous drone systems capable of performing complex tasks without direct human intervention. These advanced systems leverage artificial intelligence (AI), machine learning, and adaptive algorithms to enhance precision, efficiency, and scalability in modern farming practices. Fully automated drones can independently plan, execute, and analyze agricultural tasks by processing real-time data and adjusting operations based on environmental conditions. Equipped with AI-driven decision-making capabilities, these drones can identify crop health issues, optimize pesticide or fertilizer applications, and perform detailed field mapping. Adaptive learning allows drones to improve performance over time, refining their techniques and accuracy through continuous data analysis. One major advancement is autonomous swarm technology, where multiple drones collaborate in coordinated operations [58]. This is particularly valuable for large-scale farms, where swarms can simultaneously cover vast areas, reducing the time needed for crop monitoring and resource application. For example, a swarm can detect specific zones requiring irrigation while others conduct pesticide spraying or monitor plant health.

These systems integrate with Internet of Things (IoT) networks, enabling seamless communication with other automated devices like robotic tractors and smart irrigation systems. Through cloud-based platforms, farmers can remotely oversee and optimize operations while drones perform routine tasks such as soil analysis, disease detection, and yield estimation autonomously. The benefits of fully automated drones extend beyond productivity. They reduce labor costs, minimize human error, and enable sustainable farming by precisely applying agricultural inputs. Furthermore, these systems can operate continuously, ensuring 24/7 monitoring and response capabilities during critical growing seasons [59]. Fully automated drone systems represent the next stage of agricultural innovation, offering intelligent, adaptive, and efficient solutions that reduce manual labor while increasing the precision and sustainability of farming operations.

13.3.6 Integrating UAVs with Robotics and IoT

The integration of Unmanned Aerial Vehicles (UAVs) with robotics and the Internet of Things (IoT) is revolutionizing modern agriculture by enabling real-time data sharing, process automation, and smart decision-making. This interconnected system enhances precision farming by providing detailed insights into crop conditions and optimizing resource use through synchronized technological operations. UAVs play a critical role in data collection, capturing high-resolution images, thermal maps, and multispectral data to assess soil health, monitor crop growth, and detect pest infestations. This information is shared instantly with IoT-connected devices such as robotic tractors, automated irrigation systems, and weather sensors. These devices, in turn, can respond autonomously to changing agricultural conditions [60]. For example, drones can identify dry areas in a field and relay the data to smart irrigation systems, which then adjust water delivery accordingly.

By combining UAVs with IoT networks, farmers gain predictive analytics capabilities. Advanced algorithms analyze the continuous data flow to forecast yield outcomes, detect potential diseases, and optimize planting schedules. Additionally, these integrated systems allow for adaptive farming solutions—tailoring pesticide spraying, fertilizer application, and harvesting based on real-time crop health analysis. Robotics further enhances UAV effectiveness by automating physical tasks that UAVs cannot perform. For instance, while drones survey a field for weeds, robotic sprayers can target those areas precisely, reducing chemical waste and improving environmental sustainability [61]. This integration fosters efficient farm management by reducing manual labor, minimizing input costs, and increasing productivity.

13.3.7 Overcoming Operational Challenges

While Unmanned Aerial Vehicles (UAVs) offer transformative benefits for agriculture, their widespread adoption is hindered by operational challenges. Key obstacles include technical limitations, weather sensitivity, and high initial investment costs. Addressing these challenges is essential to ensure the effective and sustainable use of UAVs in modern farming. One major challenge lies in technical constraints such as limited battery life, payload capacity, and signal range. Agricultural UAVs often require extended flight times to cover large areas, yet many current models cannot sustain long-duration operations without frequent recharging or battery swaps. Moreover, data processing from UAVs—especially in high-resolution imaging—demands advanced computing capabilities and secure storage solutions. These technical issues necessitate continuous research to develop more efficient batteries, lightweight materials, and AI-driven systems for enhanced performance [62].

Weather conditions also pose significant challenges. UAVs are highly sensitive to rain, wind, and temperature extremes, which can disrupt flight accuracy and data collection. Adverse weather increases the risk of equipment damage and reduces operational efficiency. Solutions include developing weather-resistant drones, improving autonomous navigation in adverse conditions, and incorporating real-time weather monitoring for safer operations. Another barrier is the high initial investment required for advanced UAV technology, sensor systems, and data analysis platforms. Small-scale farmers may find these costs prohibitive, limiting access to precision agriculture benefits. Governments and industry stakeholders can address this challenge through subsidies, financing options, and community drone-sharing programs. Establishing a supportive policy framework that encourages research and provides financial assistance can lower barriers to entry and promote equitable UAV adoption [63]. Overcoming the operational challenges of UAVs in agriculture demands technological innovation, resilient infrastructure, and supportive policies. By investing in advanced research and inclusive programs, the agricultural sector can harness the full potential of UAVs to improve efficiency, sustainability, and productivity.

13.3.8 Regulatory and Legal Barriers

The adoption of Unmanned Aerial Vehicles (UAVs) in agriculture is significantly influenced by regulatory and legal frameworks, which vary across regions and present considerable challenges for drone operators. These regulations are designed to ensure airspace safety, data privacy, and environmental protection, but complex and fragmented policies can hinder the scalability and innovation of agricultural drone use. A primary regulatory challenge involves airspace restrictions. Most countries impose altitude limits, no-fly zones, and visual line-of-sight (VLOS) requirements to prevent UAV interference with manned aircraft. For instance, in the United States, the FAA Part 107 regulations limit drone flights to 400 feet above ground level and require special waivers for Beyond Visual Line of Sight (BVLOS) operations. In the European Union, the EASA mandates a UAS Operator ID and risk assessment for drones used in commercial applications, including precision agriculture [15]. Such regulations add operational complexity for farmers seeking to utilize drones over large fields or in remote locations.

Data privacy is another legal concern, particularly regarding geospatial imaging and surveillance. UAVs collect vast amounts of sensitive agricultural data, which may inadvertently capture private property or neighboring lands. Compliance with laws like the General Data Protection Regulation (GDPR) in the EU requires operators to obtain informed consent and adopt secure data handling practices. Failure to comply may lead to legal penalties and public mistrust. Navigating these barriers requires collaboration between policymakers, technology developers, and agricultural stakeholders to establish streamlined, transparent, and adaptable regulations. Simplifying licensing procedures, providing clear guidelines for UAV use in agriculture, and encouraging public–private partnerships can foster technological innovation while ensuring safety and legal compliance [64]. Overcoming regulatory and legal barriers demands harmonized policies, clear operational guidelines, and continuous dialogue between authorities and the agricultural sector. This approach can unlock the full potential of UAV technology while safeguarding public interests and airspace integrity.

13.3.9 Drone Piloting and Maintenance

Successful deployment of Unmanned Aerial Vehicles (UAVs) in agriculture depends heavily on skilled operators and rigorous maintenance practices. Trained drone pilots ensure accurate data collection, safe flight operations, and adherence to legal regulations, while regular maintenance minimizes equipment failure and extends the operational lifespan of UAV systems. Drone piloting in agriculture requires a combination of technical knowledge and field expertise. Operators must understand flight mechanics, airspace regulations, and precision agriculture techniques to effectively monitor crops, apply pesticides, and gather geospatial data [65]. Certification programs, such as the FAA Part 107 Remote Pilot Certificate in the United States or EASA's UAS Operator Certification in the European Union, are essential for legal compliance and safe operations. These programs cover topics like aerodynamics, navigation, emergency protocols, and data management, ensuring that UAVs are operated responsibly. Routine maintenance is crucial for sustaining drone performance and safety. This involves pre-flight inspections to check for mechanical wear, software updates to enhance system capabilities, and battery management to prevent power failures during flight. Preventative measures, such as cleaning sensors and calibrating cameras, preserve the accuracy of agricultural data. Post-flight diagnostics help identify technical issues early, reducing downtime and costly repairs [66].

Advanced UAVs used in large-scale farming require specialized upkeep, including propeller checks, thermal imaging sensor calibration, and GPS alignment. Manufacturers often provide maintenance schedules and service guidelines, which operators must follow to ensure optimal functionality. Investing in comprehensive training and maintenance protocols not only enhances operational efficiency but also reduces the risk of equipment failure and data loss. As drone technology evolves, continuous education and adaptive maintenance strategies will be vital for supporting the long-term integration of UAVs in agriculture. Effective drone piloting and maintenance require skilled training, regulatory compliance, and systematic care [67]. These practices ensure reliable UAV operations, extending their lifespan and maximizing their contributions to precision agriculture.

13.3.10 Innovations on the Horizon

Artificial intelligence (AI) and machine learning (ML) are transforming Unmanned Aerial Vehicle (UAV) technology, driving innovations that enhance precision agriculture and farm management. These advanced technologies empower drones to process vast amounts of data, identify patterns, and make real-time decisions, improving efficiency and accuracy in agricultural practices. AI-equipped drones use image analysis to monitor crop health, soil conditions, and pest infestations. Through advanced computer vision, UAVs can detect subtle changes in plant growth, identify disease outbreaks, and assess moisture levels across large fields. This allows for targeted interventions, such as precision spraying of fertilizers and pesticides, reducing chemical waste and minimizing environmental impact [68]. Machine learning algorithms enhance UAV capabilities by enabling predictive modeling and adaptive learning. These drones can analyze historical data to forecast crop yields, weather impacts, and optimal harvest times. With each flight, ML systems refine their analysis, improving the accuracy of future predictions and providing data-driven insights to farmers.

One breakthrough is the development of autonomous drones that can navigate and respond to changing field conditions without human intervention. AI enhances these systems by integrating sensor data and geospatial mapping, allowing UAVs to adjust flight patterns, avoid obstacles, and optimize coverage in real-time. This autonomy reduces labor demands and ensures continuous monitoring of agricultural landscapes. Additionally, AI-driven UAVs play a crucial role in sustainability. By optimizing resource allocation—such as water and fertilizer—these drones help reduce waste and environmental degradation. AI also facilitates early detection of climate-related risks, enabling farmers to adopt proactive measures and improve crop resilience. AI and machine learning are revolutionizing UAV technology in agriculture. By enabling real-time analysis, predictive modeling, and autonomous operations, these innovations enhance efficiency, sustainability, and decision-making in modern farming practices [69]. As these technologies evolve, they will continue to shape the future of precision agriculture and data-driven farm management.

13.3.11 Autonomous Drones for Scalable Agriculture

Fully autonomous drones are revolutionizing large-scale agricultural operations, offering efficient, scalable solutions for crop monitoring, resource management, and environmental conservation. These advanced UAV systems operate without human intervention, using artificial intelligence (AI), machine learning (ML), and sensor-based navigation to perform complex agricultural tasks with precision and reliability.

One of the primary advantages of autonomous drones is their ability to scale operations across large farmland. Equipped with high-resolution cameras, multispectral sensors, and geospatial mapping technology, these drones can continuously monitor crop health, soil moisture, and pest activity across vast areas. This real-time data collection allows for targeted interventions, reducing waste and improving yield outcomes [70]. In resource management, autonomous drones enhance efficiency by optimizing the use of water, fertilizers, and pesticides. For example, drones equipped with variable-rate application systems can precisely apply chemicals where needed, minimizing overuse and environmental impact. These systems also support automated irrigation monitoring, ensuring crops receive the right amount of water and promoting sustainable farming practices.

Environmental conservation is another key benefit. Autonomous drones help monitor biodiversity, detect soil erosion, and track climate changes. This data aids in developing strategies to preserve natural resources while maintaining agricultural productivity. Moreover, the use of electric and hybrid-powered autonomous drones reduces carbon emissions, contributing to eco-friendly operations. Fully autonomous drones also offer significant labor savings by reducing the need for manual oversight. With pre-programmed flight paths and AI-based decision-making, these UAVs can conduct continuous field surveillance, even in remote areas, ensuring 24/7 monitoring without human fatigue. This autonomy increases operational efficiency and allows farmers to manage larger areas with fewer resources [71]. Autonomous drones represent a transformative shift in scalable agriculture, offering precision, efficiency, and sustainability. As technology advances, these systems will play an even greater role in enhancing crop management, optimizing resource use, and promoting environmental conservation in modern farming practices.

13.4 Summary

The adoption of UAV technology in agriculture is revolutionizing farming practices through precision monitoring, resource optimization, and data-driven decision-making. Legal considerations focus on compliance with airspace regulations, flight permissions, and data protection laws, with each country having distinct guidelines governing drone use. Ethical considerations address the societal impact of UAVs, including data privacy, labor displacement, environmental sustainability, transparency, and technology access. Ensuring equitable use and responsible data handling is crucial to mitigating these ethical risks. The future of UAV technology in agriculture points toward autonomous drone systems, AI integration, and renewable energy solutions. These advancements promise improved efficiency, scalability, and environmental conservation. Challenges such as regulatory barriers, technical limitations, and maintenance must be addressed to ensure safe and effective drone operations. UAV technology holds immense potential for sustainable agriculture when aligned with legal compliance and ethical responsibility. Collaborative efforts between policymakers, researchers, and farmers will shape a future where drones contribute to food security, sustainability, and precision farming. Best practices—following regulations, respecting privacy, being transparent, minimizing harm, and ensuring equity—promote the safe, legal, and ethical use of UAV technology in agriculture while balancing technological advancement with social responsibility. Understanding regional UAV regulations, maintaining thorough documentation, ensuring proper training, and prioritizing safety are fundamental to achieving regulatory compliance in agricultural UAV use. Compliance with airspace laws, including respecting altitude limits and restricted areas, is essential for the safe and legal operation of agricultural drones. Maintaining compliance with airspace laws involves checking airspace maps, documenting flights, and staying updated on regulatory changes. Ethical implications of drone farming extend beyond technological advancements to include environmental responsibility, worker protection, fair access, and transparent data practices. By addressing these concerns, drone operators can foster sustainable, inclusive, and ethical agricultural development. Ethical drone use requires prioritizing energy efficiency, sustainable practices, and minimizing environmental harm. While drones increase efficiency in agriculture, they also pose a risk of labor displacement. Ethical drone use involves balancing automation with job preservation by promoting worker retraining and fostering inclusive access to emerging agricultural technologies. The high costs of drone technology deepen the digital divide, limiting access for small farmers and favoring wealthier producers. Ethical drone use involves reducing barriers through subsidies, community sharing, and training to ensure fair access and equitable agricultural advancement. Ensuring transparency and accountability in agricultural drone use requires clear guidelines for responsibility, public disclosure of operations, and robust regulatory frameworks. By establishing legal accountability and promoting open communication, stakeholders can protect public interests, minimize misuse, and encourage ethical and sustainable drone practices. Staying updated with emerging UAV technologies is essential for maintaining competitive and sustainable agricultural practices. Swarm drone technology enhances agricultural productivity by enabling faster, more precise, and cost-effective operations. Renewable energy-powered drones represent a promising step toward sustainable agricultural innovation, reducing carbon footprints, lowering operational costs, and providing extended surveillance capabilities. Next-generation farming leverages drones, robotics, and IoT to create a connected, automated, and sustainable agricultural future. Fully automated drone systems represent the next stage of agricultural innovation, offering intelligent, adaptive, and efficient solutions that reduce manual labor while increasing precision and sustainability of farming operations. Overcoming operational challenges of UAVs in agriculture demands technological innovation, resilient infrastructure, and supportive policies. Harmonized policies, clear operational guidelines, and continuous dialogue between authorities and the agricultural sector can unlock the full potential of UAV technology while safeguarding public interests and airspace integrity.

References

1. Singh B (2024) Unmanned aircraft systems (UAS), surveillance, risk management to cybersecurity and legal regulation landscape: unraveling the future analysis, challenges, demand, and benefits in the high sky exploring the strange new world. In: Unmanned aircraft systems. pp 313–354
2. Serafinelli E (2024) Life of drone visuals: norms, ethics, and effects. In: Theorising drones in visual culture: views from the blue. Springer, pp 119–156
3. Uddin M, Chowdhury A, Kabir MA (2024) Legal and ethical aspects of deploying artificial intelligence in climate-smart agriculture. AI Soc 39:221–234

4. Hassan M (2024) Integration of UAVs in society: opportunities and challenges from public viewpoint on UAVs in civilian airspace
5. Kemarau RA, Sakawi Z, Suab SA et al (2024) Global perspectives on unmanned aerial vehicles technology in social sciences: applications, innovations, and future research directions. Geocarto Int 39:2413547
6. Frankelius P, Norrman C, Johansen K (2019) Agricultural innovation and the role of institutions: lessons from the game of drones. J Agric Environ Ethics 32:681–707
7. Haula K, Agbozo E (2020) A systematic review on unmanned aerial vehicles in Sub-Saharan Africa: a socio-technical perspective. Technol Soc 63:101357
8. Martin P (2021) A future-focused view of the regulation of rural technology. Agronomy 11:1153
9. Rauhala A, Tuomela A, Leviäkangas P (2023) An overview of unmanned aircraft systems (UAS) governance and regulatory frameworks in the European Union (EU). In: Unmanned aerial systems in agriculture. pp 269–285
10. Senthilraj A, Rawat A, Ono D et al (2023) Farming the future: an approach to precision agriculture through UAS-based thermal scanners
11. Rushiti V, Kulakov A, Stojkoska BR (2024) An overview of UAS regulations in the European Union and the Balkan Region. IEEE, pp 977–982
12. Truxal S, Scott BI (2024) The regulation of unmanned aircraft systems in the European Union. In: Civil regulation of autonomous unmanned aircraft systems in Europe. Edward Elgar Publishing, pp 31–63
13. Ngo S, dela Rama Jr JI, Fristikawati Y et al (2022) Contemporary legal issues in the southeast asian region. Penerbit Universitas Katolik Indonesia Atma Jaya
14. Heiets I, Kuo Y-W, La J et al (2023) Future trends in UAV applications in the Australian market. Aerospace 10:555
15. Ayamga M, Tekinerdogan B, Kassahun A (2021) Exploring the challenges posed by regulations for the use of drones in agriculture in the African context. Land 10:164
16. Grote M, Pilko A, Scanlan J et al (2022) Sharing airspace with uncrewed aerial vehicles (UAVs): views of the general aviation (GA) community. J Air Transp Manag 102:102218
17. Ayamga M, Tekinerdogan B, Kassahun A, Rambaldi G (2021) Developing a policy framework for adoption and management of drones for agriculture in Africa. Technol Anal Strateg Manage 33:970–987
18. McTegg SJ, Tarsha Kurdi F, Simmons S, Gharineiat Z (2022) Comparative approach of unmanned aerial vehicle restrictions in controlled airspaces. Remote Sensing 14:822
19. Bijjahalli S, Gardi A, Pongsakornsathien N et al (2022) A unified airspace risk management framework for UAS operations. Drones 6:184
20. Del Cerro J, Cruz Ulloa C, Barrientos A, de León RJ (2021) Unmanned aerial vehicles in agriculture: a survey. Agronomy 11:203
21. Alqarni KS, Almalki FA, Soufiene BO et al (2022) Authenticated wireless links between a drone and sensors using a blockchain: case of smart farming. Wirel Commun Mob Comput 2022:4389729
22. Merz M, Pedro D, Skliros V et al (2022) Autonomous UAS-based agriculture applications: general overview and relevant European case studies. Drones 6:128
23. Mohsan SAH, Khan MA, Noor F et al (2022) Towards the unmanned aerial vehicles (UAVs): a comprehensive review. Drones 6:147
24. Asghari O, Ivaki N, Madeira H (2025) UAV operations safety assessment: a systematic literature review. ACM Comput Surv
25. Maddikunta PKR, Hakak S, Alazab M et al (2021) Unmanned aerial vehicles in smart agriculture: applications, requirements, and challenges. IEEE Sens J 21:17608–17619
26. Abro GEM, Zulkifli SAB, Masood RJ et al (2022) Comprehensive review of UAV detection, security, and communication advancements to prevent threats. Drones 6:284
27. Perz R (2024) The multidimensional threats of un-manned aerial systems: exploring biomechanical, technical, operational, and legal solutions for ensuring safety and security. Arch Transp 69:91–111

28. Wanner D, Hashim HA, Srivastava S, Steinhauer A (2024) UAV avionics safety, certification, accidents, redundancy, integrity, and reliability: a comprehensive review and future trends. Drone Syst Appl 12:1–23
29. Pathak H, Kumar G, Mohapatra S et al (2020) Use of drones in agriculture: potentials, problems and policy needs. ICAR-Natl Inst Abiotic Stress Manage 300:4–15
30. Mohsan SAH, Othman NQH, Li Y et al (2023) Unmanned aerial vehicles (UAVs): practical aspects, applications, open challenges, security issues, and future trends. Intel Serv Robot 16:109–137
31. Quamar MM, Al-Ramadan B, Khan K et al (2023) Advancements and applications of drone-integrated geographic information system technology—a review. Remote Sensing 15:5039
32. Sparrow R, Howard M (2021) Robots in agriculture: prospects, impacts, ethics, and policy. Precision Agric 22:818–833
33. Adepoju O, Esan O, Akinyomi O (2022) Food security in Nigeria: enhancing workers' productivity in precision agriculture. J Digital Food Energy Water Syst 3
34. Meng Z (2024) The role of AI in transforming local economies: exploring how AI technologies are impacting local businesses and labor markets. Appl Comput Eng 108:1–6
35. Hanrahan BV, Maitland C, Brown T et al (2021) Agency and extraction in emerging industrial drone applications: imaginaries of Rwandan farm workers and community members. Proc ACM Hum-Comput Interact 4:1–21
36. Bazargani K, Deemyad T (2024) Automation's impact on agriculture: opportunities, challenges, and economic effects. Robotics 13:33
37. Wang N, Christen M, Hunt M (2021) Ethical considerations associated with "humanitarian drones": a scoping literature review. Sci Eng Ethics 27:51
38. Hawashin D, Nemer M, Gebreab SA et al (2024) Blockchain applications in UAV industry: review, opportunities, and challenges. J Netw Comput Appl 230:103932
39. Al-Dosari K, Fetais N (2023) A new shift in implementing unmanned aerial vehicles (UAVs) in the safety and security of smart cities: a systematic literature review. Safety 9:64
40. Wang N, Christen M, Hunt M, Biller-Andorno N (2022) Supporting value sensitivity in the humanitarian use of drones through an ethics assessment framework. Int Rev Red Cross 104:1397–1428
41. Jain N, Gambhir A, Pandey M (2024) Unmanned aerial networks—UAVs and AI. In: Recent trends in artificial intelligence towards a smart world: applications in industries and sectors. Springer, pp 321–351
42. Chamuah A, Singh R (2020) Securing sustainability in Indian agriculture through civilian UAV: a responsible innovation perspective. SN Appl Sci 2:1–10
43. Chamuah A, Singh R (2022) Responsible governance of civilian unmanned aerial vehicle (UAV) innovations for Indian crop insurance applications. J Responsible Technol 9:100025
44. Renda A (2024) AI and the transformation of agricultural work: economic, social, and environmental implications. In: Handbook of artificial intelligence at work. Edward Elgar Publishing, pp 96–118
45. Dara R, Hazrati Fard SM, Kaur J (2022) Recommendations for ethical and responsible use of artificial intelligence in digital agriculture. Front Artif Intell 5:884192
46. Mathews AJ, Singh KK, Cummings AR, Rogers SR (2023) Fundamental practices for drone remote sensing research across disciplines. Drone Syst Appl 11:1–22
47. Philpott III RL (2023) Incorporating ethical, social and environmental concerns into decision making with value added modeling
48. El Alaoui M, Amraoui KE, Masmoudi L et al (2024) Unleashing the potential of IoT, artificial intelligence, and UAVs in contemporary agriculture: a comprehensive review. J Terrramech 115:100986
49. Padhiary M, Kumar R, Sethi LN (2024) Navigating the future of agriculture: a comprehensive review of automatic all-terrain vehicles in precision farming. J Inst Eng (India) Ser A 105:767–782
50. Telli K, Kraa O, Himeur Y et al (2023) A comprehensive review of recent research trends on unmanned aerial vehicles (UAVs). Systems 11:400

51. Kannan P, Abishek J, Pazhanivelan S, Firnass MMRA (2025) Farming from above: the role of UAVs in modern agricultural practices. In: Digital farming and smart agriculture for sustainable future. Chapman and Hall/CRC, pp 235–254
52. Javed S, Hassan A, Ahmad R et al (2024) State-of-the-art and future research challenges in UAV swarms. IEEE Internet Things J 11:19023–19045
53. Marek D, Paszkuta M, Szyguła J et al (2024) Swarm of drones in a simulation environment—efficiency and adaptation. Appl Sci 14:3703
54. Sarfraz S, Ali F, Hameed A et al (2023) Sustainable agriculture through technological innovations. In: Sustainable agriculture in the era of the OMICs revolution. Springer, pp 223–239
55. Rehman AU, Alamoudi Y, Khalid HM et al (2024) Smart agriculture technology: an integrated framework of renewable energy resources, IoT-based energy management, and precision robotics. Cleaner Energy Syst 9:100132
56. Louta M, Banti K, Karampelia I (2024) Emerging technologies for sustainable agriculture: the power of humans and the way ahead. IEEE Access
57. Babar AZ, Akan OB (2024) Sustainable and precision agriculture with the internet of everything (IoE). arXiv preprint arXiv:240406341
58. Hammadi M (2024) A comprehensive analysis of autonomous drone technology across multiple sectors
59. Guebsi R, Mami S, Chokmani K (2024) Drones in precision agriculture: a comprehensive review of applications, technologies, and challenges. Drones 8:686
60. Islam N, Rashid MM, Pasandideh F et al (2021) A review of applications and communication technologies for internet of things (IoT) and unmanned aerial vehicle (UAV) based sustainable smart farming. Sustainability 13:1821
61. Labib NS, Brust MR, Danoy G, Bouvry P (2021) The rise of drones in internet of things: a survey on the evolution, prospects and challenges of unmanned aerial vehicles. IEEE Access 9:115466–115487
62. Vashishth TK, Sharma V, Sharma KK et al (2024) Integration of unmanned aerial vehicles (UAVs) and IoT for crop monitoring and spraying. In: Internet of Things applications and technology. Auerbach Publications, pp 95–117
63. Sziroczak D, Rohacs D, Rohacs J (2022) Review of using small UAV based meteorological measurements for road weather management. Prog Aerosp Sci 134:100859
64. Parmaksiz O, Cinar G (2023) Technology acceptance among farmers: examples of agricultural unmanned aerial vehicles. Agronomy 13:2077
65. Joshi R, Pandey K (2024) IoT-Enabled UAV: a comprehensive review of technological change in Indian farming. In: Unmanned aircraft systems. pp 93–135
66. Cheung T (2024) Factors driving and impeding the use of unmanned aerial systems (UAS) in the Department of Transportation
67. Askerbekov D, Garza-Reyes JA, Ghatak RR et al (2024) Embracing drones and the Internet of drones systems in manufacturing—an exploration of obstacles. Technol Soc 78:102648
68. Kurunathan H, Huang H, Li K et al (2023) Machine learning-aided operations and communications of unmanned aerial vehicles: a contemporary survey. IEEE Commun Surv Tutorials 26:496–533
69. Kumar G, Altalbe A (2024) Artificial intelligence (AI) advancements for transportation security: in-depth insights into electric and aerial vehicle systems. Environ Dev Sustain 1–51
70. Sun C, Fontanesi G, Canberk B et al (2024) Advancing uav communications: A comprehensive survey of cutting-edge machine learning techniques. IEEE Open J Veh Technol
71. Mohamed M (2023) Agricultural sustainability in the age of deep learning: current trends, challenges, and future trajectories. Sustainable Machine Intelligence Journal 4:2

The manufacturer's authorised representative in the EU is Springer Nature Customer Service Centre GmbH, Europaplatz 3, 69115 Heidelberg, Germany. If you have any concerns regarding our products, please contact ProductSafety@springernature.com

Printed and bound by CPI Group (UK) Ltd, Croydon, CR0 4YY
01/10/2026
02225921-0018